S. Chowdhury

MASS SPECTROMETRY FOR DRUG DISCOVERY AND DRUG DEVELOPMENT

WILEY SERIES ON MASS SPECTROMETRY

A complete list of the titles in this series appears at the end of this volume.

MASS SPECTROMETRY FOR DRUG DISCOVERY AND DRUG DEVELOPMENT

Edited by

WALTER A. KORFMACHER

WILEY

A JOHN WILEY & SONS, INC., PUBLICATION

Published by John Wiley & Sons, Inc., Hoboken, New Jersey.
Published simultaneously in Canada.

Library of Congress Cataloging-in-Publication Data:
Mass spectrometry for drug discovery and drug development / edited by Walter A. Korfmacher.
p. ; cm.
Includes bibliographical references and index.
ISBN 978-0-470-94238-3 (cloth)
I. Korfmacher, Walter A.
[DNLM: 1. Drug Discovery. 2. Mass Spectrometry–methods. 3. Peptides–analysis. 4. Pharmaceutical Preparations–analysis. 5. Proteins–analysis. QV 745]

615.1'9–dc23

2012031775

Printed in the United States of America.

10 9 8 7 6 5 4 3 2 1

This book is dedicated to the most important people in my life:

Madeleine Korfmacher

Joseph Korfmacher

Mary McCabe

Michael McCabe

Brian McCabe

Kelly McCabe

CONTENTS

CONTRIBUTORS

Mark T. Cancilla, PhD, Merck & Co., Inc., West Point, PA

Richard M. Caprioli, PhD, Mass Spectrometry Research Center, Vanderbilt University, Nashville, TN

Guodong Chen, PhD, Bristol-Myers Squibb, Princeton, NJ

Swapan K. Chowdhury, PhD, Department of Drug Metabolism and Pharmacokinetics, Millennium Pharmaceuticals, Inc., Cambridge, MA

Filip Cuyckens, PhD, Drug Metabolism and Pharmacokinetics, Janssen Research and Development, a Division of Janssen Pharmaceutica N.V., Beerse, Belgium

Lieve Dillen, PhD, Drug Metabolism and Pharmacokinetics, Janssen Research and Development, a Division of Janssen Pharmaceutica N.V., Beerse, Belgium

Yi Du, PhD, WuXi AppTec Co., Ltd., Shanghai, China

Daniel Eikel, Dr. rer. nat., Dipl.-Chem., Advion, Ithaca, NY

William Bart Emary, PhD, Pharmacokinetics, Pharmacodynamics and Drug Metabolism, Merck & Co. Inc.,West Point, PA

W. Michael Flanagan, PhD, Genentech, Inc., South San Francisco, CA

Jack Henion, PhD, Advion, Ithaca, NY

Maarten Honing, PhD, DSM Resolve, Geleen, The Netherlands

Gérard Hopfgartner, PhD, School of Pharmaceutical Sciences, University of Lausanne, University of Geneva, Geneva, Switzerland

Benno Ingelse, PhD, MSD/Merck, Oss, The Netherlands

Mohammed Jemal, PhD, Bristol-Myers Squibb, Princeton, NJ

Walter A. Korfmacher, PhD, Genzyme, a SANOFI Company, Waltham, MA

Shuguang Ma, PhD, Department of Drug Metabolism and Pharmacokinetics Genentech, Inc., South San Francisco, CA

John Mehl, PhD, Bioanalytical Research, Bristol-Myers Squibb, Princeton, NJ

Stacey R. Oppenheimer, PhD, Pfizer, Groton, CT

Natalia Penner, PhD, Cambridge Center, Biogen Idec, Cambridge, MA

Chandra Prakash, PhD, Cambridge Center, Biogen Idec, Cambridge, MA

Birendra N. Pramanik, PhD, Parsippany, NJ

Pavlo Pristatsky, MS, Merck & Co., West Point, PA

Michael D. Reily, PhD, Bristol-Myers Squibb, Princeton, NJ

Michelle L. Reyzer, PhD, Mass Spectrometry Research Center, Vanderbilt University, Nashville, TN

Petia Shipkova, PhD, Bristol-Myers Squibb, Princeton, NJ

Wilson Z. Shou, PhD, Bristol-Myers Squibb, Wallingford, CT

Adrienne A. Tymiak, PhD, Bristol-Myers Squibb, Princeton, NJ

Hui Wei, Bristol-Myers Squibb, Princeton, NJ

Yuan-Qing Xia, MS, AB Sciex, Framingham, MA

Joanna Zgoda-Pols, PhD, Merck Research Laboratories, Rahway, NJ

Jun Zhang, PhD, Bristol-Myers Squibb, Wallingford, CT

Nanyan Rena Zhang, PhD, Pharmacokinetics, Pharmacodynamics and Drug Metabolism, Merck & Co. Inc., West Point, PA

PREFACE

This book was written as part of a series of books on the utility of mass spectrometry (MS) for various scientific fields. The emphasis for this book is the description of the application of MS to the areas of new drug discovery as well as drug development. MS is now used as the main analytical tool for all the stages of drug discovery and drug development. In many cases, the way MS is applied to these endeavors has changed significantly in recent years, so there is a need for this book in order to provide a reference to the current technology. Thus, the readers of this book would be pharmaceutical scientists including medicinal chemists, analytical chemists, and drug metabolism scientists. This book will also be of interest to any mass spectrometry scientist who wants to learn how MS is being used to support new drug discovery efforts as well as drug development applications.

The book has 15 chapters that are written by experts in the topic that is described in the chapter. The first chapter provides a current overview of the various types of MS systems that are used in new drug discovery and drug development. This chapter will be useful to those still learning about MS as well as experts who want to understand the latest MS technology. One of the major changes in the MS field has been the emergence of high-resolution mass spectrometry (HRMS) as a tool not only for qualitative analyses, but also for quantitative analyses. This change has the potential to produce a true paradigm shift. In the future, it can be predicted that many quantitative bioanalytical assays will shift from using the selected reaction monitoring (SRM) technique with high-performance liquid chromatography-tandem mass spectrometry (HPLC-MS/MS) to HPLC-HRMS. Discussions of why and how this will happen can be found in the second, third, and fourth chapters of this book. This shift from HPLC-MS/MS to HPLC-HRMS has the potential to radically change how MS is used in both new drug discovery and drug development. In addition to these three chapters, the final chapter in the book looks at the new topic of quantitative analysis of peptides and asks whether one should use SRM or HRMS for these assays.

Metabolite identification has been a major focus of MS for several decades. Chapter 5 describes the current MS technology that is used for metabolite identification including new software tools that have made this task easier. One of the newer applications of MS is the quantitative and qualitative analysis of biological drugs; this new topic is described in the sixth chapter along with a discussion of the MS analysis of proteins and peptides. Another important part of drug development is the characterization of impurities and degradation products; the utility of MS for this task is described in the seventh chapter. Medicinal chemists are at the center of all new drug discovery and drug development activities; Chapter 9 describes how MS is used to support the efforts of medicinal chemists in this effort.

An area of continuing interest is the application of MS to surface analysis in order to understand the distribution of drugs and metabolites as well as proteins and peptides on tissue slices from laboratory animal studies and sometimes human clinical tissue samples. Chapter 8 describes the new technique called liquid extraction surface analysis (LESA) that is used for tissue profiling. Chapter 10 discusses MS imaging for proteins and peptides, while Chapter 11 describes the use of MS imaging for drugs and metabolites. Together, these three chapters provide a comprehensive overview of how MS imaging is being used for various drug discovery and drug development applications.

The rest of the book covers various specific topics that are important parts of the drug discovery and drug development process. Chapter 12 deals with the important topic of screening for reactive metabolites. This topic has received increased attention in recent years because of concerns that reactive metabolites may lead to drug safety issues. Two new topics are covered in Chapters 13–14. Chapter 13 describes the use of MS for siRNA applications and Chapter 14 covers the various ways MS is used in the field of metabolomics. The last chapter in the book, Chapter 15, takes a look at the new field of quantitative analysis of peptides using MS techniques.

Overall, this book provides a comprehensive picture of the latest MS technology and how it is being used throughout the various stages of new drug discovery and drug development. I want to thank the authors of each chapter for their efforts and careful attention to detail. I also want to thank Nico Nibbering and Dominic Desiderio, the editors of this MS series, for inviting me to be the editor of this volume. Finally, I want to thank my family for their support of this effort, with special thanks going to Madeleine, my wife.

WALTER A. KORFMACHER

1

OVERVIEW OF THE VARIOUS TYPES OF MASS SPECTROMETERS THAT ARE USED IN DRUG DISCOVERY AND DRUG DEVELOPMENT

GÉRARD HOPFGARTNER

1.1 INTRODUCTION

Since J.J. Dempster published one of the first reports on the detection of volatile organic compounds using electron impact ionization in 1918, significant progress in ion sources and mass analyzers has been achieved. The aim this chapter is to focus on the most commonly used techniques in drug metabolism studies for quantitative or qualitative analysis, and also to discuss some of the "niche" techniques. In terms of the ionization techniques, atmospheric pressure ionization (API) sources including electrospray (ESI), atmospheric pressure chemical ionization (APCI), and atmospheric pressure photoionization (APPI) have revolutionized the analysis of low molecular weight compounds (LMWCs) by high-performance liquid chromatography-mass spectrometry (HPLC-MS). In addition, matrix-assisted laser desorption/ionization (MALDI) was originally developed for the characterization of biopolymers, but is also attractive for the analysis of LMWCs and for mass spectrometry imaging (MSI) of drugs and their metabolites in tissues. Ambient ionization techniques have also gained interest for the same type of applications. Finally, inductively coupled plasma (ICP) mass spectrometry has also been explored as an alternative detector to ^{14}C-labeled drug for drug metabolism studies.

Triple quadrupole MS systems have become the workhorse for quantitation and, in combination with linear ion traps (LITs), are very attractive for qualitative/quantitative workflows. Ions traps are still used as standalone mass spectrometers

Mass Spectrometry for Drug Discovery and Drug Development, First Edition. Edited by Walter A. Korfmacher.

but more and more in combination with others types of mass analyzers. A new paradigm shift will certainly come from high-resolution, accurate mass systems such as time-of-flight (TOF), ion cyclotron resonance, and Orbitraps, which will allow the application of novel approaches in mass spectrometry for drug metabolism studies. Due to the complexity of the samples, additional orthogonal separation power is always required and ion mobility mass spectrometry could play a more important role in the near future. One of the key problems in HPLC-MS is that the response is compound dependent; accelerator mass spectrometry (AMS) is one option that can be used to overcome this limitation and to provide the ultimate sensitivity in human studies.

1.2 IONIZATION TECHNIQUES

1.2.1 Electrospray

Electrospray is currently one of the most commonly used ionization techniques; in ESI, either singly or multiply charged gas phase ions are generated at atmospheric pressure by electrically charging a liquid flow. It is based on a condensed phase process where preformed solutions ions are transferred to the gas phase. ESI for mass spectrometry was developed by John Fenn and coworkers in an attempt to analyze large biomolecules by mass spectrometry [1]. Charged droplets are generated by applying a strong potential of several kilovolts (2–6 kV) to a liquid stream. An electric field gradient is generated, which induces the deformation of the liquid into a conical shape called the Taylor cone. Then the solution forms a charged aerosol. After size reduction of the droplets by evaporation at atmospheric pressure, ions escape from the droplets and are sampled into the mass analyzer. The concept of applying high potential to a metal capillary to generate ions at atmospheric pressure followed by mass spectrometric detection has also been reported by Alexandrov et al. [2, 3], and they named their method extraction of dissolved ions under atmospheric pressure (EDIAP).

The stability of the aerosol is strongly dependent on the solvent composition, the flow rate, and the applied potential; typically, electrospray works best at the flow rate of a few microliters per minute. To achieve higher flow rates, the spray formation can be assisted by a nebulizing gas (nitrogen), which has been referred to as ionspray [4] or pneumatically assisted electrospray. Most modern instruments can handle flow rates from a few nanoliters per minute to several milliliters per minute. Various atmospheric pressure ion source geometries have been developed, using in most cases some combination of nebulizing gas and heat [5]. Pneumatically assisted electrosprays are well suited as ionization sources for liquid chromatography at various flow rates. It has been stated that ion spray mass spectrometry behaves like a concentration-sensitive detector [6], where the reduction of liquid chromatography column internal diameter should result in an increase of the MS response considering that the same amount of analyte is injected. The actual behavior of ESI sources is very dependent on the ion source geometry and the instrumental settings.

ESI works best with preformed ions in solution and when preformed ions are separated from their counter ions. In 1991, Kebarle et al. [7] reported the electrophoretic nature of ESI, in which the charge balance requires the conversion of ions

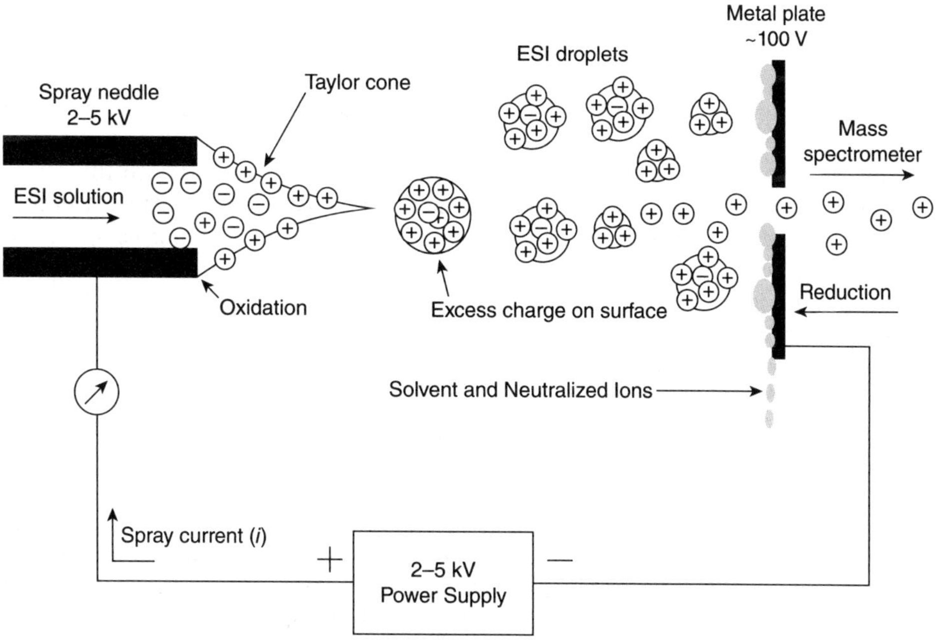

Figure 1.1 Schematic of the electrospray process (adapted with permission from Reference 136).

into electrons. Therefore, oxidation may occur at the needle (Fig. 1.1), and the interface of the mass spectrometer acts as a counter electrode.

Electrospray is particularly suitable for the analysis of inorganic ions and molecules that have acidic or basic functional groups. Organic molecules are generally observed as protonated or deprotonated molecules depending on their pKa. Bases are best detected in the positive mode, while acids give good signals in the negative mode. Therefore, for best signal, the pH of the mobile phase must be adjusted to the acidic or basic nature of the analyte. However, for peptides, it has been shown that intense signals can be observed either in the positive or in the negative mode using strongly acidic or basic solutions, respectively. These observations are reported as "wrong way round" and have been discussed by Zhou and Cook [8]. For many analytes besides the protonated or deprotonated molecules, adduct ions such as sodium or potassium adducts in the positive mode or with formate in the negative mode can be observed. Also, they can also form dimers such as $[2M+H]^+$, which are gas phase reactions [9]. Often it is almost impossible to control the intensity of sodium adducts. The formation of adducts is based on ionization by charge separation which occurs in solution and can be exploited to analyze by ESI polar compounds which are neutral or weakly acidic or basic. In the negative mode, chloride ions adducts can be formed when chlorinated solvents such as chloroform are used [10] or for the analysis of tocopherols and carotenoids where silver ions are added to form $[M+Ag]^+$ ions [11]. Analysis of analytes in highly aqueous solution is more challenging in the negative mode than in the positive mode. This is mainly due to an electrical discharge occurring at the tip of the sprayer (corona discharge) resulting in the chemical ionization of the analyte and the solvent [12, 13]. Generally, negative ESI operated at lower potential and compressed air is preferred to nitrogen as nebulizing gas.

Typical flow rates for electrospray and pneumatically assisted electrospray range from μL/mL to mL/min. Electrospray can also be operated at very low flow rates; indeed, nanoelectrospray (flow rates <500 nL/min) was developed with the intention to minimize sample consumption and maximize sensitivity [14]. The infusion of a few microliters will result in a stable signal for more than 30 min using pulled capillaries [5] or chip-based emitters [15, 16]. With the infusion signal, averaging allows one to improve the limit of detection in tandem mass spectrometry. The uniqueness of nanoelectrospray is that at nL/min flow rates the droplet sizes are in the submicron range and that the complete spray is sampled into the mass spectrometer. Nanoelectrospray has become particularly important in combination with nanoflow liquid chromatography or chip-based infusion [17]. The ionization efficiency is strongly analyte dependent. Thus, in drug metabolism studies, the relative signal intensities from the sample cannot be correlated directly to the relative abundance of the metabolites. Hop et al. [18] reported that the uniformity of the ionization response could be improved, compared with ESI, by using a chip-based nanoelectrospray source. They argue that the generation of a high electric field around the nozzles produces a large excess of protons and smaller droplets, which minimizes the differences in the ionization efficiency for the analytes.

Hirabayashi et al. [19] described an alternative to ESI called sonic spray. In their device the liquid is sprayed using a high-velocity nebulizing gas. Ions are produced without the application of heat or an electric potential typically at sonic gas velocity. For the analysis of labile compounds and noncovalent complexes the use of a cold spray ionization source was also described [20]. The solution is sprayed into a liquid nitrogen cooled electrospray source. The operating temperature is in the range (ca. –80 to 10°C) that minimizes fragmentation of the analytes compared with conventional electrospray.

The qualitative or quantitative outcome of an electrospray analysis may be strongly dependent on the settings of the experimental parameters such as solvents, flow rate, electrode, electric field, and additives, as well as the nature of the analytes (metal ions, LMWCs, polymers, oligonucleotides, peptides, or proteins). Therefore, understanding the mechanisms of how gas phase ions are formed from ions in solution is important, and reviews have been carried out by Kebarle and Verkcerk in this regard [21]. Two major mechanisms have been proposed—(1) the ion evaporation model (IEM) proposed by Iribarne and Thomson [22] and (2) the charge residue model (CRM) described by Dole et al. [23]—and have been a subject of extensive discussion [4, 24, 25].

In 1968, Dole et al.[23] reported the electrospray analysis of diluted solutions of synthetic polymers, in the negative mode, into air at atmospheric pressure, where the macroion current was detected by a Faraday cage after the light ions have been repelled from the beam by negative voltages on a repeller grid. At that time, there was no evidence of any possible solution to the "vaporization problem," for large polyatomic molecules such as proteins without extensive fragmentation and decomposition [26]. Regarding the formation of gas phase ions, Dole's proposition was that evaporation of solvent would increase the surface-charge density until it reached the Rayleigh limit at which the forces due to Coulombic repulsion and surface tension become comparable. The hydrodynamic instability results in the formation of a jet of smaller droplets repeated until complete dispersion of the liquid. Ultimately, the droplets become so small that they contain one single solute molecule, and this molecule becomes an ion, thus a "charge residue," when it retains some of the droplet charge as the last of the solvent is vaporized.

Iribarne and Thomson [22], interested in the study of small ions, proposed in 1976 the atmospheric pressure IEM consistent with the scenario described by Dole et al. [23] in that a sequence of evaporation and Coulombic explosions leads to droplets of 10^{-6} cm where charge densities are so high that the resulting electrostatic field at their surface is high enough (>10^9 V/m) to push solute ions into the gas phase. The high electric field responsible for the ion evaporation generated by the size reduction of the droplets by heat becomes competitive with further solvent evaporation. In the experiments conducted by Iribarne and Thomson, charged droplets where generated by pneumatic nebulization and the electric field was applied across the plume of evaporating spray to extract small ions and is sometimes referred to as aerospray [27]. Most published work suggests that that the two models strongly depend on the nature of the analyte and that most molecules follow the IEM proposed by Iribarne and Thomson, while large macromecules undergo mostly the charge residue mechanism. A recent study by Nguyen and Fenn [28], where the authors demonstrated the benefit of adding solvent vapor to bath gas, showed that in electrospray—at least for singly and doubly charged peptides—most gas phase ions are likely produced by the IEM rather than by the CRM. Further work indicates that CRM is preceded by IEM, in particular when buffers such as ammonium acetate or triethyl acetate are used [29], or that more nuanced emission mechanisms appear, ranging from pure ion evaporation (PIE) for small ions to pseudo-Rayleigh ion release (PRIR), a mechanism that yields charge states that are nearly indistinguishable from the CRM, for large ions [30].

Blades et al. [7, 12] showed in the early 1990s, when using stainless steel capillaries, the presence of nickel Ni(II) and iron Fe(II) ions in the electrospray solution. They compared the electrospray process with that of an electrolysis cell where an electrochemical (EC) oxidation reaction occurs at the tip of ESI capillary. In general, most of the analytes investigated are not, under standard conditions, affected by the electrochemical process that occurs in electrospray and is most pronounced at very low flow rates. Electrochemistry can be applied ion purpose for chemical derivatization, for signal enhancement, or for the oxidation and cleavage of peptides and proteins [31, 32]. EC oxidation is also of interest in the field of drug metabolism because it can mimic oxidation reactions catalyzed by the enzymes from the cytochrome P450 family [33]. EC cells (couloumetric, amperometric, or in-source) can be easily implemented online prior to MS detection with or without chromatography. In general, the conversion rate is high, but EC does not fully reflect the *in vivo* situation. However, there are two major advantages of EC over the *in vivo* technique: (1) no endogenous cofactors are needed that can affect the MS response, and (2) the direct detection of reactive metabolites becomes possible.

1.2.2 Atmospheric Pressure Chemical Ionization (APCI) and Atmospheric Pressure Photoionization (APPI)

APCI had been commercially available long before ESI but remained a niche technique for HPLC-MS [34] before its use in the quantitative analysis of pharmaceuticals in biological fluids, which started in the early 1990s [35] (Fig. 1.2).

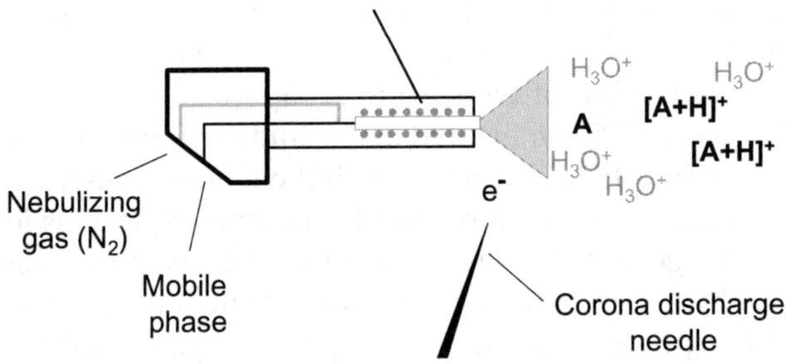

Figure 1.2 Schematic of the heated nebulizer probe.

APCI has an ionization mechanism similar to chemical ionization where gas-phase ion-molecule reactions occur between a neutral molecule and reactant ions [36]. It is a gas-phase ionization process that occurs at atmospheric pressure where the analyte is dissolved in a liquid and introduced through a heated quartz or ceramic tube (heated nebulizer probe) into the source. With the help of heat (T = 200–600°C) and a nebulizing gas (N_2) the liquid and the analyte are completely vaporized prior to being bombarded by electrons generated from a needle by a corona discharge. The discharge current is typically in the range of 1 to 5 μA. Early APCI-MS utilized a ^{63}Ni β-emitter to generate the electrons. APCI is not suited for macromolecules and forms mostly singly charged ions for molecules with a molecular weight of less than 2000 Da and which have some thermal stability. Despite the relatively high temperature of the tube, the temperature of the spray remains in the range of 120–200°C. While the typical liquid flow rates for APCI are in the range of 50–1000 μL/min for most APCI HPLC-MS applications, APCI on a microchip has also been described [37]. The reactants ions are formed through several steps. Initially, the electrons ionize the nitrogen or oxygen to form $O_2^{+\bullet}$, $N_2^{+\bullet}$ ions, which through charge-transfer mechanism form in several steps $H_3O^+(H_2O)_n$ cluster ions. In the presence of more basic molecules such as ammonia, methanol, or acetonitrile reactant ions such as $NH_4^+(H_2O)_n$ or $CH_3OH_2^+(H_2O)_n(CH_3OH)_m$ or $CH_3CNH^+(H_2O)_n(CH_3CN)_m$ will react with the neutral analytes in positive ion mode to form $[M+H]^+$ ions by proton transfer. In some cases, ammonium adduct ions $[M+NH_4]^+$ are formed. The MS response is mainly dependent on the proton affinity of the analyte gas phase acid–base chemistry. In the negative mode, ions are formed either by (1) resonance capture (AB–>AB–), (2) dissociative capture (AB–>B–), or (3) ion-molecule reaction (BH–>B–). Electron capture ionization can be exploited in APCI by derivatizing the analyte to form pentafluorobenzyl derivatives. An increase in sensitivity of 2 orders of magnitude has been demonstrated when compared with conventional APCI methodology for the analysis of steroid, steroid metabolite, prostaglandin, thromboxane, amino acid, and DNA adducts [38].

APPI was developed in an attempt to directly ionize molecules which could not ionize properly with other common API techniques. The setup for APPI [39–43] is very similar to that of APCI. The liquid phase is also vaporized by a heated pneumatic nebulizer and only the corona discharge is replaced by a gas discharge lamp

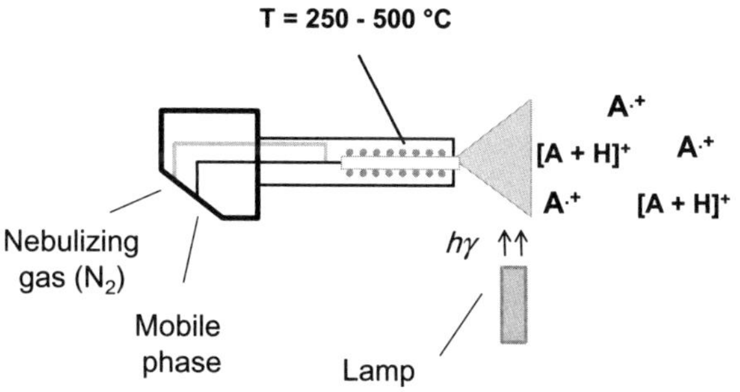

Figure 1.3 Schematic of the photoionization probe.

(Krypton, 10.0 eV) that generates ultraviolet (UV) photons in vacuum. Most analytes have ionization potentials below 10 eV, while typical HPLC solvents have higher ionization potentials (methanol 10.8 eV, acetonitrile 12.2 eV, water 12.6 eV) (Fig. 1.3).

The energy of the photon absorbed by the molecule causes the ejection of an electron and generates a radical cation. While mainly formation of even electron ions has been reported by direct photoionization, the addition of dopants such as toluene or acetone will generate protonated ions with better sensitivities. The ionization process is not fully understood, but two different mechanisms can occur: (1) dopant radical cations react with the analyte by charge transfer, or (2) the dopant radical cations can ionize the solvent molecules by proton transfer, which can then ionize the analyte. APPI can also be performed in the negative mode. As with APCI, APPI can handle a large range of analytes. Cai et al. [44] compared the APPI, APCI, and ESI performance of 106 standard compounds and 241 proprietary drug candidates that represented a wide range of chemical space. They found that APPI is an excellent complementary tool for ionizing compounds that are not ionized by ESI or APCI, and suggested that the technique may be more universal than was previously believed. The performance of APPI is dependent on the flow rate, and better sensitivities than APCI have been reported at lower flow rates. Because APCI and APPI are gas-phase ionization processes, it appears that, compared with ESI, they are less vulnerable to matrix effects [42]. APPI is a useful tool for a large variety of neutral analytes such as steroids [44].

1.2.3 Ambient Desorption Techniques

Direct analysis of solid sample or analytes present on solid surfaces without any sample preparation has always been a topic of interest but somewhat neglected with the success of HPLC-MS. Ambient ionization mass spectrometry allows the direct analysis of samples in their native state. Its major advantage over MALDI is that no matrix is needed to be deposited on the sample. Since the introduction in 2004 of desorption electrospray ionization (DESI) [45] and direct analysis in real time (DART) [46], there has been a continuous development in the field and more than 30 new techniques have been described [47, 48]. They can be classified in three

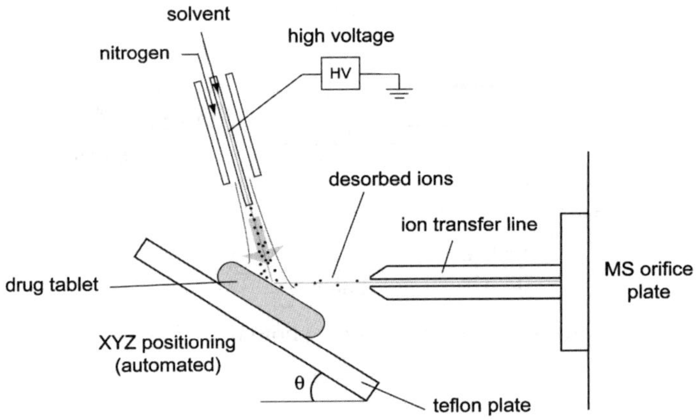

Figure 1.4 Desorption electrospray ionization interface. The sample is placed in front of the orifice and is hit by nebulized droplets. Desorbed ions are then sampled into the mass spectrometer.

groups: (1) ESI or spray-related techniques, (2) spray-based photon/energy, and (3) APCI-related techniques. For drug metabolism studies, ambient ionization mass spectrometry can be applied to analysis of pharmaceuticals and metabolites in tissues [49, 50].

1.2.3.1 DESI Desorption electrospray ionization is an atmospheric pressure ionization method based on charged liquid droplets that are directed by a high-velocity gas jet (in the order of 300 m/s) to the surface to be analyzed. Ions directly produced from the surface to be analyzed are then sampled into the mass spectrometer [51]. The incident angle of the spray plume (relative to the sample) has been extensively investigated and was found to be optimal at around 45–55° (Fig. 1.4).

Various reports on the mechanism of DESI have been published [52, 53] suggesting that both a heterogeneous charge-transfer mechanism and a droplet pick-up mechanism of ionization is occurring. Bereman and Muddiman [54] reported that the spray generates an initial wetting of the surface to analyze, allowing an extraction of the analyte into the film before a delayed formation of the droplets containing the analytes, which subsequently form ions and are sampled into the mass spectrometer. Therefore, the extraction recoveries are certainly dependent on the film properties and the impact of the spray on the sample. DESI has been demonstrated as able to analyze a large variety of analytes from pharmaceuticals to proteins, which are present in many different types of samples. Particularly attractive is the possibility of determining the active components of pharmaceutical tablets [50]. DESI has also been used to make tissue images for the detection of drugs and their metabolites [49] in tissue samples. It has also been applied to dried blood spot analysis [55].

1.2.3.2 DART Direct analysis in real time [46] is based on the reaction of corona discharge-generated metastable helium atoms with oxygen/water (negative mode) or with water clusters (positive mode). The analytes are ionized by the reactant ions either by cluster-assisted desorption or by proton exchange. Both methods generate mostly protonated or deprotonated molecular ions. Various applications of both techniques in the analysis of mass spectrometric profiling of intact biological tissue, the characterization of the active ingredients in pharmaceutical samples formulated as tablets and ointments, and the sampling of plant material have been reported. Although DART is a new development, simpler devices have been described for the analysis of volatile or semi-volatile liquid or solid materials. Atmospheric pressure solid probe analysis (ASAP) is based on APCI with corona discharge while the analytes are desorbed by a hot nitrogen gas stream (350–400°C) [56].

With all the new development in ionization techniques, it becomes quite difficult to compare them. A couple of reports stated that because negative ion (NI) APPI and DART can produce negative ions by EC, dissociative EC, proton abstraction, and halide attachment, these methods clearly ionize a wider array of compounds than NI APCI, ESI, or even chemical ionization (CI). McEwen et al. [57] investigated the API mechanism of APPI, APCI, and DART, and found that irrespective of the initial method of ionization, similar ion/electron molecule chemistries dominates. They suggest that it is not the initial method of producing the primary ions that defines the ionization efficiency, but it is the ion/electron molecule reactions. Furthermore, the sensitivity difference between the techniques should be more related to instrumental settings and configuration than to the primary ionization method.

1.2.3.3 Liquid Microjunction Surface Sampling Probe The group of van Berkel further developed a concept presented by Wachs and Henion [58] referred to as liquid microjunction surface sampling probe (LMJ-SSP) [59, 60]. Basically it is a liquid extraction probe that allows the extraction of analytes from a surface followed by electrospray analysis. Another type of liquid extraction probe, referred to as "sealing" surface sampling probe (SSSP), was introduced by Luftmann et al. [61] for the analysis of thin layer chromatography (TLC) spots. In LMJ-SSP the analyte is reconstituted from the surface by connecting the probe composed of two channels with a wall-less liquid microjunction. The liquid flow rate is regulated by the aspirating rate from the nebulizing gas of the pneumatically assisted electrospray or APCI probe. Basically any solvent combination can be applied that is compatible with the ionization method. The probe can operate in the discrete mode or in the rastering mode. The concept was also implemented to a commercially available system, the NanoMate (Advion). The chip-based infusion nano-ESI system (Advion, Ithaca, NY) allows the analysis of LMWCs from three types of surfaces: (1) stainless steel plates, (2) paper, or (3) tissues [60]. The principle of the microliquid junction remains, but the operation of the probe is somewhat different and is depicted in Figure 1.5.

A robotic arm picks up a conductive tip that is filled (typically 5 μL) with a solvent mixture and move above the specific sample to analyze. The tip is lowered to the sample and the liquid creates a junction with the target to extract the analyte. After a certain time the liquid is withdrawn and the tip is moved to the back of an ESI chip composed of 100 to 400 microfabricated nozzles. By applying a suitable potential and pressure to the tip a nanoelectrospray is generated operating at flow rates of between 20 and 500 nL/min. For each analysis a new nozzle and tip are used, allowing for carryover-free analysis.

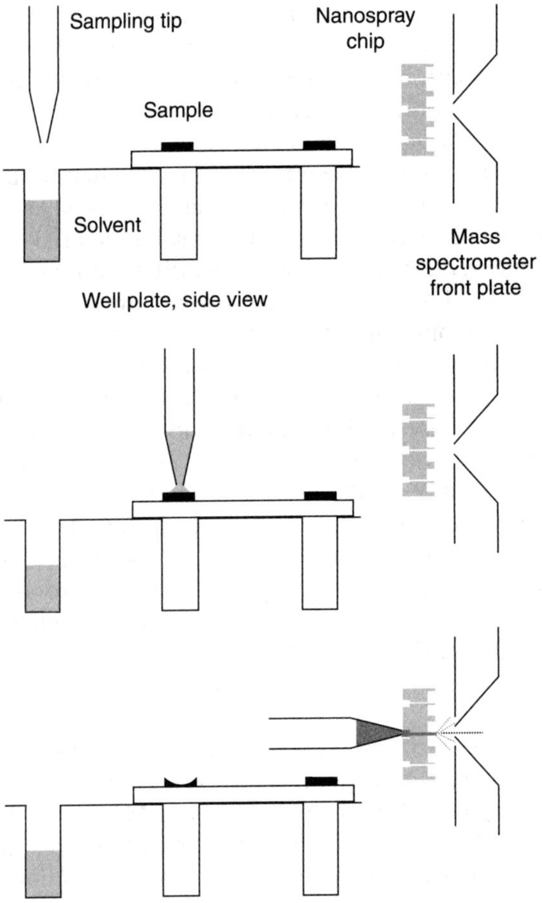

Figure 1.5 NanoMate chip-based infusion nano-ESI system.

1.2.3.4 LAESI Another ambient ionization technique strategy is based on electrospray-assisted laser desorption/ionization. The laser desorption brings the analyte in the gas phase, which is then ionized by electrospray. Basically the sample is deposited orthogonally to an electrospray source and irradiated by laser operating at a defined wavelength. Several variations of this approach, include (1) matrix-assisted laser desorption/ionization ESI (MALDIESI) with an UV laser [62], (2) the use of a nitrogen laser to desorb neutral analytes by a thermal process without matrix (ELDI) [63], and (3) the use a mid-infrared (IR) laser for water-rich samples, which has been described by Nemes and Vertes [64] as consisting of a combination of mid-IR and LAESI. The interaction with the charged droplets generated by the electrospray and neutrals emerging from the laser ablation produces some fused particles, which are the basis of the LAESI signal. The main aim of LAESI is the analysis of untreated water-rich biological samples under ambient conditions, and it was successfully applied on a variety of analytes, including pharmaceuticals, small dye molecules, peptides, proteins, explosives, synthetic polymers, and animal and plant tissues, in both positive and negative ion modes. Urine and blood samples have also been analyzed directly without any pretreatment.

1.2.4 MALDI

In the late 1960s laser ionization was mostly investigated for the analysis of organic solids and inorganic samples. In the late 1980s MALDI emerged from the efforts to analyze macromolecules by mass spectrometry. Two groups have been able to demonstrate the use of laser ionization to obtain singly charged mass spectra of proteins larger than 100 kDa [65, 66]. The group led by T. Tanaka [65] mixed the analyte in a matrix of glycerol and cobalt and used a nitrogen laser operating at 337 nm for sample ionization. The other group, formed by M. Karas and F. Hillenkampf [67], developed MALDI where originally the dissolved analyte is mixed with a matrix solution containing UV-absorbing molecules such as nicotinic acid and subjected to an Nd:YAG laser operating at 266 nm for ion generation.

In both methods, singly charged ions are formed and the laser ionization technique was coupled with a TOF mass analyzer. In MALDI, the analyte is typically diluted in a large excess of matrix. A few microliters of the solution is spotted onto a MALDI target where the sample crystallizes. After introduction of the target into the vacuum, a short pulse of a few nanoseconds from a UV laser is used to desorb and ionize the sample. The nitrogen laser emitting at 337 nm and the Nd:YAG laser emitting at 355 nm or 266 nm are the most widely used, with an operating frequency range of 20–200 Hz. More recently lasers with frequencies as high as 1000 Hz are used, in particular in MSI applications. MALDI has largely replaced former techniques such as fast-atom bombardment (FAB) for the analysis of high molecular weight compounds such as peptides and proteins [68], synthetic polymers [69], DNA [70], and lipids [71]. In most cases intact singly charged ions of the analytes or their dimers and more rarely multiply charged ions are observed. As potential matrices, many compounds have been tested randomly [72] or by design [73, 74], but only a few are now used routinely, and the most common matrix compounds are listed in Table 1.1.

The ionization mechanisms involved in MALDI are still under investigation and several of them have been proposed [75]. Unfortunately, the current understanding is not sufficient to allow the selection of the best matrix compound based only on the physicochemical properties of the analytes. In MALDI the matrix serves several functions. First, the matrix must have a strong UV chromophore to absorb the incident laser light and to allow an efficient energy transfer to the matrix causing the sample to disintegrate. Second, after ablation the matrix is vaporized together with ions and neutrals and prevents the formation of aggregates (Fig. 1.6).

Ionic compounds, which are liquids at room temperature, characterized by a very low vapor pressure, have been reported as interesting MALDI matrices because for the analysis of LMWCs the intensity of matrix-derived peaks was found to be lower than conventional matrices. These matrices are generally formed by mixing 2,5-DHB, CHCA, or sinnapinic acid (SA) with an equimolar amount of organic base such as tributylamine, pyridine, or 1-methylimidazole [76].

Matrix background can be fully eliminated by applying matrix-free laser desorption ionization where the sample is placed on a photoactive but nondesorbable support [77]. Desorption/ionization on porous silicon (DIOS) [78] without any

TABLE 1.1 Various Matrices Used for Matrix-Assisted Laser Desorption/Ionization

Matrix	Name	Comments
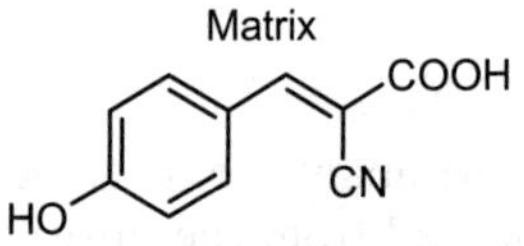	α-cyano-4-hydroxy-cinnamic acid (4-HCCA)	Peptides, low molecular weight compounds
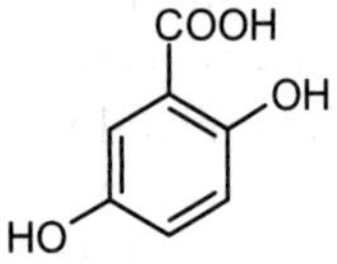	2,5-dihydroxy benzoic acid (DHB)	Proteins
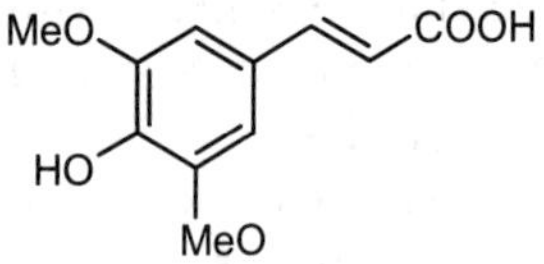	Sinnapinic acid (SA)	Proteins
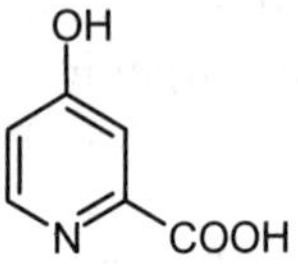	4-hydroxypicolinic acid (HPA)	Oligonucleotides
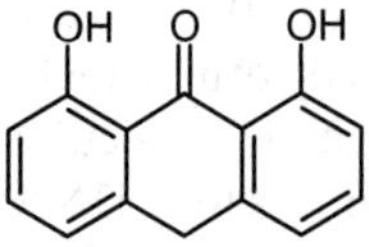	1,8,9 antrancetriol (dithranol)	Polar and apolar polymers

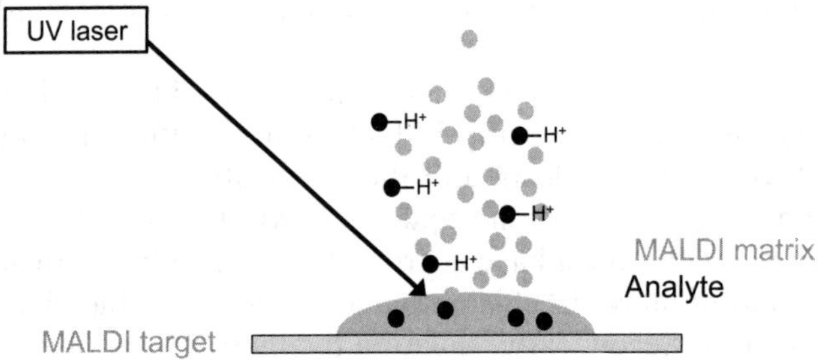

Figure 1.6 Matrix-assisted laser desorption/ionization (MALDI).

matrix has been described for the analysis of LMWCs with no chemical background [79, 80].

Compared with electrospray HPLC-MS, MALDI-MS has an intrinsic advantage in that it can achieve a higher sample throughput because the analyte separation can be decoupled from the mass spectrometric analysis and even multiplexed. The MALDI target plate can be easily archived, which allows simply the re-analysis of selected samples. One important difference between ESI and MALDI is that in ESI multiply charged ions are formed, allowing the analysis of proteins on almost any type of mass analyzer, while in MALDI a TOF mass analyzer is required in the linear mode to cover the high mass range typically needed for proteins.

The higher throughput capability of MALDI and the different ionization mechanisms make this technique also an attractive alternative to ESI for the analysis of LMWCs [81]. However, interferences of matrix ions with the protonated molecules of the LMWCs may jeopardize somewhat the larger application of MALDI-TOF despite significant improvements in the mass resolution of TOF instruments [82]. Due to the high analysis speed in the selected reaction monitoring mode (SRM), the coupling of a MALDI source with a triple quadrupole mass analyzer for quantitative analysis or for tissue imaging is particularly attractive. Combined with a triple quadrupole LIT MALDI offers interesting perspectives in the analysis of pharmaceuticals, drugs of abuse, or peptides [83, 84]. Because MALDI is a desorption technique, it is particularly suited for the analysis of surfaces such as biological tissues [85] and for the generation of mass spectrometric images of endogenous or exogenous analytes. For mass spectrometric imaging (MSI) applications, the matrix is applied on the complete surface of the tissue generally by spraying. The laser resolution is about 50–200 μm and is operated either in rastering or in discrete mode and complete analyte distribution (LMWCs, peptides, proteins) images can be recorded [86–88].

While MALDI has been widely used in the vacuum with TOF, TOF/TOF or triple quadrupole instrument its application at atmospheric pressure (API-MALDI) has also been described [89]. API-MALDI sources are commercially available and can be mounted on any type of electrospray-based instrument. Schneider et al. [90] performed a comparative study of vacuum-MALDI and API-MALDI on a triple quadrupole LIT instrument. They concluded based on the signal/background ratio that for peptides analysis both techniques provided similar results, while API-MALDI was more difficult to optimize and that thermal degradation of the analytes were observed.

Surface enhanced laser desorption/ionization (SELDI) is a distinctive form of laser desorption/ionization where the target is used in the sample preparation procedure and the ionization process [91]. The SELDI target surface can act as solid phase extraction or an affinity probe depending on the chemical or biochemical treatment of the chip surface. Chromatographic surface is used in sample fractionation and purification of biological samples prior to direct analysis by laser desorption/ionization. SELDI is mainly applied in biological fluids for protein profiling and in biomarker discovery by comparing protein profiles from control and patient groups.

1.2.5 ICP

Inductively coupled plasma mass spectrometry (ICP-MS) has gained significant interest over the last years as a sensitive technique for absolute quantitation of elements, in particular, metals. Detectable drugs or analytes are limited to halogen-, sulfur-, metal-, and metalloid-containing molecules [92]. In (ICP) ionization, the sample must be in the liquid form pumped with a peristaltic pump into a nebulizer, where it is nebulized with the help of argon gas. The fine droplets are separated from the larger ones in the spray chamber, and emerge into the ICP plasma torch.

An ionized gas is formed by the interaction of a high magnetic field, produced by a radio frequency (RF), and argon and, when bombarded with electron forms a high temperature plasma discharge. Despite the limited use of ICP-MS, it is an alternative technique for drug metabolism studies in particular since it can be combined with liquid chromatography and the response is independent from the analyte structure.

1.3 MASS ANALYZERS

1.3.1 Triple Quadrupole Systems

A two-dimensional (2D) quadrupole field is generated when applying an electric potential ($\Phi(x,y)$ within four hyperbolic or circular rods placed in parallel with identical diagonal distances from each other. The rods are electrically connected in pairs. An alternating RF potential (V) and a positive direct current (DC) potential (U) is applied on one pair of rods $\Phi(x)$ while a negative potential is applied on the other pair $\Phi(y)$. The ion trajectory is affected in the x and y directions by the total electric field composed of a quadrupolar alternating field and a constant field. When accelerated ions enter the quadrupole they maintain their velocity along the z-axis (Fig. 1.7).

The motion of ions in the quadrupole (x and y) is quite complex and its stability is defined by the solution of the Matthieu equations. The Matthieu equation contains two parameters (terms a and q), which are proportional to the RF and DC potentials, respectively. The solutions of the Matthieu equations are classified stable and unstable. An ion motion is stable when the amplitude of its oscillations never reaches the rods of the quadrupole. For a detailed description of Matthieu equations, please see the book by March and Todd [93]. To be detectable, an ion must have a stable trajectory in the x and y directions. Although there are an infinite number of

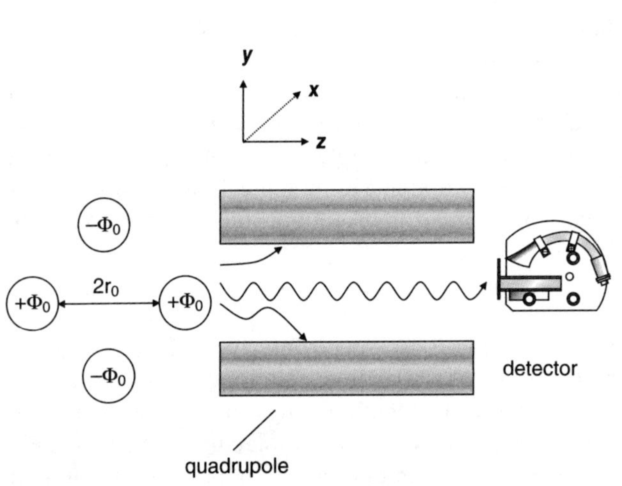

Figure 1.7 The quadrupole mass analyzer is formed by four circular or hyperbolic rods placed in parallel, Φ quadrupolar potential.

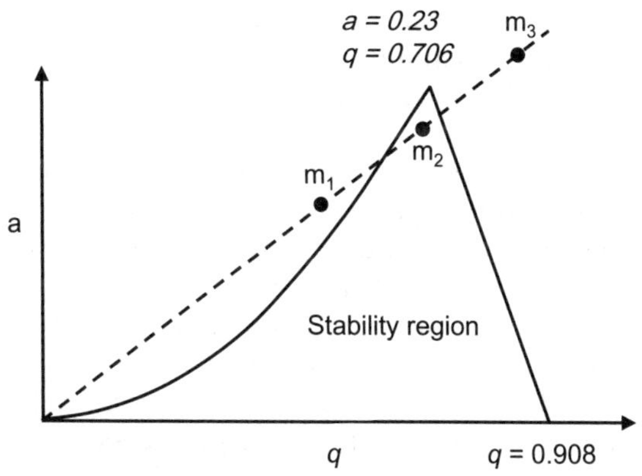

Figure 1.8 First stability region with scan line.

stability regions, most commercial quadrupoles operate in the "first" stability region, which is illustrated in Figure 1.8.

To detect an ion m_2 (Fig. 1.8) the RF and DC voltage are to be set in such a manner that m_2 is almost at the apex of the stability diagram. A lighter ion m_1 and a heavier ion m_3 with larger and smaller a and q values are outside the stability diagram. To obtain a mass spectrum, the RF and DC voltages are cramped together at a fixed ratio while their respective amplitudes are increased. In a quadrupole mass analyzer, when the DC voltage of the quadrupole is set to 0 and the RF voltage is maintained ions remain focused with no mass selectivity. Therefore, RF quadrupoles are ideal as ion guides or as collision cells. Typically, quadrupole mass analyzers operate at a unit mass resolution corresponding to a peak full width at half maximum (FWHM) of 0.6–0.7 *m/z* units. The resolution can be tuned by changing the a/q or U/V ratio. Resolution corresponding to a peak width of 0.1 *m/z* units without significant loss in sensitivity has also been reported [94, 95]. The mass range of quadrupole is typically between *m/z* 5 to *m/z* 4000 and dependent on the operating frequency (in the KHz–MHz range). The higher the frequency the lower the mass range. Often quadrupole mass analyzers are wrongly considered as slow scanning instruments. In general, most quadrupoles used for LC-MS are operated with scanning speeds of 500 to 2000 units/s, and a triple quadrupole with a scan speed of more 10,000 units/s have also been commercialized. Most common ionization sources are available on quadrupole instruments including electron ionization (EI), ESI, APCI, APPI, and MALDI.

A triple quadrupole instrument (QqQ) is a combination of two mass resolving quadrupole (tandem mass spectrometry) separated by a collision cell, which is also a quadrupole operating in the RF-only mode (Fig. 1.8). A common nomenclature is to use (Q) to describe a quadrupole that is operated in the RF/DC mode and (q) for a quadrupole that is operated in the RF-only mode. Tandem mass spectrometry or MS/MS is required to obtain structural information, in particular for soft ionization techniques such as electrospray. In a first step, a specific *m/z* ion (precursor ion) is selected in the first mass analyzer (Q1). Collision-induced dissociation (CID) occurs in the collision cell (q2) where precursor ions collided with a neutral gas such as argon or nitrogen. The fragment ions (product ions) are then sorted according to their mass-to-charge ratio in the second mass analyzer (Q3) and recorded by the detector. Because the generation of the product ions is performed sequentially, this type of CID experiment is called MS/MS in space; this is in contrast to quadrupole ion traps (QITs) where MS/MS experiments are performed in time. On triple quadrupole mass spectrometers, the potentials used to perform CID are in the range of 0–250 V. As the collision energy is defined in electrons volts (eV) it is therefore dependent on the charge of the ions. For a potential difference of 30 volts, the collision energy for a singly charged precursor ion would be 30 eV, and for a doubly charged precursor ion, 60 eV. The nature of the collision gas (N_2 or Ar) does not affect the product ion spectrum, except the energy needed to achieve similar fragment ratios. The gas pressure in the collision cell mainly influences the sensitivity, while collision energy changes the nature and intensity of the fragments (Fig. 1.9).

Quadrupole mass analyzers can be operated in transmission, scan, or fixed mode. Various types of MS and MS/MS experiments can be performed on a QqQ and are summarized in Table 1.2. A symbolism to describe various MS/MS or multistage MS^n experiments have also been proposed [96, 97].

The product ion mode is generally used to record the collision-induced fragments of a precursor ion to allow structural elucidation. The precursor ion mode is used

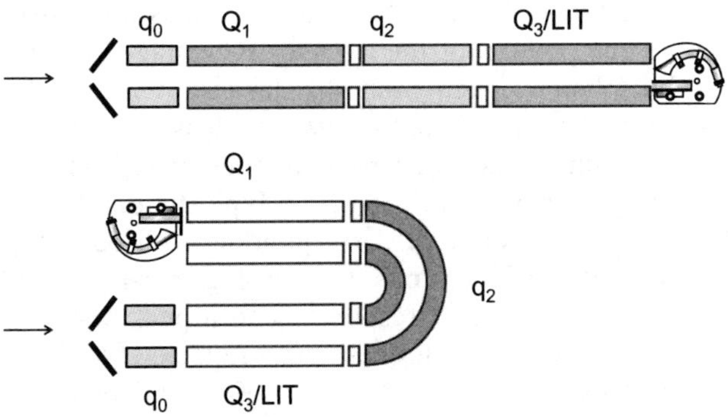

Figure 1.9 Schematic of a triple quadrupole instrument, q0 focusing quadrupole, Q1 and Q3 mass analyzing quadrupoles, q2 collision cell. In the present configuration, the collision energy (CE) is determined by the potential difference between q0 and q2.

TABLE 1.2 Settings of the Q1 and Q3 Quadrupoles for the Various Scan Modes of a Triple Quadrupole

Mode	Q1	Q3
Full Scan Q1/Single ion monitoring (SIM) Q1	Scan/fixed	RF mode
Full Scan Q3/Single ion monitoring (SIM) Q3	RF mode	Scan/fixed
Product ion scan (PIS)	Fixed	Scan
Precursor ion scan (PC)	Scan	Fixed
Neutral loss (NL)	Scan	Scan-neutral loss offset
Selected reaction monitoring (SRM)	Fixed	Fixed

to screen (m/z values corresponding generally to precursor ions) for structural analogues based on a common fragment. However, this mode requires an understanding of the fragmentation process. Neutral loss scanning, where Q3 is scanning along with Q1, but with an offset corresponding to the loss of neutral fragment, is particularly attractive to screen for phase II metabolites such a sulfates and glucuronides. Finally the SRM mode, where Q1 is fixed at the m/z of the precursor and Q3 at the m/z of a fragment, is well suited for quantitative analysis with dwell times in the range of 1 to 500 ms. Typically, a dwell time of 10 ms per transition would allow quantification of several dozen analytes simultaneously by LC-MS.

1.3.2 Ion Trap

The QIT and the related quadrupole mass filters were invented by Paul and Steinwedel in the 1950s [98]. The QIT is a device that utilizes the ion path stability for separating ions by their mass-to-charge ratio [93]. QIT mass spectrometer operates with a three-dimensional (3D) quadrupole field, while LIT operates with 2D quadrupole field. The QIT is formed by three electrodes: a donut-shaped ring electrode placed symmetrically between two end cap electrodes (Fig. 1.10).

By applying a fundamental RF potential, the QIT can be described as a small ions storage device where the ions are focused toward the center of the trap by collisional cooling with helium gas. The use of nitrogen or argon has also been described. In the QIT the x and y components of the field are combined to a single radial r component where $r^2 = x^2 + y^2$ because of the cylindrical symmetry of the trap. The motion of ions in the trap is characterized by secular frequencies, one radial and one axial. As for quadrupoles the stable motion of ions follow the solutions of Matthieu's equations (a and q). Ions can be stored in the trap in the condition that trajectories are stable in the r and z directions. Each ion at a given m/z will be trapped a certain q_z value. The lower m/z will be located at the higher q_z values, while the maximum q value is 0.908. The QIT can store only a limited number of ions before space charging occurs, and most instruments have an automatic gain control (AGC) procedure, which controls the optimal number of ions being injected. A mass spectrum can be obtained by mass-selective ejection where the amplitude of the RF potential is continuously increased at a certain rate. The mass-selective

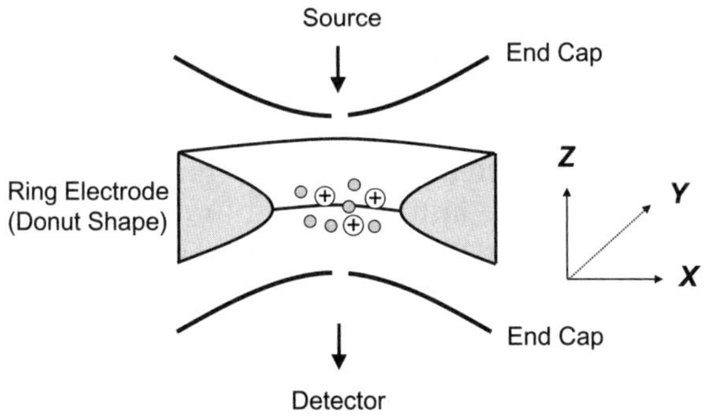

Figure 1.10 The quadrupole ion trap.

axial instability mode requires that the ions are confined at the center of the trap and at a limited mass range, typically up to *m/z* 600. To achieve a higher mass range of up to *m/z* 2000, resonant mass ejection modifies ion motion by exciting either the radial or the axial frequencies by applying a small oscillating potential at the end cap electrodes during the RF ramp. In both mass analyzing modes the resolution of the spectrum is strongly dependent on the scan speed.

The best resolution can be obtained with a slow scan speed compared with unit mass resolution. When used in the full scan mode and compared with the quadrupole, the QIT can provide higher sensitivities due to the ability to use ion accumulation in the trap before mass analysis. Rapid mass analysis with the mass instability scan allows scanning at a speed of several ten thousand *m/z* units per second. There are several important components that affect the time necessary to generate a mass spectrum (duty cycle): (1) the injection time from 0.5 to 500 ms, (2) the scan speed (in the range of 250 to 20,000 units/s), and (3) isolation of the precursor ion and fragmentation in tandem MS or MS^n. Contrarily to the triple quadrupole, MS/MS is not performed in *space* but in *time*. Another significant difference is the use of helium as collision gas. Because the trap is permanently filled with gas, the instrument can switch very rapidly from single MS to MS/MS mode. High sensitivity can be achieved in the QIT because of selective accumulation of the precursor. After injection the precursor ion is isolated and then excited while fragments are trapped. The next MS^3 step would be to isolate a fragment ion to perform isolation and CID fragmentation. Because MS/MS is performed in time in the same physical device, the operation can be repeated several times. Most commercial instruments can perform MS^n up to 10 or 11 times. A challenge in trap MS/MS is to excite the precursor ions efficiently and trap the product ions in the same device. Generally, solely the precursor is excited in a specific window corresponding to 1 to 4 *m/z* units. The consequence is that fragment ions are not further excited and cannot produce second-generation fragments. In many cases MS^2 trap CID generates mass spectra similar to quadrupole CID, but there are many cases where the mass spectra differ significantly. Greater similarity can be achieved when combining MS^2 and MS^3 trap spectra.

For molecules that can easily lose water or ammonia, the most abundant fragment observed in MS^2 will be M-18 or M-17, which is not very informative. To overcome this limitation, wide band excitation (range 20 *m/z* units) can be applied. Another difference compared to QqQ is that QIT have a low mass cutoff of about one third of the mass of the precursor ion.

Due to their high sensitivity, QITs are mainly used for qualitative analysis in drug metabolism and proteomics studies. QITs can achieve similar sensitivies to QqQ but at the expense of precision and accuracy. A major difference is the number of transitions that can be monitored at the same time. While several hundred SRM transitions can be recorded on a QqQ with dwell times of 1 ms, this number is much lower with a QIT (generally 4 to 8 transitions). Ion traps have mass ranges (up to 50,000) higher than quadrupole instruments but lower than TOF mass analyzers. Most commercial instruments use two mass ranges: (1) *m/z* 50 to 2000–3000 with a mass resolution of 0.6 *m/z* units or higher, and (2) *m/z* 200 to 4000–6000 with a mass resolution of 2–4 *m/z* units.

1.3.2.1 LIT Over the last years LITs have gained much interest for various applications, either as a standalone mass analyzer or coupled with a Fourier

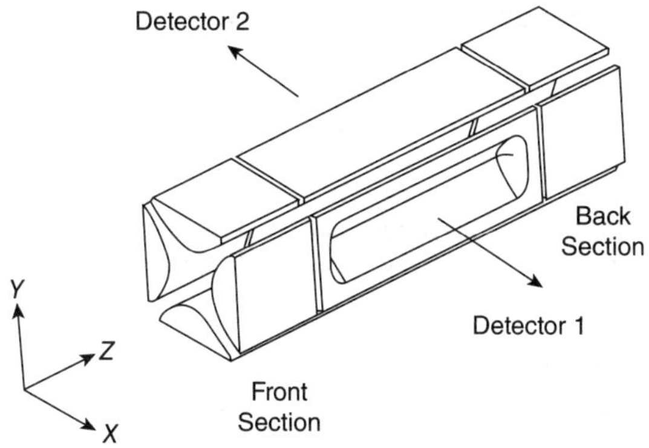

Figure 1.11 Standalone linear ion trap. Because the ions are ejected radially, two detectors are required for best sensitivity (with permission from Reference 100).

transform ion cyclotron, 3D IT, TOF, or an Orbitrap mass analyzer [99]. In a LIT, the ions are confined radially by a 2D RF field. To avoid that, the ions escape axially when a DC potential is applied to the end of the quadrupoles. Trapping ions in a 2D trap has several advantages over using 3D traps: (1) there is no quadrupole field along the *z*-axis, (2) trapping efficiency is enhanced, (3) more ions can be stored before space charging effects occur, and (4) strong focusing of ions can be done along the center line instead of focusing ions to a point. Schwartz et al. have [100] described a standalone LIT where mass analysis is performed by ejecting the ions radially through slits of the rods using the mass instability mode (Fig. 1.11). To maximize sensitivity, the detection is performed by two detectors placed axially on either side of the rods.

LITs have been successfully coupled to TOF (LIT-TOFMS) [101] and Fourier transform-ion cyclotron resosnance (FT-ICR)-MS [102, 103]. Such hybrid instruments are designed to combine ion accumulation and MS^n features with the superior mass analysis (accuracy and resolution) and the high sensitivity of TOF-MS or FTICR-MS. The ions stored in the trap are then axially ejected in a nonmass-dependent fashion to the mass analyzer.

1.3.2.2 Triple Quadrupole LIT Basically one can operate a quadrupole in the classical quadrupole ion mode by applying RF/DC potentials or in the LIT mode by applying an RF potential. In a LIT, the most efficient way of performing mass analysis is to eject ions radially. Hager [104] has demonstrated that by using the fringe field effects, ions can be mass selectively ejected in the axial direction. There are several benefits of axial ejection compared with radial ejection: (1) it does not require open slits in the quadrupole, and (2) the device can be operated either as a regular quadrupole or as a LIT using one detector. A hybrid mass spectrometer was developed based on a triple quadrupole platform where Q3 can be operated either in normal RF/DC mode or in the LIT ion trap mode, named the QTRAP. Linear and curve geometries have been also commercialized.

In the QTRAP, tandem MS is performed in space where the LIT serves only as trapping and mass analyzing device. While quadrupole CID spectra are obtained in MS^2, MS^3 are typical trap CID spectra. MS^3 is performed in the following manner:

TABLE 1.3 Modes of Operation of the Triple Quadrupole Linear Ion Trap (QqQ_{LIT})

Mode of Operation	Q1	q2	Q3
Q1 Scan	Resolving (Scan)	RF-only	RF-only
Q3 Scan	RF-only	RF-only	Resolving (Scan)
Product ion scan (PI)	Resolving (Fixed)	Fragment	Resolving (Scan)
Precursor ion scan (PC)	Resolving (Scan)	Fragment	Resolving (Fixed)
Neutral loss scan (NL)	Resolving (Scan)	Fragment	Resolving (Scan Offset)
Selected reaction monitoring mode (SRM)	Resolving (Fixed)	Fragment	Resolving (Fixed)
Enhanced Q3 single MS (EMS)	RF-only	No fragment	Trap/scan
Enhanced product ion (EPI)	Resolving (Fixed)	Fragment	Trap/scan
MS^3	Resolving (Fixed)	Fragment	Isolation/fragment trap/scan
Time delayed fragmentation (TDF)	Resolving (Fixed)	Trap/No fragment	Fragment/trap/ scan
Enhanced resolution Q3 single MS (ER)	RF-only	No fragment	Trap/scan
Enhanced multiply charged (EMC)	RF-only	No fragment	Trap/scan

the first stage of fragmentation is accomplished by accelerating the precursor ions chosen by Q1 into the pressurized collision cell, q2. The fragments and residual precursor ions are transmitted into the Q3 LIT mass spectrometer and are cooled for approximately 10 ms. The next generation precursor ion is isolated within the LIT by application of resolving DC near the apex of the stability diagram. The ions are then excited by a single frequency of 85 kHz auxiliary signal and fragmented. The particularity of the QqQ_{LIT} is that the instrument can be operated in various ways as described in Table 1.3 [105, 106]. MS^2 spectra are obtained in the quadrupole CID mode, while MS^3 spectra are obtained in the trap CID mode.

The major advantage of this instrument is that both qualitative and qualitative analysis can be performed in the same LC-MS run. As an example in a data-dependent experiment, the selected reaction monitoring mode can be used as a survey scan and the enhanced product ion (EPI) as a dependent scan. The consequence is that for each quantified analyte a confirmatory MS/MS spectrum can be obtained.

1.3.3 TOF

Time-of-flight may be the simplest way to perform mass spectrometric analysis (Fig. 1.12). TOF is the measure of the time that ions need to cross in a field free tube of a length of about 1 m [107, 108]. It is a pulsed technique and requires a starting point or event. At the starting point all ions get the same kinetic energy kick. As the motion of an ion is characterized by its kinetic energy $E_c = 1/2\ mv^2$ (m = mass, v = speed), the speed of ions or the time to fly through the tube is proportional to

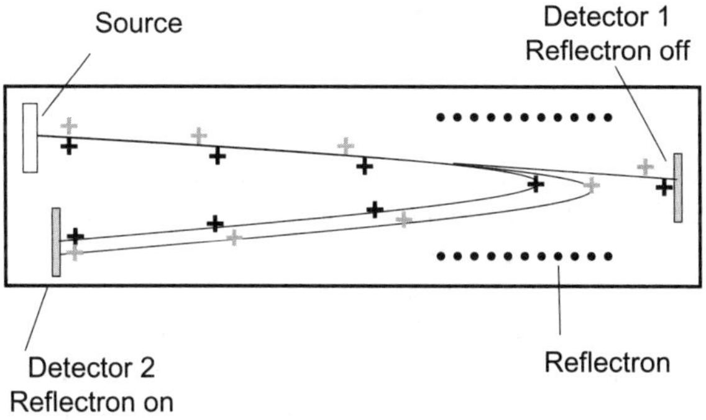

Figure 1.12 Schematic of a time-of-flight mass spectrometer equipped with a reflectron. The instrument can be operated in the linear mode (reflectron off) or in the reflectron mode (reflector on).

their $\sqrt{m/z}$ value. After ionization the velocity of the ions is generally low, and they need to be accelerated by strong electric fields (2–30 KV). Low mass ions reach the detector sooner than high mass ions. Due to the short fly time (50–100 µns) and the good transmission, a mass spectrum can be generated within 100 m over an almost unlimited mass range. Detection of the ions is performed with either an analog or a digital multichannel plate detector (MCP), which has, in general, a relatively small dynamic range (two to three orders of magnitude).

With ionization techniques such as MALDI, singly charged ions of several hundred thousands of m/z can be routinely detected. The mass range is mainly limited by the fact that the detector response decreases with the increase in m/z of the ions. At first, the resolution of a TOF mass analyzer were relatively poor (unit mass resolution or less) because there are factors that create a distribution in the flight time of ions with the same m/z. There are two ways to increase the resolution: increase the length of flight tube or reduce the kinetic energy spread of the ions leaving the source.

The resolution can be improved significantly by applying an electrostatic mirror (mass reflectron) placed in the drift region of ions (Fig. 1.12). Briefly, ions with higher energy penetrate deeper into the ion mirror region than those with the same m/z at lower energy. Because the different trajectories in the reflectron, all ions of the same m/z reach the detector at the same time. Thus, all ions of the same m/z have a much lower energy dispersion. With the reflectron the flight path is also increased without changing the physical size of the instrument. In the reflectron mode a typical mass resolving power of 10,000 is standard, while instruments with resolving power of 30,000 and higher have become commercially available. TOF instruments can detect simultaneously a large mass range of ions resulting in an increased sensitivity compared with scanning instruments.

In general the commercial MALDI-TOF instruments have two detectors: one for the linear mode and one for the reflectron mode. A TOF detector must have a fast response, and physically it must be relatively large because the ion beam should be several centimeters large. TOF instruments are equipped with microchannel plate

(MCP) electron multipliers [109], which are flat arrays of micrometer-sized channels acting as a single electron multiplier.

The combination of MALDI with TOF is ideal because both are pulsed techniques. However, it is also possible to utilize continuous beams as generated by ESI. For that purpose orthogonal acceleration was developed [110]. The ion beam is introduced perpendicularly to the TOF and packets are accelerated orthogonally (oa-TOF) at similar frequencies, improving the sensitivity. While a packet of ions is analyzed, a new beam is formed in the orthogonal acceleration.

TOF instrument have been mainly used for qualitative analysis. With MALDI one of the main applications is the identification of proteins by analyzing their peptides after trypsin digestion (peptide mass finger print [PMF]). For TOF, the acquisition rate is in the range of 10 to 100 Hz; therefore these mass analyzers are best suited for the interfacing of fast liquid chromatographic separations or capillary electrophoresis with ESI. Due to the fast acquisition rates and high-resolution capabilities of TOF mass analyzers, they are often used as the last mass analyzing stage in hybrid tandem mass spectrometers (QqTOF or QIT-TOF).

Full scan tandem mass spectrometry with triple quadrupole (QqQ) instruments suffers from a relatively poor sensitivity and limited mass resolution. The replacement of the last quadrupole section (Q3) of a triple quadrupole by a TOF analyzer to form a hybrid quadrupole-time-of-flight instrument (QqTOF) represents a powerful combination in terms of mass range (*m*/*z* 5 to *m*/*z* 40,000), mass resolution (resolving power of 10,000 or higher), and sensitivity [111, 112]. In single MS mode, the quadrupoles (q0, Q1, and q2) serve as RF ion guides and the mass analysis is performed in the TOF. After the collision cell, which can be filled with collision gas, the ions are reaccelerated and refocused before they enter the ion modulator. A pulsed field is applied in the ion modulator to push the ions orthogonally to their initial direction into the TOF analyzer. In the RF-only mode, quadrupoles have a high mass cutoff and ions with high *m*/*z* values can be lost in any of the three quadrupole regions during spectrum acquisition. To extend the transmission window in q0, Q1, and q2, the RF voltage is modulated during full scan spectrum acquisition but at the cost of duty cycle (Fig. 1.13).

In the tandem MS mode, because the product ions are recorded with the same TOF mass analyzer as in the full scan mode, the same high resolution and mass accuracy is obtained. Isolation of the precursor ion can be performed either at unit mass resolution or at 2–3 *m*/*z* units for multiply charged ions. Accurate mass measurements of the elemental composition of product ions greatly facilitate spectral interpretation. The main applications are peptide analysis and metabolite identification using ESI [113]. In TOF mass analyzers, accurate mass determination can be affected by various parameters such as (1) ion intensities, (2) room temperature, and (3) detector dead time. Interestingly, the mass spectrum can be recalibrated post-acquisition using the mass of known ions (lock masses). The lock mass can be a cluster ion in full scan mode or the residual precursor ion in product ion mode. For LC-MS analysis a dual spray (LockSpray, Waters, Milford, MA) source has been described, which allows the continuous introduction of a reference analyte into the mass spectrometer for improved accurate mass measurements [114].

The versatile precursor ion scan, another specific feature of the triple quadrupole, is maintained in the QqTOF instrument. However, in the precursor scan mode, sensitivity is lower in QqTOF than in QqQ instruments. The low duty cycle quadrupole

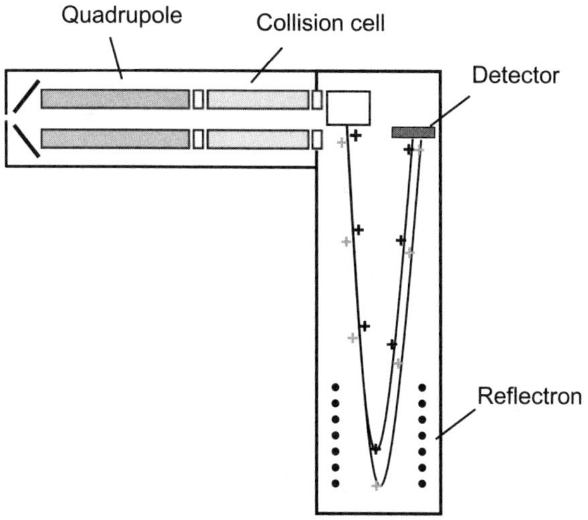

Figure 1.13 Schematic of a quadrupole time-of-flight instrument. Quadrupole q0 is used for collisional cooling and ion focusing. Nitrogen or argon is generally used as the collision gas. The ion modulator pushes the ions orthogonally to their initial direction into the TOF analyzer.

represents a significant part of these losses. To improve the sensitivity in the precursor ion mode, ions can be trapped in q2 collision cell and released as a short burst into the ion modulator. The lack of good quality product ion spectra on a conventional MALDI-TOF instrument made the use of MALDI on a QqTOF instrument an interesting alternative for the sequencing of peptides. Also MALDI is complementary to electrospray and shows better tolerance to sample contamination. As for electrospray TOF, in the case of the QqTOF, the MALDI ion production process needs to be decoupled from the mass measurement step. The technique to interface MALDI with QqTOF is named orthogonal MALDI (o-MALDI) TOF with collisional cooling. With o-MALDI the pulse is almost converted in a continuous beam equivalent to that originated from an electrospray source.

The TOF mass analyzer has a low duty cycle, and the combination with an ion accumulation device such as an ion trap is therefore very advantageous. It also offers MS^n capabilities with accurate mass measurement. In all acquisition modes, the ions are accelerated into the TOF for mass analysis. Both the QIT [115] and LIT [99] have been combined with TOF either with MALDI or ESI sources.

Various tandem TOF combination (TOF/TOF) have been described and some of the commercial ones are presented in Figure 1.14 [116]. Vestal and Campbell [117] have developed an instrument allowing high energy CID MALDI TOF/TOF. Basically, the instrument is composed by two time-of flight sections connected through a collision cell as depicted in Figure 1.14a. The first TOF operates in a linear mode, while the second TOF is equipped with a reflectron for improved resolution.

In the TOF/TOF instrument, selection of the precursor ion is performed by a timed ion selector (TIS). The TIS is basically an ion gate where ions outside the

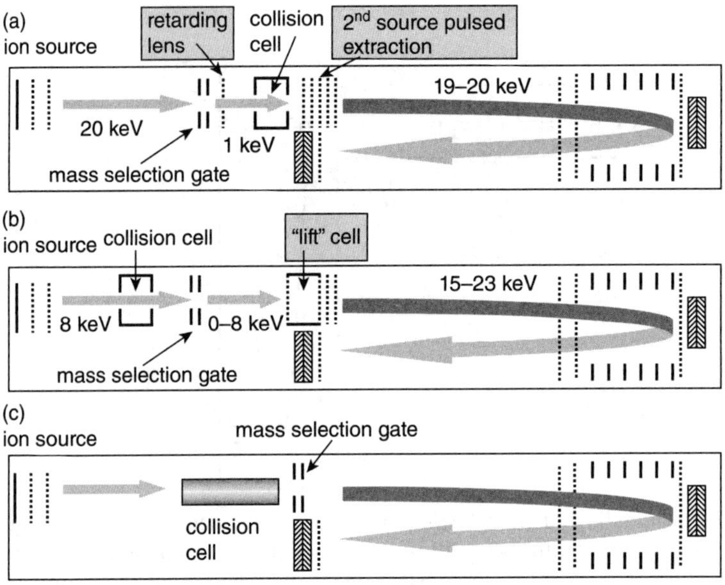

Figure 1.14 Schematic comparison of various commercial tandem TOF/TOF geometry. (a) AB Sciex, Toronto, Canada; (b) Bruker, Bremen, Germany; (c) Shimadzu, Manchester, UK (adapted with permission from Reference 116).

selected mass window are deflected. The isolation window of the precursor ions is typically in the range 3–7 *m*/*z* units. CID energy is defined by the potential difference between the source potential and collision cell, and various gases such as argon, neon, and nitrogen can be used. Under standard operation conditions of 1 kV both low-energy and high-energy fragments can be observed.

A different configuration has been described by Suckau et al. [118] that allows laser-induced dissociation (LID) MS/MS or CID MS/MS, providing high energy CID for the analysis of peptides. In the mass spectrometer (Fig. 1.14b) the collision cell is placed prior to the TIS. The "LIFT" device is the heart of the system where precursor and fragments ions are post-accelerated to a kinetic energy in the range of 19 to 27 kV. High energy CID allows differentiation of isobaric leucine and isoleucine amino acid residues by the side-chain fragmentation. In the last configuration (Fig. 1.14c) the mass selection gate is placed after the collision cell.

1.3.4 Fourier Transform Mass Spectrometry (FT-MS)

1.3.4.1 FT-ICR The main components of an FT-MS are a superconducting magnet and a cubic or cylindrical cell (Fig. 1.15). Typically, field strengths (B) are in the range of 3 to 12 Tesla. Ions are stored in the cell according to their cyclotronic motion arising from the interaction of an ion with the unidirectional constant homogenous magnetic field. A static magnetic field applied on the z direction confines ions in the x- and y-directions according the cyclotronic motion. To avoid

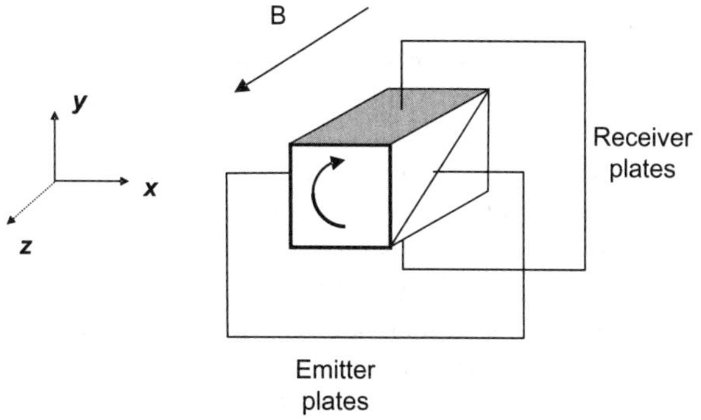

Figure 1.15 Diagram of an ion cyclotron resonance instrument. The magnetic field is oriented along the *z*-axis and ions are trapped according the same axis. Due to the cyclotronic motion, the ions rotate around the *z*-axis in the *x-y* plane.

escaping of ions along the *z*-axis, a low electrostatic potential is applied to the endcap electrodes [119].

The trapping of ions generates a further fundamental motion of ions called magnetron motion. Magnetron frequencies are independent of m/z of the ions and are much lower frequencies (1–100 Hz) than cyclotron motion (5 KHz–5 MHz). Cyclotron motion is characterized by its cyclotron frequency (f), which depends on (1) the magnetic field (B), (2) charge of the ion (z), and (3) and the mass of the ion m.

Unlike other types of mass spectrometers, the detection is performed in a nondestructive way. The ions are detected by excitation applying a coherent broadband excitation on the excitation plates. The ions undergo cyclotron motion as a packet with a larger radius. When the ion packet approaches the detection plates it generates an alternating current named image current. Ions of any mass can be detected simultaneously with FTMS. The resulting signal is called a free induction decay, transient, or interferogram, which consists of a superposition of sine waves and is converted by applying a Fourier transform.

Mass resolution is best with high field strength; it decreases when the mass increases and it is dependent on acquisition time. The resolution (R) is strongly dependent on the length of the transient time (T). Typical transient times are in the range of 0.1 to 2 s. With commercial instruments, resolving power of 100,000 or more can be routinely achieved. CID can also be performed in the FT-ICR cell. The transient signal decreases with collision of ions and neutral gas molecules. It is therefore essential to work at very high vacuum (10^{-10} Torr). The dynamic range of FT-ICR is relatively poor because the instrument suffers from the fact that the number of ions in the trap must be in a specified range. Over- or under-filling of the trap results in mass shifts toward high and low values, respectively. To have a better control of the ion population in the cell, a hybrid instrument was developed that combined LIT and FT-ICR [103]. Because the LIT is equipped with two detectors, data can be acquired simultaneously in the ion trap and in the FT-ICR. In this way, the FT-ICR operates only as a high-resolution detector for MS while MS^n experiments are performed in the LIT.

1.3.4.2 Orbitrap Mass analyzers based on Paul's or Penning's traps have become popular devices for MS analysis in part due to their high sensitivity. The major limitation of QIT is insufficient mass accuracy, while FT-ICR MS remains a complex and expensive device to operate. Makarov [120] invented a novel type of mass spectrometer based on the orbital trapping of ions around a central electrode using electrostatic fields named Orbitrap. Kingdon [121] has already described the orbiting of ions around a central electrode using electrostatic fields in 1923, but the device had been only used for ion capturing and not as a mass analyzer. The Orbitrap is formed by a central spindle-like electrode surrounded by a barrel-like electrode. The m/z is reciprocal proportionate to the frequency (ω) of the ions oscillating along the z-axis. The stability of trajectories in the Orbitrap is achieved only if the ions have sufficient tangential velocity to collide with the inner electrode and short injection times are used [122].

There is no collisional cooling inside the Orbitrap, which operates at very high vacuum (2×10^{-10} mbar). Detection is performed by measuring the current image of the axial motion of the ions around the inner electrode. The mass spectrum is obtained after Fourier transformation of the image current. The mass resolving power depends on the time constant of the decay transient. The Orbitrap provides resolution exceeding 100,000 (FWHM) and a mass accuracy better than 3 ppm. To be operational as a mass spectrometer, the Orbitrap requires external ion accumulation, cooling, and fragmentation. A commercial setup (LTQ-Orbitrap) is depicted in Figure 1.16. The instrument consists of a linear ion trap with two detectors connected to the Orbitrap via a C-trap. On the right side of the C-trap, a high energy collision cell is mounted. While with the LIT trap CID experiments can be performed, the high energy collision cell (HCD) allows higher fragmentation energies and quadrupole CID-type collisions. When the Orbitrap is used as a detector the ions are transferred into the C-trap, where they are collisionally damped by nitrogen at low pressure. Injection into the Orbitrap is then performed with short pulses of high voltages.

The uniqueness of the LTQ–Orbitrap instrument is the independent operation of the Orbitrap and the LIT. Because high resolution requires longer transient time,

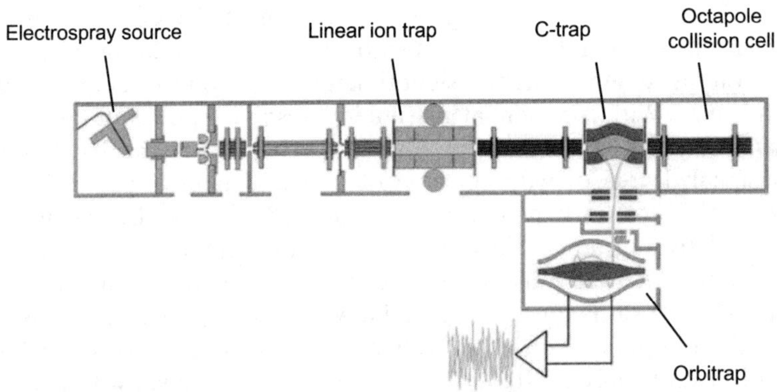

Figure 1.16 Schematic of the LTQ-Orbitrap. One of the features of the system is that linear ion trap (LIT) has two detectors. Therefore, the LIT can perform various experiments at the same time (adapted with permission from Reference 137).

further data can already be collected in the LIT at the same time. As an example, accurate mass measurements of the precursor ion can be performed in the Orbitrap while MS^2 and MS^3 spectra are recorded with the LIT. High resolution and accurate mass have become essential for structure elucidation. In 2011, to allow fast selection of precursor ions prior to HCD fragmentation, an instrument was developed (Q Exactive) where the LIT is replaced by a quadrupole mass analyzer [123].

While FT-MS can achieve the highest resolving power, it is often at the cost of time. For example, to achieve a resolving power of 100,000 the Obitrap requires a duty cycle of more than one second, which is not compatible with fast or ultra-high pressure liquid chromatography. Recent improvements in the resolving power of TOF systems make them attractive alternatives when fast acquisition times are needed when combining ultra-HPLC with mass spectrometry.

1.3.5 Ion Mobility

Ion mobility mass spectrometry (IMS) separates ions on the basis of their size/charge ratios as well as their interactions with a buffer gas (nitrogen, helium). It has gained interest in recent years to improve selectivity or to obtain further structural information of the analytes. In case of drug metabolism, IMS could add a significant new dimension in the separation of structural isomers that have the same mass-to-charge ratio. Different types of IMS have been developed including time drift, aspiration, and differential and traveling wave, which can be interfaced with many types of mass analyzers. The IMS can operate either at atmospheric pressure or reduced pressure conditions [124].

One of the simplest designs of ion mobility at atmospheric pressure is to have ions passed through to flat parallel electrodes or two concentric cylinder electrodes with a flow buffer gas. An alternating electric field is applied between the two electrodes so that the ions move perpendicular to the flow direction in alternating low field and high field directions. It is the difference in mobility in the two directions that separates ions. Generally, the dispersion voltage is fixed and a compensation voltage (CoV) is ramped to separate the different species. Field asymmetric waveform ion mobility (FAIMS) [125] and differential ion mobility (DMS) [126] are the most common names for this type of ion mobility. FAIMS has been applied to improve the selectivity is LC-MS/MS quantitative assays. More recently, the use of modifiers such as isopropanol in the buffer was shown to improve the separation power of DMS significantly [127]. A novel method of IMS has been described and commercialized under the name traveling-wave ion mobility mass spectrometry (TWIMS). In this device the ions are moved through the cell in a pulse of waves. The cell is mounted in a hybrid quadrupole/traveling-wave ion mobility separator/orthogonal acceleration TOF instrument [128].

1.3.6 AMS

Accelerator mass spectrometry is an isotope ratio mass spectrometry-based technique originally developed in the late 1970s for measuring isotopic ratios in environmental and archeological studies [129, 130]. Carbon (^{14}C) remains the most important element where AMS can measure precisely the amount of ^{12}C, ^{13}C, and ^{14}C, but several other elements that have long half-lives can be considered. While in the early days most applications of AMS were focused on environmental studies, in recent years, interest has grown in the use of AMS in the biomedical field, in particular drug quantitation after microdosing [131], as well as metabolite profiling in human studies [132, 133]. In biomedical applications, radioactivity is typically measured by liquid scintillation counting (LSC). AMS offers unmatched sensitivity for measuring ^{14}C amounts in a sample, and it is about 10^6 times more sensitive than LSC [134].

An AMS system includes the following components [135]: (1) a negative ion source, (2) various electric and magnetic fields used as filters, (3) various counting detectors for the various isotopes on elements, (4) an accelerator that accelerates ions to high energies to several millions volts, and (5) a collision cell where ions collide with argon, resulting in a charge stripping from negative to positive charge.

1.4 FUTURE TRENDS

Over the last decades significant progress has been achieved in the field of mass spectrometry regarding ionization techniques and mass analyzers, but there are remaining challenges that require improvements in instrument performance. Although with modern instruments good resolving power can be achieved for the analysis of LMWCs, better sensitivity and high dynamic range are still a concern. High-resolution mass spectrometry does not only improve MS/MS spectra interpretation, but it will also play a major role in the future for integrated QUAL/QUAN workflows. Data-dependent acquisition and MS^{All} approaches are hampered by several restrictions such as the loss of information or the limited selectivity regarding dirty samples. The use of unsupervised MS and MS/MS acquisition may become essential, and approaches such as global precursor scan or sequential window acquisition of all theoretical fragment-ion spectra (SWATH) based on multiple precursor windows and collision energy ramping have shown interesting potential. Certainly, these MS and MS/MS acquisition scheme will require new software development to fully extract the relevant information. CID spectra for many metabolites are not sufficiently informative for many metabolites where only a few fragments are generated. Gas phase chemistry or alternative fragmentation process would be worth exploring to improve fragmentation efficiency.

With many API techniques such as electrospray the MS response factor is analyte dependent and unpredictable. Furthermore, matrix effects can severely affect ionization performance and challenge sample comparison. Alternative ionization methods need to be developed to overcome these limitations.

Miniaturization of mass spectrometers is certainly highly desirable and several approaches have already been described and further improvements are expected for the future. Ion mobility should attract interest, in particular because of its ability to separate isobaric compounds or to improve assay selectivity. It could also be considered an additional orthogonal separation dimension in LC-MS analysis.

REFERENCES

[1] Fenn JB, Mann M, Meng CK, Wong SF, Whitehouse CM. Electrospray ionization—principles and practice. Mass Spectrom Rev 1990;9:37–70.

[2] Alexandrov ML, Gall LN, Krasnov NV, Nikolaev VI, Pavlenko VA, Shkurov VA. Ion extraction from solutions at atmospheric-pressure—a method of mass-spectrometric analysis of bioorganic substances. Dokl Akad Nauk SSSR 1984;277:379–83.

[3] Alexandrov ML, Gall LN, Krasnov NV, Nikolaev VI, Pavlenko VA, Shkurov VA. Extraction of ions from solutions under atmospheric pressure as a method for mass spectrometric analysis of bioorganic compounds. Rapid Commun Mass Spectrom 2008;22:267–70.

[4] Bruins AP, Covey TR, Henion JD. Ion spray interface for combined liquid chromatography/atmospheric pressure ionization mass-spectrometry. Anal Chem 1987;59:2642–6.

[5] Covey TR, Thomson BA, Schneider BB. Atmospheric pressure ion sources. Mass Spectrom Rev 2009;28:870–97.

[6] Hopfgartner G, Bean K, Henion J, Henry R. Ion spray mass spectrometric detection for liquid chromatography: a concentration- or a mass-flow-sensitive device? J Chromatogr 1993;647:51–61.

[7] Blades AT, Ikonomou MG, Kebarle P. Mechanism of electrospray mass-spectrometry—electrospray as an electrolysis cell. Anal Chem 1991;63:2109–14.

[8] Zhou S, Cook KD. Protonation in electrospray mass spectrometry: wrong-way-round or right-way-round? J Am Soc Mass Spectrom 2000;11:961–6.

[9] Tong H, Bell D, Tabei K, Siegel MM. Automated data massaging, interpretation, and e-mailing modules for high throughput open access mass spectrometry. J Am Soc Mass Spectrom 1999;10:1174–7.

[10] Cole RB, Zhu JH. Chloride anion attachment in negative ion electrospray ionization mass spectrometry. Rapid Commun Mass Spectrom 1999;13:607–11.

[11] Rentel C, Strohschein S, Albert K, Bayer E. Silver-plated vitamins: a method of detecting tocopherols and carotenoids in LC/ESI-MS coupling. Anal Chem 1998;70:4394–400.

[12] Ikonomou MG, Blades AT, Kebarle P. Electrospray ion spray—a comparison of mechanisms and performance. Anal Chem 1991;63:1989–98.

[13] Ikonomou MG, Blades AT, Kebarle P. Electrospray mass-spectrometry of methanol and water solutions suppression of electric-discharge with SF_6 gas. J Am Soc Mass Spectrom 1991;2:497–505.

[14] Wilm M, Mann M. Analytical properties of the nanoelectrospray ion source. Anal Chem 1996;68:1–8.

[15] Zhang S, Van Pelt CK. Chip-based nanoelectrospray mass spectrometry for protein characterization. Expert Rev Proteomics 2004;1:449–68.

[16] Koster S, Verpoorte E. A decade of microfluidic analysis coupled with electrospray mass spectrometry: an overview. Lab Chip 2007;7:1394–412.

[17] Sikanen T, Franssila S, Kauppila TJ, Kostiainen R, Kotiaho T, Ketola RA. Microchip technology in mass spectrometry. Mass Spectrom Rev 2010;29:351–91.

[18] Hop CE, Chen Y, Yu LJ. Uniformity of ionization response of structurally diverse analytes using a chip-based nanoelectrospray ionization source. Rapid Commun Mass Spectrom 2005;19:3139–42.

[19] Hirabayashi A, Sakairi M, Koizumi H. Sonic spray ionization method for atmospheric-pressure ionization mass-spectrometry. Anal Chem 1994;66:4557–9.

[20] Yamaguchi K. Cold-spray ionization mass spectrometry: principle and applications. J Mass Spectrom 2003;38:473–90.

[21] Kebarle P, Verkcerk UH. Electrospray: from ions in solution to ions in the gas phase, what we know now. Mass Spectrom Rev 2009;28:898–917.

[22] Iribarne JV, Thomson BA. On the evaporation of small ions from charged droplets. J Chem Phys 1976;64:2287–94.

[23] Dole M, Mack LL, Hines RL. Molecular beams of macroions. J Chem Phys 1968; 49:2240–50.

[24] Fenn JB, Mann M, Meng CK, Wong SF, Whitehouse CM. Electrospray ionization for mass-spectrometry of large biomolecules. Science 1989;246:64–71.

[25] Kebarle P, Peschke M. On the mechanisms by which the charged droplets produced by electrospray lead to gas phase ions. Anal Chim Acta 2000;406:11–35.

[26] Fenn J. Electrospray ionization mass spectrometry: how it all began. J Biomol Tech 2002;13:101–18.

[27] Fenn JB. Ion formation from charged droplets—roles of geometry, energy, and time. J Am Soc Mass Spectrom 1993;4:524–35.

[28] Nguyen S, Fenn JB. Gas-phase ions of solute species from charged droplets of solutions. Proc Natl Acad Sci U S A 2007;104:1111–7.

[29] Hogan CJ, Carroll JA, Rohrs HW, Biswas P, Gross ML. Charge carrier field emission determines the number of charges on native state proteins in electrospray ionization. J Am Chem Soc 2008;130:6926–7.

[30] Labowsky M. A model for solvated ion emission from electrospray droplets. Rapid Commun Mass Spectrom 2010;24:3079–91.

[31] Roeser J, Permentier HP, Bruins AP, Bischoff R. Electrochemical oxidation and cleavage of tyrosine- and tryptophan-containing tripeptides. Anal Chem 2010;82:7556–65.

[32] Permentier HP, Bruins AP. Electrochemical oxidation and cleavage of proteins with on-line mass spectrometric detection: development of an instrumental alternative to enzymatic protein digestion. J Am Soc Mass Spectrom 2004;15:1707–16.

[33] Baumann A, Karst U. Online electrochemistry/mass spectrometry in drug metabolism studies: principles and applications. Expert Opin Drug Metab Toxicol 2010;6:715–31.

[34] Thomson BA, Danylewychmay L, Henion JD. LC MS MS with an atmospheric-pressure chemical ionization source. Abstr Pap Am Chem Soc 1983;186:19–1A.

[35] Fouda H, Nocerini M, Schneider R, Gedutis C. Quantitative-analysis by high-performance liquid-chromatography atmospheric-pressure chemical ionization mass-spectrometry—the determination of the renin inhibitor Cp-80,794 in human serum. J Am Soc Mass Spectrom 1991;2:164–7.

[36] Carroll DI, Dzidic I, Stillwell RN, Haegele KD, Horning EC. Atmospheric pressure ionization mass spectrometry. Corona discharge ion source for use in a liquid chromatograph-mass spectrometer-computer analytical system. Anal Chem 1975;47: 2369–72.

[37] Ostman P, Marttila SJ, Kotiaho T, Franssila S, Kostiainen R. Microchip atmospheric pressure chemical ionization source for mass spectrometry. Anal Chem 2004; 76:6659–64.

[38] Singh G, Gutierrez A, Xu KY, Blair IA. Liquid chromatography/electron capture atmospheric pressure chemical ionization/mass spectrometry: analysis of pentafluorobenzyl derivatives of biomolecules and drugs in the attomole range. Anal Chem 2000; 72:3007–13.

[39] Bos SJ, van Leeuwen SM, Karst U. From fundamentals to applications: recent developments in atmospheric pressure photoionization mass spectrometry. Anal Bioanal Chem 2006;384:85–99.

[40] Hanold KA, Fischer SM, Cormia PH, Miller CE, Syage JA. Atmospheric pressure photoionization. 1. General properties for LC/MS. Anal Chem 2004;76:2842–51.

[41] Raffaelli A, Saba A. Atmospheric pressure photoionization mass spectrometry. Mass Spectrom Rev 2003;22:318–31.

[42] Robb DB, Blades MW. State-of-the-art in atmospheric pressure photoionization for LC/MS. Anal Chim Acta 2008;627:34–49.

[43] Robb DB, Covey TR, Bruins AP. Atmospheric pressure photoionization: an ionization method for liquid chromatography-mass spectrometry. Anal Chem 2000;72:3653–9.

[44] Cai Y, Kingery D, McConnell O, Bach AC, 2nd. Advantages of atmospheric pressure photoionization mass spectrometry in support of drug discovery. Rapid Commun Mass Spectrom 2005;19:1717–4.

[45] Takats Z, Wiseman JM, Gologan B, Cooks RG. Mass spectrometry sampling under ambient conditions with desorption electrospray ionization. Science 2004;306:471–3.

[46] Cody RB, Laramee JA, Durst HD. Versatile new ion source for the analysis of materials in open air under ambient conditions. Anal Chem 2005;77:2297–302.

[47] Weston DJ. Ambient ionization mass spectrometry: current understanding of mechanistic theory; analytical performance and application areas. Analyst 2010;135:661–8.

[48] Ifa DR, Wu CP, Ouyang Z, Cooks RG. Desorption electrospray ionization and other ambient ionization methods: current progress and preview. Analyst 2010;135:669–81.

[49] Wiseman JM, Ifa DR, Song QY, Cooks RG. Tissue imaging at atmospheric pressure using desorption electrospray ionization (DESI) mass spectrometry. Angew Chem Int Ed Engl 2006;45:7188–92.

[50] Leuthold LA, Mandscheff JF, Fathi M, Giroud C, Augsburger M, Varesio E, et al. Desorption electrospray ionization mass spectrometry: direct toxicological screening and analysis of illicit Ecstasy tablets. Rapid Commun Mass Spectrom 2006;20:103–10.

[51] Cooks RG, Ouyang Z, Takats Z, Wiseman JM. Detection technologies. Ambient mass spectrometry. Science 2006;311:1566–70.

[52] Takats Z, Wiseman JM, Cooks RG. Ambient mass spectrometry using desorption electrospray ionization (DESI): instrumentation, mechanisms and applications in forensics, chemistry, and biology. J Mass Spectrom 2005;40:1261–75.

[53] Nefliu M, Smith JN, Venter A, Cooks RG. Internal energy distributions in desorption electrospray ionization (DESI). J Am Soc Mass Spectrom 2008;19:420–7.

[54] Bereman MS, Muddiman DC. Detection of attomole amounts of analyte by desorption electrospray ionization mass spectrometry (DESI-MS) determined using fluorescence spectroscopy. J Am Soc Mass Spectrom 2007;18:1093–6.

[55] Wiseman JM, Evans CA, Bowen CL, Kennedy JH. Direct analysis of dried blood spots utilizing desorption electrospray ionization (DESI) mass spectrometry. Analyst 2010;135:720–5.

[56] McEwen CN, McKay RG, Larsen BS. Analysis of solids, liquids, and biological tissues using solids probe introduction at atmospheric pressure on commercial LC/MS instruments. Anal Chem 2005;77:7826–31.

[57] McEwen CN, Larsen BS. Ionization mechanisms related to negative ion APPI, APCI, and DART. J Am Soc Mass Spectrom 2009;20:1518–21.

[58] Wachs T, Henion J. Electrospray device for coupling microscale separations and other miniaturized devices with electrospray mass spectrometry. Anal Chem 2001;73:632–8.

[59] Van Berkel GJ, Sanchez AD, Quirke JME. Thin-layer chromatography and electrospray mass spectrometry coupled using a surface sampling probe. Anal Chem 2002; 74:6216–23.

[60] Kertesz V, Van Berkel GJ. Fully automated liquid extraction-based surface sampling and ionization using a chip-based robotic nanoelectrospray platform. J Mass Spectrom 2010;45:252–60.

[61] Luftmann H. A simple device for the extraction of TLC spots: direct coupling with an electrospray mass spectrometer. Anal Bioanal Chem 2004;378:964–8.

[62] Sampson JS, Hawkridge AM, Muddiman DC. Generation and detection of multiply-charged peptides and proteins by matrix-assisted laser desorption electrospray ionization (MALDESI) Fourier transform ion cyclotron resonance mass spectrometry. J Am Soc Mass Spectrom 2006;17:1712–6.

[63] Huang MZ, Hsu HJ, Wu CI, Lin SY, Ma YL, Cheng TL, et al. Characterization of the chemical components on the surface of different solids with electrospray-assisted laser desorption ionization mass spectrometry. Rapid Commun Mass Spectrom 2007; 21:1767–75.

[64] Nemes P, Vertes A. Laser ablation electrospray ionization for atmospheric pressure, *in vivo*, and imaging mass spectrometry. Anal Chem 2007;79:8098–106.

[65] Tanaka K, Waki H, Ido Y, Akita S, Yoshida Y, Yoshida T. Protein and polymer analysis up to *m/z* 100,000 by laser ionization time-of-flight mass spectrometry. Rapid Commun Mass Spectrom 1988;2:151.

[66] Karas M, Bahr U, Hillenkamp F. UV laser matrix desorption ionization mass-spectrometry of proteins in the 100,000 Dalton range. Int J Mass Spectrom Ion Process 1989;92:231–42.

[67] Karas M, Hillenkamp F. Laser desorption ionization of proteins with molecular masses exceeding 10,000 daltons. Anal Chem 1988;60:2299–301.

[68] Rappsilber J, Moniatte M, Nielsen ML, Podtelejnikov AV, Mann M. Experiences and perspectives of MALDI MS and MS/MS in proteomic research. Int J Mass Spectrom 2003;226:223–37.

[69] Nielen M. MALDI time-of-flight mass spectrometry of synthetic polymers. Mass Spectrom Rev 1999;18:309–44.

[70] Gut IG. DNA analysis by MALDI-TOF mass spectrometry. Hum Mutat 2004; 23:437–41.

[71] Fuchs B, Suss R, Schiller J. An update of MALDI-TOF mass spectrometry in lipid research. Prog Lipid Res 2010;49:450–75.

[72] Fitzgerald MC, Parr GR, Smith LM. Basic matrices for the matrix-assisted laser-desorption ionization mass-spectrometry of proteins and oligonucleotides. Anal Chem 1993;65:3204–11.

[73] Porta T, Grivet C, Knochenmuss R, Varesio E, Hopfgartner G. Alternative CHCA-based matrices for the analysis of low molecular weight compounds by UV-MALDI-tandem mass spectrometry. J Mass Spectrom 2011;46:144–52.

[74] Jaskolla TW, Lehmann WD, Karas M. 4-Chloro-alpha-cyanocinnamic acid is an advanced, rationally designed MALDI matrix. Proc Natl Acad Sci U S A 2008;105: 12200–5.

[75] Knochenmuss R. Ion formation mechanisms in UV-MALDI. Analyst 2006;131: 966–86.

[76] van Kampen JJ, Burgers PC, de Groot R, Gruters RA, Luider TM. Biomedical application of MALDI mass spectrometry for small-molecule analysis. Mass Spectrom Rev 2011;30:101–20.

[77] Peterson DS. Matrix-free methods for laser desorption/ionization mass spectrometry. Mass Spectrom Rev 2007;26:19–34.

[78] Wei J, Buriak JM, Siuzdak G. Desorption-ionization mass spectrometry on porous silicon. Nature 1999;399:243–6.

[79] Go EP, Prenni JE, Wei J, Jones A, Hall SC, Witkowska HE, et al. Desorption/ionization on silicon time-of-flight/time-of-flight mass spectrometry. Anal Chem 2003;75:2504–6.

[80] Lewis WG, Shen Z, Finn MG, Siuzdak G. Desorption/ionization on silicon (DIOS) mass spectrometry: background and applications. Int J Mass Spectrom 2003;226:107–16.

[81] Cohen LH, Gusev AI. Small molecule analysis by MALDI mass spectrometry. Anal Bioanal Chem 2002;373:571–86.

[82] Persike M, Karas M. Rapid simultaneous quantitative determination of different small pharmaceutical drugs using a conventional matrix-assisted laser desorption/ionization time-of-flight mass spectrometry system. Rapid Commun Mass Spectrom 2009;23: 3555–62.

[83] Kovarik P, Grivet C, Bourgogne E, Hopfgartner G. Method development aspects for the quantitation of pharmaceutical compounds in human plasma with a matrix-assisted laser desorption/ionization source in the multiple reaction monitoring mode. Rapid Commun Mass Spectrom 2007;21:911–9.

[84] Lesur A, Varesio E, Hopfgartner G. Protein quantification by MALDI-selected reaction monitoring mass spectrometry using sulfonate derivatized peptides. Anal Chem 2010;82:5227–37.

[85] Caprioli RM, Farmer TB, Gile J. Molecular imaging of biological samples: localization of peptides and proteins using MALDI-TOF MS. Anal Chem 1997;69:4751–60.

[86] Reyzer ML, Hsieh Y, Ng K, Korfmacher WA, Caprioli RM. Direct analysis of drug candidates in tissue by matrix-assisted laser desorption/ionization mass spectrometry. J Mass Spectrom 2003;38:1081–92.

[87] Rohner TC, Staab D, Stoeckli M. MALDI mass spectrometric imaging of biological tissue sections. Mech Ageing Dev 2005;126:177–85.

[88] Hopfgartner G, Varesio E, Stoeckli M. Matrix-assisted laser desorption/ionization mass spectrometric imaging of complete rat sections using a triple quadrupole linear ion trap. Rapid Commun Mass Spectrom 2009;23:733–6.

[89] Laiko VV, Baldwin MA, Burlingame AL. Atmospheric pressure matrix assisted laser desorption/ionization mass spectrometry. Anal Chem 2000;72:652–7.

[90] Schneider BB, Lock C, Covey TR. AP and vacuum MALDI on a QqLIT instrument. J Am Soc Mass Spectrom 2005;16:176–82.

[91] Tang N, Tornatore P, Weinberger SR. Current developments in SELDI Affinity technology. Mass Spectrom Rev 2003;23:34–44.

[92] Gammelgaard B, Hansen HR, Sturup S, Moller C. The use of inductively coupled plasma mass spectrometry as a detector in drug metabolism studies. Expert Opin Drug Metab Toxicol 2008;4:1187–207.

[93] March RE, Todd JFJ. Quadrupole Ion Trap Mass Spectrometry. Hoboken, NJ: Wiley-Interscience, 2005.

[94] Yang L, Amad M, Winnik WM, Schoen AE, Schweingruber H, Mylchreest I, et al. Investigation of an enhanced resolution triple quadrupole mass spectrometer for high-throughput liquid chromatography/tandem mass spectrometry assays. Rapid Commun Mass Spectrom 2002;16:2060–6.

[95] Jemal M, Xia YQ. LC-MS development strategies for quantitative bioanalysis. Curr Drug Metab 2006;7:491–502.

[96] de Hoffmann E. Tandem mass spectrometry: a primer. J Mass Spectrom 1996;31: 129–37.

[97] Schwartz JC, Wade AP, Enke CG, Cooks RG. Systematic delineation of scan modes in multidimensional mass spectrometry. Anal Chem 1990;62:1809–18.

[98] Paul W, Steinwedel H. A new mass spectrometer without magnetic field. Z Naturforsch 1953;8a:448–50.

[99] Douglas DJ, Frank AJ, Mao D. Linear ion traps in mass spectrometry. Mass Spectrom Rev 2005;24:1–29.

[100] Schwartz JC, Senko MW, Syka JEP. A two-dimensional quadrupole ion trap mass spectrometer. J Am Soc Mass Spectrom 2002;13:659–69.

[101] Collings BA, Campbell JM, Mao D, Douglas DJ. A combined linear ion trap time-of-flight system with improved performance and MS^n capabilities. Rapid Commun Mass Spectrom 2001;15:1777–95.

[102] Belov ME, Nikolaev EN, Alving K, Smith RD. A new technique for unbiased external ion accumulation in a quadrupole two-dimensional ion trap for electrospray ionization Fourier transform ion cyclotron resonance mass spectrometry. Rapid Commun Mass Spectrom 2001;15:1172–80.

[103] Syka JEP, Marto JA, Bai DL, Horning S, Senko MW, Schwartz JC, et al. Novel linear quadrupole ion Trap/FT mass spectrometer: performance characterization and use in the comparative analysis of histone H3 post-translational modifications. J Proteome Res 2004;3:621–26.

[104] Hager JW. A new linear ion trap mass spectrometer. Rapid Commun Mass Spectrom 2002;16:512–26.

[105] Hopfgartner G, Varesio E, Tschäppät V, Grivet C, Emmanuel Bourgogne E, Leuthold LA. Triple quadrupole linear ion trap mass spectrometer for the analysis of small molecules and macromolecules. J Mass Spectrom 2004;39:845–55.

[106] Hopfgartner G, Zell M. Q Trap MS: a new tool for metabolite identification. In Korfmacher W, editor. Using Mass Spectrometry for Drug Metabolism Studies. Boca Raton, FL: CRC Press, 2005, pp. 277–304.

[107] Mamyrin BA. Time-of-flight mass spectrometry (concepts, achievements, and prospects). Int J Mass Spectrom 2001;206:251–66.

[108] Uphoff A, Grotemeyer J. The secrets of time-of-flight mass spectrometry revealed. Eur J Mass Spectrom 2003;9:151–64.

[109] Birkinshaw K. Fundamentals of focal plane detectors. J Mass Spectrom 1997;32: 795–806.

[110] Guilhaus M, Selby D, Mlynski V. Orthogonal acceleration time-of-flight mass spectrometry. Mass Spectrom Rev 2000;19:65–107.

[111] Morris HR, Paxton T, Dell A, Langhorne J, Berg M, Bordoli RS, et al. High sensitivity collisionally-activated decomposition tandem mass spectrometry on a novel quadrupole/orthogonal-acceleration time-of-flight mass spectrometer. Rapid Commun Mass Spectrom 1996;10:889–96.

[112] Chernushevich IV, Loboda AV, Thomson BA. An introduction to quadrupole-time-of-flight mass spectrometry. J Mass Spectrom 2001;36:849–65.

[113] Hopfgartner G, Chernushevich IV, Covey T, Plomley JB, Bonner R. Exact mass measurement of product ions for the structural elucidation of drug metabolites with a tandem quadrupole orthogonal-acceleration time-of-flight mass spectrometer. J Am Soc Mass Spectrom 1999;10:1305–14.

[114] Wolff JC, Eckers C, Sage AB, Giles K, Bateman R. Accurate mass liquid chromatography/mass spectrometry on quadrupole orthogonal acceleration time-of-flight mass analyzers using switching between separate sample and reference sprays. 2. Applications using the dual-electrospray ion source. Anal Chem 2001;73:2605–12.

[115] Martin RL, Brancia FL. Analysis of high mass peptides using a novel matrix-assisted laser desorption/ionisation quadrupole ion trap time-of-flight mass spectrometer. Rapid Commun Mass Spectrom 2003;17:1358–65.

[116] Cotter RJ, Griffith W, Jelinek C. Tandem time-of-flight (TOF/TOF) mass spectrometry and the curved-field reflectron. J Chromatogr B 2007;855:2–13.

[117] Vestal ML, Campbell JM. Tandem time-of-flight mass spectrometry. Meth Enzymol 2005;402:79–108.

[118] Suckau D, Resemann A, Schuerenberg M, Hufnagel P, Franzen J, Holle A. A novel MALDI LIFT-TOF/TOF mass spectrometer for proteomics. Anal Bioanal Chem 2003;376:952–65.

[119] Marshall AG, Hendrickson CL, Jackson GS. Fourier transform ion cyclotron resonance mass spectrometry: a primer. Mass Spectrom Rev 1998;17:1–35.

[120] Makarov A. Electrostatic axially harmonic orbital trapping: a high-performance technique of mass analysis. Anal Chem 2000;72:1156–62.

[121] Kingdon KH. A method for the neutralization of electron space charge by positive ionization at very low gas pressures. Phys Rev 1923;21:408–18.

[122] Perry RH, Cooks RG, Noll RJ. Orbitrap mass spectrometry: instrumentation, ion motion and applications. Mass Spectrom Rev 2008;27:661–99.

[123] Michalski A, Damoc E, Hauschild JP, Lange O, Wieghaus A, Makarov A, et al. Mass spectrometry-based proteomics using q exactive, a high-performance benchtop quadrupole orbitrap mass spectrometer. Mol Cell Proteomics 2011;10:1–11.

[124] Kanu AB, Dwivedi P, Tam M, Matz L, Hill HH, Jr. Ion mobility-mass spectrometry. J Mass Spectrom 2008;43:1–22.

[125] Guevremont R. High-field asymmetric waveform ion mobility spectrometry: a new tool for mass spectrometry. J Chromatogr A 2004;1058:3–19.

[126] Krylov EV, Nazarov EG. Electric field dependence of the ion mobility. Int J Mass Spectrom 2009;285:149–56.

[127] Schneider BB, Covey TR, Coy SL, Krylov EV, Nazarov EG. Chemical effects in the separation process of a differential mobility/mass spectrometer system. Anal Chem 2010;82:1867–80.

[128] Pringle SD, Giles K, Wildgoose JL, Williams JP, Slade SE, Thalassinos K, et al. An investigation of the mobility separation of some peptide and protein ions using a new hybrid quadrupole/travelling wave IMS/oa-ToF instrument. Int J Mass Spectrom 2007;261:1–12.

[129] Bennett CL, Beukens RP, Clover MR, Elmore D, Gove HE, Kilius L, et al. Radiocarbon dating with electrostatic accelerators—dating of milligram samples. Science 1978;201: 345–7.

[130] Nelson DE, Korteling RG, Stott WR. C-14—direct detection at natural concentrations. Science 1977;198:507–8.

[131] Tozuka Z, Kusuhara H, Nozawa K, Hamabe Y, Ikushima I, Ikeda T, et al. Microdose study of C-14-Acetaminophen with accelerator mass spectrometry to examine pharmacokinetics of parent drug and metabolites in healthy subjects. Clin Pharmacol Ther 2010;88:824–30.

[132] Arjomand A. Accelerator mass spectrometry-enabled studies: current status and future prospects. Bioanalysis 2010;2:519–41.

[133] Lappin G, Stevens L. Biomedical accelerator mass spectrometry: recent applications in metabolism and pharmacokinetics. Expert Opin Drug Metab Toxicol 2008;4:1021–33.

[134] Lappin G, Garner RC. Current perspectives of C-14-isotope measurement in biomedical accelerator mass spectrometry. Anal Bioanal Chem 2004;378:356–64.

[135] Litherland AE, Zhao XL, Kieser WE. Mass spectrometry with accelerators. Mass Spectrom Rev 2011;30:1037–72.

[136] Cech NB, Enke CG. Practical implications of some recent studies in electrospray ionization fundamentals. Mass Spectrom Rev 2001;20:362–87.

[137] Olsen JV, Macek B, Lange O, Makarov A, Horning S, Higher-Energy MM. C-trap dissociation for peptide modification analysis. Nat Methods 2007;4:709–12.

2

UTILITY OF HIGH-RESOLUTION MASS SPECTROMETRY FOR NEW DRUG DISCOVERY APPLICATIONS

William Bart Emary and Nanyan Rena Zhang

2.1 INTRODUCTION

The pharmaceutical industry has used high-resolution mass spectrometry (HRMS) for many years for the determination of the molecular formula of a compound, and this has been critical to the research and development process. Instrument companies have made significant technological advancements in HRMS over the past two decades, and this has created much excitement in the world of bioanalytical chemistry. New systems can perform more sophisticated scans and are easier to use because of more stable calibration and new software tools. Instrument costs of the latest systems are decreasing and becoming more competitive with triple quadrupole mass spectrometers (TQMS), which only provide unit mass resolution. Quantitation with a TQMS is typically carried out using selected reaction monitoring (SRM), which requires optimization of each analyte.

Drugs are extraordinarily expensive to discover and bring to market, in part due to the high attrition rate of compounds during the multi-year testing process [1]. The inflation rate of drugs cannot continue to increase at the same pace and pharmaceutical companies must find ways to increase the probability of success. Scientists create data and convert it into knowledge. Analytical laboratories are at the center of this effort and these laboratories need to undergo continuous process improvement. The latest generation of HRMS provides more information with less effort than TQMS. In the future, many SRM assays will be performed instead on an HRMS system.

Mass Spectrometry for Drug Discovery and Drug Development, First Edition. Edited by Walter A. Korfmacher.

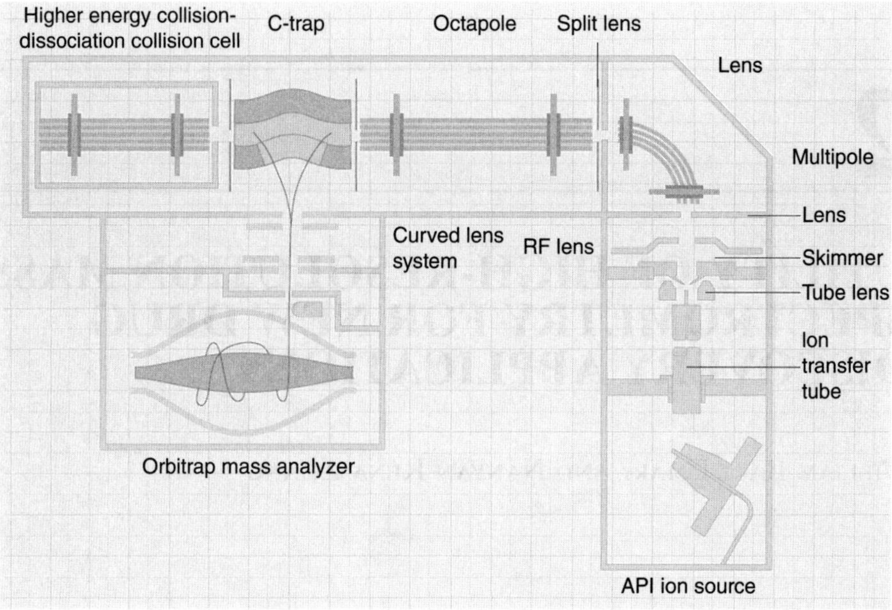

Figure 2.1 Orbitrap mass spectrometer.

2.2 QUALITATIVE/QUANTITATIVE WORKFLOW IN DRUG METABOLISM

2.2.1 Instrumentation

While instruments based on magnetic and electric sectors dominated HRMS for decades, two other types of systems supplanted them since about 2000. Makarov introduced the Orbitrap, which is based on electrostatically trapped ions circulating between two oval electrodes (Fig. 2.1) [2]. Each m/z has a characteristic harmonic oscillation frequency that induces a current on a detection electrode. The total detector signal is complex and a summation of all frequencies and intensities of the trapped ions. An inverse Fourier transform is performed on the total current signal, generating a frequency to ion intensity spectrum, which can then be converted to a mass spectrum. The Orbitrap mass analyzer is frequently preceded by a linear quadrupole ion trap for the purpose of conducting collision-induced dissociation (CID). The Orbitrap analyzer is stable enough to use for accurate mass work for weeks without needing recalibration.

Wong examined the effects of parameters such as resolution, maximum injection time, and automatic gain control on discovery assay quantitative performance [3]. While the lower limit of quantitation (LLOQ) could be decreased by increasing the length of time that ions were injected into the analyzer, this came at a sacrifice of linearity at the high end of the concentration curve. The LLOQ for most compounds in mouse plasma was in the 1–3 nM range and the dynamic range was three to four orders of magnitude. A resolution of 20,000 full width at half maximum (FWHM) was found to strike the best balance of signal to noise. The results were competitive

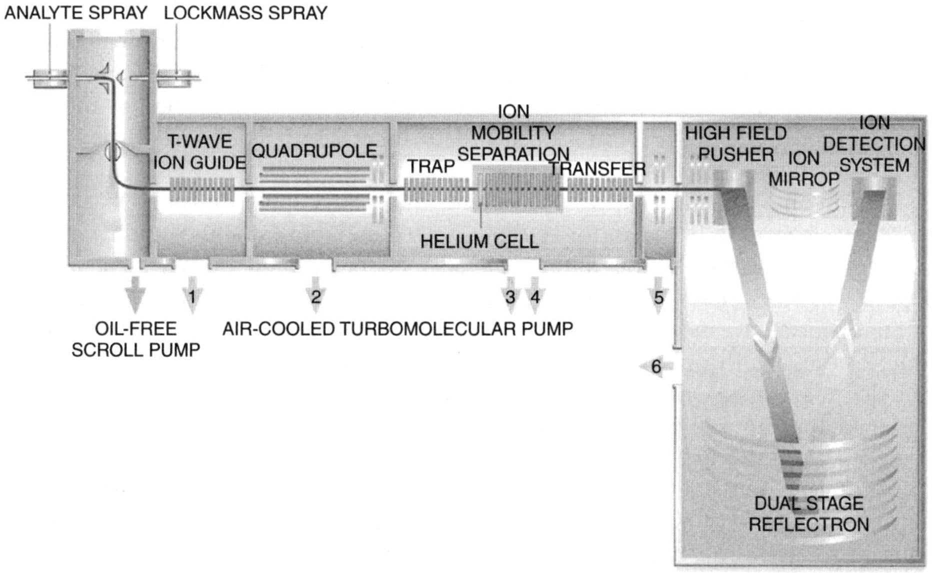

Figure 2.2 Synapt G2™ (Waters Corporation, Milford, MA) ion mobility-TOF mass spectrometer.

to many triple quadrupoles but illustrated the need for thoughtful consideration of parameter values to meet study objectives.

The second type of HRMS system commonly used today is based on the time-of-flight (TOF) design. Similar to the Orbitrap, the high-resolution device can be preceded with a quadrupole for CID or other mass filter purposes. Researchers have used TOF for many years with pulsed ion sources but the resolution of such systems was relatively low until the 1990s [4]. The combination of a delayed extraction pulse of ions in the source plus a reflectron (Fig. 2.2) significantly increased resolution, and 30,000 FWHM is readily achievable today on commercial instruments [5]. Although the calibration must be carried out more frequently than with an Orbitrap, TOF is automated and, in general, seamlessly integrated into the unknown sample run batches. Tandem TOF/TOF systems have also been used to provide simultaneous high-resolution analysis of both parent and fragment ions [6].

2.2.2 Applications

Adoption of high-performance liquid chromatography-mass spectrometry (HPLC-MS) occurred quickly after the introduction of robust atmospheric pressure ionization (API) sources in the early 1990s. This advancement occurred at a time when pharmaceutical companies were building better tools to improve the accuracy of human pharmacokinetic (PK) predictions and reduce the chance of drug–drug interactions. Larger organizations' workload increased exponentially and their laboratories specialized in either quantitative or qualitative experiments. We are now beginning to see a reversal of this trend because better hardware and software tools for both qualitative and quantitative data allow us to consider high-level process changes to increase efficiency. Hardware ruggedness and software ease of use are critical features for laboratories that analyze thousands of samples each day. Likewise, scientists charged with metabolite characterization rely on sophisticated tools so that they can rapidly provide feedback to medicinal chemists regarding metabolic soft spots. Scientists' energy must be focused toward problem solving as much as possible and neither specialist can afford to take a step backwards when these workflows merge together.

Implementation of HRMS to combine qualitative/quantitative (Qual/Quan) workflows has already begun for routine *in vitro* metabolic stability and characterization experiments. Although this can also be performed on a low-resolution TQMS, it is less efficient because it requires compound tuning for the SRM portion of the experiment, and multiple injections to provide similar levels of information. State-of-the-art ultra-performance liquid chromatography (UPLC)-HRMS systems can do this with a single experiment and injection [7]. If the experiments are designed appropriately, data can be reprocessed later to answer new questions that come up after the first examination of the data. Post-acquisition data processing software using mass defect and biotransformation filters allow for the discovery of new metabolites that might have initially been missed, and even estimation of their concentrations [8]. Several laboratories have expanded the concept to *in vivo* studies [9, 10]. Ramanathan used a Q-TOF for routine discovery PK profiling of 20 proprietary compounds [11]. The LLOQ of the parent compound and PK parameters were comparable to SRM, with no loss of metabolite information. The authors asked, "Is it time for a paradigm shift from SRM to HRMS?"

This is indeed happening in several laboratories and one must ask, "To what extent will this occur for all routine drug metabolism discovery assays?" While the transformation should begin for many assays, there is still anecdotal information that indicates that for some compounds, SRM results in lower LLOQs. Maximum sensitivity is needed sometimes in drug discovery for a variety of reasons. For example, drug potency is continuously being reduced to minimize risk of adverse events in humans. Chemists are always challenging bioanalytical groups to increase compound PK screening throughput, and for some laboratories *in vivo* cassette dosing is an essential tool. Researchers typically dose five to seven compounds per cassette and a proportional reduction in dose of each compound is warranted as the number of compounds increase. Instrumentation is constantly evolving and expensive. Each analytical laboratory will be challenged to decide the best mix of systems depending on their circumstances and budgets. The laboratory of tomorrow will likely include both HRMS and TQMS systems. Analytical scientists will also leverage other skills in their toolbox, namely extraction and chromatography.

Xia et al. performed extensive experiments and looked at parameter effects on quantitative performance measures for over 153 drugs in human plasma. They found a resolution of 20,000 and extracted ion chromatogram (XIC) less than +20 ppm generally provided good conditions for quantitation [12]. The number of endogenous peaks and interferences were characterized at different resolution and extracted ion chromatogram (EIC) windows. Their findings demonstrated a continued need for chromatography to complement HRMS (Fig. 2.3). Kaufmann measured concentrations of over 100 veterinary drugs in animal tissues using an Orbitrap and compared this with the European Union method using SRM detection. Results (precision, accuracy, and dynamic range) were comparable, although in a few cases, SRM was more sensitive [13]. In many cases, one wants to screen for as many drugs as possible

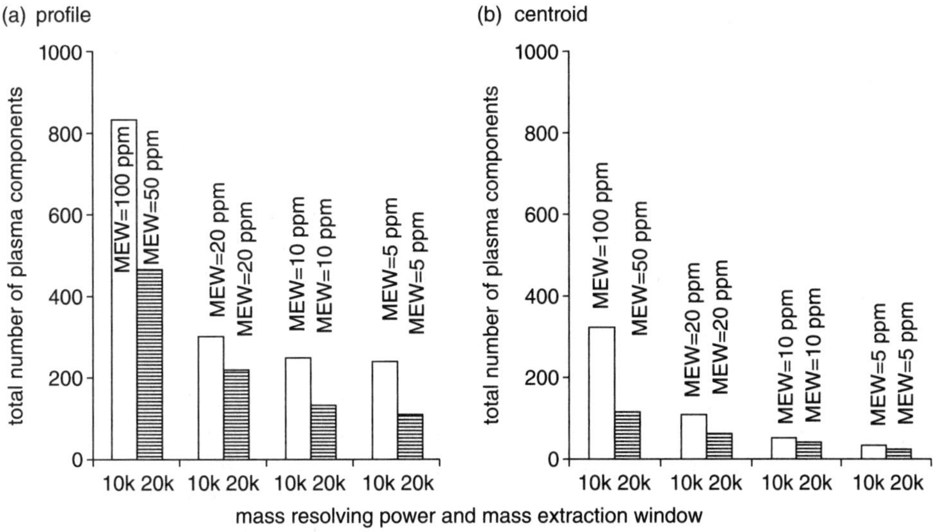

Figure 2.3 Total number of endogenous chromatographic peaks in human plasma detected in the extracted ion chromatogram windows of 153 drugs. Data analysis in (a) profile mode; (b) centroid mode.

in the unknown. In this case, HRMS is superior because it requires only one injection and did not require tuning and SRM transitions for all possible drugs.

Hopfgartner reported a new Q-TOF technique called sequential window acquisition of all theoretical fragment-ion spectra (SWATH). He used it to improve performance and quality of data for Qual/Quan experiments of drugs (e.g., talinolol, a beta-blocker) [14]. In SWATH, Q is set to band pass a narrow mass range approximately 20 Da wide. Low and high collision energy spectra are generated over 25 ms and combined into a single data scan providing a spectrum that contains significant parent ion intensity plus fragment ions. By limiting the parent ion mass window to a small range, correlation of fragments from parents are much easier to make. This facilitates structure elucidation of metabolites.

The first example using HRMS for regulated analysis was recently published for prednisone and prednisolone in human plasma [15]. The LLOQ of 5 ng/mL (resolution 20,000) meets most needs but is still not as good as several reported SRM methods. Signal to noise is good even with a relatively wide mass filter window of +40 ppm. The Jetstream™ spray device increased ion signal strength by focusing ions using a supersonic nitrogen stream. This increased sensitivity approximately 10-fold. Henry recently developed clinical methods for 17 drugs on both the Orbitrap and TQMS in the SRM mode [16]. Performance measures were similar for both systems, and they highlighted the benefit of full scan MS to diagnose problems such as standard curve nonlinearity.

2.3 BIOMARKERS

Biomarkers encompass physiological (e.g., blood pressure) and biochemical (e.g., cholesterol) domains. Undesirable biochemical changes are the root of disease, and also the cause of drug side effects. In a recent survey of development candidates that failed during testing, over 66% were due to lack of efficacy and 21%, to unacceptable safety margins [1]. The high rate of failure years after first synthesis is the impetus for biomarker research. Measurement of biochemical biomarkers is being utilized earlier in research since changes can occur before positive or negative effects manifest themselves. There are thousands of potential chemical biomarkers and this, in combination with natural variation and response to other stimuli, makes identification of relevant ones very difficult. Bioanalytical labs are increasingly asked to add one or more biomarker concentration measurements to their PK drug candidate assays for the purpose of differentiating between several interesting leads. The transition of work between biomarker discovery and its application to selection of drug candidates must be seamless.

Examples of HRMS for biomarker discovery and compound assessment are beginning to appear in the literature. Differential mass spectrometry (DMS) has been developed as a tool within Merck to locate potential Alzheimer's biomarkers in preclinical models [17]. Multiple samples were taken from each subject and served as their own control. The strategy utilized in-house software routines to map spectral components between samples, and those that changed significantly were identified using statistical analysis tools. It did not require stable label tools, which are frequently used to measure protein concentration changes. Tests were conducted using secretase inhibitors dosed to rhesus monkeys and confirmed changes in A-beta, a known biomarker of the disease. The findings by DMS correlated well with enzyme-linked immunosorbent assay (ELISA).

Drug companies stop development of many compounds after good laboratory practice (GLP) preclinical multi-day toxicity testing. These studies are relatively late in the discovery process and are expensive. The industry recognizes that to reduce attrition rates as early as possible, it is critical to identify important safety genomics and metabonomics biomarkers that can be implemented earlier [18]. For example, muscle toxicity was observed after the treatment of ibipinabant to dogs. Zhang et al. used UPLC-HRMS and a plasma pooling strategy to find several statistically significant metabolome biomarkers in dogs after 6 weeks of dosing. Sample cleanup was straightforward and the chromatographic run times were relatively short. In-house software was created to carefully compare control and study sample spectral intensities. Changes in several acylcarnitine concentrations were identified [19]. The importance of continued research like this to better understand toxicity and link them to biochemical biomarkers cannot be understated.

In the small-molecule lead optimization world, the challenge we are facing is "How do we measure concentrations of drug candidates and an increasing array of biomarkers efficiently?" In the past, we have optimized our processes around molecules in a relatively small physical–chemical space first highlighted by Lipinski [20]. Biomarkers significantly expand this space and create new challenges. In order to build the best possible PK/pharmacodynamic (PD) correlations, we must generate information from the same subject/sample. Many transgenic animal models are in mouse, which limits blood volumes after serial sampling. Different diseases, and complexity of biochemical events, may dictate that many simultaneous biomarker measurements are necessary to achieve acceptable correlation of PD to drug PK [21]. For these applications, HRMS becomes increasingly attractive compared with SRM in order to maintain simplicity and to allow post-acquisition data mining.

2.4 TISSUE

Since the majority of pharmacological targets are in tissues, concentrations of drugs and metabolites are important to help understand toxicity and efficacy or lack thereof. One important evolving MS field is the use of ion or laser beams to directly probe tissue and provide complementary information to radiolabel drug disposition studies [22, 23]. It can even supplant radiolabel to answer early discovery questions before radiolabel is available. Ubiquitous chemical noise makes interpretation of low-resolution MS data challenging. Most reports mapping drug and metabolite to tissue location use SRM to provide good signal to noise, but this limits the information in the raw data set. HRMS will play a critical role because full scan MS data will be generated with a high level of specificity. Spengler et al. developed a high spatial and mass resolution system demonstrating the utility [24]. They were able to distinguish between different types of endogenous compounds in kidney tissue, separating isobaric phospholipid ions, both from themselves and matrix ions.

Mann et al. recently reported measurement of 30 colon cancer biomarkers from tissue using laser capture microdissection followed by LC-HRMS. They were able to monitor several thousand proteins in a single analysis from as few as 1000 cells [25]. Addition of polyethylene glycol or dextran carriers increased peptide yield critical to reduce cell count. These carriers add complexity to spectra, but this can be compensated with high mass resolution. Similar strategies may be employed in the future for preclinical oncology models to better understand their relevance to humans, and to improve tools for compound selection.

2.5 siRNA AND PROTEIN THERAPIES

While most drugs are small organic molecules, drug companies are developing an increasing number of medicines with molecular weights (MWs) greater than 1000. The hormone Epogen (Amgen, Thousand Oaks, CA) and oncology antibody Herceptin (Genentech, South San Francisco, CA) are examples of protein therapies. Researchers use ELISA for most protein quantitative assays because it provides high throughput and extremely low LLOQ. It requires time-consuming antibody preparation, and assays can sometimes suffer from specificity problems if cross reactivity occurs with metabolites. MS is now used to provide early candidate disposition assessment in preclinical studies and before investment in ELISA is warranted. Protein drugs are challenging to work with from an MS perspective because they are large and multiply charged, and their glycosylation is heterogeneous [26]. In order to improve PK properties, chemical moieties such as polyethylene glycol polymers are added and this only makes MS spectra more complex. Much of this work is carried out with digestions and sample clean-up followed by SRM. While chromatography is critical, HRMS allows the work to be carried out faster because of greater specificity, diverse acquisition modes, and powerful data processing tools [27]. Ruan reported that 60-K mass resolution was needed to provide good selectivity for measurement of intact protein with MWs of 15 k Dalton in plasma (Fig. 2.4) [28]. The effect of different extracted ion windows on isobaric analyte and endogenous interference signal is shown in Figure 2.5.

Researchers are testing another class of high MW therapies made of small interfering RNA (siRNA). While small-molecule drugs bind reversibly to receptor

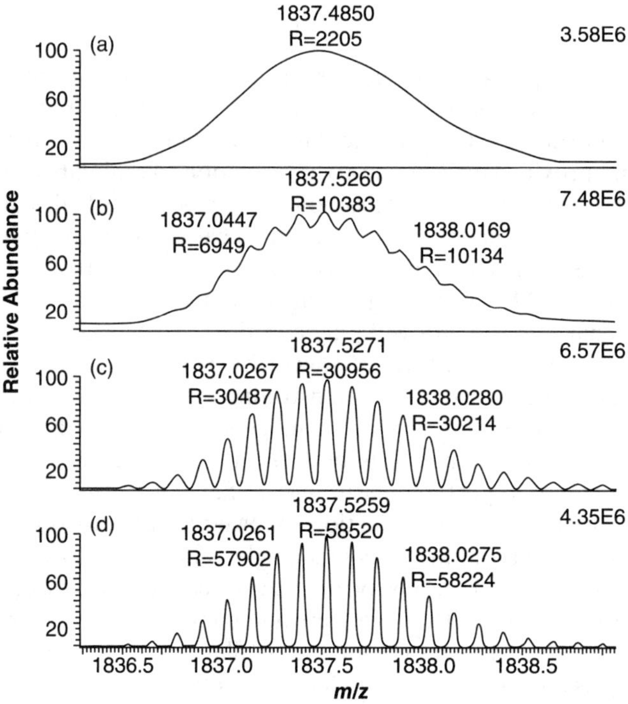

Figure 2.4 Effect of resolving power on separation of isotopic peaks of octuply charged lysozymes: (A) resolving power of 2 k, (B) resolving power of 10 k, (C) resolving power of 30 k, and (D) resolving power of 60 k.

proteins to cause the desired outcome, siRNA bind to expressed messenger RNA so that it cannot be transcribed into the protein associated with the disease. The MWs of siRNA duplexes are usually 10,000 to 14,000 Da. Robust HPLC-MS assays are challenging to develop in a high-throughput format because unique analytical conditions are required for the passenger and guide stands, which are negatively and multiply charged due to the phosphate backbone of nucleotides. As with other drugs, it is necessary to characterize the siRNA molecules to determine structure and stability. It is also important to understand their PK and metabolism. Q-RT-PCR methods are the standard for siRNA quantification, but they cannot always distinguish small metabolic changes from parent molecule, and require reporter probe reagent preparation [29]. Li showed that HPLC-HRMS and MS/MS complemented qPCR and used it in PK studies to confirm parent concentration, and also provided metabolic information of the siRNA candidates [30]. Zou differentiated between two potential siRNA metabolites with molecular mass difference less than 1 Da [31]. Takami characterized metabolites in plasma, and then used a linear ion trap quadrupole–Fourier transform (LTQ-FT) to quantify both parent and the metabolite concentrations [32]. We have also used HRMS to confirm siRNA qPCR concentrations and also to optimize PK of lipid nanopartilces (LNP), which deliver the active molecules to the liver target (unpublished) [33]. LNPs are complex,

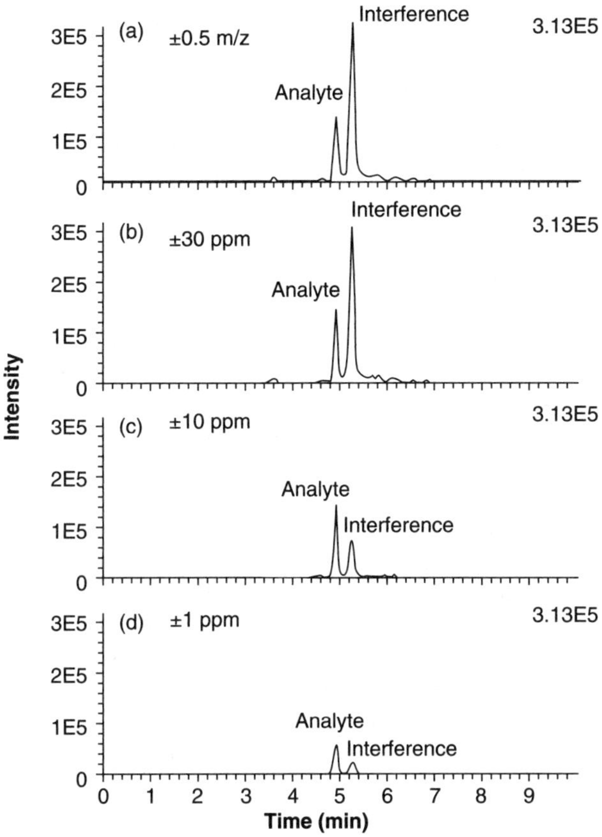

Figure 2.5 Effect of extracted ion chromatogram (EIC) window on assay selectivity and sensitivity for lysozome in human plasma: (A) ±0.5 *m/z*, (B) ±30 ppm, (C) ±10 ppm, (D) ±1 ppm.

multicomponent vehicles that prevent the active siRNA molecule from being degraded by nucleases during transport into the cells [34]. Understanding their PK and distribution is important to maximize target engagement.

2.6 DRUG METABOLITE IDENTIFICATION USING MASS DEFECT FILTERS

In order to more fully assess safety risks to humans, increasing scrutiny is being paid to the level of exposure of metabolites in preclinical GLP studies [35]. The science is still evolving and events leading to manifestation of toxicity are complex. For example, factors such as projected human dose level need to be accounted for to increase the confidence level of toxicity prediction and to avoid unnecessarily eliminating good drug candidates [36]. Many organizations are implementing new and more sophisticated HRMS hardware and software technologies in order to move the metabolite characterization process further upstream in the research process, and to do it efficiently. Mass resolution greater than 30,000 and accuracy of ±5 ppm from current high-resolution mass spectrometers allow unambiguous assignment of elemental composition. Rapid characterization of drug candidate metabolites is critical in order to successfully influence new compound synthesis and to improve the PK and structure–activity relationships (SAR). Chemists need to make decisions quickly, and will do so based on whatever timely information is available.

Characterization of metabolites early in discovery is typically carried out without the aid of radiolabels. For many years, researchers have used advanced SRM systems with intelligent software to control scan functions and automate manual steps in metabolite characterization. They require some knowledge of fragmentation patterns of the parent compound and common metabolic pathways. Mortishire-Smith was among the earliest researchers to characterize metabolites in a new way, using HRMS with mass defect filtering (MDF). Minute mass differences of drug and metabolites can be used to distinguish themselves from isobaric endogenous compounds. Mortishire-Smith et al. studied commercial and proprietary compounds in metabolic stability studies and were able to automate data interrogation routines and fully characterize drugs with less work [37]. MDF also allowed them to identify less common metabolic products. Bateman expanded upon this work and generated parent and fragment information in a single injection on a Q-TOF MS [38]. The quadrupole was used as a band pass filter and interleaved scans at low and high energy produced mass spectra that contained either parent or a combination of parent and fragment ions. This provided more structure information in addition to the molecular formula.

Once putative metabolites are characterized, the focus turns to the concentration level of more abundant ones because this may impact selection between candidates or, at a minimum, affect design of future studies. These questions arise before authentic standards are available, and while selectivity is a hallmark strength of MS, it can also be viewed as a weakness because endogenous compounds eluting near the metabolite can cause suppression or enhancement of ionization and skew the measured concentration. Yang et al. attempted to correct for the weakness by coupling an ultraviolet (UV) detector to HRMS [39]. UV detectors at low wavelengths (200–230 nm) tend to show a more uniform response for different compounds. They reduced the error of metabolite concentration from eight- to two- to threefold.

Zhang et al. increased metabolite hit quality by adding control sample background subtraction routines to MDF. They applied it to highly complex *in vivo* samples such as bile, urine, and plasma (Fig. 2.6) [40]. The tool was used to characterize metabolites of troglitazone, a compound withdrawn from the market after idiosyncratic hepatotoxicity was observed [41]. The combination of MDF with improved subtraction routines provided complete characterization of all metabolites, including critical ones implicated in the toxicity. They have continued making improvements to detect low concentration metabolites by adding more sophisticated scan techniques on the latest hardware [42].

Glutathione (GSH) drug conjugates are phase 2 metabolic products, typically generated by the organism to protect itself from xenobiotic electrophilic species. Some electrophiles may be associated with idiosyncratic drug reactions, causing medicines to be removed from the market after approval. While GSH conjugates

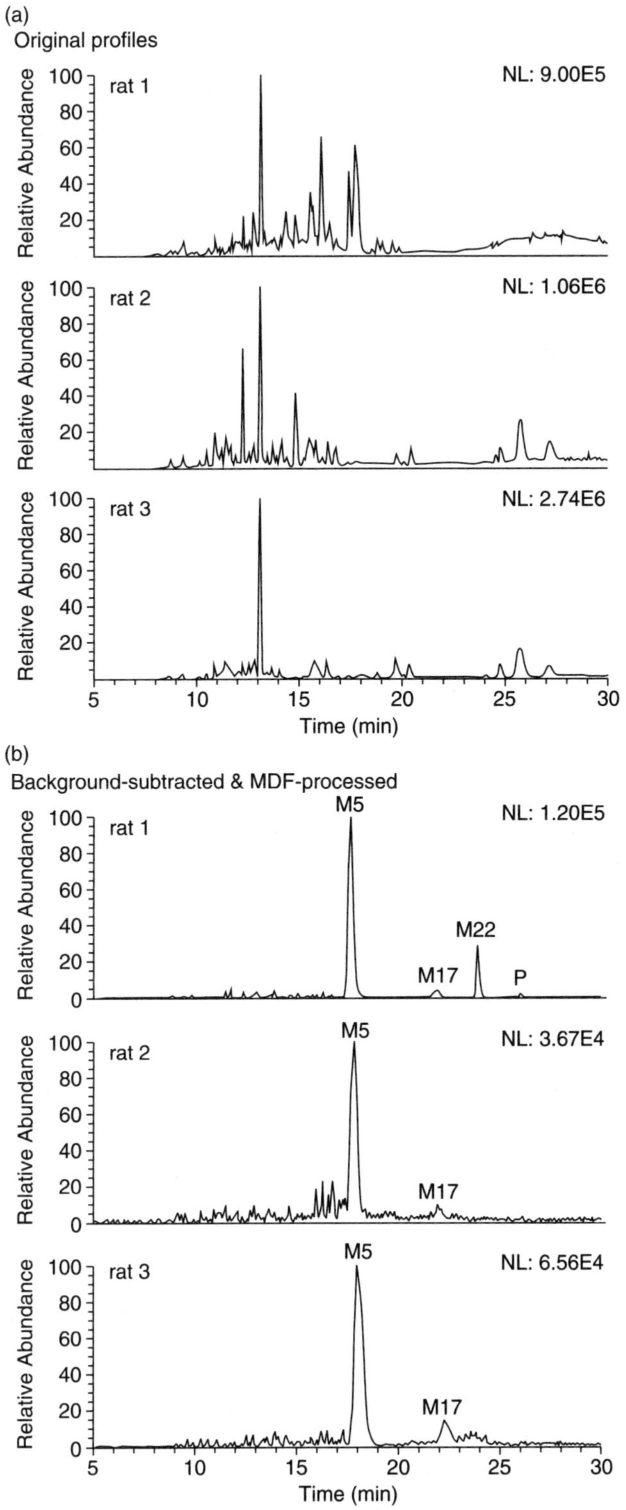

Figure 2.6 Troglitazone metabolites in rat urine: (a) full scan raw total ion chromatogram data, (b) full scan extracted ion chromatogram using mass defect filtering and background subtraction of two control samples.

are present in many safe medicines, they are of heightened interest during drug development. Phase 2 metabolites such as glucuronide and sulfate conjugates can easily be detected using neutral loss scans on quadrupole instruments. In contrast, GSH metabolites do not always fragment with the same neutral loss (e.g., 129), making it more challenging to find all of them on TQMS. Chen et al. used background subtraction and MDF tools to investigate *in vitro* formation rates of GSH conjugates from a series of compounds containing a thiophene moiety, a known structural alert [43]. Their results demonstrated the utility of MDF for rapid identification of electrophiles.

The process of identifying a metabolite is not complete until the biological sample is tested along with an authentic, synthetic standard. Discovery programs move quickly, and some degree of uncertainty is inevitable before a standard is available. Holman characterized metabolites of a drug candidate that underwent oxidation of a sulfur atom [44]. One particular metabolite underwent unusual intramolecular rearrangement after CID on a TQMS. This initially led to an incorrect assignment of the metabolite structure. Holman performed HRMS and hydrogen/deuterium exchange experiments that showed unresolved fragment ions and ultimately led to a change in the proposed metabolite structure. This is a clear demonstration of improved data quality in early preclinical space.

2.7 FUTURE TRENDS

Improvements in software are at the forefront of current HRMS research. Manufacturers are attempting to bring together the best of both qualitative and quantitative packages of the past. Different types of ion filtering devices are also being introduced into newer instruments. Orthogonal gas phase separation can be done by coupling an ion mobility separation (IMS) to the TOF (Fig. 2.2) [45]. This is a powerful combination, and in some cases, cross-sectional differences in the drift tube allow rapid (MS) separation of isobaric ions such as isomeric peptides (Fig. 2.7) [46]. Dwivedi built an IMS-TOF device and then used it to analyze the metabonome in human blood. Figure 2.8 shows two-dimensional mobility-*m/z* plots and the segregation of the various classes of compounds found in blood. These new, powerful techniques should find utility identifying biomarkers [47]. New fragmentation tools have also been added to the Orbitrap. Electron capture dissociation (ECD) and electron transfer dissociation (ETD) have demonstrated benefits in peptide sequencing compared with CID, and preserved post-translational modifications of proteins [48].

Most routine PK and metabolite characterization work is done using 2.1–4.6 mm inner diameter (ID) columns with flow rates greater than 0.5 mL/min. The metabonomics and proteomics communities have used lower flows and nano-liquid chromatography and ionization sources for many years. They had to use these systems because some compound concentrations were very low, creating a demand for maximum sensitivity. Drug metabolism scientists do not use the systems routinely for PK and drug metabolism studies because the number of analyzed samples is larger and the flow paths become blocked more frequently. Instrument companies have modified ion source designs to bridge the gap between nano- and conventional flow rates. Ramanathan used the CaptiveSpray™ (Bruker-Michrom, Auburn, CA)

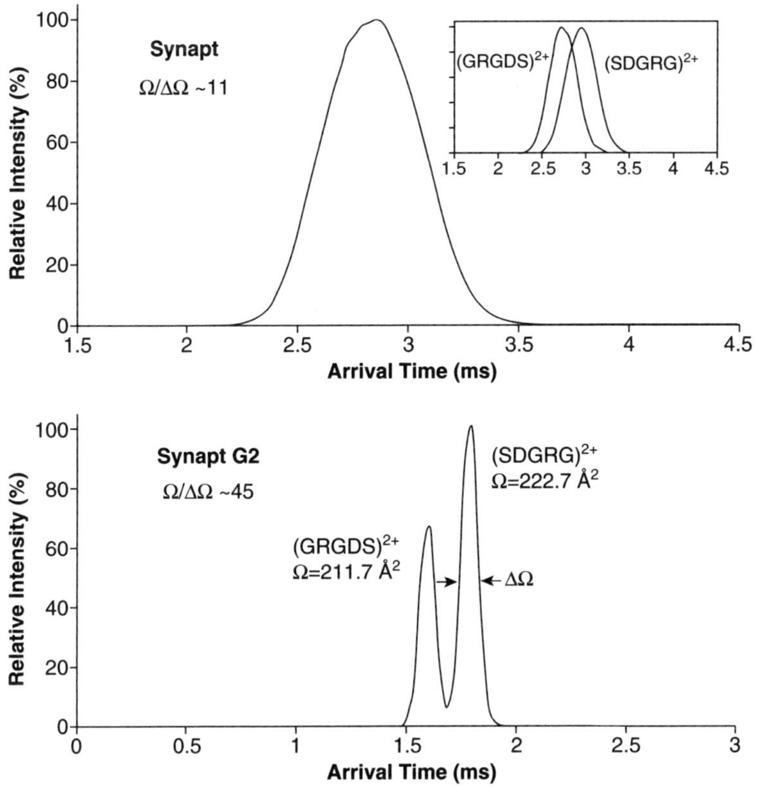

Figure 2.7 Ion mobility separation of isobaric peptide ions.

ion source because it provided benefits associated with nano-spray, but at higher flow rates [49]. Ramanathan et al. did routine bile metabolic profiling using a 75 micron ID column running at 0.3 μL/min [50]. Samples were diluted up to 1000-fold, but this was counterbalanced by a gain in sensitivity of about 1000-fold due to the smaller column and spray device. No special workup was used other than high speed centrifugation at 13,000 RPM. They found that lower RPM was insufficient for sample cleanup. Comparable metabolic profile data were generated with significantly less sample and solvent consumption. Ramanathan found that columns lasted for months, which meets the need of most laboratories.

2.8 CONCLUSIONS

The creation of new analytical instrumentation is one of the keys to opening new doors in science [51]. Instrumentation companies and academic and industrial analytical scientists have worked well together to develop and use new systems in the pursuit of new medicines. Pharmaceutical scientists are making high-resolution mass spectrometers one of their most important tools in drug discovery.

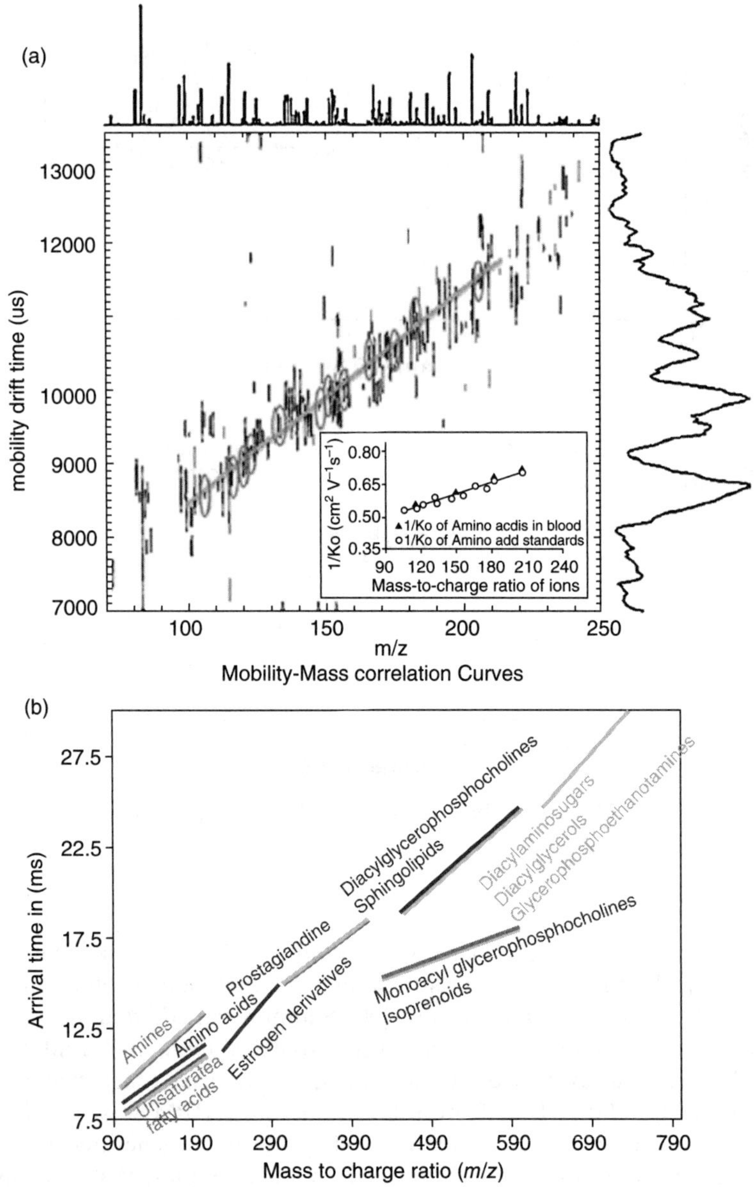

Figure 2.8 Ion mobility-TOF two-dimensional plot of the metabonome from human blood.

ACKNOWLEDGMENTS

We acknowledge the following for providing their original artwork: Drs. Prabha Dwivedi, Michaela Scigelova, Qin Ji, Mohammed Jemal, Mark Wrona, Haiying Zhang, and Iain Campuzano. We also acknowledge Drs. Walter Korfmacher, Kathy Cox, and Lucinda Cohen for critical reading of the manuscript.

REFERENCES

[1] Arrowsmith J. TRIAL WATCH Phase III and submission failures: 2007–2010. Nat Rev Drug Discov 2011;10:1.

[2] Scigelova M, Makarov A. Advances in bioanalytical LC-MS using the Orbitrap (TM) mass analyzer. Bioanalysis 2009;1:741–54.

[3] Wong RL, Xin BM, Olah T. Optimization of Exactive Orbitrap (TM) acquisition parameters for quantitative bioanalysis. Bioanalysis 2011;3:863–71.

[4] Korfmacher W. High-resolution mass spectrometry will dramatically change our drug-discovery bioanalysis procedures. Bioanalysis 2011;3:1169–71.

[5] Cotter RJ. The new time-of-flight mass spectrometry. Anal Chem 1999;71:445A–51A.

[6] Nagao H, Shimma S, Hayakawa S, Awazu K, Toyoda M. Development of a tandem time-of-flight mass spectrometer with an electrospray ionization ion source. J Mass Spectrom 2010;45:937–43.

[7] Bateman KP, Kellmann M, Muenster H, Papp R, Taylor L. Quantitative-qualitative data acquisition using a benchtop Orbitrap mass spectrometer. J Amer Soc Mass Spectrom 2009;20:1441–50.

[8] Zhang ZP, Zhu MS, Tang W. Metabolite identification and profiling in drug design: current practice and future directions. Curr Pharm Des 2009;15:2220–35.

[9] Zhang NR, Yu S, Tiller P, Yeh S, Mahan E, Emary WB. Quantitation of small molecules using high-resolution accurate mass spectrometers—a different approach for analysis of biological samples. Rapid Commun Mass Spectrom 2009;23:1085–94.

[10] Du F, Liu T, Shen T, Zhu F, Xing J. Qualitative-(semi)quantitative data acquisition of artemisinin and its metabolites in rat plasma using an LTQ/Orbitrap mass spectrometer. J Mass Spectrom 2012;47:246–52.

[11] Ramanathan R, Jemal M, Ramagiri S, Xia YQ, Humpreys WG, Olah T, Korfmacher WA. It is time for a paradigm shift in drug discovery bioanalysis: from SRM to HRMS. J Mass Spectrom 2011;46:595–601.

[12] Xia YQ, Lau J, Olah T, Jemal M. Targeted quantitative bioanalysis in plasma using liquid chromatography/high-resolution accurate mass spectrometry: an evaluation of global selectivity as a function of mass resolving power and extraction window, with comparison of centroid and profile modes. Rapid Commun Mass Spectrom 2011;25:2863–78.

[13] Kaufmann A, Butcher P, Maden K, Walker S, Widmer M. Quantitative and confirmative performance of liquid chromatography coupled to high-resolution mass spectrometry compared to tandem mass spectrometry. Rapid Commun Mass Spectrom 2011;25: 979–92.

[14] Hopfgartner G, Tonoli D, Wagner-Rousset E. High-resolution mass spectrometry for integrated qualitative and quantitative analysis of pharmaceuticals in biological matrices. Anal Bioanal Chem 2012;402(8):2587–96.

[15] Fung EN, Xia YQ, Aubry AF, Zeng JN, Olah T, Jemal M. Full-scan high resolution accurate mass spectrometry (HRMS) in regulated bioanalysis: LC-HRMS for the quantitation of prednisone and prednisolone in human plasma. J Chromatogr B Analyt Technol Biomed Life Sci 2011;879:2919–27.

[16] Henry H, Sobhi HR, Scheibner O, Bromirski M, Nimkar SB, Rochat B. Comparison between a high-resolution single-stage Orbitrap and a triple quadrupole mass spectrometer for quantitative analyses of drugs. Rapid Commun Mass Spectrom 2012;26: 499–509.

[17] Paweletz CP, Wiener MC, Bondarenko AY, Yates NA, Song QH, Liaw A, Lee AYH, Hunt BT, Henle ES, Meng FY, Sleph HF, Holahan M, Sankaranarayanan S, Simon AJ, Settlage RE, Sachs JR, Shearman M, Sachs AB, Cook JJ, Hendrickson RC. Application of an end-to-end biomarker discovery platform to identify target engagement markers in cerebrospinal fluid by high resolution differential mass spectrometry. J Proteome Res 2010;9:1392–1401.

[18] Przybylak R, Cronin M. In silico models for drug-induced liver injury—current status. Expert Opin Drug Metab Toxicol 2012;8:201–7.

[19] Zhang HY, Petrone L, Kozlosky J, Tomlinson L, Cosma G, Horvath J. Pooled sample strategy in conjunction with high-resolution liquid chromatography-mass spectrometry-based background subtraction to identify toxicological markers in dogs treated with ibipinabant. Anal Chem 2010;82:3834–9.

[20] Lipinski CA, Lombardo F, Dominy BW, Feeney PJ. Experimental and computational approaches to estimate solubility and permeability in drug discovery and development settings. Adv Drug Deliv Rev 1997;23:3–25.

[21] Daniels LB. Combining multiple biomarkers for cardiovascular risk assessment: more is usually better—up to a point. Bioanalysis 2011;3:1679–82.

[22] Eikel D, Vavrek M, Smith S, Bason C, Yeh S, Korfmacher WA, Henion JD. Liquid extraction surface analysis mass spectrometry (LESA-MS) as a novel profiling tool for drug distribution and metabolism analysis: the terfenadine example. Rapid Commun Mass Spectrom 2011;25:3587–96.

[23] Li FB, Hsieh YS, Kang L, Sondey C, Lachowicz J, Korfmacher WA. MALDI-tandem mass spectrometry imaging of astemizole and its primary metabolite in rat brain sections. Bioanalysis 2009;1:299–307.

[24] Rompp A, Guenther S, Schober Y, Schulz O, Takats Z, Kummer W, Spengler B. Histology by mass spectrometry: label-free tissue characterization obtained from high-accuracy bioanalytical imaging. Angew Chem Int Ed Engl 2010;49:3834–8.

[25] Wisniewski JR, Ostasiewicz P, Mann M, High recovery FASP applied to the proteomic analysis of microdissected formalin fixed paraffin embedded cancer tissues retrieves known colon cancer markers. J Proteome Res 2011;10:3040–9.

[26] Beck A, Wurch T, Bailly C, Corvaia N. Strategies and challenges for the next generation of therapeutic antibodies. Nature Rev Immunol 2010;10:345–52.

[27] McAlister GC, Phanstiel D, Wenger CD, Lee MV, Coon JJ. Analysis of tandem mass spectra by FTMS for improved large-scale proteomics with superior protein quantification. Anal Chem 2010;82:316–22.

[28] Ruan Q, Ji QC, Arnold ME, Humphreys WG, Zhu MS. Strategy and its implications of protein bioanalysis utilizing high-resolution mass spectrometric detection of intact protein. Anal Chem 2011;83:8937–44.

[29] Shi B, Keough E, Matter A, Leander K, Young S, Carlini E, Sachs AB, Tao WK, Abrams M, Howell B, Sepp-Lorenzino L. Biodistribution of small interfering RNA at the organ and cellular levels after lipid nanoparticle-mediated delivery. J Histochem Cytochem 2011;59:727–40.

[30] Li G, Crowe D, Beverly M. LC-MS analysis of therapeutic siRNA in mouse liver, American Society Mass Spectrometry Conference, Denver, 2011.

[31] Zou Y, Tiller P, Chen IW, Beverly M, Hochman J. Metabolite identification of small interfering RNA duplex by high-resolution accurate mass spectrometry. Rapid Commun Mass Spectrom 2008;22:1871–81.

[32] Takami T, Nishida Y, Shioyama S, Goto R. Quantitative and qualitative analysis of an siRNA and its degradation products in plasma by LC-MS/MS. Poster 311 (636), 59th ASMS Conference on Mass Spectrometry and Allied Topics. Denver, CO, June 5–9, 2011.

[33] Xu Y, Zhang N, Koeplinger K, Mahah E, Yeh S, Cancilla M. Unpublished, 2012. Unpublished Work.

[34] Zhang JT, Fan HH, Levorse DA, Crocker LS. Interaction of cholesterol-conjugated ionizable amino lipids with biomembranes: lipid polymorphism, structure-activity relationship, and implications for siRNA delivery. Langmuir 2011;27:9473–83.

[35] Baillie TA, Cayen MN, Fouda H, Gerson RJ, Green JD, Grossman SJ, Klunk LJ, LeBlanc B, Perkins DG, Shipley LA. Drug metabolites in safety testing. Toxicol Appl Pharm 2002;182:188–96.

[36] Nakayama S, Atsumi R, Takakusa H, Kobayashi Y, Kurihara A, Nagai Y, Nakai D, Okazaki O. A zone classification system for risk assessment of idiosyncratic drug toxicity using daily dose and covalent binding. Drug Metab Dispos 2009;37:1970–7.

[37] Mortishire-Smith RJ, O'Connor D, Castro-Perez JM, Kirby J. Accelerated throughput metabolic route screening in early drug discovery using high-resolution liquid chromatography/quadrupole time-of-flight mass spectrometry and automated data analysis. Rapid Commun Mass Spectrom 2005;19:2659–70.

[38] Bateman KP, Castro-Perez J, Wrona M, Shockcor JP, Yu K, Oballa R, Nicoll-Griffith DA. MSE with mass defect filtering for *in vitro* and *in vivo* metabolite identification. Rapid Commun Mass Spectrom 2007;21:1485–96.

[39] Yang YO, Grubb MF, Luk CE, Humphreys WG, Josephs JL. Quantitative estimation of circulating metabolites without synthetic standards by ultra-high-performance liquid chromatography/high resolution accurate mass spectrometry in combination with UV correction. Rapid Commun Mass Spectrom 2011;25:3245–51.

[40] Zhang HY, Ma L, He K, Zhu MS. An algorithm for thorough background subtraction from high-resolution LC/MS data: application to the detection of troglitazone metabolites in rat plasma, bile, and urine. J Mass Spectrom 2008;43:1191–200.

[41] Zhu MS, Ma L, Zhang DL, Ray K, Zhao WP, Humphreys WG, Skiles G, Sanders M, Zhang HY. Detection and characterization of metabolites in biological matrices using mass defect filtering of liquid chromatography/high resolution mass spectrometry data. Drug Metab Dispos 2006;34:1722–33.

[42] Zhang HY, Zhang DL, Ray K, Zhu MS. Mass defect filter technique and its applications to drug metabolite identification by high-resolution mass spectrometry. J Mass Spectrom 2009;44:999–1016.

[43] Chen WQ, Caceres-Cortes J, Zhang HY, Zhang DL, Humphreys WG, Gan JP. Bioactivation of substituted thiophenes including alpha-chlorothiophene-containing compounds in human liver microsomes. Chem Res Toxicol 2011;24:663–9.

[44] Holman SW, Wright P, Langley GJ. High-throughput approaches towards the definitive identification of pharmaceutical drug metabolites. 2. An example of how unexpected dissociation behaviour could preclude correct assignment of sites of metabolism. Rapid Commun Mass Spectrom 2009;23:2017–25.

[45] Wallace A. A high-resolution ion mobility mass spectrometry platform for breakthrough discoveries in life science research and the pharmaceutical industry. Am Lab 2010;42:13–7.

[46] Giles K, Williams JP, Campuzano I. Enhancements in travelling wave ion mobility resolution. Rapid Commun Mass Spectrom 2011;25:1559–66.

[47] Dwivedi P, Schultz AJ, Hill HH. Metabolic profiling of human blood by high-resolution ion mobility mass spectrometry (IM-MS). Int J Mass Spectrom 2010;298:78–90.

[48] Syka JEP, Coon JJ, Schroeder MJ, Shabanowitz J, Hunt DF. Peptide and protein sequence analysis by electron transfer dissociation mass spectrometry. Proc Natl Acad Sci U S A 2004;101:9528–33.

[49] Liu J, Zhao Z, Teffera Y. Application of on-line nano-liquid chromatography/mass spectrometry in metabolite identification studies. Rapid Commun Mass Spectrom 2012;26: 320–6.

[50] Ramanathan R, Raghavan N, Comezoglu SN, Humphreys WG. A low flow ionization technique to integrate quantitative and qualitative small molecule bioanalysis. Int J Mass Spectrom 2011;301:127–35.

[51] Cooks RG. Creativity through instrumentation. Anal Chem 1985;57:823A–43A.

3

QUANTITATIVE MASS SPECTROMETRY CONSIDERATIONS IN A REGULATED ENVIRONMENT

Mohammed Jemal and Yuan-Qing Xia

3.1 INTRODUCTION

Systematic evaluation of pharmacokinetics and metabolism has been recognized to be essential in reducing failure in drug discovery and development. Consequently, appropriately designed bioanalytical methods are critical in providing support for such studies. In general, bioanalytical method attributes include accuracy, precision, selectivity, sensitivity, reproducibility, and stability [1, 2]. High-performance liquid chromatography-tandem mass spectrometry (HPLC-MS/MS) has been widely used for the quantification of drugs and their metabolites in biological matrices due to high degree of specificity provided by MS/MS [3–11]. In this chapter, we highlight themes that have recently emerged as central to building quality during the development of HPLC-MS/MS bioanalytical methods. Use of quality methods developed in this manner will reduce bioanalytical risk and avoid the costly and nerve-racking experience of finding out that an invalid method, although "officially validated," has been used to support pivotal clinical and nonclinical studies. Accordingly, we discuss in some detail avoidance of the matrix effect, elimination of interference, and enhancing chromatography and mass spectrometric detection. After presenting the chromatographic, mass spectrometric, and extraction behaviors of phospholipids, we recommend a dual approach that involves selective sample extraction and chromatographic separation, for incorporation during method development in order to avoid phospholipids-related bioanalytical risks. For interference, we discuss potential pitfalls in HPLC-MS/MS bioanalysis and the means to avoid them. Such pitfalls

Mass Spectrometry for Drug Discovery and Drug Development, First Edition. Edited by Walter A. Korfmacher.

may occur due to mass spectral interference from metabolites or prodrugs, or due to the use of inappropriate calibration standard and quality control (QC) samples for analysis involving unstable drugs or metabolites. For enhancing chromatography, the significance of systematic selection of HPLC columns and mobile phases in HPLC-MS/MS bioanalysis is discussed. Looking to the future, we briefly present our perspective on the use of full scan high-resolution accurate mass spectrometry (HRAMS) for quantification of drugs and metabolites. The importance of using incurred sample (postdose sample) during method development is discussed. The significance of using incurred sample during method development cannot be overemphasized. This is essential in order to ensure that metabolites and other components present only in the incurred samples, and not in QC samples, do not interfere with the accurate quantitation of the drug or any particular metabolite of interest. This is different from the concept of incurred sample reanalysis (ISR), which is currently being widely practiced during the application of a validated method. ISR tests only reproducibility and may not have any relevance to detecting potential metabolite interference. In the end, we put forth an outline of a protocol for a systematic approach to method development in order to reduce bioanalytical risks.

3.2 CONSIDERATIONS OF AVOIDING POTENTIAL PITFALLS IN LC-MS/MS BIOANALYSIS

3.2.1 Biological Sample Collection, Storage, and Preparation

3.2.1.1 Plasma Collection and Storage The complex nature of biological matrices requires that sample preparation be an integral part of bioanalytical methods. One of the challenges faced in plasma sample collection, storage, and extraction is the potential instability of drugs, metabolites, and prodrugs in biological samples. Compound stability in plasma may be affected by enzymes and/or pH of the biological samples, anticoagulants, storage temperature, and freeze–thaw cycles [12–28]. While the degradation of a drug during sample collection and storage causes underestimation of the drug concentration, the degradation of the metabolite or prodrug may cause overestimation of the drug concentration. Acylglucuronides are probably the most commonly encountered problematic metabolites in bioanalysis. Acylglucuronides tend to be unstable and hydrolyze to release the original aglycone (the drug) under neutral and alkaline conditions and elevated temperatures [12, 13], although different acylglucuronides have been shown to have different rates of hydrolysis [13]. The mildly acidic pH of 3–5 tends to be the most desirable pH region for minimizing the hydrolysis of acylglucuronides in biological samples. The common recommendation is that blood samples should be immediately cooled on ice after collection, and then plasma should be separated using a cooled centrifuge within 10 min after blood collection. The plasma samples are then stored frozen at −70°C. During sample processing for analysis, the plasma samples should be kept cooled on ice and the aliquotted portions immediately buffered to lower the pH to 3–5. The lactone is another commonly encountered metabolite functional group that could be bioanalytically challenging since the lactone may be converted to its open-ring hydroxy acid drug. Systematic studies of the effects of pH and temperature on the stability of the lactone metabolites of hydroxy acid drugs show that the sample pH should be adjusted to 3–5 in order to minimize the hydrolysis of the lactone metabolite back to the drug or vice versa [14, 15].

It should be noted that if a drug, metabolite or prodrug is very unstable, it may be necessary to add a stabilizing reagent directly to the blood before obtaining plasma samples. It has thus been reported that citric acid added directly to the blood, in the amount of 5 mg/mL of blood, provides the lower pH required for stabilization of pH-sensitive labile compounds without causing the undesirable gelling of the blood [16]. Another important finding is that the pH of the biological sample may change during storage or processing and thus cause unexpected degradation of drugs or metabolites [17, 18]. Fura et al. found that the pH of plasma samples may be significantly higher than the normally expected value of approximately 7.4 depending on the conditions of sample storage and processing [17]. Initial addition of appropriate amounts of citrate or phosphate buffers to the plasma sample can be used to maintain the pH of plasma during storage or processing [17].

Metabolites with O-methyloximes [19] and carbon–carbon double bonds [20] can undergo E to Z isomerization due to nonoptimal pH or exposure to light. Epimerization is another potential cause for sample instability [21, 22]. A sample that contains a thiol drug and its disulfide metabolite may also cause an analytical challenge due to the potential for the conversion of the thiol to the disulfide or vice versa [23].

The type of anticoagulant used during blood sample collection may affect the stability of drugs or their metabolites. Evans et al. compared the stability of hormones in human blood samples collected with ethylenediaminetetraacetic acid (EDTA), lithium heparin, sodium fluoride, and potassium oxalate [24]. They showed that most of the hormones were stable in blood collected with EDTA or fluoride at 4°C. A variety of chemical agents, such as methyl acrylate, sulphate, thiosulphate, fluoride, borate, phosphate, paraoxon, eserine, and organophosphate reagents, have been used to stabilize analytes in biological matrices [13, 18, 25–28].

Another important aspect of plasma sample storage is the minimization of clot formation that could impede the facile aliquotting of the plasma sample for analysis. The clogging of pipetting tips is mainly caused by the formation of fibrinogen clots when the plasma samples are stored at –20°C and undergo freeze–thaw cycles [29]. The problem of fibrinogen clots can be overcome by storing plasma samples at –80°C [29]. Berna et al. employed an interesting approach to solving the clot formation problem by the use of 96-well polypropylene filter plates to collect, store frozen, and then filter plasma samples prior to bioanalysis [30]. Sadagopan et al. reported that the failure rate for transfer of EDTA plasma, by automated workstation or manually, is less than that of heparinized plasma [31]. During analysis of a large number of samples, the task of manually uncapping and recapping the sample tubes could take a significant amount of time and cause physical stress. As a solution to this problem, Teitz et al. reported the direct transfer of plasma samples from pierceable capped tubes into a 96-well plate using a robotic system [32].

3.2.1.2 Offline Plasma Extraction Offline sample extraction methods, namely solid phase extraction (SPE), liquid–liquid extraction (LLE), and protein precipitation (PPT), are widely used in quantitative HPLC-MS bioanalysis [33–47]. The multiple steps of the three modes of offline extraction have been automated in 96-well format to different degrees using robotic liquid handlers, as described in detail in an excellent book solely dedicated to bioanalytical sample preparation [34].

Using modular 96-well SPE plates populated with sorbent cartridges of different chemistries, Jemal et al. reported a strategy for the fast development of SPE procedures in a standardized, comprehensive manner [35]. Classical LLE methods, while considered to be less automation-friendly than SPE, may yield cleaner extracts as evidenced by less matrix effect and less tendency for clogging HPLC columns. A variant of LLE, solid-supported LLE, which is more automation-friendly, has also been used [34]. PPT, considered as a "dilute-and-shoot" technique, is one of the simplest sample preparation methods in bioanalysis. The other attraction to the PPT technique is its universality since the same basic procedure can be applied to extract almost any analyte. However, the disadvantage of this technique is that the extract obtained is not as clean as that obtained with SPE and LLE as may be evidenced by the matrix effect and soiling of the HPLC system and mass spectrometer. Therefore, automated SPE, LLE, and online extraction (described below) are considered to be the preferred techniques, especially at a stage when a drug candidate enters the development phase, where sensitive, accurate, precise, and rugged bioanalytical methods are required for routine analyses of large batches of clinical samples.

3.2.1.3 Online Plasma Extraction Online sample preparation, like PPT, is considered as another "dilute-and-shoot" method since there is no sample preparation except for aliquotting the samples, adding the internal standard (IS), and centrifugation. However, the main difference between these two methodologies is that online sample extraction provides a cleaner extract with reduced chance for matrix effect and less soiling of the HPLC-MS system. The unique feature of probably the most commonly used online technique is the use of an online extraction column packed with large particles of a stationary phase material (typically >20 μm), which is used in conjunction with a very high flow rate of the mobile phase (typically 3–5 mL/min). This translates to a very fast linear speed of 10–17 cm/s when a narrow bore column (typically 1 mm i.d.) is used. The combination of the fast flow and large particle sizes allows for the rapid removal of proteins with simultaneous retention of the small-molecule analyte of interest. The principles and numerous applications of this fast-flow online extraction technique have been presented in a number of publications [3, 48–58], including an excellent review article [50]. Both homemade and commercially available online extraction systems have been used. Early on, a simple, homemade online system, based on a single extraction column for both extraction and analysis, was used [48, 51]. However, this approach gives little or no chromatographic separation. Then a system that incorporated an analytical column in-line with the extraction column was used to achieve adequate chromatographic separation [51, 52]. Further improvement to this type of homemade configuration was made by using two parallel extraction columns as well as two parallel analytical columns [53]. Large particle-size extraction columns commonly used include polymeric and silica-based sorbents, which work based on reversed-phase, ion exchange, or mixed (combination of ion exchange and reversed phase) mode of separation. Fully integrated commercial systems are available that provide very desirable features such as the ability to achieve isocratic focusing while eluting from the extraction column in order to retain the eluted analyte as a sharp and well-defined plug at the head of the analytical column [57, 58]. Large particle-size extraction columns commonly used include polymeric and silica-based sorbents, which work based on reversed-phase, ion exchange, or mixed (combination of ion exchange and reversed-phase) mode of separation.

A second category of extraction columns consists of columns that are packed with sorbent materials known as restricted access media (RAM) [50, 59–64]. RAM designates a family of sorbents that allows direct injection of biological fluids by limiting the accessibility of interaction sites within the pores to small molecules only. Macromolecules are excluded and interact only with the outer surface of the particle support coated with hydrophilic groups, which minimizes the adsorption of matrix proteins. RAM can be classified according to the protein exclusion mechanism. Macromolecules can be excluded by a physical barrier due to the pore diameter or by a diffusion barrier created by a protein (or polymer) network at the outer surface of the particle. RAM can further be classified by subdividing RAM sorbents with respect to their surface chemistry. RAM-based online extractions have been used either in single-column mode, with the RAM column serving both as the extraction and analytical column, or in conjunction with a second analytical column. The main advantage of the one-column system is simplicity. Hsieh et al. reported the application of such a system using a single RAM column at a regular flow rate (1.0 mL/min for a column of 4.6 × 50 mm, 5 μm) for drug analysis in plasma [61]. A disadvantage of a one-column system is that only limited chromatographic separation of analytes can be achieved. On the other hand, a dual-column system consisting of one extraction column and one analytical column allows the achievement of superior chromatographic resolution by the use of analytical columns of various modes of chromatographic separation. For example, separation and quantification of enantiomers can be achieved by simply coupling a chiral column to the extraction column [62–64].

Online extraction HPLC-MS systems using monolithic alkyl-bonded silica rod columns for quantitative analysis have also been reported [65–67]. Successful quantitative bioanalyses have been accomplished by using the monolithic column in a single-column mode, where the column serves as both the extraction and analytical column, or by using it as an analytical column in conjunction with a different extraction column. Although monolithic column based online extraction looks promising, there is a need for more practical experience with this approach, and for the availability of more variety of monolithic columns, both in terms of column dimensions and in terms of stationary phase types.

3.2.2 Matrix Effect

3.2.2.1 Background The performance of HPLC-MS methods in electrospray (ESI) and atmospheric pressure chemical ionization (APCI) is hampered by matrix effects, which result mostly in ion suppression and sometimes in ion enhancement. Kebarle and Tang first observed the matrix effect phenomena in ESI where the analyte response decreased in the presence of other organic compounds [68]. Several publications describe the mechanism of the matrix effects, methodologies to detect the phenomena, and biological sample preparation procedures to minimize the effects [68–82]. In ESI, droplets with a surface excess charge are created. Enke introduced a predictive model based on competition among the ions in the solution for the limited number of excess charge sites [69]. Thus, at low concentrations of the analyte, the response–concentration relation is linear. However, at higher concentrations, the response becomes independent of the analyte concentration but highly affected by the presence of other analytes [69]. Pan and McLuckey studied the effects of small cations on the positive ESI responses of proteins at low pH. They concluded that the extent to which ions concentrate on the droplet surface is the major factor in determining the ion suppression efficiency [70]. King et al. investigated the mechanism of ion suppression in ESI and found out that the gas-phase reaction leading to the loss of net charge on the analyte is not the important process that causes ion suppression [71]. However, the presence of nonvolatile solute is much more important since this changes the droplet solution properties.

A number of approaches to assess ion suppression or enhancement have been proposed in the literature. Among them, a post-column infusion of analyte of interest while injecting the blank matrix [72] has been used widely. This system allows the bioanalyst to determine the extent of the effect of endogenous components in blank plasma on the analyte response as a function of chromatographic retention time. Ion suppression or enhancement is illustrated as a valley or a hill, respectively, in the otherwise flat response–time trace [72]. Kaufmann and Butcher reported the use of segmented post-column analyte addition to visualize and compensate signal suppression/enhancement [73]. Matuszewski et al. reported a practical approach for the quantitative assessment of the absolute and relative matrix effect [74] as part of formal validation of bioanalytical HPLC-MS/MS methods.

In addition to the endogenous components in biological samples, dosing vehicles, such as PEG400, propylene glycol, Tween 80, hydroxypropyl-β-cyclodextrin, and N,N-dimethylacetamide, have also been found to cause matrix effect to varying degrees [75, 76]. The postdose samples, especially the early time-point samples, may contain high concentrations of these dosing vehicles. Thus, the drug concentrations in such samples could be underestimated if lower analyte response is obtained due to the matrix effect caused by the dosing vehicle. Another potential source of matrix effect is the containers used for sample storage or sample processing [77, 78]. It is very important to recognize that matrix effect can also be caused by co-eluting IS, including stable-isotope labeled analogs [79, 80]. This phenomenon can be explained by competition between the analyte and IS for the droplet surface charge [69, 81]. It should be noted that analyte suppression is flow rate-dependent and is practically absent at flow rates below 20 nL/min [82].

3.2.2.2 Association of Plasma Phospholipids with Matrix Effect Association of plasma phospholipids with ionization suppression or enhancement (matrix effect) has come to the fore in the bioanalytical community thanks to the pioneering work by Bennett, Van Horne, Ahnoff, et al. [83–85]. It is thus important to monitor the fate of phospholipids during method development and application. As shown in Figure 3.1, plasma phospholipids belong to one of two structural classes of phospholipids: glycerophospholipids and sphingomyelins (SM) [86–89]. Glycerophospholipids are made up of a three-carbon glycerol backbone with different substitutions at the three carbon (C) positions. As shown in Figure 3.1, the C-1 position contains a fatty acid ester, an alkyl ether, or a vinyl ether, giving rise to phosphatidyl, plasmanyl, and plasmenyl (plasmalogen) glycerophospholipids, respectively. The C-2 position may contain a fatty-acid ester group, or the C-2 hydroxyl group (OH) may remain free without being esterified, the latter giving rise

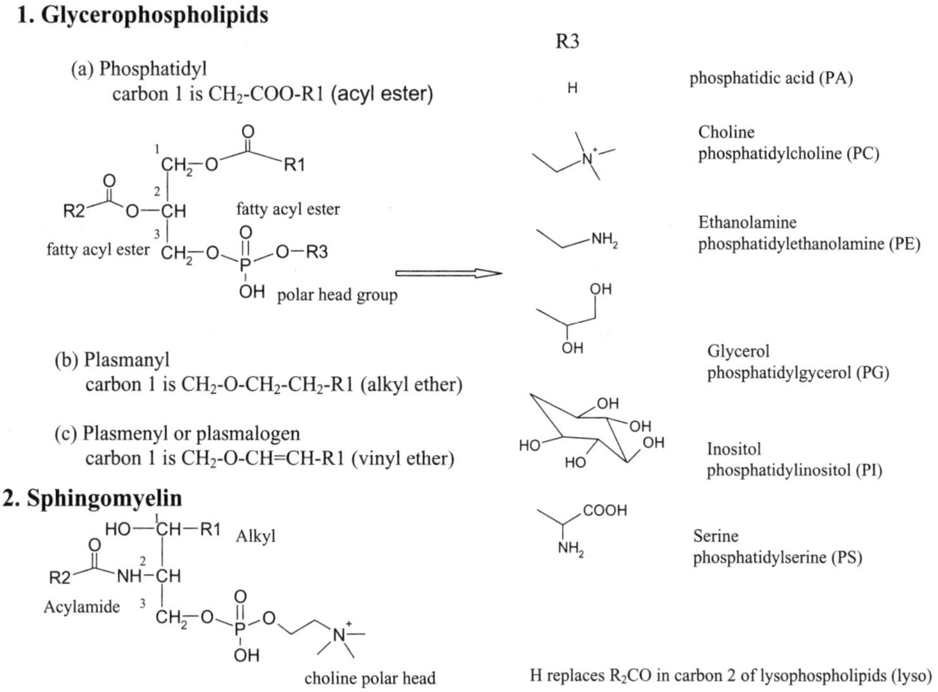

Figure 3.1 Phospholipids structure. Reproduced from Xia and Jemal, *Rapid Commun. Mass Spectrom.* **2009**, *23*, 2125–2138, with permission from John Wiley & Sons, Ltd.

to the lysophospholipids (lyso) class. The C-3 position OH is esterified with a phosphoric acid, which in turn is esterified to create different polar head groups. Depending on the polar head groups at the C-3 position, glycerophospholipids are classified into different classes: phosphatidic acid (PA), phosphatidylcholine (PC), phosphatidylethanolamine (PE), phosphatidylglycerol (PG), phosphatidylinositol (PI), and phosphatidylserine (PS). On the other hand, as shown in Figure 3.1, SMs contain an alky chain at the C-1 position, with the OH remaining free. While the C-2 position contains a fatty acid amide, the C-3 position contains the same group found in the PC class of glycerophospholipids. The total concentration of all phospholipids in human plasma is about 1.6–3.0 mg/mL [90]. The PC class of phospholipids accounts for about 60–70% of the total phospholipids found in plasma [91, 92]. The other abundant classes are SM, PE, PI, lyso-PC, PS, and PG listed in the order of decreasing concentration [92].

3.2.2.3 Mass Spectrometric Monitoring of Phospholipids A timely and well-received technique for monitoring phospholipids during HPLC-MS/MS bioanalytical method development and application has been reported [93]. The technique is based on using positive electrospray (ESI) selected reaction monitoring (SRM) of *m/z* 184→*m/z* 184, with the precursor ion (*m/z* 184) generated in the source due to collision-induced dissociation (CID) of phospholipids. This technique can be used to monitor PC, lyso-PC, and SM phospholipids. It cannot be used to monitor the other phospholipids, such as PE, PI, PG, PS, and PA, since these phospholipids do not generate the *m/z* 184 ion in the source. Xia and Jemal later evaluated three tandem mass spectrometry (MS/MS) techniques to monitor phospholipids, using positive and negative ESI, and we finally recommended the following all-inclusive technique [86]: positive precursor ion scan of *m/z* 184 for the detection of all the PC, lyso-PC, and SM phospholipids; positive neutral loss scan of 141 Da for the detection of lysoPE and PE; and negative precursor ion scan of *m/z* 153 for the detection of the remaining phospholipids. As illustrated below, not only can this technique detect all the classes of phospholipids, but it also possesses other distinct advantages compared with the *m/z* 184→*m/z* 184 SRM technique.

The application of the all-inclusive technique in the detection of phospholipids in human plasma was demonstrated by injecting the supernatant of acetonitrile-precipitated plasma [86]. As shown in Figure 3.2c, using the positive precursor ion scan of *m/z* 184, a number of intensive chromatographic peaks were obtained in the elution window between 2–10 min. Post-acquisition interrogation of the data enables the identification of the positive precursor ions corresponding to the chromatographic peaks. The results confirmed the presence of the previously reported [92, 94–96] major lyso-PC phospholipids (*m/z* 496, *m/z* 520, *m/z* 522, *m/z* 524, and *m/z* 544), PC phospholipids (*m/z* 758, *m/z* 760, *m/z* 786, and SM phospholipids (*m/z* 701 and *m/z* 703) in human plasma. The SM phospholipids can be easily distinguished from PC phospholipids as the protonated molecule of an SM is expected to be an odd number, whereas the protonated molecule of a PC is expected to be an even number (nitrogen rule). The lyso-PC phospholipids eluted in the 2–5 min region and the PC and SM phospholipids eluted in the 5–10 min region, with the *m/z* 760 and *m/z* 786 PC phospholipids being among the late eluting components. The phospholipid profile obtained using the SRM of *m/z* 184→*m/z* 184. (Fig. 3.2d) is, in general, similar to that obtained using the positive precursor ion scan of *m/z* 184 (Fig. 3.2c). A very important advantage of the technique based on the precursor ion scan of *m/z* 184 over the *m/z* 184→*m/z* 184 technique is that the former not only detects all the phospholipids that have the choline polar head, but it also identifies the precursor ions corresponding to each chromatographic peak, which is not the case with the latter. This feature enables the facile distinction between the PC, lyso-PC, and SM phospholipids. This is very important when working under different chromatographic conditions where the elution order of the different phospholipids is not known a priori. The results obtained using the positive neutral loss scan of 141 Da (Fig. 3.2b) confirmed the presence of previously reported [92, 94, 97] PE phospholipids (*m/z* 742, *m/z* 744, *m/z* 768) in human plasma. The results obtained using the negative precursor ion scan of *m/z* 153 are shown in Figure 3.2a. The results obtained using the SRM transitions of the individual phospholipids species selected to represent the different classes of phospholipids are shown in Figure 3.2e.

3.2.2.4 Chromatographic Elution Behavior of Phospholipids It is very important to evaluate the elution behavior of the phospholipids in order to establish a strategy for eliminating bioanalytical risk due to phospholipids that have not eluted completely during the analytical run time. The results obtained in the authors' laboratory [86] using different mobile phases in reversed-phase chromatography, the most commonly used mode of chromatography in LC-MS/MS bioanalysis, are

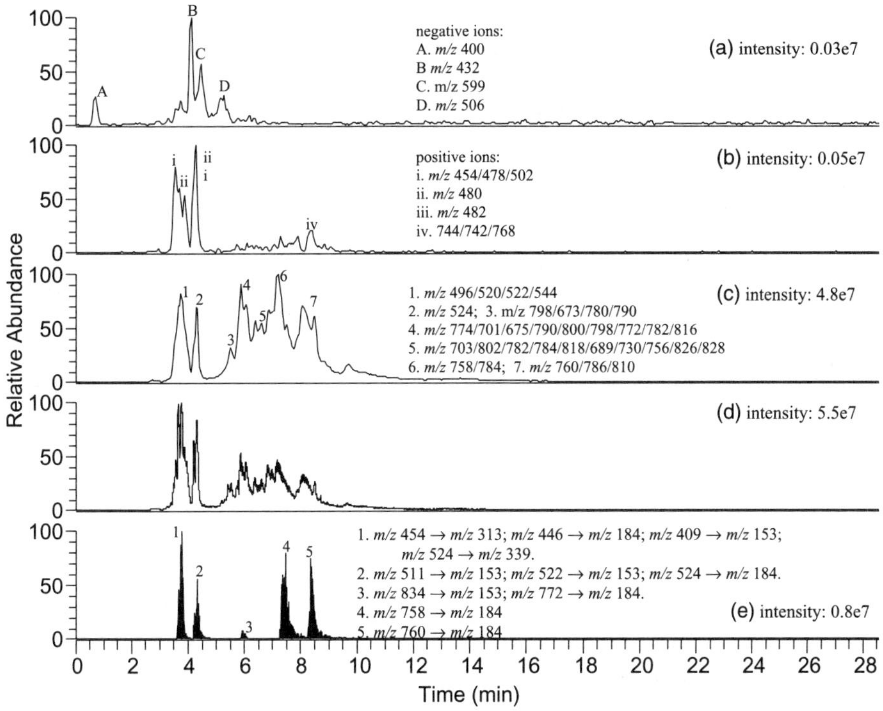

Figure 3.2 Comparison of three techniques used to monitor human plasma phospholipids from acetonitrile-precipitated human plasma injected into LC-MS/MS system using a gradient elution. Chromatograms obtained using the technique based on the precursor ion/neutral loss scans: (a) negative precursor-ion scan of *m/z* 153; (b) positive neutral loss of 141 Da; (c) positive precursor ion scan of *m/z* 184; (d) chromatogram obtained using the technique based on the in-source-CID *m/z* 184 → *m/z* 184; (e) chromatogram obtained using a technique based on the using SRM transitions of phospholipids species each representing a class of phospholipids. XBridge C18 (2.1 × 50 mm, 3.5 μm); flow rate: 0.3 mL/min; column temperature: 30°C. Gradient elution with eluent A (5 mM ammonium formate and 5 mM formic acid in water, pH 3.2) and eluent B (acetonitrile): start at 30% B and hold for 0.5 min; increase % B to 95% in 4.5 min; hold for 23 min. Reproduced from Xia and Jemal, *Rapid Commun. Mass Spectrom.* **2009**, *23*, 2125–2138, with permission from John Wiley & Sons, Ltd.

summarized in Figure 3.3. Under isocratic conditions, using a mobile phase consisting of 5 mM ammonium formate/5 mM formic acid in water as the aqueous eluent and acetonitrile as the organic eluent in different proportions, phospholipids eluted faster as the acetonitrile percentage was increased (Fig. 3.3a–d). A total removal of phospholipids required about 80 min for 65% acetonitrile (Fig. 3.3a), 25 min for 75% acetonitrile (Fig. 3.3b), 10 min for 85% acetonitrile (Fig. 3.3c), and 5 min for 95% acetonitrile (Fig. 3.3d). The results obtained using 20% of the same aqueous eluent (5 mM ammonium formate/5 mM formic acid) and 80% of acetonitrile, methanol, or isopropyl alcohol as the organic eluent are shown in Figure 3.3e–g. Under these conditions, the total removal of plasma phospholipids required about 20 min for acetonitrile (Fig. 3.3e), 205 min for methanol (Fig. 3.3f), and 3 min for isopropyl

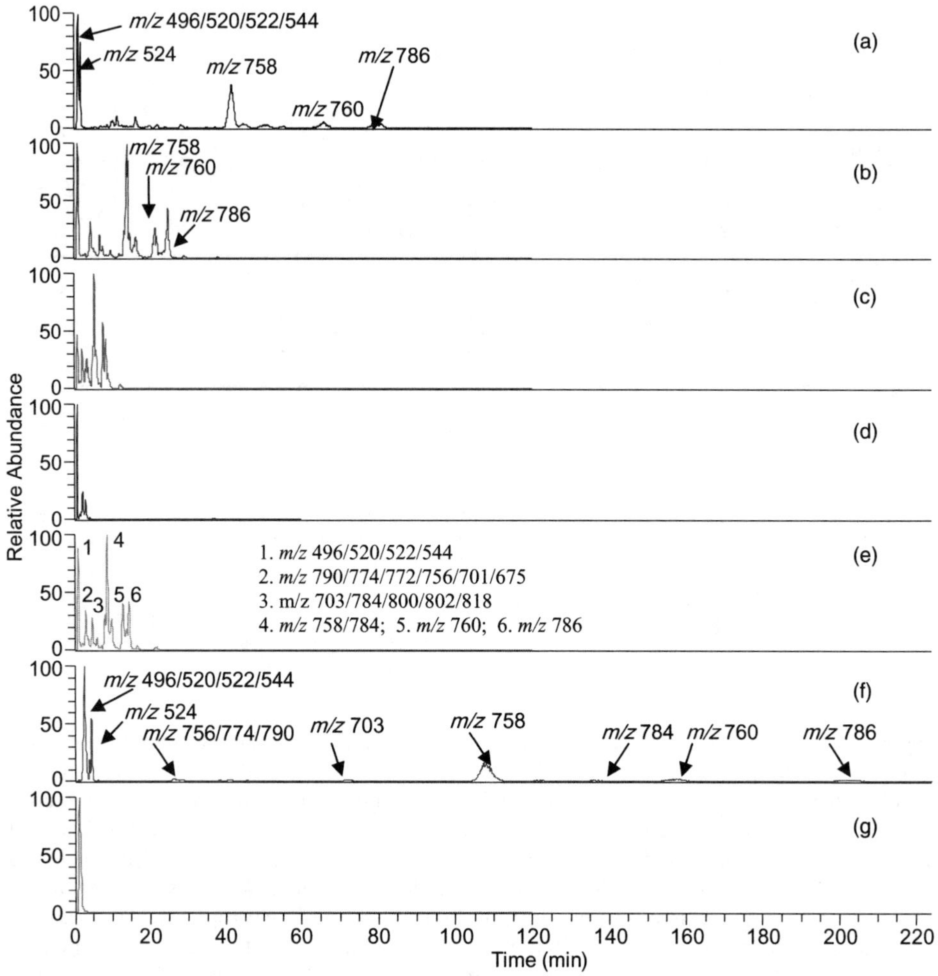

Figure 3.3 Phospholipids chromatographic elution profiles illustrated with the LC-MS/MS chromatograms obtained from acetonitrile-precipitated human plasma using the positive precursor ion scan of *m/z* 184 for detection of phospholipids, with an isocratic elution using the following mobile phases made up of an aqueous eluent (5 mM ammonium formate/5 mM formic acid in water) and an organic eluent: (a) 65% acetonitrile, (b) 75% acetonitrile, (c) 85% acetonitrile, (d) 95% acetonitrile, (e) 80% acetonitrile, (f) 80% methanol, (g) 80% isopropyl alcohol. Reproduced from Xia and Jemal, *Rapid Commun. Mass Spectrom.* **2009**, *23*, 2125–2138, with permission from John Wiley & Sons, Ltd.

alcohol (Fig. 3.3g). Thus, the most effective organic eluent to remove phospholipids from the reversed-phase column used was isopropyl alcohol, followed by acetonitrile and methanol, which is in contrast with a previous report [93], where methanol was indicated to be a stronger eluent than acetonitrile. The difference in the acetonitrile and methanol elution order seen between the two laboratories could be due to the different HPLC column used in the previous work. The work in the authors' laboratory [86] showed that, using the same organic eluent, different aqueous eluents did not make a significant difference in the elution behavior of phospholipids, which is in contrast with the dramatic effect seen when changing the type and/or percentage of the organic eluent used.

Under the reversed-phase chromatographic conditions used for HPLC-MS/MS bioanalytical methods, typically a flow rate of 0.3–0.6 mL/min for a 2 × 50 mm column with a mobile phase consisting of less than 80% methanol or acetonitrile, the lysophospholipids, such as lyso-PC, will elute near the analytes (drugs, metabolites, or biomarkers), while the phospholipids, such as PC, will elute much later. This reality puts into question the perceived advantage of column washing via increasing the organic component of the mobile phase to 100% following the elution of the analyte, unless the washing procedure is conducted long enough to remove all the phospholipids. Inadequate column washing may be worse than not washing at all since the washing procedure would cause the otherwise late eluting phospholipids to elute sooner as narrower peaks with enhanced probability to cause ionization suppression or enhancement.

3.2.2.5 Strategy of Avoiding Phospholipids via Chromatographic Separation of the Analyte from Phospholipids It is important to adopt the strategy of chromatographically separating the analyte from phospholipids during quantitative HPLC-MS/MS method development for drugs and their metabolites in plasma. Experience in the authors' laboratory with a large number of drugs and metabolites having diverse physicochemical properties shows that, using reversed-phase chromatography, appropriate column–mobile phase combinations can be found to achieve the separation of the analytes from all the phospholipids, including the early eluting lyso phospholipids, with the analyte eluting before the lyso phospholipids and the larger phospholipids eluting after the lyso phospholipids [4]. Figure 3.4 illustrates the effect of the HPLC column using the same mobile phase in achieving the separation of the analyte from the phospholipids [4]. Under the same chromatographic conditions, the analyte was separated from plasma phospholipids on column 1, while the analyte co-eluted with the phospholipids on column 2.

It is highly desirable that the phospholipids elute after the analyte of interest since this would allow, if so desired, changing the chromatographic conditions to rapidly elute the phospholipids off the column after the analyte elution. When analyzing a large batch of samples, there are different possible approaches to dealing with phospholipids eluting slowly after the analyte elution. One approach would be to completely elute all the phospholipids off the column before the next injection of a sequence consisting of a large batch of samples. Following the elution of the analyte, this can be accomplished in a rapid manner by increasing the flow rate and/or raising the organic component of the mobile phase to 100%, or by introducing another organic solvent of greater eluting strength, such as isopropanol. Another approach would be to wait until the completion of the batch analysis and then conduct the column washing. This is illustrated in Figure 3.5 for a plasma sample extracted using LLE, which removed almost all the phospholipids from the extract. There were no phospholipid peaks detected in the blank plasma LLE extract in the first injection (Fig. 3.5a) using isocratic elution. On the 201th injection (Fig. 3.5b), using the same isocratic elution, there were no phospholipid peaks detected from the same LLE extract; however, the overall baseline signal intensity significantly increased, indicating continuous, slow phospholipids bleeding off the column. On

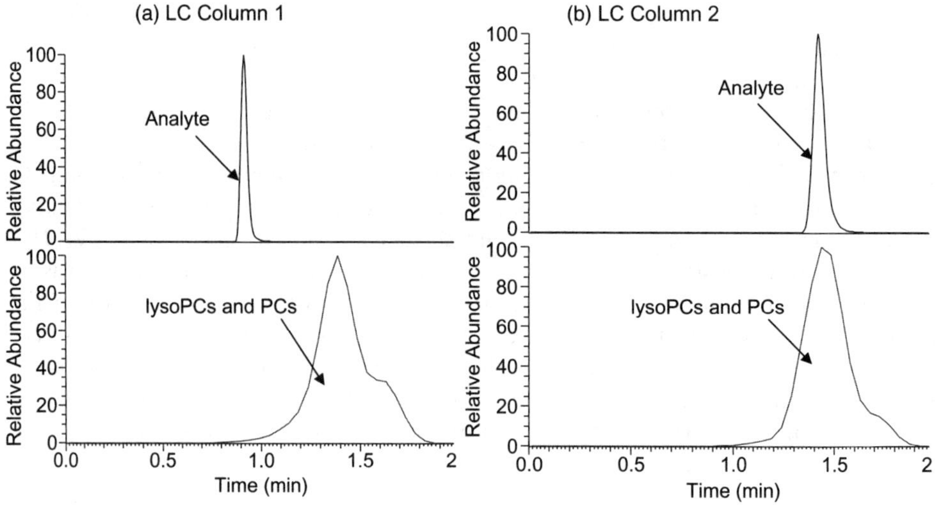

Figure 3.4 LC column effect on the separation of an analyte from plasma phospholipids with the same mobile phase: (a) column 1, (b) column 2. In each of panel (a) or (b), the top trace depicts the SRM transition of the analyte, and the bottom trace depicts phospholipids monitoring using the positive precursor ion scan of *m/z* 184 (+Pre *m/z* 184). Reproduced from Jemal et al., *Biomed. Chromatogr.* **2010**, *24*, 2–19, with permission from John Wiley & Sons, Ltd.

the 202nd injection (Fig. 3.5c), a high organic wash step was applied after the point of the analyte elution. As shown, the wash step achieved the removal of the accumulated phospholipids as indicated by the presence of the *m/z* 758 and *m/z* 786 ions. On the other hand, there was no accumulation of lyso phospholipids.

3.2.2.6 Strategy of Avoiding Phospholipids via Removal of Phospholipids during Sample Extraction The experience in the authors' laboratory show that the technique of LLE for the extraction of small-molecule drugs and metabolites from plasma samples compares favorably with the other techniques of sample extraction, namely SPE, PPT, and online extraction. In general, LLE, if properly optimized, gives very clean extracts as gauged by the absence of endogenous peaks interfering with the selected reaction quantitation of the analyte even at the level of the lower limit of quantitation (LLOQ) [4]. The selectivity achieved involving the extraction of a number of drugs and metabolites was, in general, as good as or better than that obtained with SPE. The run-to-run assay reproducibility obtained with LLE was very good, with no untoward effects observed arising from lot-to-lot variation that may occasionally be observed with SPE products. LLE-based methods are easy to transfer from one laboratory to another and are relatively of low cost. With the advent of 96-well format devices, automation of LLE has become practical and user-friendly. Traditionally, the extraction of analytes using LLE has been performed by conducting the extraction after adjusting the plasma pH to a value equal to pKa minus 2 for acidic analytes and to pKa plus 2 for basic analytes, assuming that the nonionized analyte species have a better extractability behavior than the ionized species. While this may hold true in general, the experience in the

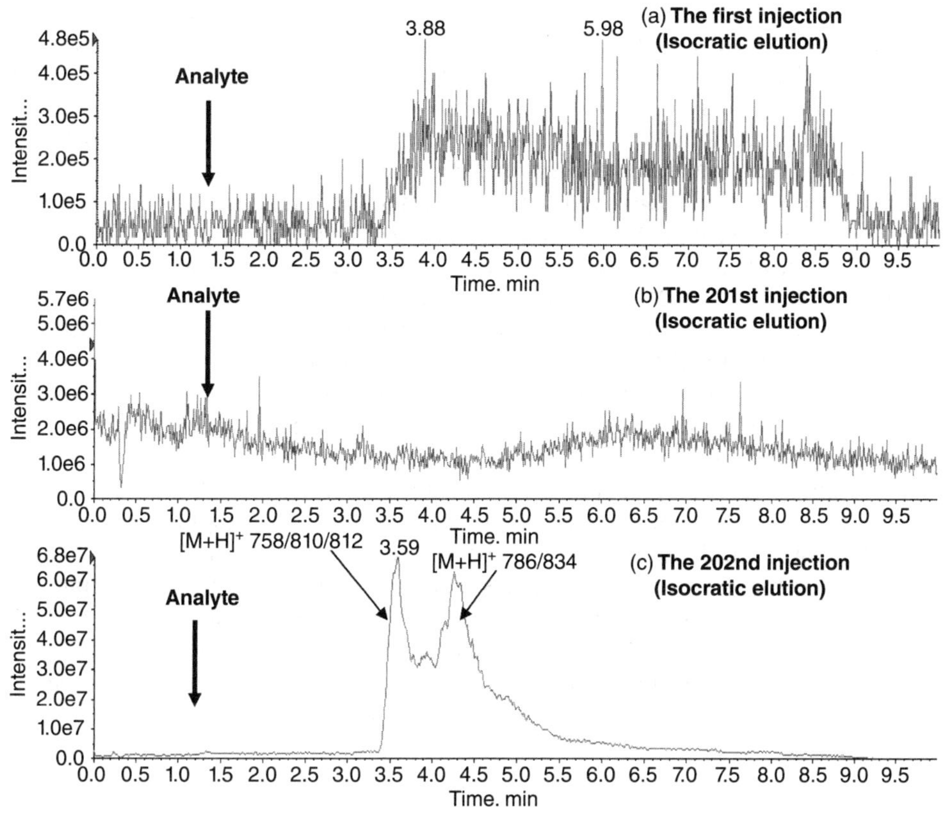

Figure 3.5 The chromatographic fate of phospholipids, monitored using the precursor ion scan of *m/z* 184, in a blank plasma extract, obtained via liquid–liquid extraction (LLE), during 202 runs (injections), all under isocratic conditions except for the final run where isocratic elution was used for the first three minutes and then a steep gradient scheme was applied as a wash step: (a) first injection with isocratic elution, (b) 201st injection with isocratic elution, (c) 202nd injection with isocratic elution for the first 3 min followed by steep gradient elution. Reproduced from Jemal et al., *Biomed. Chromatogr.* **2010**, *24*, 2–19, with permission of John Wiley & Sons, Ltd.

authors' laboratory show that, for some analytes, good extraction efficiency could be achieved under conditions where the analyte is apparently largely ionized. A paper published recently [98], which gives an excellent theoretical treatise of extraction efficiency vis-à-vis analyte pKa, confirms the authors' experience.

Because of the recent enhanced awareness about the bioanalytical risks posed by the plasma phospholipids, a study was conducted on the fate of phospholipids during the different LLE procedures normally used for the extraction of drugs and metabolites from plasma. In general, it was found that a number of organic solvents commonly used in LLE, such as *n*-butyl chloride (1-chlorobutane) and methyl-*tert*-butyl ether (MTBE), and solvent combinations, such as hexane/ethyl acetate and hexane/2-methyl-1-butanol, are very selective in extracting the analyte and leaving the phospholipids behind [99]. Figure 3.6 shows the relative amounts of a lyso phosphatidylcholine (C16:0 lyso-PC) contained in LLE plasma extracts obtained using *n*-butyl chloride, MTBE, and ethyl acetate compared with a PPT extract as a benchmark. The extraction of the C16:0 lyso-PC into *n*-butyl chloride, MTBE, and ethyl acetate was less than 0.1%, 1%, and 15%, respectively. While evaluating the extraction of phospholipids into the different organic solvents used in LLE, limited evaluation was conducted on the extraction of the phospholipids using SPE under commonly used conditions and using recently commercialized phospholipids removal plates [100, 101]. As shown in Figure 3.6, the amount of the C16:0 lyso-PC in the SPE extracts, under the conditions used, was larger than that seen in the *n*-butyl chloride and MTBE LLE extracts. On the other hand, the phospholipids removal plates were as good as *n*-butyl chloride LLE in providing extracts free of the C16:0 lyso-PC. In LLE, it can be generally concluded that the extraction of the PC and lyso-PC species increases with the increase in the polarity of the organic solvent and the lyso-PC species are extracted to a lesser degree compared with the PC species. It can also be concluded that, considering the minimal extraction of phospholipids, especially the lyso species, and the likelihood of obtaining a decent LLE recovery for the commonly encountered small-molecule drugs and metabolite, *n*-butyl chloride and MTBE are the best single-component solvents for LLE. Although the lyso phospholipids are less abundant in plasma than the phospholipids, the former are more important in HPLC-MS/MS bioanalysis since they elute earlier and closer to the commonly encountered small-molecule drugs and the associated metabolites under the normally used conditions of reversed-phase

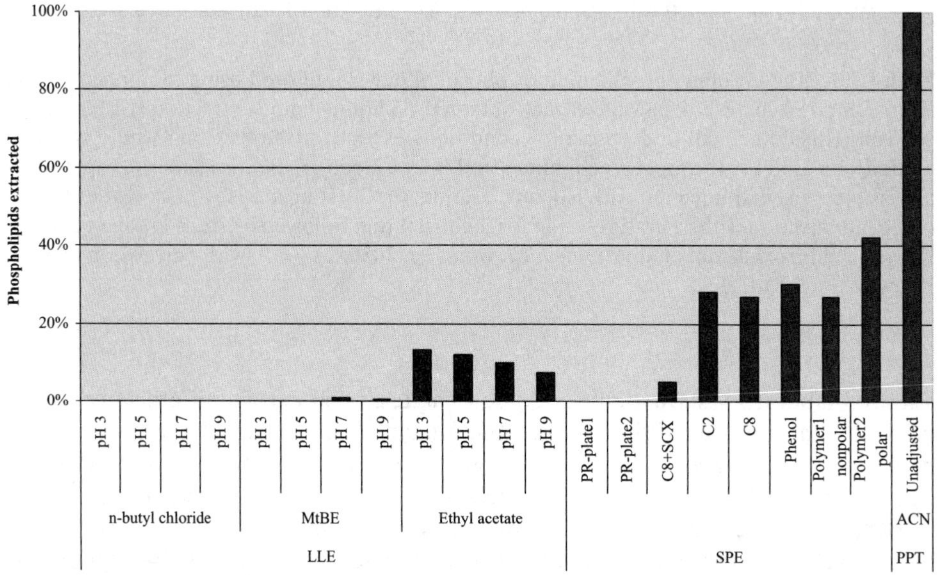

Figure 3.6 Extraction of C16:0 lyso phosphatidylcholine (C16:0 lyso-PC) from human plasma using liquid–liquid extraction (LLE) with three different solvents under different pHs, compared with typical solid-phase extraction (SPE) and two commercialized phospholipids removal sorbents (PR-plate1 and PR-plate2). The lyso-PC was monitored using selected reaction monitoring (SRM) of *m/z* 496 → *m/z* 184. Reproduced from Jemal et al., *Biomed. Chromatogr.* **2010**, 24, 2–19, with permission from John Wiley & Sons, Ltd.

chromatography. Hence, it is important that LLE using *n*-butyl chloride and MTBE is especially efficient in removing the lyso phospholipids.

Depending on the solvent used for the preparation of an IS working solution used for adding the IS to a plasma sample, a substantial amount of an organic solvent may be introduced to the plasma sample. This could affect the amount of the phospholipids extracted, as illustrated in Figure 3.7 for C16:0 lyso-PC using an acetonitrile IS solution and LLE with MTBE. While the effect of adding acetonitrile equal to 5% and 10% of the plasma volume was negligible, the amount of the lyso-PC extracted into MTBE increased nearly 10-fold, compared with the 0% acetonitrile addition, when the added acetonitrile was equal to 20% of the plasma volume. Thus, it is important to be cognizant of the effect of the organic solvent introduced during the IS addition. Consequently, the IS solution used for spiking a plasma sample should contain as little organic solvent as possible, considering the solubility and stability of the compound.

3.2.3 Interference

3.2.3.1 Metabolite Mass Spectrometric Interference Although HPLC-MS/MS has excellent selectivity, bioanalytical specificity may be hampered by interference from a co-eluting metabolite that can undergo in-source CID conversion to the molecular ion of the drug molecule itself, thereby providing the same SRM transition as that used for the drug and thus interfering with the quantitation of the drug. This type of interference was illustrated in an original article on this subject in 1999 [102], and since then there have been a number of publications dealing with the same

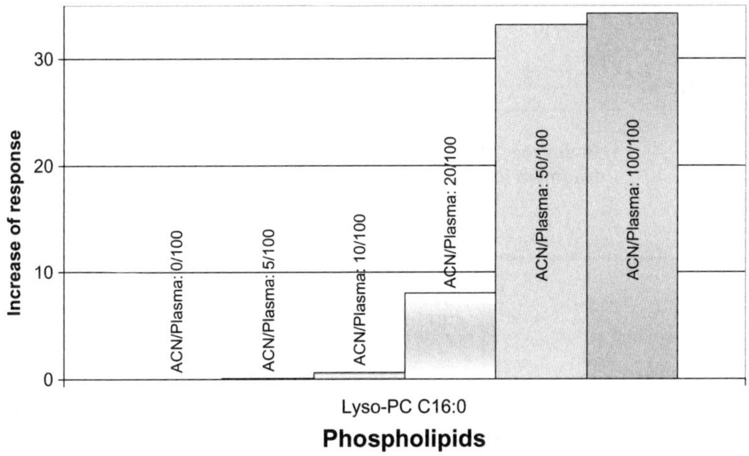

Figure 3.7 The effect of acetonitrile added to human plasma on the extraction of C16:0 lyso phosphatidylcholine (C16:0 lyso-PC) by liquid–liquid extraction (LLE) with methyl-*tert*-butyl ether (MTBE). The acetonitrile : plasma volume ratios are: 0:100; 5:100; 10:100; 20:100; 50:100; and 100:100. The extracted C16:0 lyso-PC increased by as much as 35-fold depending on the volume of the acetonitrile added. The lyso-PC was monitored using selected reaction monitoring (SRM) of *m/z* 496 → *m/z* 184. Reproduced from Jemal et al., *Biomed. Chromatogr.* **2010**, *24*, 2–19, with permission from John Wiley & Sons, Ltd.

theme [103–108]. Drugs producing metabolites that can potentially interfere in this manner include those containing lactone, carboxylic acid, and sulfhydryl (thiol) functional groups [102, 104]. The corresponding metabolites, which have hydroxy acid, acylglucuronide, and disulfide functional groups, undergo in-source conversion to generate the molecular ion of the corresponding parent drug. This is illustrated in Figures 3.8 and 3.9. Other published examples include analysis of a drug in the presence of its N-oxide metabolite [109] and analysis of a primary amine containing drug in the presence of its carbamoyl glucuronide [110]. A summary of putative metabolites of drugs of different chemical structures and the associated SRM transitions is presented in Table 3.1 [111].

For a hydroxy acid drug that produces a corresponding lactone metabolite, chromatographic separation between the drug and the metabolite may be needed even in the absence of in-source generation of the drug entity [112]. Such a situation arises when both the metabolite and the drug form the $[M + H]^+$ and $[M + NH_4]^+$ ions and the quantitation of the drug is based on its $[M + H]^+$ ion being used as the precursor ion for the SRM transition. The interference arises because of the M + 1 isotopic contribution of the $[M + NH_4]^+$ ion of the lactone metabolite, which is lower than the $[M + H]^+$ ion of the hydroxy acid by only one mass-unit. This occurrence

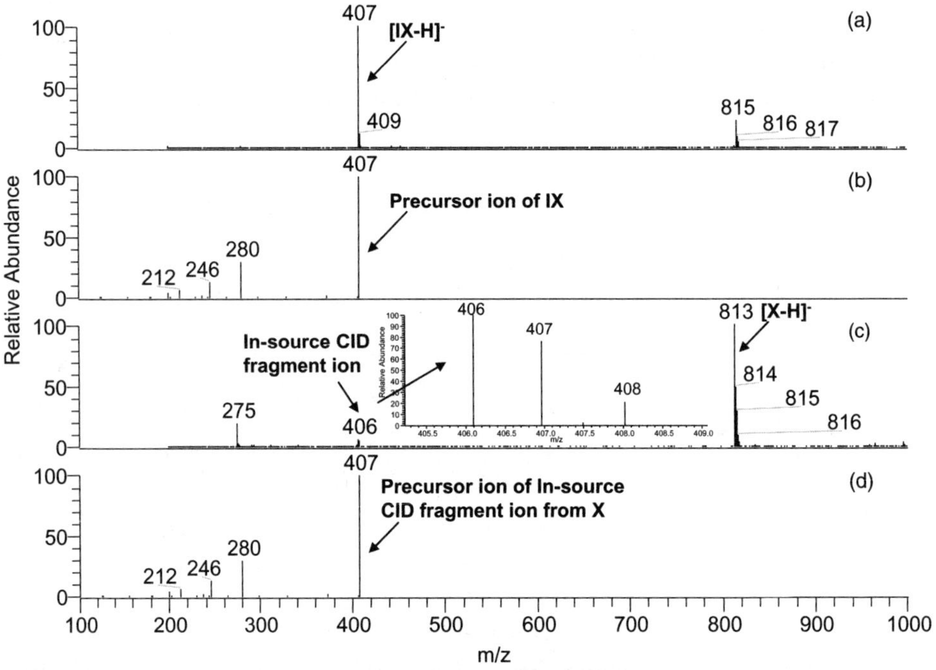

Figure 3.8 (a) and (b) show negative full scan MS and product ion spectrum using precursor ion *m/z* 407 for thiol drug IX; (c) shows negative full scan MS of the disulfide metabolite, X, and in-source collision-induced dissociation (CID) generated thiol drug IX ion of *m/z* 407 from X; (d) shows the product ion spectrum using the CID generated ion of *m/z* 407 from X. Reproduced from Jemal and Xia, *Rapid Commun. Mass Spectrom.* **1999**, *13*, 97–106 with permission from John Wiley & Sons, Ltd.

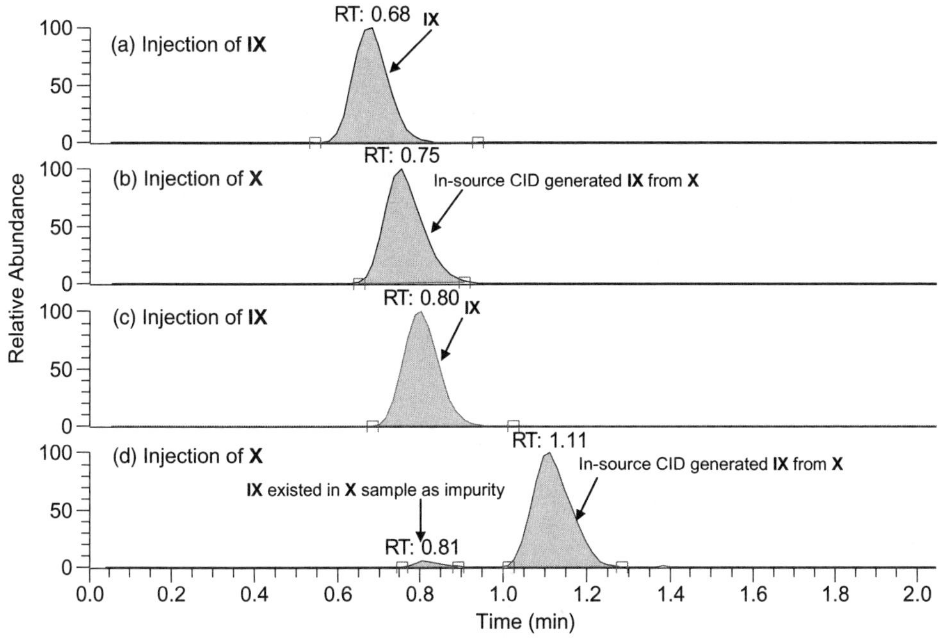

Figure 3.9 HPLC-SRM (*m/z* 407 →*m/z* 280) chromatograms of a thiol drug R-SH (IX, $[IX-H]^-$ = 407) and its disulfide metabolite R-S-S-H (X, $[X-H]^-$ = 813). (a) and (b) show SRM chromatograms of IX and X obtained using an isocratic elution of 45% aqueous ammonium acetate (5 mM) and 55% acetonitrile; the peak detected in B was generated from X due to the in-source collision-induced dissociation (CID); clearly IX and X were not separated under this low k′ value of 0.43 and 0.56, respectively. On the other hand, (c) and (d) show that IX and X were well resolved under isocratic elution of 55% aqueous ammonium acetate (5 mM) and 45% acetonitrile, with k′ values of 0.68 and 1.3, respectively. The small peak (0.81 min) detected in (d) was trace amount of IX contained in X sample as an impurity. Reproduced from Jemal and Xia, *Rapid Commun. Mass Spectrom.* **1999**, *13*, 97–106 with permission from John Wiley & Sons, Ltd.

is due to the unique feature of a lactone and the corresponding hydroxy acid in that there is an 18-mass-unit difference between the two compounds, which is only one mass-unit different from the 17-mass-unit difference that exists between the $[M + H]^+$ and $[M + NH_4]^+$ ions of any one compound.

Another type of mass spectrometric interference arises from isomeric metabolites such as the Z-isomeric metabolite of a drug containing a methyloxime group of E-configuration [102, 104]. Epimeric, diastereomeric, and enantiomeric metabolites [21, 113] also cause the same kind of interference since such metabolites would obviously interfere with the SRM transition used for the quantitation of the drug.

A third type of mass spectrometric interference arises from other metabolites that are isobaric with the drug. A phosphate prodrug, which is used as a drug delivery strategy for enhancing the solubility of drug candidates containing alcohol or phenol groups, hydrolyzes *in vivo* to form the parent drug, which may further undergo conjugation to form a sulfate metabolite [114]. Since the addition of the

TABLE 3.1 Metabolite or Prodrug Interference Due to In-Source Conversion for Drugs of Different Functional Groups

Drug	SRM from Drug	Metabolite or Prodrug	SRM from Metabolite or Prodrug
E-isomer Methyloxime 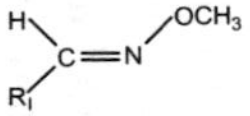	m/z 327 → m/z 97[a]	Z-isomer Methyloxime 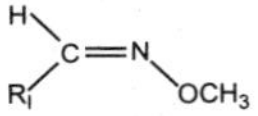	m/z 327 → m/z 97[a]
Lactone 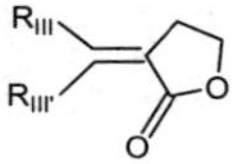	m/z 363 → m/z 285[a]	Carboxylic acid 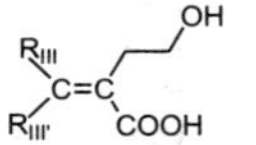	m/z 381 → m/z 363[b] m/z 363 → m/z 285[a]
Carboxylic acid 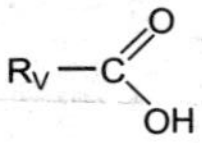	m/z 441 → m/z 423[a]	Acylglucuronide 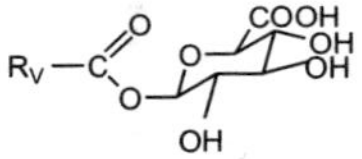	m/z 617 → m/z 441[b] m/z 441 → m/z 423[a]
γ or δ hydroxy carboxylic acid	$[M + H]^+ \rightarrow P^+$[a]	Lactone	$[M + H\text{-}18]^+ \rightarrow [M + H]^+$[b] $[M + H]^+ \rightarrow P^+$[a]
Phenol 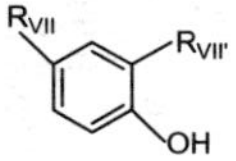	m/z 369 → m/z 229[a]	Prodrug 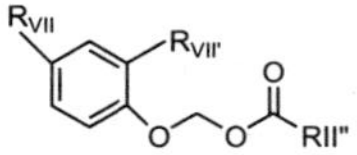	m/z 646 → m/z 369[b] m/z 369 → m/z 229[a]
Thiol R_{IX}-SH	m/z 407 → m/z 280[a]	Disulfide R_{IX}-S-S-R_{IX}	m/z 813 → m/z 407[b] m/z 407 → m/z 280[a]
Fosinoprilat	m/z 436 → m/z 390[a]	Fosinopril	m/z 564 → m/z 436[b] m/z 436 → m/z 390[a]
Simvastatin	m/z 419 → m/z 285[a]	Simvastatin acid	m/z 437 → m/z 419[b] m/z 419 → m/z 285[a]
Hydroxyl OH or phenolic OH	$[M + H]^+ \rightarrow P^+$[a]	O-glucuronide	$[M + H + 176]^+ \rightarrow [M + H]^+$[b] $[M + H]^+ \rightarrow P$[a]
Hydroxyl OH or phenolic OH	$[M + H]^+ \rightarrow P^+$[a]	O-sulfate	$[M + H + 80]^+ \rightarrow [M + H]^+$[b] $[M + H]^+ \rightarrow P^+$[a]
Amine	$[M + H]^+ \rightarrow P^+$[a]	N-glucuronide	$[M + H + 176]^+ \rightarrow [M + H]^+$[b] $[M + H]^+ \rightarrow P^+$[a]
Amine	$[M + H]^+ \rightarrow P^+$[a]	N-oxide	$[M + H + 16]^+ \rightarrow [M + H]^+$[b] $[M + H]^+ \rightarrow P^+$[a]

aSRM adopted for drugs;
bin-source conversion of metabolite or prodrug to the parent drug. Reproduced from Jemal and Xia, *Current Drug Metabolism*, **2006**, *7*, 491–502, with permission from Bentham Science Publishers, Ltd.

sulfate or phosphate group increases the mass of the parent drug by the same 80.0 Da, a phosphate prodrug and its sulfate metabolite are isobaric. It should be noted that the sulfate metabolite, if not chromatographically separated, would interfere not only with the isobaric phosphate prodrug but also with the parent drug since the sulfate metabolite would undergo in-source conversion to produce the same protonated molecule as the parent drug [114]. A drug containing a methyl ether group (RCH_2OCH_3) is likely to produce RCOOH metabolite, containing the COOH group instead of the CH_2OCH_3 group in the parent drug, with the two groups being isobaric. Chromatographic separation between the drug and the metabolite would be required to avoid the interference. In the absence of prior information about this type of metabolite, there is a chance that the metabolite would co-elute with the drug, especially when using an acidic mobile phase and rapid chromatography with a low resolving power.

Metabolites with masses that are lower by 1 or 2 Da than the corresponding parent drugs will cause interference due to M + 1 or M + 2 isotopic contributions of the metabolites, with the M + 2 contributions enhanced for chlorine or bromine containing drugs. Metabolites lower by 1 Da (compared with the parent drug) could originate via oxidative deamination of the drug ($R_1R_2CHNH_2$) to form a ketone metabolite (R_1R_2CO). Metabolites lower by 2 Da could originate via hydroxylation followed by dehydration ($-H_2$), oxidation of primary alcohol to aldehyde ($-H_2$), and oxidation of secondary alcohol to ketone ($-H_2$). While metabolites higher by 1 or 2 Da would not interfere with the drug quantitation when at least a unit-mass resolution is maintained, it should be noted that the quantitation of such metabolites would be interfered with by the isotopic contribution of the parent drug. Metabolites higher by 1 Da could originate via oxidative deamination of the drug to form an alcohol metabolite, and hydrolysis of an $RCONH_2$ drug to an RCOOH metabolite. Metabolites higher by 2 Da could originate via the reduction of ketone or aldehyde to alcohol. As described above, achieving chromatographic separation between a drug and its metabolite is essential when the metabolite causes mass spectrometric interference with the quantitation of the drug.

3.2.3.2 Metabolite Interference due to Conversion to the Parent Drug during Sample Processing An unstable metabolite may have an untoward effect on the accurate quantitation of the parent drug since it may convert to the drug during the multiple steps of sample collection, handling, and preparation that precede the HPLC-MS/MS analysis. Acylglucuronides are such metabolites and tend to be unstable, especially under alkaline conditions and elevated temperatures, degrading to generate the parent drug [12, 13]. Mildly acidic conditions of pH 3–5 tend to be the most desirable pH region for minimizing the hydrolysis of acylglucuronides in biological samples or during the numerous steps involved during sample analysis. Lactone metabolites of hydroxy acid drugs are also prone to degradation to generate the parent drug. Combinations of low temperature and mildly acidic pH of 3–5 have been used to minimize the hydrolysis of the lactone metabolite back to the drug or vice versa [112]. Other metabolite functional groups may undergo E to Z isomerization (or vice versa) due to exposure to light or undesirable pH conditions to generate the parent drug. Such functional groups include O-methyloximes [19] and carbon–carbon double bonds [20]. Metabolites could also undergo epimerization to generate the drug [21, 22].

For the quantitation of a drug in the presence of its unstable metabolite, which potentially converts to the parent drug, conditions must be optimized to minimize such a conversion. However, even the optimal conditions adopted may not totally prevent conversion of the metabolite to the drug. It is thus essential to design method development appropriately in order to minimize the adverse effects of such a conversion on the accuracy and precision of the method. The significance of proper method design was systematically illustrated using two compounds, pravastatin (a hydroxy acid) and pravastatin lactone, which would undergo interconversion to different degrees depending on the conditions used [115]. The important attribute of the method design was the use of the appropriate ratios of the concentrations of pravastatin and pravastatin lactone in the QC samples. Methods that implemented such an approach of bioanalytical method design include those used for the simultaneous quantitation of simvastatin and simvastatin acid [115].

Methanol is a commonly used organic solvent in different steps of HPLC-MS/MS bioanalysis, including standard stock preparation, washing steps of SPE, reconstitution following evaporation, and chromatography. Metabolites containing ester groups, including acylglucuronide metabolites, may react with methanol to produce the corresponding methyl ester, especially under basic conditions [12, 112]. Thus, the use of methanol should be avoided with a drug containing methyl ester group ($RCOOCH_3$) since such a drug could produce the corresponding carboxylic acid metabolite (RCOOH) and the corresponding acylglucuronide (RCOO-acylglucuronide) metabolite. In the presence of methanol, especially under basic conditions, the acylglucuronide could react with methanol to produce $RCOOCH_3$, which is the drug. Under such conditions, the measured concentration of the drug would be highly inflated if the acylglucuronide concentration is large compared with the drug concentration. On the other hand, it should be noted that for a drug that contains an ester group other than a methyl ester, such as an ethyl ester, a reaction with methanol would cause underestimation of the measured drug concentration since methanol would react with the drug to form the methyl ester analog of the drug [116].

3.2.4 Enhancing Chromatography

3.2.4.1 Column Stationary Phase and Mobile Phase Screening Reversed-phase (RP) chromatography remains the technique of choice for the analysis of pharmaceutical compounds. Manufacturers are continuously introducing new RP stationary phases to meet the demand for columns that give retention selectivity and symmetrical, sharp peaks. Differences in the performance of seemingly similar commercial columns may lie in both the nature of the silica support and the technique used to produce the bonded phase. Factors that can influence retention selectivity and peak shape include surface area, pore size, trace metal activity, bonded phase surface activity, bonding chemistry, and silica deactivation process. Extensive efforts have been expended characterizing silica-based RP columns in terms of their surface coverage, hydrophobic selectivity, shape selectivity, hydrogen bonding capacity, and ion-exchange capacity at different pH values and with different mobile phases. The objective of such characterizations, using appropriate acidic, basic and neutral probe analytes of different polarities, is to aid in selecting the appropriate column for a given application. Despite such tests and the claims made by the column manufacturers, it is still difficult to predict a priori the right column without some level of experimentation. Commercially available RP column types include those packed with alkyl, cyano, phenyl, perfluorinated, polar embedded, polar/hydrophilic endcapped, and a variety of novel phases. The different modifications of the RP phases not only provide different retention behaviors but also extend the usable pH range and allow the use of highly aqueous mobile phases. For certain analytes, perfluorinated phases have been shown to exhibit not only RP characteristics but also normal-phase-like characteristics, especially in a mobile phase of high organic percentage.

In HPLC-MS/MS bioanalysis, the selection of columns and mobile phases is further complicated since a mobile phase could significantly affect the nature and intensity of the analyte mass spectrometric response obtained using the same column [3, 5, 117–123], as illustrated below for negative and both ESI. Although not as widely reported, there have also been reports of observations that the type/brand of the column used could affect the analyte response using the same mobile phase [117, 124]. Thus, during method development, it is prudent to incorporate some level of column and mobile phase screening in order to obtain appropriate mass spectrometric response, in addition to achieving the requisite chromatographic separation with symmetrical and efficient peaks. Such screening and optimization could be achieved efficiently using automated systems as illustrated in Figure 3.10 [117].

The effect of mobile phase on ESI efficiency is not well understood and hence the behavior of a compound under a set of LC mobile phase conditions cannot be routinely predicted. In a study of negative ESI response of a carboxylic acid compound, both formic acid and ammonium formate in a water/acetonitrile mobile phase decreased analyte response, but ammonium formate caused the more severe decrease [122]. Acidification of the mobile phase with formic acid also had the added benefit of maintaining a reasonably high retention factor (k) for the analyte even at a relatively high acetonitrile concentration. A concentration of 1 mM formic acid in the mobile phase was found to be optimal as it achieved reproducibility, elongated retention time with only 60% loss in response. Dalton et al. investigated the effects of various mobile-phase modifiers on the negative ESI response of several model compounds (all without carboxylic acid or any other strongly acidic group). They found that acetic, propionic, and butyric acid at low concentrations improved the responses of the analytes to varying degrees [125]. On the other hand, formic acid decreased response, as did neutral salts (ammonium formate and ammonium acetate) and bases (ammonium hydroxide and triethylamine).

A dramatic difference in the positive ESI response was found when acetonitrile was substituted by methanol in the mobile phase [123]. The compound in a water/acetonitrile mobile phase gave only a weak ESI response in the positive ion mode with formic acid and/or ammonium acetate. The compound in a water/methanol mobile phase, in contrast, gave a significantly higher response, approximately 25-fold, with formic acid and/or ammonium acetate. On the other hand, for the same analyte, acetonitrile and methanol mobile phases gave about the same response in the negative ion mode.

3.2.4.2 Specialized Reversed-Phase Columns for Polar Analytes Retention of polar analytes in reversed-phase chromatography often requires a highly aqueous mobile phase to achieve retention, which can cause a number of problems such as the collapse of the stationary phase due to pore dewetting and decreased sensitivity

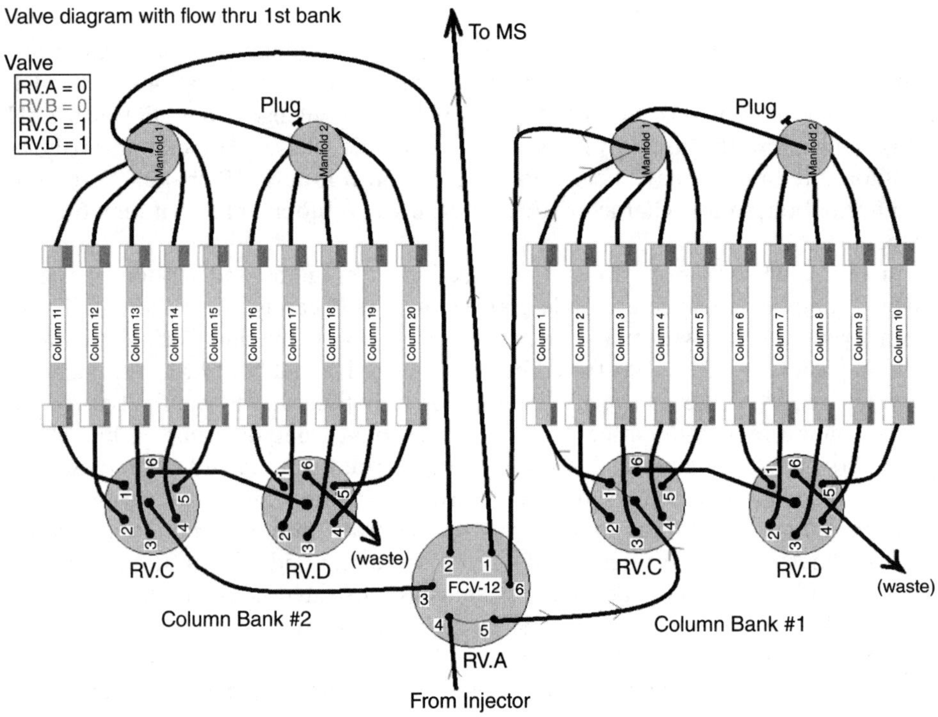

Figure 3.10 Multicolumn screening system for HPLC-MS/MS bioanalytical optimization.

in ESI detection. The result of phase collapse is that chromatography becomes more problematic with retention loss, retention irreproducibility, increased tailing, and long gradient regeneration times. To avoid the phase collapse that may be experienced during high-aqueous reversed-phase chromatography for polar compounds using silica-based alkyl bonded phases, specialized phases have been developed, including polar-embedded and polar-endcapped phases. These phases are modifications of classical alkyl-bonded silica chemistry (typically C_{18}) with the addition of a polar functional group, such as an amide or a carbamate group, within the alkyl chain itself, or with a polar functional group used as an endcapping agent. Compared with the use of the classical alkyl-bonded phases, these modifications prevent dewetting of the alkyl-bonded phases, provide increased retention for polar analytes, give good peak shapes, and allow higher sample throughput because of faster mobile phase re-equilibration.

Another alternative for achieving retention and separation of polar analytes is the use of a special packing material known as porous graphitic carbon (PGC) [113, 126]. PGC generally provides markedly greater retention and selectivity for polar compounds than alkyl-bonded silica columns. Solvents and additives that are used for elution from PGC columns are similar to those used in traditional reversed-phase chromatography, with most of them being MS compatible. Xing et al. used the PGC column to resolve very polar mixtures containing nucleosides and their mono-, di-, and triphosphates under conditions suitable for HPLC-MS [126]. A water/acetonitrile mobile phase containing ammonium acetate and diethylamine as modifiers was used. The ammonium acetate proved to be critical for retention and diethylamine was found to improve the peak shapes of di- and triphosphates.

Hydrophilic interaction chromatography (HILIC) is another alternative for achieving retention and separation of polar analytes. HILIC is conducted on polar stationary phases such as bare (nonbonded, underivatized) silica, amino, diol, and cyclodextrin-based packings [127, 128]. A high-organic, low-aqueous mobile phase is used to retain polar analytes. Retention is proportional to the polarity of the solute and inversely proportional to the polarity of the mobile phase [9]. There are several retention mechanisms on underivatized silica. A combination of hydrophilic interaction, ion exchange and reversed-phase retention results in unique selectivity. The hydrophilic interaction mechanism involves partitioning between the adsorbed polar component of the mobile phase and the remaining hydrophobic component of the mobile phase. As the aqueous content of the mobile phase is increased, the observed retention time of the analyte will decrease due to the strong elution strength of water in the HILIC mode. Thus, greater retention is achieved when the organic content of the mobile phase is high, typically greater than 70%. This high organic mobile phase is usually ideal for analyte desolvation and ionization, which leads to enhanced response in MS detection, compared with traditional reversed-phase methods. Because water is the strongest elution solvent in HILIC, samples dissolved in high organic content, such as samples extracted via SPE, can be injected directly onto the HILIC column, obviating the need for evaporation followed by reconstitution [129]. The backpressure experienced with HILIC using underivatized silica is significantly lower than that experienced with reversed-phase chromatography for the same column dimension and flow rates. Hence, the use of higher flow rates is feasible, resulting in significant time saving.

3.2.5 Enhancing Mass Spectrometric Detection

3.2.5.1 High-Resolution Accurate Mass Spectrometry For nearly two decades, HPLC-MS/MS, using triple quadrupole platforms in the selected reaction monitoring (SRM) mode, has been the technique of choice for quantitation of small-molecule drugs and metabolites in plasma and other biological samples obtained from animals and human subjects following the administration of drugs/drug candidates [3–10]. The advantages of the HPLC-SRM bioanalytical methods used for the analysis of such postdose samples are well known, which include: (1) excellent selectivity vis-à-vis biological endogenous components, metabolites, and formulation ingredients; (2) great sensitivity with or without extensive cleanup of the biological samples; (3) good assay precision and accuracy; (4) great dynamic range, easily achieving a four-order linear regression; (5) fast turnaround analysis due to fast chromatography and rapid SRM acquisition speed; (6) straightforward post-acquisition data processing and quantitation; and (7) excellent long-term reproducibility and ruggedness, with ease of method transferability from one laboratory to another [130]. Consequently, appropriately enough, the LC-SRM technique is well entrenched in the bioanalytical laboratories, equipped exclusively with triple quadrupole mass spectrometers. However, with the recent introduction of reasonably priced high-resolution accurate mass spectrometers with enhanced mass resolving power, mass accuracy, acquisition speed, and dynamic range, there is currently a growing interest in the bioanalytical community to explore the utility of high-performance liquid chromatography coupled with full scan high-resolution accurate mass spectrometry (HPLC-HRAMS) for quantitation of drugs/drug candidates and metabolites in postdose samples. Recent publications demonstrated proof of principle for quantitation of small-molecule drugs/drug candidates in plasma samples using Orbitrap-based HPLC-HRAMS [130–132].

There are basically two platforms that can be used for HPLC-HRAMS, namely, those based on the Orbitrap technology and those based on the time-of-flight (TOF) technology. Whether an Orbitrap or a TOF mass spectrometer is used, it is important to examine the pros and cons of using HPLC-HRAMS vis-à-vis HPLC-SRM, the current workhorse used for quantitation of target drugs/drug candidates and metabolites. The most distinct feature of HPLC-SRM is that only those precursor ions and their selected product ions, as specified by the SRM transitions used, are detected and thus no information about the other ions in the injected sample is acquired and recorded. This feature possesses the desirable attributes of enhanced selectivity and sensitivity arising from increased the duty cycle, which is important for the accurate and precise quantitation of the target analyte. The drawback is the absence of information about any other compounds for conducting post-acquisition queries to answer questions that may arise related to the presence/absence of metabolites and other components that may be co-eluting and causing unexpectedly low or high response of the target analyte.

On the other hand, the most distinct feature of HPLC-HRAMS is its ability to detect not only all the ions (protonated, sodium, or ammonium adducted, etc.) generated from the target analyte but also ions generated from other compounds in the injected sample. Accordingly, post-acquisition data queries can be conducted if the need arises to obtain information about components other than the target analyte. Thus, qualitative or quantitative information about metabolites, biomarkers, phospholipids, and formulation agents can be obtained from the same sample injection used for the quantitation of the target analyte. This feature of HPLC-HRAMS obviously offers a significant advantage over HPLC-SRM.

Under the circumstance, before HPLC-HRAMS can be fully embraced by the bioanalytical community, the main issue to address is whether or not bioanalytical methods can be developed and routinely applied for the quantitation of a target drug, drug candidate, or metabolite with the sensitivity, selectivity, accuracy, precision, and ruggedness approaching those obtained using HPLC-SRM. Limited, preliminary comparison conducted in the authors' and other laboratories show that HPLC-SRM methods are, for most compounds, more sensitive than HPLC-HRAMS methods. However, TOF-based HPLC-HRAMS methods have been shown to achieve an LLOQ of 1 ng/mL and 5 ng/mL for at least 50% and 90%, respectively, of model pharmaceutical compounds in plasma, following a simple sample extraction using plasma protein precipitation [133]. This level of sensitivity should be adequate for most of the pharmacokinetic studies normally conducted. As the HRAMS systems continue to improve, it is expected that sensitivity will be further enhanced. So, in general, sensitivity should not be an impediment in adopting HPLC-HRAMS for target analyte quantitation.

It is also important to make an evaluation of the mass resolving power required for HPLC-HRAMS bioanalytical methods to provide the selectivity needed for pharmacokinetic studies. For most HPLC-SRM bioanalytical methods, a unit-mass resolution, with a full width at half maximum (FWHM) of 0.7 Da, is normally used at both mass analyzers, providing a good balance between selectivity/sensitivity and the resolving power that can be reasonably achieved on the state-of-the-art triple quadrupole systems. Similarly, a determination should be made regarding the HRAMS resolving power deemed reasonable, balancing the selectivity/sensitivity achieved against the equipment cost and ease of applicability. It should be noted that accuracy of mass assignment, which is crucial in HPLC-HRAMS, is affected by the resolving power depending on the complexity of the matrix.

Another important aspect in HPLC-HRAMS method development is the effect of the mass extraction window (MEW) used in the post-acquisition data processing to obtain the extracted ion chromatogram (EIC) for the target analyte from the total ion chromatogram (TIC). The accuracy of the area value obtained for the target analyte chromatographic peak in the EIC is dependent on using the optimum MEW. Thus, understanding this step of method development and selecting the optimum MEW, which is related to the resolving power used during the TIC acquisition, is crucial in developing a sound and rugged HPLC-HRAMS bioanalytical method. This post-acquisition step makes HPLC-HRAMS method development more complicated than HPLC-SRM method development, where there is no such post-acquisition step involved. On the other hand, HPLC-HRAMS method development does not involve the extra step of MS/MS optimization, which is needed for HPLC-SRM method development [130].

The authors' in-depth evaluation of the different parameters of HPLC-HRAMS for quantitative bioanalysis in plasma show that a resolving power of 20 k is generally adequate to achieve accurate and precise quantitation even at low analyte concentrations [130]. The relation between the resolving power and the MEW required to achieve optimum EIC selectivity is in line with the theoretical expectation and the optimum MEW, expressed in ppm, is only a small fraction of the FWHM, expressed in ppm. The relation between analyte EIC response and MEW is also as expected from theory in that, for a standard prepared in a matrix-free solvent, the peak response does not change with change in MEW using the centroid data, whereas the response in the profile mode increases with increase in MEW until the maximum MEW is reached, which is approximately equal to FWHM. The observed apparent selectivity is better in the centroid mode compared with the profile mode. On the other hand, there is a higher probability of reporting a false negative, especially when using a narrow MEW, in the centroid mode. The use of centroid data allows the application of a simple rule, namely the analyte EIC response does not change with change in MEW, in assessing the presence/absence of a co-eluting interference at a given concentration and MEW. This simple rule aids in selecting optimum MEW and establishing the LLOQ in quantitative bioanalysis. In general, a narrower MEW should be used in the profile mode compared with the centroid mode.

In conclusion, if a systematic method development is adopted, HPLC-HRAMS can be used for accurate and precise quantitation of drugs and/or metabolites as an alternative to HPLC-MS/MS. The bonus that comes with using HPLC-HRAMS for target quantitation is that the TIC obtained for the quantitative analysis can be subsequently queried to look for information about other compounds of interest in the sample (e.g., metabolites) and for troubleshooting analytical problems.

3.2.5.2 High-Field Asymmetric Waveform Ion Mobility Spectrometry High-field asymmetric waveform ion mobility spectrometry (FAIMS) has been demonstrated to reduce chemical noise, enhance analyte detection, and eliminate some interfering metabolites and prodrugs in HPLC-MS/MS bioanalysis [103, 134–140]. FAIMS separation is based on the differences in ion mobility at high versus low electric fields, and the separation occurs in an atmospheric pressure gas-phase environment. A FAIMS system that is physically located between the sprayer and the orifice of a mass spectrometer controls the type of ions entering the orifice of the mass spectrometer, thereby discarding the unwanted ions. Thus, a FAIMS device, when used in conjunction with HPLC-MS/MS, acts as a post-column, pre-MS ion filter in which only a subset of the ions formed are transmitted. The ion selection in FAIMS is achieved by applying a compensation voltage (CV) that is specific to the ion of interest. Ideally, when the selected CV is optimum for the drug, the protonated molecules of the drug pass through the FAIMS system into the orifice of a mass spectrometer, while the protonated molecules of the metabolite are filtered away before reaching the orifice. Consequently, FAIMS provides separation between the drug and the metabolite even in the absence of chromatographic separation. Work by Kapron et al. demonstrated the advantage of using FAIMS in a situation where the selected reaction monitoring (SRM) transition used for a drug exhibited interference due to in-source conversion of its N-oxide metabolite to generate the same protonated molecule as the drug [103]. The FAIMS device removed the metabolite ion before entrance to the mass spectrometer. In addition, the separation of an acylglucuronide metabolite from its parent drug using FAIMS has been demonstrated in the authors' laboratory [140].

3.3 APPLICATION OF INCURRED SAMPLES IN METHOD DEVELOPMENT

There are two types of risk associated with metabolites contributing to the measured drug concentration in incurred biological samples. One type of interference is due to MS in-source conversion of metabolites to the parent drug molecule to cause inaccurate quantitative determination of the drug. The second type is due to the generation of the drug from its metabolite that undergoes degradation during the sample collection, storage, extraction, and analysis.

The accepted practice of developing and validating HPLC-MS/MS bioanalytical methods is based on using calibration standards and QC samples prepared by spiking the drug into a blank biological matrix, as described in the 2001 Food and Drug Administration (FDA) guidance for bioanalytical method validation [1], and the conference report of the third American Association of Pharmaceutical Scientists (AAPS)/FDA bioanalytical workshop held in 2006 [141]. However, metabolites found in incurred samples could affect the accuracy and precision of the drug measurement. Ideally, the QC samples used for method development and validation should contain the metabolites in addition to the drug. However, it is not practical to obtain the reference standards of all metabolites at the time of method development and validation, especially for drugs in early development. In 2002, a strategy was put forth [111] based on using incurred sample, known to contain the drug metabolites, to challenge a method previously validated in the traditional manner using spiked QC samples. It was proposed that the strategy be applied to methods used for the analysis of samples from first-in-human (FIH) studies. As soon as the FIH samples from high dose panels become available, a pooled incurred sample would be prepared to test the method for metabolite interference. If the method were found not to be valid due to metabolite interference, the method was to be reoptimized and revalidated. Consequently, all the samples previously analyzed by the invalid method were to be reanalyzed with the newly validated method shown to be free of metabolite interference. It should be noted that the reanalysis may not be feasible if the presence of an unstable metabolite would have required stabilization of the sample at collection.

A current strategy is based on using a pooled incurred sample during method development, not based on using a pooled incurred sample to challenge an already validated method as proposed in 2002 [111]. The new strategy is to be used in late drug discovery or early drug development. For the development of methods intended to support animal studies, a pooled incurred sample is obtained from earlier animal discovery studies conducted for pharmacokinetics and other discovery studies. For methods intended to support FIH studies, incurred sample is obtained from selected pre-investigational new drug (IND) animal studies. Later, when appropriate human incurred samples are available, the FIH method is to be retested using a human pooled incurred sample. The proposed use of a pooled incurred sample for method development is completely different from the ISR currently implemented in bioanalytical laboratories providing data for submission to regulatory agencies. The concept of ISR was adopted in the third AAPS/FDA bioanalytical workshop held in 2006, as described in the conference report [141], and the details of how to conduct ISR were discussed in a workshop dedicated to ISR in 2008, as described in the conference report [142]. The stated purpose of ISR is to demonstrate assay reproducibility by analyzing a set of incurred samples on two different occasions using the same validated method in exactly the same manner. Unfortunately, an invalid method, namely a method that does not accurately measure the drug analyte due to the metabolite interference described above, may exhibit excellent ISR reproducibility. A clear case in point is an incurred sample containing a stable metabolite that causes mass spectrometric interference and co-elutes with the drug. Thus, it is essential to build quality during method development before the initiation of method validation. One important way of incorporating quality is through the use of a pooled incurred sample during method development. There are two steps in the strategy of using a pooled incurred sample for method development. For both categories, a pooled incurred sample is prepared by taking aliquots from high-dose-panel samples, including different animals or human subjects and different post-dosing time points, in order to obtain an adequate volume of the pooled sample.

In step 1, the objective is to ensure the chromatographic separation of a drug from its metabolites that show response in the SRM channel of the drug. First, a portion of the pooled incurred sample is extracted using acetonitrile precipitation. The extract is then analyzed using the chromatographic conditions of the method under development using the SRM transitions of known and potential metabolites, in addition to that of the drug. This procedure is then repeated using modified chromatographic conditions that give a larger retention factor (k), using a weaker mobile phase by lowering the organic amount, or using a shallower gradient. Third, the procedure is repeated using a different column–mobile phase combination to achieve orthogonal chromatography. The purpose of using the modified chromatographic conditions is to test if an additional peak would appear in the drug SRM channel, which would indicate the presence of a metabolite that is not separated from the drug under the initially used chromatographic conditions. With isomeric metabolites such as, diastereomers, epimers, and E/Z isomers, unlike metabolites that undergo in-source conversion to produce the protonated molecules of the drug, there are no additional SRM channels that can distinguish the metabolites from the drug and hence chromatographic separation is essential [19, 21, 113]. It is impossible to know the chromatographic conditions or chromatographic run times required to achieve the chromatographic separation of the drug from such metabolites in the absence of authentic reference standards. In general, it is recommended that a retention factor of at least 5 be achieved for the drug. It should be mentioned that this does not address the problem caused by the presence of an enantiomeric metabolite. It should also be noted that an incurred sample may contain a metabolite that has the same SRM transition as the IS, especially when a structural analog of the drug is used as the IS [143]. Therefore, it is prudent to analyze a pooled incurred sample without the IS and show that the IS SRM channel is clean at the retention time of the IS.

In step 2, the objective is to ensure that the drug is not generated from its metabolites during sample handling, storage, preparation, and extraction. First, a 24-h room temperature stability evaluation of the incurred sample is conducted vis-à-vis a QC sample treated in the same manner. The incurred and QC samples are then extracted identically and analyzed by the HPLC-MS/MS method under development. Second, the effect of the conditions used for extraction on the incurred samples is evaluated. For an LLE extraction procedure, this would involve adding the requisite buffer and keeping the sample at room temperature for at least 1 h before conducting the LLE extraction. Third, the stability of the extract, obtained from the incurred sample, is tested for at least 24 h.

3.4 A PROTOCOL FOR SYSTEMATIC METHOD DEVELOPMENT

Herein, an outline of a protocol is given for the systematic development of an HPLC-MS/MS bioanalytical method in plasma in order to reduce bioanalytical risk [4]. This protocol would ideally be applicable in late drug discovery or early drug development, where substantial information on the physicochemical properties and the metabolism of the drug candidate is already available.

Before initiating method development, a variety of information is assembled from appropriate groups working on the drug candidate of interest. Thus, information related to different physicochemical properties are obtained, such as stability in aqueous solutions at different pHs, photostability, aqueous solubility at different pHs, solubility in selected water-miscible and water-immiscible solvents, pKa, log D, and log P. Information is also obtained on *in vitro* and *in vivo* metabolism, serum protein binding, and partitioning between plasma and red blood cells. It is also essential to have detailed information related to any bioanalytical methods previously used in support of any earlier discovery studies so that any lessons learned can be incorporated into the new method under development. It is also important to obtain the realistic LLOQ required for the planned nonclinical and clinical studies.

Method development starts with the optimization of the acquisition of the positive and negative ESI full scan spectra of the drug candidate and any metabolites for which reference standards are available. These experiments are conducted by infusing each analyte prepared in water/acetonitrile (50/50, v/v) into a mobile phase flowing into the mass spectrometer. Typically, a mobile phase consisting of 40% water (with 0.1% formic acid) and 60% acetonitrile is used for positive ESI, and a mobile phase consisting of 40% water (with 0.005% formic acid) and 60% acetonitrile is used for negative ESI. Using the protonated or deprotonated molecule as the precursor ion, product ion spectra are obtained and then SRM transitions are selected. At this stage of method development, it is essential to utilize at least two SRM transitions since a co-eluting metabolite or an endogenous component may interfere with one or more of the selected SRM transitions. The MS parameters, such as sprayer voltage, capillary temperature, tube lens potential, skimmer voltage, collision energy, sheath gas, and auxiliary gas pressure are optimized using the selected SRM transitions.

Step 2 of method development involves the screening of LC mobile phases and stationary phases using an automated system consisting of multiple mobile phases and columns packed with different stationary phases. Typical aqueous eluents used include 0.1% formic acid (pH 2.7), 10 mM ammonium formate with 0.1% formic acid (pH 3.2), 10 mM ammonium acetate with 0.05% acetic acid (pH 4.2), and 10 mM ammonium bicarbonate with 0.1% ammonium hydroxide (pH 9.5). The organic eluents are acetonitrile and methanol, each used in conjunction with the aqueous eluents. A variety of RP columns from different manufacturers are used including those packed with C_8, C_{18}, phenyl, perfluorinated, polar endcapped, and polar embedded stationary phases. Typically, 2 × 50 mm columns are used and the particle sizes range from sub-2 μm (1.7–1.9 μm) to 3.5 μm, depending on the brand of the column used. The analyte solution, prepared in water/acetonitrile or water/methanol (70/30, v/v), is injected under isocratic or gradient conditions into the screening system for detection by the SRM transitions selected above. Based on the analyte peak shape (symmetry), efficiency (N), response (peak height), retention factor, and resolution between critical pairs (if known at this stage), two or three column/mobile phase combinations are selected.

In step 3, the selected column/mobile phase combinations are tested for the separation of the analytes from plasma phospholipids under the isocratic or gradient conditions used above. This is achieved by injecting the supernatant of acetonitrile-precipitated plasma and monitoring phospholipids by the all-inclusive technique described earlier, which involves the positive precursor ion scan of *m*/*z* 184, positive neutral loss scan of 141 Da and negative precursor ion scan of *m*/*z* 153.

Step 4 involves the use of incurred samples to test the specificity of the method in the presence of drug-related metabolites, endogenous plasma components, and possibly formulation agents. For this purpose, portions of plasma samples from high-dose panels of one or more *in vivo* studies are combined in order to obtain an adequate volume of a pooled incurred sample. The pooled sample should include aliquots from at least three regions of the pharmacokinetic profile, including the early and late time points and around the maximum concentration region. Ideally, the pooled sample should be from a species known to contain a high concentration of potentially problematic metabolites. If high-concentration plasma samples are not available, urine, bile, and *in vitro* samples, such as liver microsomes and hepatocytes, may be used for spiking plasma to generate a metabolite-rich plasma sample. An aliquot of the pooled incurred plasma sample is then extracted using acetonitrile precipitation and the supernatant is used for the analysis. A large number of SRM transitions, as many as the mass spectrometric system allows, are created to cover all the potential metabolites of the drug candidate. The SRM transitions of all known and expected metabolites are included. This can be easily accomplished if the mass spectrometric system is equipped with a metabolite SRM table builder; otherwise, the table has to be created manually. It is very important to include the SRM transitions of all known and potential phase 1 and 2 metabolites that may undergo in-source CID and hence give response in the SRM channels used for the drug. It is also important to include metabolites with masses lower than that of the parent drug by one or two Daltons. The incurred sample extract is analyzed using the created metabolite SRM table along with the drug SRM transitions using the isocratic or gradient elution conditions of the method. The chromatograms corresponding to the SRM transitions used for the drug are then carefully examined for the presence/absence of chromatographic peaks at different retention times. If these chromatograms show any peaks eluting before or after the drug peak, a determination should be made whether these peaks originate from the metabolites by examining the chromatograms corresponding to the metabolite SRM channels. Obviously, it is essential to make a determination of whether a metabolite SRM chromatogram shows a peak at the retention time of the drug. To this end, the analysis of the incurred sample extract is repeated using the same column–mobile phase combination with a shallower gradient scheme (for a gradient method) or with a lower organic percentage mobile phase (for an isocratic method). The purpose of this analysis is to demonstrate that the single peak attributed to the drug under the faster elution conditions still remains as a single peak, confirming that the absence of another component co-eluting with the drug. For the same purpose, the analysis is repeated using a column–mobile phase combination orthogonal to the first column–mobile phase combination.

After the completion of steps 1–4, the column–mobile phase combination and the gradient or isocratic conditions are fixed for the method. In step 5, different selective plasma sample extraction procedures are evaluated in the presence of an IS, preferably a stable isotope labeled analog of the analyte. The analyte extraction recoveries of the different extraction procedures are then compared. An extraction procedure that gives adequate analyte recovery, depending on the sensitivity required, and at the same time achieves efficient removal of the plasma phospholipids is selected. The use of LLE make steps 5 relatively easy since the phospholipids removal information of the different organic solvents used in LLE can be easily obtained and documented under the relatively limited number of the buffer–pH combinations normally used. Thus, if adequate analyte recovery, say 40%, is obtained with n-butyl chloride extraction at pH 9.0, this extraction procedure would be selected forthright since it has been documented that n-butyl chloride extracts are nearly 100% free of phospholipids.

In the final step, a few more experiments are conducted before the method is considered suitable for a formal validation. Under the chromatographic conditions selected and using the selected extraction procedure, post-column infusion of the analyte is conducted with the injection of a blank plasma extract to demonstrate the absence of ion suppression/enhancement at or the retention time of the analyte.

In addition, the effect of different reconstitution solvents (injection solvents) on peak shape and peak response is evaluated. A 24-h bench-top room-temperature stability of the incurred sample is also conducted and compared with a zero-time incurred sample. As a control, a QC sample is also subjected to the same stability testing regimen. If the analyte response in the incurred sample shows a significant increase with time, the presence of a conjugate metabolite that is degrading to generate the parent drug is indicated. A similar test is conducted to test the stability of the incurred sample in the aqueous buffer used for buffering the plasma sample during extraction. The stability of the extract obtained from the incurred sample should also be evaluated. Another quick test conducted is the evaluation of the effect of the IS equilibration time in plasma. Finally, method qualification is conducted to gauge the accuracy, precision, and ruggedness of the method. For accuracy and precision determination, a set of calibration standards and QC samples are analyzed. Ruggedness is determined by performing at least 200 injections, consecutively, from a single pooled QC sample extract, at a concentration within the first quartile of the standard curve range, portioned into several wells of a 96-well plate. Ruggedness is gauged by the relative standard deviation (RSD) of the peak area ratios (analyte/IS) across the entire ruggedness run, as well as the RSDs of the analyte and IS peak areas. If the assay qualification shows the method will meet the required assay performance criteria for sensitivity, accuracy, precision, ruggedness, and linearity, the method is deemed ready for a formal validation.

3.5 CONCLUSIONS

The relatively new concepts and technologies discussed in this review article attest to the fact that practical research activities still abound in quantitative HPLC-MS/MS bioanalysis, although it has been over 15 years since HPLC-MS/MS has been embraced as the technique of choice for bioanalysis of small molecules. The concepts and technologies covered in this review article can be used to enhance HPLC-MS/MS bioanalytical method development and to minimize the pitfalls that may be encountered during analysis of postdose study samples. It cannot be overemphasized that well thought-out concepts and strategies are essential in enhancing quantitative HPLC-MS/MS bioanalysis and that new technologies alone are not a substitute for bioanalytical thinking.

In order to avoid the costly and nerve-racking experience of finding out by chance that an HPLC-MS/MS bioanalytical method, validated as per the currently accepted industry practice, is invalid after it has been used to support a number of pivotal clinical and nonclinical studies, it is important to incorporate quality during method development. The protocol for systematic method development recommended herein, which is based on the use of incurred sample, column–mobile phase optimization and phospholipids avoidance, goes a long way in enhancing the quality of the developed method.

REFERENCES

[1] U.S. Department of Health and Human Services, Food and Drug Administration, Center for Drug Evaluation and Research, Center for Veterinary Medicine. Guidance

for Industry: Bioanalytical Method Validation. USA Administration, CDER; Rockville, MD, 2001.

[2] Shah VP, Midha KK, Findlay JWA, Hill HM, Hulse JD, McGilveray IJ, McKay G, Miller KJ, Patnaik RN, Powell ML, Tonelli A, Viswanathan CT, Yacobi A. Bioanalytical method validation—A revisit with a decade of progress. Pharm Res 2000;17:1551–7.

[3] Jemal M, Xia Y-Q. LC-MS development strategies for quantitative bioanalysis. Curr Drug Metab 2006;7:491–502.

[4] Jemal M, Ouyang Z, Xia Y-Q. Systematic LC-MS/MS bioanalytical method development. Biomed Chromatogr 2010;24:2–19.

[5] Jemal M. High-throughput quantitative bioanalysis by LC/MS/MS. Biomed Chromatogr 2000;14:422–9.

[6] Korfmacher WA. Principles and applications of LC-MS in new drug discovery. Drug Discov Today 2005;10:1357–67.

[7] Hopfgartner G, Bourgogne E. Quantitative high-throughput analysis of drugs in biological matrices by mass spectrometry. Mass Spectrom Rev 2003;22:195–214.

[8] Berna MJ, Ackermann BA, Murphy AT. High-throughput chromatographic approaches to liquid chromatographic/tandem mass spectrometric bioanalysis to support drug discovery and development. Anal Chim Acta 2004;509:1–9.

[9] Naidong W. Bioanalytical liquid chromatography tandem mass spectrometry methods on underivatized silica columns with aqueous/organic mobile phases. J Chromatogr B 2003;796:209–24.

[10] Cech NB, Enke CG. Practical implications of some recent studies in electrospray ionization fundamentals. Mass Spectrom Rev 2001;20:362–87.

[11] Mansoori BA, Volmer DA, Boyd RK. "Wrong-way-round" electrospray ionization of amino acids. Rapid Commun Mass Spectrom 1997;11:1120–30.

[12] Khan S, Teitz DS, Jemal M. Kinetic analysis by HPLC-electrospray mass spectrometry of the pH-dependent acyl migration and solvolysis as the decomposition pathways of ifetroban 1-O-acyl glucuronide. Anal Chem 1998;70:1622–8.

[13] Shipkova M, Armstrong VW, Oellerich M, Wieland E. Acyl glucuronide drug metabolites: toxicological and analytical implications. Ther Drug Monit 2003;25:1–16.

[14] Jemal M, Ouyang Z, Chen BH, Teitz D. Quantitation of the acid and lactone forms of atorvastatin and its biotransformation products in human serum by high-performance liquid chromatography with electrospray tandem mass spectrometry. Rapid Commun Mass Spectrom 1999;13:1003–15.

[15] Jemal M, Rao S, Salahudeen I, Chen BH, Kates R. Quantitation of cerivastatin and its seven acid and lactone biotransformation products in human serum by liquid chromatography-electrospray tandem mass spectrometry. J Chromatogr B 1999;736:19–41.

[16] Ong VS, Stamm GE, Menacherry ES, Chu S-Y. Quantitation of TNP-470 and its metabolites in human plasma: sample handling, assay performance and stability. J Chromatogr B 1998;710:173–82.

[17] Fura A, Harper TW, Zhang H, Fung L, Shyu WC. Shift in pH of biological fluids during storage and processing: effect on bioanalysis. J Pharm Biomed Anal 2003;32:513–22.

[18] Boink ABTJ, Buckley BM, Christiansen TF, Covington AK, Maas AHJ, Müller-Plathe O, Sachs C, Siggaard-Anderson O. International Federation of Clinical Chemistry (IFCC) Scientific Division. IFCC recommendation—recommendation on sampling, transport and storage for the determination of concentration of ionized calcium in whole blood, plasma and serum. Clin Chim Acta 1991;202:S13–22.

[19] Xia Y-Q, Whigan DB, Jemal M. A simple liquid–liquid extraction with hexane for low-picogram determination of drugs and their metabolites in plasma by high-performance liquid chromatography with positive ion electrospray tandem mass spectrometry. Rapid Commun Mass Spectrom 1999;13:1611–21.

[20] Wang CJ, Pao LH, Hsiong CH, Wu CY, Whang-Peng JJK, Hu OYP. Novel inhibition of cis/trans retinoic acid interconversion in biological fluids—an accurate method for determination of trans and 13-cis retinoic acid in biological fluids. J Chromatogr B 2003;796:283–91.

[21] Testa B, Carrupt PA, Gal J. The so-called "interconversion" of stereoisomeric drugs: an attempt at clarification. Chirality 1993;5:105–11.

[22] Won CM. Epimerization and hydrolysis of dalvastatin, a new hydroxymethylglutaryl coenzyme A (HMG-CoA) reductase inhibitor. Pharm Res 1994;11:165–70.

[23] Gilbert HF. Thiol/disulfide exchange equilibria and disulfide bond stability. Methods Enzymol 1995;251:8–29.

[24] Evans MJ, Livesey JH, Ellis MJ, Yandle TG. Effect of anticoagulants and storage temperatures on stability of plasma and serum hormones. Clin Biochem 2001;34: 107–12.

[25] Testa M. Hydrolysis in Drug and Prodrug Metabolism. Zurick: Wiley-VCH, 2003.

[26] Jemal M, Khan S, Teitz DS, McCafferty JA, Hawthorne DJ. LC/MS/MS determination of omapatrilat, a sulfhydryl-containing vasopeptidase inhibitor, and its sulfhydryl- and thioether-containing metabolites in human plasma. Anal Chem 2001;73:5450–6.

[27] Redinbo MR, Potter PM. Mammalian carboxylesterases: from drug targets to protein therapeutics. Drug Discov Today 2005;10:313–25.

[28] Satoh T, Taylor P, Bosron WF, Sanghani SP, Hosokawa M, La Du BN. Current progress on esterases: from molecular structure to function. Drug Metab Dispos 2002; 30:488–93.

[29] Watt AP, Morrison D, Locker KL, Evans DC. Higher throughput bioanalysis by automation of a protein precipitation assay using a 96-well format with detection by LC-MS/MS. Anal Chem 2000;72:979–84.

[30] Berna M, Murphy AT, Wilken B, Ackermann B. Collection, storage, and filtration of *in vivo* study samples using 96-well filter plates to facilitate automated sample preparation and LC/MS/MS analysis. Anal Chem 2002;74:1197–201.

[31] Sadagopan NP, Li W, Cook JA, Galvan B, Weller DL, Fountain ST, Cohen LH. Investigation of EDTA anticoagulant in plasma to improve the throughput of liquid chromatography/tandem mass spectrometric assays. Rapid Commun Mass Spectrom 2003;17:1065–70.

[32] Teitz DS, Khan S, Powell ML, Jemal M. An automated method of sample preparation of biofluids using pierceable caps to eliminate the uncapping of the sample tubes during sample transfer. J Biochem Biophys Methods 2000;45:193–204.

[33] Jemal M, Teitz D, Ouyang Z, Khan S. Comparison of plasma sample purification by manual liquid-liquid extraction, automated 96-well liquid-liquid extraction and automated 96-well solid-phase extraction for analysis by high-performance liquid chromatography with tandem mass spectrometry. J Chromatogr B 1999;732:501–8.

[34] Wells DA. High Throughput Bioanalytical Sample Preparation: Methods and Automation Strategies. Amsterdam: Elsevier, 2003.

[35] Ouyang Z, Khan S, Jemal M. 53rd ASMS Conference on Mass Spectrometry and Allied Topics. San Antonio, TX, 2005.

[36] Mallet CR, Lu Z, Fisk R, Mazzeo JR, Neue UD. Performance of an ultra-low elution-volume 96-well plate: drug discovery and development applications. Rapid Commun Mass Spectrom 2003;17:163–70.

[37] AbuRuz S, Millership J, McElnay J. Determination of metformin in plasma using a new ion pair solid phase extraction technique and ion pair liquid chromatography. J Chromatogr B 2003;798:203–09.

[38] Wachs T, Henion J. A device for automated direct sampling and quantitation from solid-phase sorbent extraction cards by electrospray tandem mass spectrometry. Anal Chem 2003;75:1769–75.

[39] Palandra J, Weller D, Hudson G, Li J, Osgood S, Hudson E, Zhong M, Buchholz L, Cohen LH. Flexible automated approach for quantitative liquid handling of complex biological samples. Anal Chem 2007;79:8010–5.

[40] Bolden RD, Hoke SH, II, Eichhold TH, McCauley-Myers DL, Wehmeyer KR. Semi-automated liquid–liquid back-extraction in a 96-well format to decrease sample preparation time for the determination of dextromethorphan and dextrorphan in human plasma. J Chromatogr B 2002;772:1–10.

[41] Eerkes A, Shou WZ, Naidong W. Liquid/liquid extraction using 96-well plate format in conjunction with hydrophilic interaction liquid chromatography-tandem mass spectrometry method for the analysis of fluconazole in human plasma. J Pharm Biomed Anal 2003;31:917–28.

[42] Xu N, Kim GE, Gregg H, Wagdy A, Swaine BA, Chang MS, El-Shourbagy TA. Automated 96-well liquid–liquid back extraction liquid chromatography–tandem mass spectrometry method for the determination of ABT-202 in human plasma. J Pharm Biomed Anal 2004;36:189–95.

[43] Ji QC, Reimer MT, El-Shourbagy TA. 96-Well liquid–liquid extraction liquid chromatography-tandem mass spectrometry method for the quantitative determination of ABT-578 in human blood samples. J Chromatogr B 2004;805:67–75.

[44] Xue Y-J, Pursley J, Arnold ME. A simple 96-well liquid-liquid extraction with a mixture of acetonitrile and methyl t-butyl ether for the determination of a drug in human plasma by high-performance liquid chromatography with tandem mass spectrometry. J Pharm Biomed Anal 2004;34:369–78.

[45] Zhang N, Yang A, Rogers JD, Zhao JJ. Quantitative analysis of simvastatin and its β-hydroxy acid in human plasma using automated liquid–liquid extraction based on 96-well plate format and liquid chromatography-tandem mass spectrometry. J Pharm Biomed Anal 2004;34:175–87.

[46] O'Connor D, Clarke DE, Morrison D, Watt AP. Determination of drug concentrations in plasma by a highly automated, generic and flexible protein precipitation and liquid chromatography/tandem mass spectrometry method applicable to the drug discovery environment. Rapid Commun Mass Spectrom 2002;16:1065–71.

[47] Polson C, Sarkar P, Incledon B, Raguvaran V, Grant R. Optimization of protein precipitation based upon effectiveness of protein removal and ionization effect in liquid chromatography–tandem mass spectrometry. J Chromatogr B 2003;785:263–75.

[48] Ayrton J, Dear GJ, Leavens WJ, Mallett DN, Plumb RS. The use of turbulent flow chromatography/mass spectrometry for the rapid, direct analysis of a novel pharmaceutical compound in plasma. Rapid Commun Mass Spectrom 1997;11:1953–8.

[49] Ayrton J, Dear GJ, Leavens WJ, Mallett DN, Plumb RS. Optimisation and routine use of generic ultra-high flow-rate liquid chromatography with mass spectrometric detection for the direct on-line analysis of pharmaceuticals in plasma. J Chromatogr A 1998;828:199–207.

[50] Souverain S, Rudaz S, Veuthey J-L. Restricted access materials and large particle supports for on-line sample preparation: an attractive approach for biological fluids analysis. J Chromatogr B 2004;801:141–56.

[51] Jemal M, Xia Y-Q, Whigan DB. The use of high-flow high performance liquid chromatography coupled with positive and negative ion electrospray tandem mass spectrometry for quantitative bioanalysis via direct injection of the plasma/serum samples. Rapid Commun Mass Spectrom 1998;12:1389–99.

[52] Jemal M, Ouyang Z, Xia Y-Q, Powell ML. A versatile system of high-flow high performance liquid chromatography with tandem mass spectrometry for rapid direct injection analysis of plasma samples for quantitation of a b-Lactam drug candidate and its open-ring biotransformation product. Rapid Commun Mass Spectrom 1999;13:1462–77.

[53] Xia Y-Q, Hop CECA, Liu DQ, Vincent SH, Chiu S-HL. Parallel extraction columns and parallel analytical columns coupled with liquid chromatography tandem mass spectrometry for on-line simultaneous quantification of a drug candidate and its six metabolites in dog plasma. Rapid Commun Mass Spectrom 2001;15:2135–44.

[54] Mallett DN, Dear GJ, Plumb RS. Direct analysis of a polar pharmaceutical compound in plasma using ultra-high flow rate liquid chromatography/mass spectrometry with a mixed-mode column. Rapid Commun Mass Spectrom 2001;15:2526–9.

[55] Wu J-T, Zeng H, Qian M, Brogdon BL, Unger SE. Direct plasma sample injection in multiple-component LC-MS-MS assays for high-throughput pharmacokinetic screening. Anal Chem 2000;72:61–7.

[56] Ramos L, Bringnol N, Bakhtir R, Ray T, McMahon LM, Tse PSE. High-throughput approaches to the quantitative analysis of ketoconazole, a potent inhibitor of cytochrome P450 3A4, in human plasma. Rapid Commun Mass Spectrom2000;14:2282–93.

[57] Herman JL. Generic method for on-line extraction of drug substances in the presence of biological matrices using turbulent flow chromatography. Rapid Commun Mass Spectrom 2002;16:421–6.

[58] Herman JL. The use of turbulent flow chromatography and the isocratic focusing effect to achieve on-line cleanup and concentration of neat biological samples for low-level metabolite analysis. Rapid Commun Mass Spectrom 2005;19:696–700.

[59] Chiap P, Rbeida O, Christiaens B, Hubert PH, Lubda D, Boos K-S, Crommen J. Use of a novel cation-exchange restricted-access material for automated sample clean-up prior to the determination of basic drugs in plasma by liquid chromatography. J Chromatogr A 2002;975:145–55.

[60] Papp R, Mullett WM, Kwong E. A method for the direct analysis of drug compounds in plasma using a single restricted access material (RAM) column. J Pharm Biomed Anal 2004;36:457–64.

[61] Hsieh Y, Brisson J-M, Ng K, Korfmacher WA. Direct simultaneous determination of drug discovery compounds in monkey plasma using mixed-function column liquid chromatography/tandem mass spectrometry. J Pharm Biomed Anal 2002;27: 285–93.

[62] Xia Y-Q, Liu DQ, Bakhtiar R. Use of online-dual-column extraction in conjunction with chiral liquid chromatography tandem mass spectrometry for determination of terbutaline enantiomers in human plasma. Chirality 2002;14:742–49.

[63] Xia Y-Q, Bakhtiar R, Franklin RB. Automated online-dual-column extraction coupled with teicoplanin stationary phase for simultaneous determination of (*R*)- and (*S*)-propranolol in rat plasma using liquid chromatography/tandem mass spectrometry. J Chromatogr B 2003;788:317–29.

[64] Wu ST, Xing J, Apedo A, Wang-Iverson DB, Olah TV, Tymiak AA, Zhao N. High-throughput chiral analysis of albuterol enantiomers in dog plasma using on-line sample extraction/polar organic mode chiral liquid chromatography with tandem mass spectrometric detection. Rapid Commun Mass Spectrom 2004;18:2531–6.

[65] Plumb R, Dear G, Mallett D, Ayrton J. Direct analysis of pharmaceutical compounds in human plasma with chromatographic resolution using an alkyl-bonded silica rod column. Rapid Commun Mass Spectrom 2001;15:986–93.

[66] Hsieh Y, Wang G, Wang Y, Chackalamannil S, Korfmacher WA. Direct plasma analysis of drug compounds using monolithic column liquid chromatography and tandem mass spectrometry. Anal Chem 2003;75:1812–8.

[67] Zeng H, Deng Y, Wu J-T. Fast analysis using monolithic columns coupled with high-flow on-line extraction and electrospray mass spectrometric detection for the direct and simultaneous quantitation of multiple components in plasma. J Chromatogr B 2003;788: 331–7.

[68] Kebarle P, Tang L. From ions in solution to ions in the gas phase: the mechanism of electrospray mass spectrometry. Anal Chem 1993;65:972A–86A.

[69] Enke CG. A predictive model for matrix and analyte effects in electrospray ionization of singly-charged ionic analytes. Anal Chem 1997;69:4885–93.

[70] Pan P, McLuckey SA. The effect of small cations on the positive electrospray responses of proteins at low pH. Anal Chem 2003;75:5468–74.

[71] King R, Bonfiglio R, Fernandez-Metzler C, Miller-Stein C, Olah T. Mechanistic investigation of ionization suppression in electrospray ionization. J Am Soc Mass Spectrom 2000;11:942–50.

[72] Bonfiglio R, King RC, Olah TV, Merkle K. The effects of sample preparation methods on the variability of the electrospray ionization response for model drug compounds. Rapid Commun Mass Spectrom 1999;13:1175–85.

[73] Kaufmann A, Butcher P. Segmented post-column analyte addition; a concept for continuous response control of liquid chromatography/mass spectrometry peaks affected by signal suppression/enhancement. Rapid Commun Mass Spectrom 2005;19:611–7.

[74] Matuszewski BK, Constanzer ML, Chavez-Eng CM. Strategies for the assessment of matrix effect in quantitative bioanalytical methods based on HPLC-MS/MS. Anal Chem 2003;75:3019–30.

[75] Tong XS, Wang J, Zheng S, Pivnichny JV, Grifin PR, Shen X, Donnelly M, Vakerich K, Nunes C, Fenyk-Melody J. Effect of signal interference from dosing excipients on pharmacokinetic screening of drug candidates by liquid chromatography/mass spectrometry. Anal Chem 2002;74:6305–13.

[76] Shou WZ, Naidong W. Post-column infusion study of the "dosing vehicle effect" in the liquid chromatography/tandem mass spectrometric analysis of discovery pharmacokinetic samples. Rapid Commun Mass Spectrom 2003;17:589–97.

[77] Mei H, Hsieh Y, Nardo C, Xu X, Wang S, Ng K, Korfmacher WA. Investigation of matrix effects in bioanalytical high-performance liquid chromatography/tandem mass spectrometric assays: application to drug discovery. Rapid Commun Mass Spectrom 2003;1: 97–103.

[78] Xia Y-Q, Patel S, Bakhtiar R, Franklin RB, Doss GA. Identification of a new source of interference leached from polypropylene tubes in mass-selective analysis. J Am Soc Mass Spectrom 2005;16:417–21.

[79] Liang HR, Foltz RL, Meng M, Bennett P. Ionization enhancement in atmospheric pressure chemical ionization and suppression in electrospray ionization between target drugs and stable-isotope-labeled internal standards in quantitative liquid chromatography/tandem mass spectrometry. Rapid Commun Mass Spectrom 2003;17:2815–21.

[80] Sojo LE, Lum G, Chee P. Internal standard signal suppression by co-eluting analyte in isotope dilution LC-ESI-MS. Analyst 2003;128:51–4.

[81] Cech NB, Enke CG. Effect of affinity for droplet surfaces on the fraction of analyte molecules charged during electrospray droplet fission. Anal Chem 2001;73:4632–9.

[82] Schmidt A, Karas M, Dülcks T. Effect of different solution flow rates on analyte ion signals in nano-ESI MS, or: when does ESI turn into nano-ESI? J Am Soc Mass Spectrom 2003;14:492–500.

[83] Bennett PK, Van Horne KC. Identification of the major endogenous and persistent compounds in plasma, serum and tissue that cause matrix effects with electrospray LC/MS techniques. American Association of Pharmaceutical Scientists Annual Meeting and Exposition. Salt Lake City, UT, 2003.

[84] Van Horne KC, Bennett PK. Preventing matrix effects by using new sorbents to remove phospholipids from biological samples. American Association of Pharmaceutical Scientists Annual Meeting and Exposition. Salt Lake City, UT, 2003.

[85] Ahnoff M, Wurzer A, Lindmark B, Jussila R. Characterisation of serum albumin and lysoPCs as major contributors to plasma sample matrix effects on electrospray ionisation efficiency. 51st ASMS Conference on Mass Spectrometry. Montreal, Canada, 2003.

[86] Xia Y-Q, Jemal M. Phospholipids in liquid chromatography/mass spectrometry bioanalysis: comparison of three tandem mass spectrometric techniques for monitoring plasma phospholipids, the effect of mobile phase composition on phospholipids elution and the association of phospholipids with matrix effects. Rapid Commun Mass Spectrom 2009;23:2125–38.

[87] Murphy RC, Fiedler J, Hevko J. Analysis of nonvolatile lipids by mass spectrometry. Chem Rev 2001;101:479–526.

[88] Pulfer M, Murphy RC. Electrospray mass spectrometry of phospholipids. Mass Spectrom Rev 2003;22:332–64.

[89] Perterson BL, Cummings BS. A review of chromatographic methods for the assessment of phospholipids in biological samples. Biomed Chromatogr 2006;20:227–43.

[90] Lehninger AL. Principles of Biochemistry. New York: Worth Publishers, 1982.

[91] Schwarz HP, Dahlke MB, Drelsbach L. Phospholipid composition of blood plasma, erythrocytes, and “ghosts” in sickle cell disease. Clin Chem 1977;23:1548–50.

[92] Pang LQ, Liang QL, Wang YM, Ping L, Luo GA. Simultaneous determination and quantification of seven major phospholipid classes in human blood using normal-phase liquid chromatography coupled with electrospray mass spectrometry and the application in diabetes nephropathy. J Chromatogr B 2008;869:118–25.

[93] Little JL, Wempe MF, Buchanan CM. Liquid chromatography–mass spectrometry/mass spectrometry method development for drug metabolism studies: examining lipid matrix ionization effects in plasma. J Chromatogr B 2006;833:219–30.

[94] Uran S, Larsen A, Jacobsen PB, Skotland T. Analysis of phospholipid species in human blood using normal-phase liquid chromatography coupled with electrospray ionization ion-trap tandem mass spectrometry. J Chromatogr B 2001;758:265–75.

[95] Wang C, Xie S, Yang J, Yang Q, Xu G. Structural identification of human blood phospholipids using liquid chromatography/quadrupole-linear ion trap mass spectrometry. Anal Chim Acta 2004;525:1–10.

[96] Takatera A, Takeuchi A, Saiki K, Morisawa T, Yokoyama N, Matsuo M. Quantification of lysophosphatidylcholines and phosphatidylcholines using liquid chromatography–tandem mass spectrometry in neonatal serum. J Chromatogr B 2006;838:31–6.

[97] Taguchi R, Houjou T, Nakanishi H, Yamazaki T, Ishida M, Imagawa M, Shimizu T. Focused lipidomics by tandem mass spectrometry. J Chromatogr B 2005;823:26–36.

[98] Hendriks G, Uges DRA, Franke JP. Reconsideration of sample pH adjustment in bioanalytical liquid–liquid extraction of ionisable compounds. J Chromatogr B 2007;853: 234–41.

[99] Ouyang Z, Wu S, Jemal M. Biological sample phospholipids clean up: a comparison of sample preparation techniques. 57th ASMS Conference on Mass Spectrometry. Philadelphia, PA, 2009.

[100] Aurand C, Trinh A, Brandes HK, Bell DS, Increased YM. Bioanalytical throughput using selective phospholipid depletion. 57th ASMS Conference on Mass Spectrometry. Philadelphia, PA, 2009.

[101] Yong B, Hudson W, Jones D, Chen Y-L. A simple way to remove phospholipids from bioanalytical samples. 57th ASMS Conference on Mass Spectrometry. Philadelphia, PA, 2009.

[102] Jemal M, Xia Y-Q. The need for adequate chromatographic separation in the quantitative determination of drugs in biological samples by high-performance liquid chromatography with tandem mass spectrometry. Rapid Commun Mass Spectrom 1999;13: 97–106.

[103] Kapron J, Jemal M, Duncan G, Kolakowski B, Purves R. Removal of metabolite interference during liquid chromatography/tandem mass spectrometry using high-filed asymmetric waveform ion mobility spectrometry. Rapid Commun Mass Spectrom 2005;19:1979–83.

[104] Jemal M. Pitfalls in Quantitative LC-MS/MS: Metabolites Contributing to Measured Drug Concentrations in "Identification and Quantification of Drugs, Metabolites and Metabolizing Enzymes by LC-MS". In: Chowdhury SK, editor. Amsterdam: Elsevier, 2005.

[105] Vanderhoeven SJ, Lindon JC, Troke J, Nicholson JK, Wilson ID. NMR spectroscopic studies of the transacylation reactivity of ibuprofen 1-β-O-acyl glucuronide. J Pharm Biomed Anal 2006;41:1002–6.

[106] Xue YJ, Simmons NJ, Liu J, Unger SE, Anderson DF, Jenkins RG. Separation of a BMS drug candidate and acyl glucuronide from seven glucuronide positional isomers in rat plasma via high-performance liquid chromatography with tandem mass spectrometric detection. Rapid Commun Mass Spectrom 2006;20:1776–86.

[107] Schwartz MS, Desai RB, Bi S, Miller AR, Matuszewski BK. Determination of a prostaglandin D2 antagonist and its acyl glucuronide metabolite in human plasma by high performance liquid chromatography with tandem mass spectrometric detection—A lack of MS/MS selectivity between a glucuronide conjugate and a phase I metabolite. J Chromatogr B 2006;837:116–24.

[108] Xue YJ, Akinsanya B, Raghavan N, Zhang D. Optimization to eliminate the interference of migration isomers for measuring 1-O-α-acyl glucuronide without extensive chromatographic separation. Rapid Commun Mass Spectrom 2008;22:109–20.

[109] Ramanathan R, Su AD, Alvarez N, Blumenkranz N, Chowdhury SK, Alton K, Patrick J. Liquid chromatography/mass spectrometry methods for distinguishing N-oxides from hydroxylated compounds. Anal Chem 2000;72:1352–9.

[110] Liu DQ, Pereira T. Interference of a carbamoyl glucuronide metabolite in quantitative liquid chromatography/tandem mass spectrometry. Rapid Commun Mass Spectrom 2002;16:142–6.

[111] Jemal M, Ouyang Z, Powell M. A strategy for a post-method-validation use of incurred biological samples for establishing the acceptability of a liquid chromatography/tandem mass-spectrometric method for quantitation of drugs in biological samples. Rapid Commun Mass Spectrom 2002;16:1538–47.

[112] Jemal M, Ouyang Z. The need for chromatographic and mass Resolution in liquid chromatography-tandem mass spectrometric methods used for quantitation of lactones and corresponding hydroxy acids in biological samples. Rapid Commun Mass Spectrom 2000;14:1757–65.

[113] Xia Y-Q, Jemal M, Zheng N, Shen X. Utility of porous graphitic carbon stationary phase in quantitative LC-MS/MS bioanalysis: quantitation of diastereomers in plasma. Rapid Commun Mass Spectrom 2006;20:1831–7.

[114] Wu ST, Cao K, Bonacorsi SJ, Jr., Zhang H, Jemal M. Distinguishing a phosphate ester prodrug from its isobaric sulfate metabolite by mass spectrometry without the metabolite standard. Rapid Commun Mass Spectrom 2009;23:3107–13.

[115] Jemal M, Xia Y-Q. Bioanalytical method validation design for the simultaneous quantitation of analytes that may undergo interconversion during analysis. J Pharm Biomed Anal 2000;22:813–27.

[116] Ferreirós N, Dresen S, Alonso RM, Weinmann W. Hydrolysis and transesterification reactions of candesartan cilexetil observed during the solid phase extraction procedure. J Chromatogr B 2007;855:134–8.

[117] Xia Y-Q, Ouyang Z, Jemal M. Automated column/mobile phase screening system to achieve optimum chromatographic separation and sensitivity during LC-MS/MS bioanalytical method development. 54th ASMS Conference on Mass Spectrometry. Seattle, WA, 2006.

[118] Delatour C, Leclercq L. Positive electrospray liquid chromatography/mass spectrometry using high-pH gradients: a way to combine selectivity and sensitivity for a large variety of drugs. Rapid Commun Mass Spectrom 2005;19:1359–62.

[119] Patring JDM, Jastrebova JA. Application of liquid chromatography–electrospray ionisation mass spectrometry for determination of dietary folates: effects of buffer nature and mobile phase composition on sensitivity and selectivity. J Chromatogr A 2007;1143:72–82.

[120] Grujic S, Vasiljevic T, Lausevic M, Ast T. Study on the formation of an amoxicillin adduct with methanol using electrospray ion trap tandem mass spectrometry. Rapid Commun Mass Spectrom 2008;22:67–74.

[121] Peng L, Farkas T. Analysis of basic compounds by reversed-phase liquid chromatography–electrospray mass spectrometry in high-pH mobile phases. J Chromatogr A 2008;1179:131–44.

[122] Jemal M, Almond R, Ouyang Z, Teitz D. Negative ion electrospray high-performance liquid chromatography-mass spectrometry method development for determination of a highly polar phosphonic acid/sulfonic acid compound in plasma. Optimization of ammonium acetate concentration and in-source collision-induced dissociation. J Chromatogr B 1997;703:167–75.

[123] Jemal M, Hawthorne DJ. Effect of high performance liquid chromatography mobile phase (methanol versus acetonitrile) on the positive and negative ion electrospray response of a compound that contains both an unsaturated lactone and a methyl sulfone group. Rapid Commun Mass Spectrom 1999;13:61–6.

[124] Ouyang Z, Jemal M, Shen X. Does LC column type/brand affect LC/MS response? 53th ASMS Conference on Mass Spectrometry. San Antonio, TX, 2005.

[125] Wu Z, Gao W, Phelps MA, Wu D, Miller DD, Dalton JT. Favorable effects of weak acids on negative-ion electrospray ionization mass spectrometry. Anal Chem 2004;76:839–47.

[126] Xing J, Apedo A, Tymiak A, Zhao N. Liquid chromatographic analysis of nucleosides and their mono-, di- and triphosphates using porous graphitic carbon stationary phase

coupled with electrospray mass spectrometry. Rapid Commun Mass Spectrom 2004;18: 1599–606.

[127] Guo Y, Gaiki S. Retention behavior of small polar compounds on polar stationary phases in hydrophilic interaction chromatography. J Chromatogr A 2005;1074:71–80.

[128] Grumbach ES, Wagrowski-Diehl DM, Mazzeo JR, Alden B, Iraneta PC. Hydrophilic interaction chromatography using silica columns for the retention of polar analytes and enhanced ESI-MS sensitivity. LCGC North America 2004;22(10):1010–23.

[129] Li AC, Junga H, Shou WZ, Bryant MS, Jiang X-Y, Naidong W. Direct injection of solid-phase extraction eluents onto silica columns for the analysis of polar compounds isoniazid and cetirizine in plasma using hydrophilic interaction chromatography with tandem mass spectrometry. Rapid Commun Mass Spectrom 2004;18:2343–50.

[130] Xia Y-Q, Lau J, Olah T, Jemal M. Targeted quantitative bioanalysis in plasma using liquid chromatography-high resolution accurate mass spectrometry: an evaluation of global selectivity as a function of mass resolving power and extraction window, with comparison of centroid and profile modes. Rapid Commun Mass Spectrom 2011;25: 2863–78.

[131] Bateman KP, Kellmann M, Muenster H, Papp R, Taylor L. Quantitative–qualitative data acquisition using a benchtop orbitrap mass spectrometer. J Am Soc Mass Spectrom 2009;20:1441–50.

[132] Zhang NR, Yu S, Tiller P, Yeh S, Mahan E, Emary WB. Quantitation of small molecules using high-resolution accurate mass spectrometers—a different approach for analysis of biological samples. Rapid Commun Mass Spectrom 2009;23:1085–94.

[133] Fung NKE, Jemal M, Lau J, Olah T, Xia Y-Q. Comparison of full scan high resolving power MS and triple quadrupole SRM in quantitative bioanalysis. 58th ASMS Conference on Mass Spectrometry and Allied Topics. Salt Lake City, UT, 2010.

[134] Guevremont R. Understanding and designing field asymmetric waveform ion mobility spectrometry separations in gas mixtures. J Chromatogr A 2004;1058:3–19.

[135] Shvartsburg AA, Tang K, Smith RD. Understanding and designing field asymmetric waveform ion mobility spectrometry separations in gas mixtures. Anal Chem 2004;76: 7366–74.

[136] Shvartsburg AA, Tang K, Smith RD. Optimization of the design and operation of FAIMS analyzers. J Am Soc Mass Spectrom 2005;16:2–12.

[137] Kolakoswski BM, McCooeye MA, Mester Z. Compensation voltage shifting in high-field asymmetric waveform ion mobility spectrometry-mass spectrometry. Rapid Commun Mass Spectrom 2006;20:3319–29.

[138] Wu ST, Xia Y-Q, Jemal M. LC-ESI-FAIMS-MS/MS multi-component bioanalytical method development, performance evaluation and demonstration of the constancy of the compensation voltage with change of mobile phase composition or flow rate. Rapid Commun Mass Spectrom 2007;21:3667–76.

[139] Xia Y-Q, Wu ST, Jemal M. LC-FAIMS-MS/MS for quantification of a peptide in plasma and evaluation of FAIMS global selectivity from plasma components. Anal Chem 2008;80:7133–43.

[140] Xia Y-Q, Jemal M. High-field asymmetric waveform ion mobility spectrometry for determining the location of in-source collision-induced dissociation in electrospray ionization mass spectrometry. Anal Chem 2009;81:7839–43.

[141] Viswanathan CT, Bansal S, Booth B, DeStefano AJ, Rose MJ, Sailstad J, Shah VP, Skelly JP, Swann PG, Weiner R. Workshop/conference report—quantitative bioanalytical methods validation and implementation: best practices for chromatographic and ligand binding assays. AAPS J 2007;9(1):E30–42.

[142] Fast DM, Kelley M, Viswanathan CT, O'Shaughnessy J, King SP, Chaudhary A, Weiner R, DeStefano AJ, Tang D. Workshop report and follow-up—AAPS workshop on current topics in GLP bioanalysis: assay reproducibility for incurred samples—implications of Crystal City recommendations. AAPS J 2009;11:238–41.

[143] Matuszewski BK, Chavez-Eng CM, Constanzer ML. Development of high-performance liquid chromatography-tandem mass spectrometric methods for the determination of a new oxytocin receptor antagonist (L-368,899) extracted from human plasma and urine: a case of lack of specificity due to the presence of metabolites. J Chromatogr B 1998;716:195–208.

4

MASS SPECTROMETRY FOR QUANTITATIVE *IN VITRO* ADME ASSAYS

JUN ZHANG AND WILSON Z. SHOU

4.1 INTRODUCTION

In vitro absorption, distribution, metabolism, and excretion (ADME) assays are routinely conducted throughout drug discovery and development to characterize pharmacokinetics-(PK) and toxicity-related properties of new molecular entities (NMEs). These assays usually have demonstrated reasonable *in vitro–in vivo* correlations (IVIVCs), and can serve a number of purposes for *in vivo* studies such as assessing potential liabilities, selecting/prioritizing compounds to advance, troubleshooting unexpected results, and predicting parameters in preclinical/clinical studies [1]. Since *in vitro* experiments are generally less expensive to perform than animal studies, and are also far more amenable to miniaturization and automation in a high-density plate format, they have been utilized extensively in drug discovery to guide lead optimization effort by quickly providing structure–liability relationship (SLR) in parallel to potency optimization efforts.

The most important and therefore frequently performed *in vitro* ADME assays include metabolic stability, permeability, transporters, drug–drug interaction (DDI), physicochemical property, and cardio- and hepato-toxicity assays (Table 4.1). Briefly, metabolic stability assays are performed by incubating test compounds with an *in vitro* preparation such as liver microsomes or hepatocytes to assess the metabolic liability of these compounds and also provide prediction of *in vivo* clearance. Permeability assays measure the permeability of discovery compounds across *in vitro* membranes such as parallel artificial membranes and human colonic carcinoma

Mass Spectrometry for Drug Discovery and Drug Development, First Edition. Edited by Walter A. Korfmacher.

TABLE 4.1 Commonly Performed *in vitro* ADME Assays. Reprinted with Permission from Reference [10]

Assay Categorization	Assay names and References	Objectives and Application	Experimental
Metabolism	• Metabolic stability • Metabolic soft-spot identification • Reactive metabolites	• Assessing *in vitro* metabolic liabilities by drug-metabolism enzymes such as CYP • Predicting *in vivo* clearance	• Assay: • Liver microsomes from various species • NADPH as co-factor • 0.5–5 μM incubation concentration • Analysis: • Protein precipitation followed by LC-MS/MS or online SPE-MS/MS
Absorption	• PAMPA • Permeability (Caco-2, MDCK cells, etc.) • Bi-direction permeability	• Assessing *in vitro* membrane permeability • Predicting *in vivo* absorption	• Assay: • Dual-chamber "sandwich" setup using either artificial or live cell membranes • 3–50 μM compound concentration • Analysis: • Protein precipitation followed by LC-MS/MS
Drug–drug Interaction	• CYP inhibition (rCYP, HLM) • CYP induction (PXR-TA, Fa2N-4, hepatocytes) • CYP reaction phenotyping	• Assessing *in vitro* CYP inhibition and induction potentials • Predicting *in vivo* drug–drug interactions	• Assay (HLM CYP inhibition): • Human live microsomes • 0.5–100 μM probe concentration depending on each isozyme • Analysis: • Protein precipitation followed by LC-MS/MS or online SPE-MS/MS
Transporters	• P-gp assay • Other transporter assays	• Assessing transporter involvement in *in vitro* absorption experiments • Predicting *in vivo* transporter-related liabilities	• Assay: • Bi-directional permeability assays using Caco-2 or other cell lines • 3–50 μM compound concentration • Analysis: • Protein precipitation followed by LC-MS/MS
Property	• Structural integrity • Solubility (nephelometry) • Stability (buffer, plasma) • Serum protein binding	• Assessing physicochemical properties • Providing context to *in vivo* data interpretation	• Assay (protein binding): • Serum from multiple species • 5–10 μM compound concentration • Ultrafiltration or equilibrium dialysis in 96-well format • Analysis: • Samples in both buffer and serum matrices • Protein precipitation followed by LC-MS/MS

CYP, cytochrome P450; PAMPA, parallel artificial membrane permeability assay; Caco-2, human colonic adenocarcinoma cell; MDCK, Madin Darby canine kidney cell; rCYP, recombinant cytochrome P450; HLM, human liver microsomes; PXR-TA, pregnane X receptor transactivation; Fa2N-4, Fa2N-4 cryopreserved immortalized hepatocytes; P-gp, P-glycoprotein.

(Caco-2) cell monolayers, in order to predict NMEs *in vivo* absorption across intestinal epithelium layer. In addition, permeability assays with various cell lines can also be used to assess the involvement of influx and efflux membrane transporters in the absorption of NMEs and predict transporter-related liabilities. *In vitro* assays with liver microsomes and hepatocytes respectively are used extensively to assess the inhibition and induction of cytochrome P450 (CYP) enzyme activities by NMEs, and predict *in vivo* DDI potentials. A growing number of *in vitro* assays such as hERG, sodium and calcium channel assays, as well as human hepatotoxicity assay are being used to assess and predict toxicity. Lastly, physicochemical property assessments such as buffer/plasma solubility/stability and serum protein binding assay are important to provide context in interpreting both *in vitro* and *in vivo* data. Detailed discussions of the rationale, conduct, and utilities of these *in vitro* ADME assays are outside the scope of this chapter, and therefore interested readers are encouraged to consult with a number of excellent articles on the subject [2–6].

The analyses of *in vitro* ADME assay samples require the specific, sensitive, and fast detection of analytes with a lower limit of quantitation (LLOQ) typically in the low nM range. Depending on the analysis required, these assays can be categorized into two types, namely "compound-specific" and "probe-specific." For assays requiring probe-specific analysis, a single "probe" analyte is monitored for all samples—even though there is usually a different NME in the same assay well, its concentration is normally not measured. Examples of probe-specific assays include DDI, transporter inhibition, and toxicity assays. For assays requiring compound-specific analysis, each sample requires the analysis of a specific NME; therefore it is possible that samples from up to several dozens of different NMEs need to be analyzed from a single ADME assay batch. Such assays include metabolic stability, permeability, transporter substrate, and protein binding assays. For probe-specific assays, a variety of bioanalytical techniques including fluorescence, luminescence, and radioactivity measurement in a plate-reader format are commonly used for sample analysis in a very high throughput (minutes per plate), with the caveat being that the analyte typically needs to be labeled with fluorescent/luminescent/radioactive moieties, which may prevent a native, physiologically relevant substrate from being used in the assay. These techniques, however, are obviously not applicable to compound-specific assays since a different analyte needs to be measured for each sample.

Mass spectrometry is currently the only practical method of detection with adequate sensitivity, specificity, and speed for both assay types without the requirement of having a labeled analyte (label-free) in the assays, and naturally it has been the enabling sample analysis method of choice for a majority of *in vitro* ADME assay support [7–10]. As in other areas of bioanalysis, *in vitro* ADME bioanalysis follows the same process including method development, sample preparation, sample analysis, and data review and reporting, although the sample preparation (usually a simple protein precipitation) is typically incorporated into the assay automation. However, in comparison to other bioanalytical areas, there are several characteristics unique to bioanalytical mass spectrometry (MS) in support of *in vitro* ADME. First, the number of samples generated by the highly automated *in vitro* ADME assays ranges from hundreds to thousands on a daily basis; therefore the bioanalytical speed and turnaround time are of extremely high importance in order to generate quality data in a timely fashion. Second, for compound-specific assay support, the large number of structurally diverse compounds in each *in vitro* ADME screen poses a significant challenge for high-performance liquid chromatography-tandem mass spectrometry (HPLC-MS/MS) method development to achieve the selective and sensitive detection of each of the compounds. Third, samples from *in vitro* ADME assays are usually cleaner than their *in vivo* counterparts, and a fit-for-purpose approach is usually used to balance quality with throughput and capacity requirements. Lastly, as a result of these characteristics and requirements, automation and software tools are extensively used in order to maximize productivity and reduce errors. In this chapter, we will divide the discussion of quantitative *in vitro* ADME support based on the type of mass spectrometric techniques being used, into the following sections: (1) LC coupled with triple quadrupole mass spectrometers, (2) LC coupled with high-resolution mass spectrometers, and (3) direct mass spectrometric analysis without chromatographic separation.

4.2 HPLC-MS/MS WITH TRIPLE QUADRUPOLE MASS SPECTROMETERS

As in other areas of bioanalysis, LC coupled with a triple quadrupole mass spectrometer equipped with an atmospheric pressure ionization (API) source (either electrospray [ESI] or atmospheric pressure chemical ionization [APCI]), operated under selected reaction monitoring (SRM) mode has been the predominant technique for *in vitro* ADME bioanalysis due to its combination of specificity, sensitivity, speed, robustness, and its general applicability to the majority of discovery compounds. A prerequisite for performing SRM analysis for an analyte is the method development (referred to as optimization) effort to find the most suitable MS/MS conditions including ionization polarity, interface region lens voltage, precursor to product ion transitions, and corresponding collision energies. This method development process has to be performed for each analyte, and is typically conducted manually through infusing compound solution directly into the mass spectrometer. However, compound-specific *in vitro* ADME experiments are usually performed for several dozens of compounds at a time, and conducting manual optimization for that many compounds would be too time-consuming and also error-prone. To address this challenge in ADME bioanalysis, researchers and vendors developed automated hardware and software solutions to perform automated SRM method development. Earlier efforts include QuanOptimise™ (Waters, Milford, MA) and Automaton™ (AB Sciex, Framingham, MA). In both of these approaches, flow injection analysis (FIA) of compound solutions is performed with an LC system without a column, while the mass spectrometer automatically ramps its parameters to obtain the optimized conditions. These FIA-based method development approaches have been used extensively by bioanalytical groups supporting *in vitro* ADME assays [11–13].

Recently, several even more sophisticated solutions for automated SRM development have been developed that offer high-quality, manual infusion-like optimization results in addition to the speed enhancement. For example, QuickQuan™ (Thermo Scientific, Waltham, MA) uses a CTC autosampler to perform automated infusion for SRM method development (Fig. 4.1), and is capable of performing faster optimization (2 min per compound) while obtaining high-quality, reproducible results similar to those obtained by manual infusion [14, 15]. AB Sciex's new DiscoveryQuant™ Optimize software still uses an FIA mode for automated

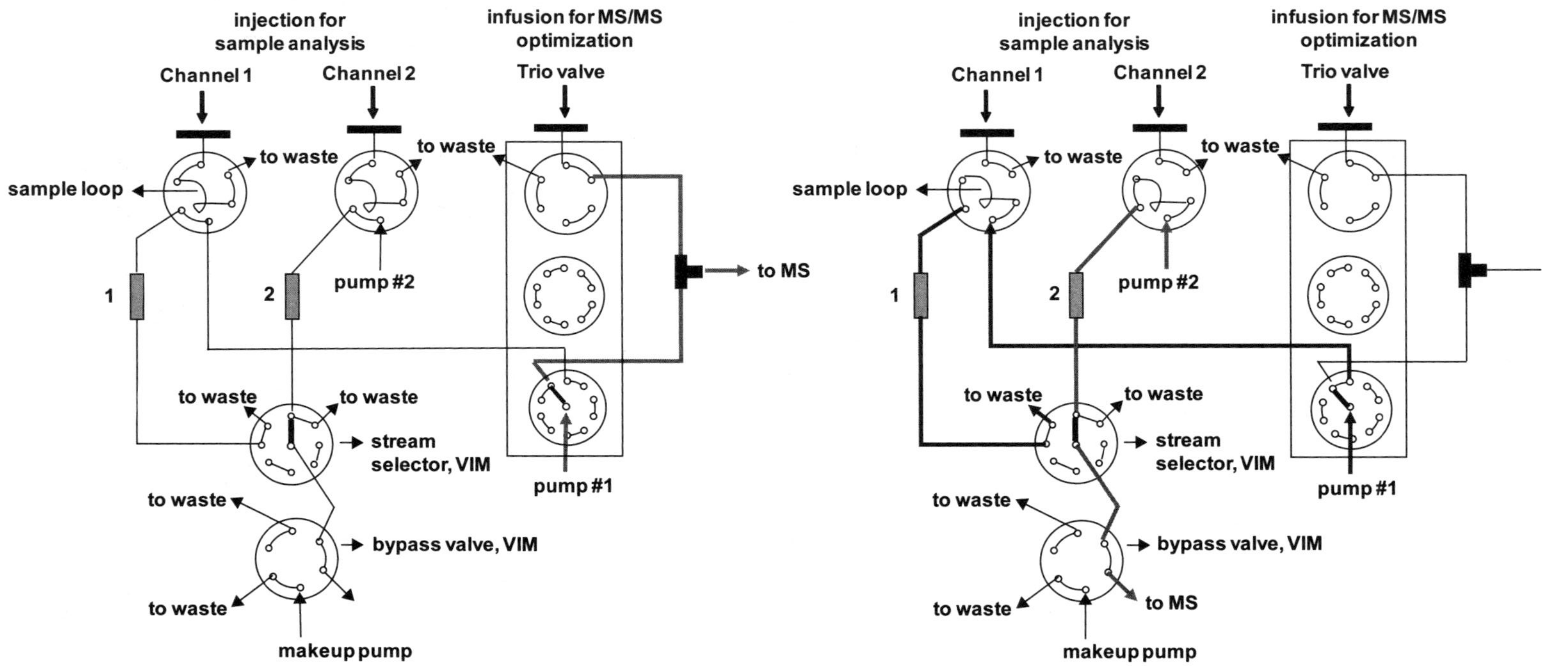

Figure 4.1 CTC autosampler valve configurations for (A) MS/MS method optimization by automated infusion; (B) sample analysis with multiplexed LC-MS/MS. The active flow path is highlighted in a dark color. Reprinted with permission from Reference 21.

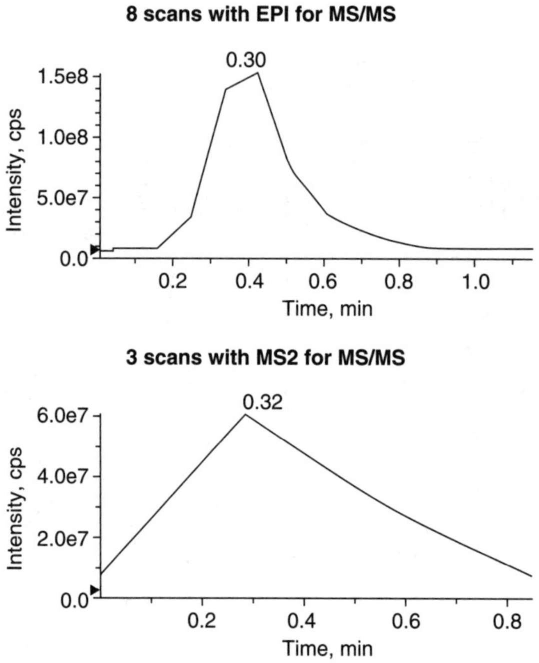

Figure 4.2 Comparison of scan numbers across the flow injection analysis (FIA) peak during SRM optimization by DiscoveryQuant™ Optimize: using trap scan (enhanced product ion [EPI]) for MS/MS optimization (top); and using quadrupole scan (product ion scan) for MS/MS optimization (bottom).

optimization; however, utilizing the fast-scanning ability of its QTrap instruments to perform all MS and MS/MS optimization in one injection provides more number of scans across an FIA peak (Fig. 4.2), and therefore better data quality than using a triple quadrupole MS. The software also offers users the option of performing a further "fine tune" with smaller parameter ramping steps to obtain the most optimized SRM conditions [9]. Furthermore, both QuickQuan™ and DiscoveryQuant™ have an elaborate global database feature, which allows multiple users to upload, query and share SRM conditions over network from many locations across the world.

In terms of chromatographic separation methods prior to SRM mass spectrometric analysis of *in vitro* samples, high-performance liquid chromatography (HPLC) or ultra-high pressure liquid chromatography (UHPLC), is still the prevalent technique in ADME support, as in other areas of bioanalysis. A generic, ballistic gradient running on a short (<5 cm) column is a common practice for *in vitro* ADME bioanalysis [16, 17]. However, in order to meet the requirement of a very high sample load (hundreds to even thousands of samples per day), a number of additional technologies have been developed and implemented specifically for *in vitro* ADME support. The first of such technologies is multiplexed LC, which involves performing multiple HPLC runs in parallel and directing the eluents into a single mass spectrometer. In the most popular multiplexing configuration, namely the "staggered

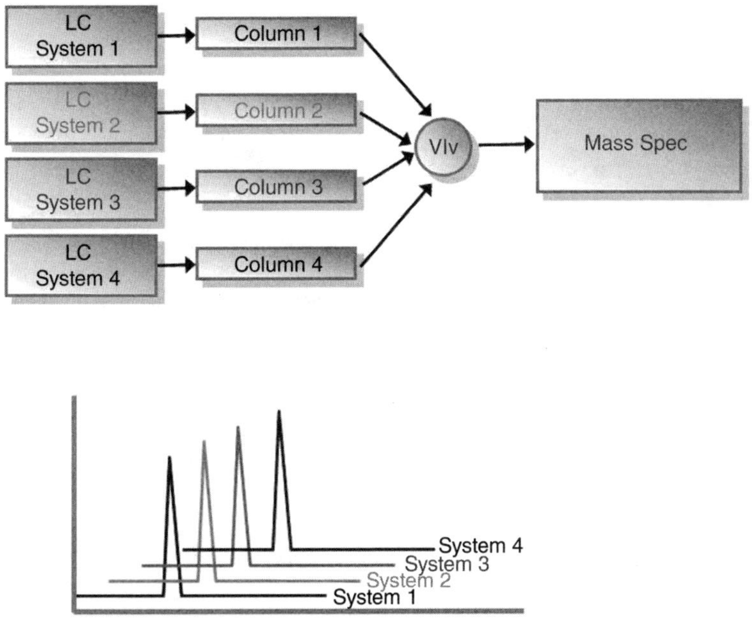

Figure 4.3 Schematic of a four-channel, staggered parallel multiplexed HPLC-MS/MS system. Reprinted with permission from Reference 10.

parallel" approach (Fig. 4.3), up to four independent (U)HPLC systems are connected to a single MS through a selector valve, with staggered injections made to each LC and the selector valve programmed to direct alternating HPLC streams into the MS [18]. With this multiplexing approach, injection-to-injection cycle time can be significantly reduced by minimizing "dead time" such as gradient equilibration and autosampler overhead, whereas at the same time the integrity of the HPLC separation is maintained. Multiplexed HPLC-MS/MS has been used to support various *in vitro* ADME assays [19–21], and an injection cycle time of as fast as 15 s per sample has been reported using this approach [22].

Another high-throughput separation approach commonly employed for *in vitro* sample analysis is direct online solid-phase extraction (SPE) followed by MS detection. Since many *in vitro* samples are in "lighter" matrices such as buffers and microsomes where ionization matrix effect is not very severe, online SPE with direct elution into the mass spectrometer can be an adequate alternative to LC separation while offering potentially much higher speed. First pioneered by Janiszewski et al. [23], this method has been used extensively in the analysis of *in vitro* ADME samples using home-built systems [24, 25]. More recently, two commercial direct online SPE systems, namely Agilent's RapidFire (Santa Clara, CA) [26] and Apricot Design's ADDA (Covina, CA) [27], have been introduced based on the same concept, with both offering a cycle time of less than 10 s per sample. A number of groups have reported the use of RapidFire system (Fig. 4.4) for the ultra-fast analysis of *in vitro* ADME samples [28–31]. When using this direct online SPE approach (as well as other high-speed separation mode such as multiplexed HPLC-MS/MS), it is common to acquire multiple injections into the same MS data file (Fig. 4.5), in

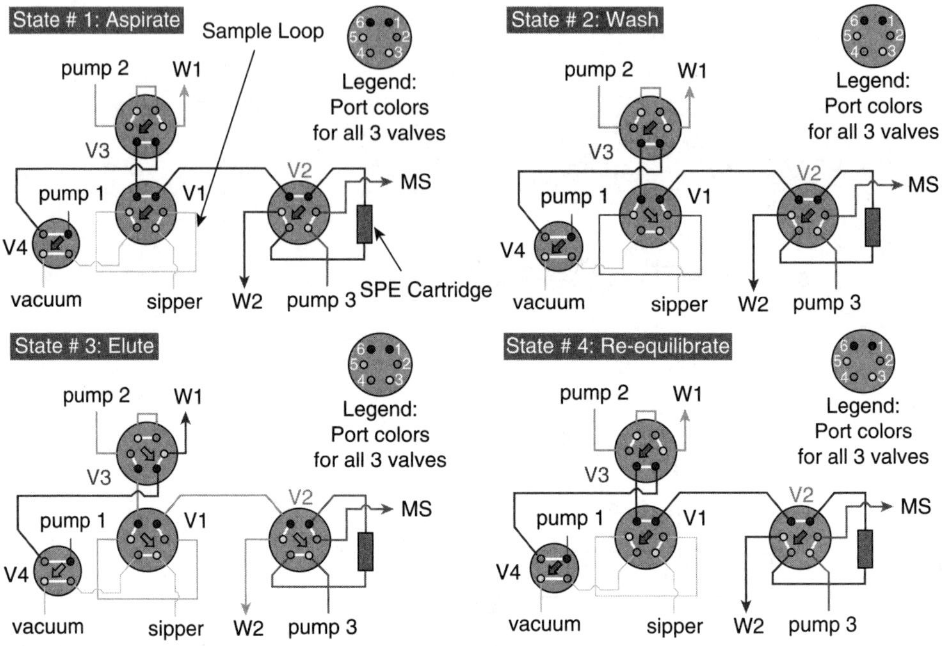

Figure 4.4 Schematic of the RapidFire direct online SPE-MS/MS system. Reprinted with permission from Agilent.

order to minimize the time required for the MS instrument software to download acquisition methods. These multiple injections are then "dissected" by custom software tools to perform peak integration and results review.

A third throughput-enhancing approach commonly used for *in vitro* ADME is sample reduction through the use of either cassette incubation or cassette analysis. Cassette incubation is performing assays for multiple NMEs in the same well, therefore reducing the number of sample generated for analysis by multiple folds. Several conditions have to be met in order for the cassette incubation approach to be effective: first the assay itself has to be free from potential drug–drug interaction problems within the cassette; second the compound selection needs to eliminate potential isobaric interferences between compounds in the same cassette during bioanalysis; and lastly a single SRM method and the corresponding data processing method containing all the cassette compounds can be generated with reasonable ease (i.e., automatically) [12, 32–34]. Cassette analysis, also called "sample pooling," does not incubate multiple compounds in the same well. Instead, samples from discrete assay wells for different compounds are combined after assays and analyzed together with HPLC-MS/MS [35, 36]. Cassette analysis eliminates the assay interaction issue; however sample dilution during pooling requires a more sensitive bioanalytical method in order to analyze the compounds in the pooled samples accurately and reproducibly. For both of these cassette approaches it is typical to combine 4–8 compounds together (and therefore reduce the sample number by 4–8 folds), since beyond these numbers too many SRM transitions would have to be built into a single method, which could result in too long a scan time to be compatible with fast

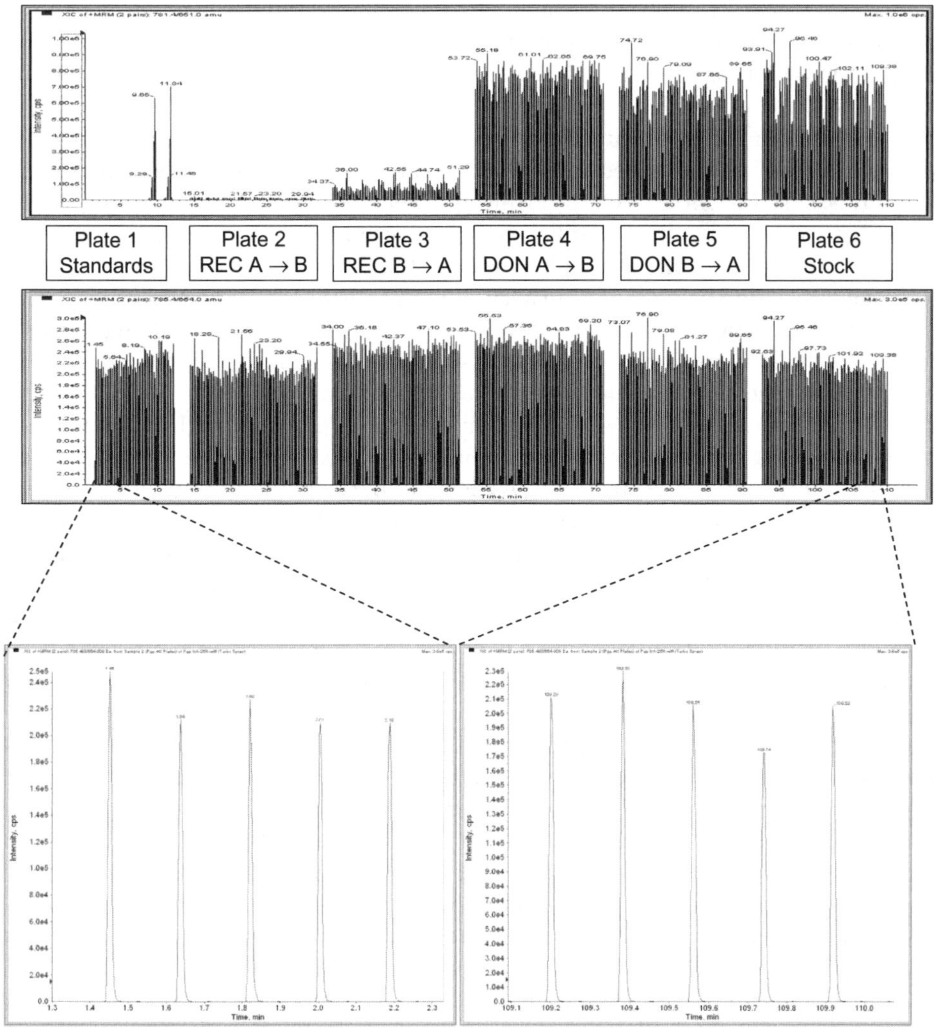

Figure 4.5 Mass chromatogram from the RF-MS/MS analysis of a typical P-gp inhibition assay: top panel is the digoxin trace and the bottom panel is the internal standard trace. Reprinted with permission from Reference 30.

gradients typically employed in sample analysis. A solution to this problem is to use full scan accurate mass for MS analysis instead of SRM on triple quadrupole MS, which will be described in the next section.

4.3 HPLC-MS WITH HIGH-RESOLUTION MASS SPECTROMETERS

Bioanalysis with full scan acquisition on high-resolution mass spectrometers has been previously explored using time-of-flight (TOF) instruments [37]. In comparison to SRM-based quantitation on triple quads, this approach could potentially offer a couple of distinctive advantages, namely the complete elimination of SRM method development and the ability to simultaneously collect quantitative and qualitative information, both of which could be very attractive for *in vitro* ADME bioanalysis. However, those efforts have not resulted in successful implementation due to the limitation in sensitivity, linearity, robustness, and the high cost of the previous generation TOF instruments. Recently, a new generation of high-resolution mass spectrometers, based on either TOF or Orbitrap™ technology, featuring high sensitivity, wide dynamic range, robust stability, good user-friendliness, and good affordability has become commercially available. As a result, there has been a renewed interest in bioanalysis using high-resolution accurate MS (HRAM) to fully realize its potentials in various drug discovery and development drug metabolism and pharmacokinetics (DMPK) applications including *in vitro* ADME support [38, 39].

O'Connor and coworkers [40] used a UPLC coupled with a Waters QTOF instrument to support a metabolic stability assay and simultaneously perform metabolite identification, at a run time of 2.5–3.5 min per sample. Temesi et al. [41] described the use of a newer TOF instrument (Agilent) to perform the bioanalysis of a high-throughput metabolic stability assay in hepatocytes. They demonstrated that the results for more than 1000 compounds obtained using LC-HRAM were similar to those from LC-SRM from triple-quads, and the data acquisition time was reduced by 20% due to the elimination of SRM method optimization. Using a benchtop Orbitrap instrument, Bateman and coworkers [42] demonstrated the feasibility of simultaneous quantitation of metabolic stability samples and the qualitative assessment of potential metabolites. High-speed separation methods mentioned previously, such as multiplexed LC and direct online SPE, have been both successfully used with high-resolution mass spectrometry (HRMS) as the detection method. For instance, Agilent recently launched the RapidFire 360 system, which combines direct online SPE with a TOF mass analyzer to achieve ultrafast ADME sample analysis without the need to perform SRM method optimization. Similarly, Murphy and coworkers reported the high-throughput quantitation of peptides using multiplexed LC coupled with high-resolution MS on a bench top Orbitrap instrument (Q Exactive) with an 18-s injection-to-injection cycle time [43].

Another advantage of full scan HRAM-based bioanalysis is that theoretically an unlimited number of analytes can be monitored simultaneously, since each scan takes the same time regardless of analyte numbers. Therefore, the common problem of "running out of dwell time" encountered on triple quadrupole-based bioanalysis of multiple analytes can be effectively resolved by using full scan HRAM. Zhang and coworkers [44] demonstrated this utility by performing cassette incubation of up to 32 compounds and analyzing the resulting samples with full scan HRAM (Fig. 4.6). Similar bioanalytical and biological results for parallel artificial membrane permeability assay (PAMPA) and protein binding assays were obtained using this cassette incubation/HRAM analysis approach to those obtained from discrete incubation with triple quadrupole MS SRM analysis (Fig. 4.7). In addition, a significant reduction in sample analysis time, as well as cost savings due to the reduced reagent usage, was also reported.

With the development in excellent high-resolution MS hardware and the demonstration of full scan bioanalytical feasibility for *in vitro* ADME support, there has been a pressing need for the development of corresponding software tools for data review and interpretation. While no SRM method development is required upfront

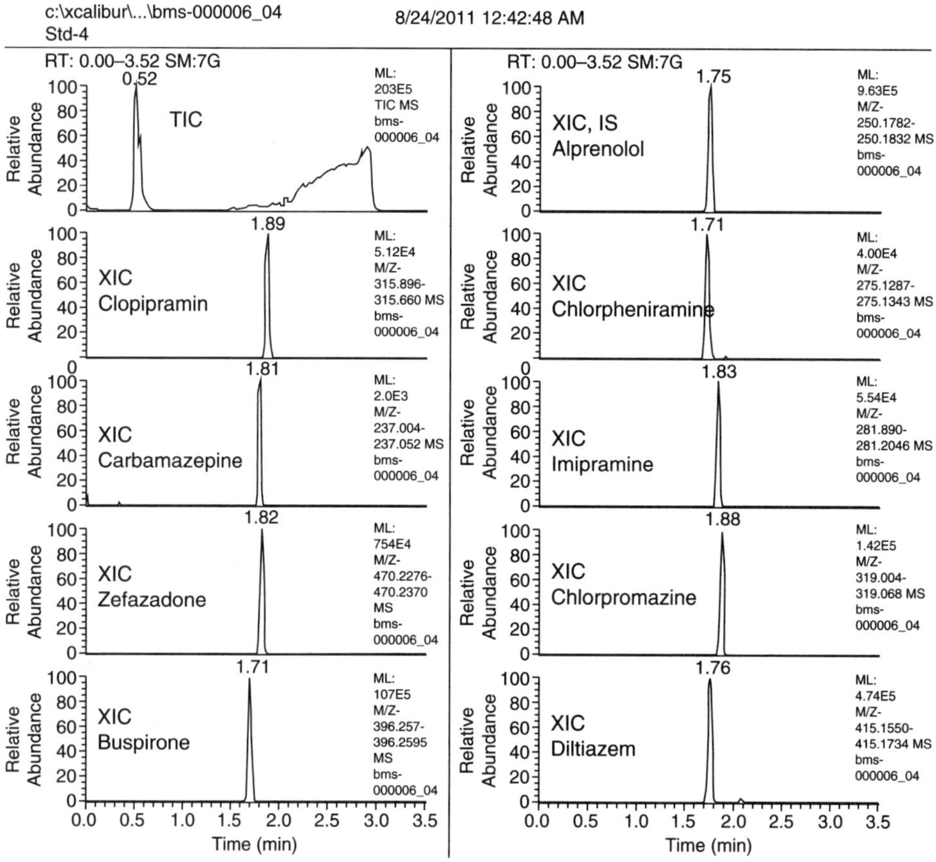

Figure 4.6 Extracted ion chromatograms of 50 nM standard in n = 8 cassette PAMPA assay. Mass extraction window ±5 ppm. Reprinted with permission from Reference 44.

for sample analysis using HRAM, post-acquisition extraction of accurate masses of analytes is required for quantitation. This step would be especially tedious if carried out manually with instrument software for *in vitro* ADME assay samples, due to the large number of samples from many different compounds in need of peak extraction and integration [41]. Zhang reported the use of a third-party software tool (GMSU/QC, Gubbs, Alpharetta, GA) to perform automated accurate mass extraction and peak integration from full scan high-resolution data acquired on an Orbitrap instrument [44]. Separately, AB Sciex has developed MultiQuant™ software that provides quantitation support for full scan data acquired on its QTOF instruments.

In addition to automated tools to process high-resolution MS data and generate quantitative results, software that facilitates the (preferably automated) data interpretation of full scan data to provide qualitative information of the sample (such as metabolites) is also in high demand. For *in vitro* metabolic stability assays, it has long been the "Holy Grail" of drug metabolism and pharmacokinetics/bioanalytical (DMPK/BA) scientists to perform a single incubation/analysis and obtain both the half-life of parent compound and the identities of metabolites. With the development

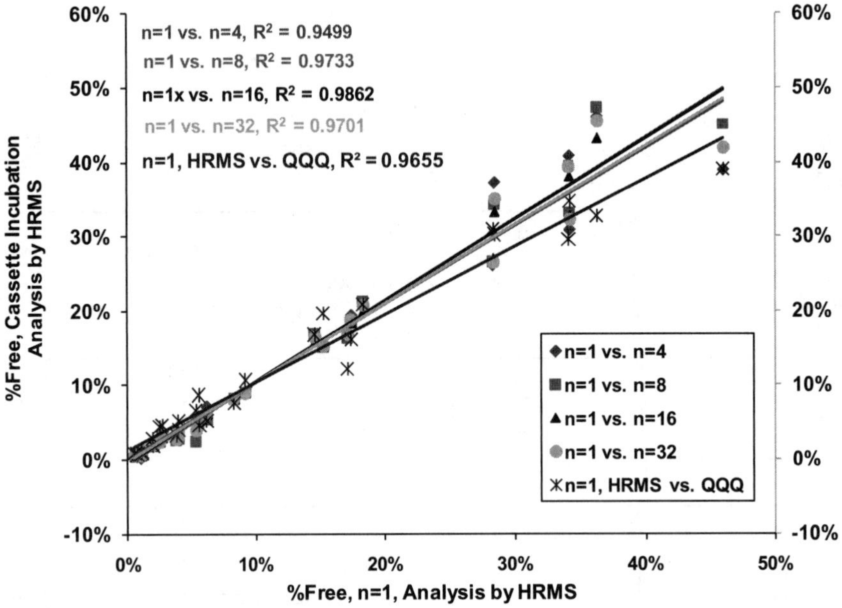

Figure 4.7 Correlation of %Free in protein binding assay between results obtained from cassette incubation and high-resolution mass spectrometric (HRMS) analysis and those from discrete incubation and analysis with selected reaction monitoring (SRM) on a triple quadrupole (QQQ) mass spectrometer. Reprinted with permission from Reference 44.

of generic data acquisition modes such as MS^E [45] and information-dependent analysis (IDA) [46] on newer high-resolution mass spectrometers, it has become possible to capture the necessary information for parent/metabolite quantitation as well as metabolite identification in the data set within a single injection. However, the automated identification of metabolites and subsequent assignment of their structures based on MS (/MS) spectra remain a challenge. Software development in this area has been very active, and included recently the introduction of a number of tools that combine intelligent metabolism prediction (including dealkylation), data reduction (mass defect filter, isotopic pattern matching, etc.), and structural assignment tools to perform metabolite identification and soft spot localization in a semi-automated fashion. Examples of such tools include MetabolitePilot™ from AB Sciex, MetaboLynx XS with MassFragment™ from Waters, and Mass-MetaSite [47] from Molecular Discovery (Perugia, Italy).

4.4 DIRECT MS ANALYSIS WITHOUT CHROMATOGRAPHIC SEPARATION

Direct analysis of samples without any chromatography or sample cleanup is another bioanalytical topic with extensive research over the years, due to its potential to achieve a very high analysis speed, which is attractive to sample-heavy applications such as *in vitro* ADME support. Earlier efforts include the direct introduction of samples into API sources with flow injection [48] or automated infusion through the use of Nanomate™ instruments (Advion, Ithaca, NY) [49, 50]. However, ionization suppression and interferences from matrix components in the ESI and atmospheric pressure chemical ionization (APCI) sources are common problems, unless extensive sample preparation approaches such as liquid–liquid extraction is performed prior to sample analysis. As a result, a number of alternative ion sources have been explored to perform the direct MS analysis of *in vitro* ADME samples.

The development of matrix-assisted laser desorption ionization (MALDI) paralleled that of API sources, and naturally attempts have been made to perform direct bioanalysis of biological samples to take advantage of the speed of MALDI analysis. Gobey et al. [51] used MALDI coupled with a triple quadrupole MS to perform sample analysis for a metabolic stability assay, and achieved a speed of 7 s per sample (limited by the speed of sample stage movement). Similarly, Rathore and coauthors [52] reported the use of MALDI/SRM at a speed of 1.2 s per sample to support an *in vitro* high-throughput screen of acetylcholinesterase inhibition. Despite the speed advantage, ionization suppression, and interferences from MALDI matrices were also observed. To address these issues, matrix-free desorption ionization techniques using various inert surfaces are currently being explored [53, 54].

Also recently, a number of ambient sampling/ionization techniques have been developed for MS-based direct analysis. These techniques are usually a two-step ionization process, with the usually solid samples first being introduced into gas phase for ionization, followed by the ambient ionization through an atmospheric pressure ionization source (ESI and APCI) [55]. Desorption electrospray ionization (DESI) [56] and direct analysis in real time (DART) [57] are the first two widely accepted ambient desorption/ionization techniques with many reported applications, and the utility of DART for *in vivo* bioanalysis has been evaluated with somewhat mixed results [58, 59]. Several other ambient ionization techniques use laser-based sampling methods such as laser ablation or desorption to introduce samples into gas phase, followed by ESI or APCI ionization and mass analysis. These techniques include laser ablation electrospray ionization (LAESI) [60] and laser diode array thermal desorption (LDTD), and LDTD has been demonstrated for *in vitro* CYP inhibition assay support at a speed of up to 18–28 s per sample [61, 62]. All the ambient sampling/ionization techniques discussed herein are compatible with both triple quadrupole and high-resolution mass spectrometric detections.

It is worth pointing out that while the aforementioned direct analysis techniques could potentially achieve a throughput approaching that of a plate reader (<1 s/sample), they are mostly solid-state ionization methods, which require the samples to be deposited on some types of solid support and dried prior to analysis. This requirement not only increases cost, but more importantly makes it difficult to integrate them easily with *in vitro* ADME assay automation, since the assays are usually performed in high-density plate format and generate liquid-phase samples. Therefore, corresponding developments in liquid handling that enable the fast, parallel transfer of incubation samples from plates to support materials would be required to truly realize the high-throughput potentials of these MS-based direct analysis methods for *in vitro* ADME support.

4.5 CONCLUSIONS

MS-based quantitation is the method of choice to support both probe-specific and compound-specific bioanalysis of *in vitro* ADME assay samples. While HPLC-MS/

MS on triple quadrupole mass spectrometers is utilized extensively for quantitative *in vitro* ADME support as in other areas of bioanalysis, technologies such as automated SRM MS/MS optimization, multiplexed LC separation, and direct online SPE with corresponding software tools have been developed and implemented to address its unique requirement of quickly analyzing large number of *in vitro* samples from numerous structurally diverse compounds. Emerging technologies such as high-resolution MS and ambient sampling/ionization are continuously being explored to achieve an even higher throughput in support of *in vitro* ADME profiling, and at the same time provide both quantitative and qualitative information to better serve the needs of drug discovery.

REFERENCES

[1] Kerns EH. Editorial: high throughput *in vitro* ADME/tox profiling for drug discovery. Curr Drug Metab 2008;9:845–6.

[2] Li AP. Screening for human ADME/Tox drug properties in drug discovery. Drug Discov Today 2001;6:357–66.

[3] Herbst JJ, Dickinson K. Automated high-throughput ADME-Tox profiling for optimization of preclinical candidate success. Am Pharm Rev 2005;8:96–101.

[4] Hop CE, Cole MJ, Davidson RE, Duignan DB, Federico J, Janiszewski JS, et al. High throughput ADME screening: practical considerations, impact on the portfolio and enabler of in silico ADME models. Curr Drug Metab 2008;9:847–53.

[5] Wang J, Urban L, Bojanic D. Maximising use of *in vitro* ADMET tools to predict *in vivo* bioavailability and safety. Expert Opin Drug Metab Toxicol 2007;3:641–65.

[6] Wan H, Holmén AG. High throughput screening of physicochemical properties and *in vitro* ADME profiling in drug discovery. Combin Chem High Throughput Screen 2009; 12:315–29.

[7] Shou WZ, Zhang J. Recent development in high-throughput bioanalytical support for *in vitro* ADMET profiling. Expert Opin Drug Metab Toxicol 2010;6:321–36.

[8] Carlson TJ, Fisher MB. Recent advances in high throughput screening for ADME properties. Comb Chem High Throughput Screen 2008;11:258–64.

[9] Janiszewski JS, Liston TE, Cole MJ. Perspectives on bioanalytical mass spectrometry and automation in drug discovery. Curr Drug Metab 2008;9:986–94.

[10] Jian W, Shou WZ, Edom RW, Weng N, Zhu M. LC-MS in drug metabolism and pharmacokinetics: a pharmaceutical industrial perspective. In Lee MS, editor. Mass Spectrometry Handbook. Hoboken, NJ: Wiley, 2012, p. 119.

[11] Fung EN, Chu I, Li C, Liu T, Soares A, Morrison R, et al. Higher-throughput screening for Caco-2 permeability utilizing a multiple sprayer liquid chromatography/tandem mass spectrometry system. Rapid Commun Mass Spectrom 2003;17:2147–52.

[12] Fung EN, Chen YH, Lau YY. Semi-automatic high-throughput determination of plasma protein binding using a 96-well plate filtrate assembly and fast liquid chromatography-tandem mass spectrometry. J Chromatogr B Analyt Technol Biomed Life Sci 2003;795: 187–94.

[13] Chovan LE, Black-Schaefer C, Dandliker PJ, Lau YY. Automatic mass spectrometry method development for drug discovery: application in metabolic stability assays. Rapid Commun Mass Spectrom 2004;18:3105–12.

[14] Kieltyka K, Zhang J, Li S, Vath M, Baglieri C, Ferraro C, et al. A high-throughput bioanalytical platform using automated infusion for tandem mass spectrometric method

optimization and its application in a metabolic stability screen. Rapid Commun Mass Spectrom 2009;23:1579–91.

[15] Smalley J, Xin B, Olah TV. Increasing high-throughput Discovery bioanalysis using automated selected reaction monitoring compound optimization, ultra-high-pressure liquid chromatography, and single-step sample preparation workflows. Rapid Commun Mass Spectrom 2009;23:3457–64.

[16] Plumb RS, Potts Iii WB, Rainville PD, Alden PG, Shave DH, Baynham G, et al. Addressing the analytical throughput challenges in ADME screening using rapid ultra-performance liquid chromatography/tandem mass spectrometry methodologies. Rapid Commun Mass Spectrom 2008;22:2139–52.

[17] Rainville PD, Wheaton JP, Alden PG, Plumb RS. Sub one minute inhibition assays for the major cytochrome P450 enzymes utilizing ultra-performance liquid chromatography/tandem mass spectrometry. Rapid Commun Mass Spectrom 2008;22:1345–50.

[18] Wu J. The development of a staggered parallel separation liquid chromatography/tandem mass spectrometry system with on-line extraction for high-throughput screening of drug candidates in biological fluids. Rapid Commun Mass Spectrom 2001;15:73–81.

[19] Lindqvist A, Hilke S, Skoglund E. Generic three-column parallel LC-MS/MS system for high-throughput *in vitro* screens. J Chromatogr A 2004;1058:121–6.

[20] Briem S, Pettersson B, Skoglund E. Description and validation of a four-channel staggered LC-MS/MS systems for high-throughput *in vitro* screens. Anal Chem 2005;77: 1905–10.

[21] Zhang J, Shou WZ, Vath M, Kieltyka K, Maloney J, Elvebak L, et al. An integrated bioanalytical platform for supporting high-throughput serum protein binding screening. Rapid Commun Mass Spectrom 2010;24:3593–601.

[22] Peltier JM. Challenges and opportunities in adapting LC/MS/MS to high-throughput screening. Proceedings—58th ASMS Conference on Mass Spectrometry and Allied Topics. Salt Lake City, UT, 2010.

[23] Janiszewski JS, Rogers KJ, Whalen KM, Cole MJ, Liston TE, Duchoslav E, et al. A high-capacity LC/MS system for the bioanalysis of samples generated from plate-based metabolic screening. Anal Chem 2001;73:1495–501.

[24] Kerns EH, Kleintop T, Little D, Tobien T, Mallis L, Di L, et al. Integrated high capacity solid phase extraction-MS/MS system for pharmaceutical profiling in drug discovery. J Pharm Biomed Anal 2004;34:1–9.

[25] Yan Z, Lu C, Wu JT, Elvebak L, Brockman A. Validation of a high-throughput absorption, distribution, metabolism, and excretion (ADME) system and results for 60 literature compounds. Rapid Commun Mass Spectrom 2005;19:1191–9.

[26] Miller VP. SPE/MS analysis of ADME assays: a tool to increase throughput and steamline workflow. Bioanalysis 2012;4:1111–21.

[27] Janiszewski J. Next generation sample delivery platform for HT-LC/MS/MS. Proceedings—59th ASMS Conference on Mass Spectrometry and Allied Topics. Denver, CO, 2011.

[28] Lim KB, Özbal CC, Kassel DB. Development of a high-throughput online solid-phase extraction/tandem mass spectrometry method for cytochrome P450 inhibition screening. J Biomol Screen 2010;15:447–52.

[29] Luippold AH, Arnhold T, Jörg W, Süssmuth RD. An integrated platform for fully automated high-throughput LC-MS/MS analysis of *in vitro* metabolic stability assay samples. Int J Mass Spectrom 2010;296:1–9.

[30] Wagner AD, Kolb JM, Özbal CC, Herbst JJ, Olah TV, Weller HN, et al. Ultrafast mass spectrometry based bioanalytical method for digoxin supporting an *in vitro* P-glycoprotein (P-gp) inhibition screen. Rapid Commun Mass Spectrom 2011;25:1231–40.

[31] Luippold AH, Arnhold T, Jörg W, Krüger B, Süssmuth RD. Application of a Rapid and Integrated Analysis System (RIAS) as a high-throughput processing tool for *in vitro* ADME samples by liquid chromatography/tandem mass spectrometry. J Biomol Screen 2011;16:370–7.

[32] Bu HZ, Poglod M, Micetich RG, Khan JK. High-throughput Caco-2 cell permeability screening by cassette dosing and sample pooling approaches using direct injection/on-line guard cartridge extraction/tandem mass spectrometry. Rapid Commun Mass Spectrom 2000;14:523–8.

[33] Youdim KA, Lyons R, Payne L, Jones BC, Saunders K. An automated, high-throughput, 384 well Cytochrome P450 cocktail IC50 assay using a rapid resolution LC-MS/MS end-point. J Pharm Biomed Anal 2008;48:92–9.

[34] Zhao SX, Forman D, Wallace N, Smith BJ, Meyer D, Kazolias D, et al. Simple strategies for reducing sample loads in *in vitro* metabolic stability high-throughput screening experiments: a comparison between traditional, two-time-point and pooled sample analyses. J Pharm Sci 2005;94:38–45.

[35] Halladay JS, Wong S, Jaffer SM, Sinhababu AK, Khojasteh-Bakht SC. Metabolic stability screen for drug discovery using cassette analysis and column switching. Drug Metab Lett 2007;1:67–72.

[36] Xu R, Manuel M, Cramlett J, Kassel DB. A high throughput metabolic stability screening workflow with automated assessment of data quality in pharmaceutical industry. J Chromatogr A 2010;1217:1616–25.

[37] Williamson LN, Bartlett MG. Quantitative liquid chromatography/time-of-flight mass spectrometry. Biomed Chromatogr 2007;21:567–76.

[38] Ramanathan R, Jemal M, Ramagiri S, Xia YQ, Humpreys WG, Olah T, et al. It is time for a paradigm shift in drug discovery bioanalysis: from SRM to HRMS. J Mass Spectrom 2011;46:595–601.

[39] Korfmacher W. High-resolution mass spectrometry will dramatically change our drug-discovery bioanalysis procedures. Bioanalysis 2011;3:1169–71.

[40] O'Connor D, Mortishire-Smith R, Morrison D, Davies A, Dominguez M. Ultra-performance liquid chromatography coupled to time-of-flight mass spectrometry for robust, high-throughput quantitative analysis of an automated metabolic stability assay, with simultaneous determination of metabolic data. Rapid Commun Mass Spectrom 2006;20:851–7.

[41] Temesi DG, Martin S, Smith R, Jones C, Middleton B. High-throughput metabolic stability studies in drug discovery by orthogonal acceleration time-of-flight (OATOF) with analogue-to-digital signal capture (ADC). Rapid Commun Mass Spectrom 2010;24:1730–6.

[42] Bateman KP, Kellmann M, Muenster H, Papp R, Taylor L. Quantitative-Qualitative Data Acquisition Using a Benchtop Orbitrap Mass Spectrometer. J Am Soc Mass Spectrom 2009;20:1441–50.

[43] Murphy K, Bennett PK, Duczak N. High throughput quantitation of large molecules using Multiplexed Chromatography and High Resolution/Accurate Mass LC/MS. Bioanalysis 2012;4:1013–24.

[44] Zhang J, Maloney J, Drexler D, Cai X, Stewart J, Mayer C, et al. Cassette incubation followed by bioanalysis using high resolution mass spectrometry for *in vitro* ADME screening assays. Bioanalysis 2012;4:581–93.

[45] Bateman KP, Castro-Perez J, Wrona M, Shockcor JP, Yu K, Oballa R, et al. MSE with mass defect filtering for *in vitro* and *in vivo* metabolite identification. Rapid Commun Mass Spectrom 2007;21:1485–96.

[46] Ruan Q, Peterman S, Szewc MA, Li M, Cui D, Humphreys WG, et al. An integrated method for metabolite detection and identification using a linear ion trap/Orbitrap mass

spectrometer and multiple data processing techniques: application to indinavir metabolite detection. J Mass Spectrom 2008;43:251–61.

[47] Bonn B, Leandersson C, Fontaine F, Zamora I. Enhanced metabolite identification with MSE and a semiautomated software for structural elucidation. Rapid Commun Mass Spectrom 2010;24:3127–38.

[48] Wang T, Zeng L, Strader T, Burton L, Kassel DB. A new ultra-high throughput method for characterizing combinatorial libraries incorporating a multiple probe autosampler coupled with flow injection mass spectrometry analysis. Rapid Commun Mass Spectrom 1998;12:1123–9.

[49] Balimane PV, Pace E, Chong S, Zhu M, Jemal M, Van Pelt CK. A novel high-throughput automated chip-based nanoelectrospray tandem mass spectrometric method for PAMPA sample analysis. J Pharm Biomed Anal 2005;39:8–16.

[50] Van Pelt CK, Zhang S, Fung E, Chu I, Liu T, Li C, et al. A fully automated nanoelectrospray tandem mass spectrometric method for analysis of Caco-2 samples. Rapid Commun Mass Spectrom 2003;17:1573–8.

[51] Gobey J, Cole M, Janiszewski J, Covey T, Chau T, Kovarik P, et al. Characterization and performance of MALDI on a triple quadrupole mass spectrometer for analysis and quantification of small molecules. Anal Chem 2005;77:5643–54.

[52] Rathore R, Corr JJ, Lebre DT, Seibel WL, Greis KD. Extending matrix-assisted laser desorption/ionization triple quadrupole mass spectrometry enzyme screening assays to targets with small molecule substrates. Rapid Commun Mass Spectrom 2009;23: 3293–300.

[53] Peterson DS. Matrix-free methods for laser desorption/ionization mass spectrometry. Mass Spectrom Rev 2007;26:19–34.

[54] Greving MP, Patti GJ, Siuzdak G. Nanostructure-initiator mass spectrometry metabolite analysis and imaging. Anal Chem 2011;83:2–7.

[55] Huang MZ, Cheng SC, Cho YT, Shiea J. Ambient ionization mass spectrometry: a tutorial. Anal Chim Acta 2011;702:1–15.

[56] Takáts Z, Wiseman JM, Cooks RG. Ambient mass spectrometry using desorption electrospray ionization (DESI): instrumentation, mechanisms and applications in forensics, chemistry, and biology. J Mass Spectrom 2005;40:1261–75.

[57] Cody RB, Laramée JA, Durst HD. Versatile new ion source for the analysis of materials in open air under ambient conditions. Anal Chem 2005;77:2297–302.

[58] Zhao Y, Lam M, Wu D, Mak R. Quantification of small molecules in plasma with direct analysis in real time tandem mass spectrometry, without sample preparation and liquid chromatographic separation. Rapid Commun Mass Spectrom 2008;22:3217–24.

[59] Yu S, Crawford E, Tice J, Musselman B, Wu JT. Bioanalysis without sample cleanup or chromatography: the evaluation and initial implementation of direct analysis in real time ionization mass spectrometry for the quantification of drugs in biological matrixes. Anal Chem 2009;81:193–202.

[60] Nemes P, Vertes A. Laser ablation electrospray ionization for atmospheric pressure, *in vivo*, and imaging mass spectrometry. Anal Chem 2007;79:8098–106.

[61] Wu J, Hughes CS, Picard P, Letarte S, Gaudreault M, Lévesque JF, et al. High-throughput cytochrome P450 inhibition assays using laser diode thermal desorption-atmospheric pressure chemical ionization-tandem mass spectrometry. Anal Chem 2007;79:4657–65.

[62] Beattie I, Smith A, Weston DJ, White P, Szwandt S, Sealey L. Evaluation of laser diode thermal desorption (LDTD) coupled with tandem mass spectrometry (MS/MS) for support of *in vitro* drug discovery assays: increasing scope, robustness and throughput of the LDTD technique for use with chemically diverse compound libraries. J Pharm Biomed Anal 2012;59:18–28.

5

METABOLITE IDENTIFICATION USING MASS SPECTROMETRY IN DRUG DEVELOPMENT

NATALIA PENNER, JOANNA ZGODA-POLS, AND CHANDRA PRAKASH

5.1 INTRODUCTION

High-performance liquid chromatography coupled with mass spectrometry (HPLC-MS) is the key analytical technique for metabolite characterization and quantitative analysis of drugs in modern drug discovery and development. The availability of various scan types (product ion, MS^n, neutral loss, and precursor ion scans), as well as accurate mass measurements, depending on the design of a mass spectrometer, allows the identification and characterization of putative and unexpected metabolites with no or little prior knowledge of biotransformation pathways of a given drug molecule.

In the last decade, there has been an enormous increase in the popularity and availability of mass spectrometric instrumentation. The instruments currently available on the market are equipped with a variety of mass analyzers, including an ion trap (IT), triple quadrupole (QQQ), time-of-flight (TOF), Fourier transform-ion cyclotron resonance (FT-ICR), and the newest mass analyzer, the FT-Orbitrap. Each type of MS has its advantages and disadvantages in terms of speed, sensitivity, ease of use and robustness, mass accuracy, mass resolution, and cost to own and operate. Until recently, IT and QQQ mass spectrometers were routinely utilized, while higher mass accuracy–higher cost TOFs, Orbitraps, and especially FT-ICR instruments were used to answer specific or critical questions. However, the clear trend in the biotransformation area has been to switch to high-resolution mass spectrometers (HRMS) with high mass accuracy for improved metabolite detection and structural elucidation.

In general, metabolite profiling and identification in discovery and development stages is performed using similar instrumentations and analytical strategies.

Mass Spectrometry for Drug Discovery and Drug Development, First Edition. Edited by Walter A. Korfmacher.

However, these strategies are applied to address different questions. While in discovery metabolite identification is focused on major metabolites and metabolic liabilities, development work is designed to ensure that all human metabolites have been adequately assessed in safety testing in animals. A number of compounds going through discovery and development as well as significantly different regulatory expectations affect analytical methods and conditions used for metabolite profiling. Discovery metabolite identification is mostly performed in high-throughput screening mode with short generic methods, and if a radiolabel is used, it is typically ^{3}H rather than ^{14}C, due to the ease of incorporating ^{3}H into a molecule, and even if tritiated water loss is observed [1, 2], this is acceptable for early decision making in the discovery stage. In development, when radiolabeled drug (usually ^{14}C) is available, metabolite profiling is conducted under conditions optimized for a given compound, which often translates into longer HPLC separation methods. There have been a number of publications and reviews on the use of mass spectrometry in metabolite identification in both drug discovery and drug development setting [3–9]. This chapter is focused mostly on current trends and novel approaches in the area of drug development.

Once a decision is made to progress a compound from discovery to development, the focus of pharmaceutical research shifts to characterizing the safety of a drug candidate, which is of primary concern to regulatory authorities around the world. Metabolism plays an important role in the overall safety profile of a drug; therefore, regulatory authorities and the pharmaceutical industry have paid much attention to how and when characterization of human metabolites should occur. The Food and Drug Administration (FDA) final guidance [10] on metabolites in safety testing (MIST) encourages the identification of differences in drug metabolism between animals used for long-term safety assessments and humans as early as possible during the drug development process. The guidance also warns the industry that the discovery of disproportionate human drug metabolites late in drug development can potentially cause development and marketing delays, which is generally understood to mean that all metabolism-related data should be available before initiation of large-scale clinical trials. The updated ICH: M3(R2) guidance for industry on Nonclinical Safety Studies for the Conduct of Human Clinical Trials and Marketing Authorization for Pharmaceuticals [11] established the following qualification regarding human metabolite testing: nonclinical characterization of a human metabolite(s) is only warranted when that metabolite(s) is observed at exposures greater than 10% of total drug-related exposure and at significantly greater levels in humans than the maximum exposure seen in the toxicity studies using the highest safe dose. In addition, the FDA MIST guidance also requires comparison of steady-state exposure of drug-related components across species as a more representative means of evaluating disproportionate human metabolites.

Approaches to mitigate MIST-related risks and to avoid delays in the late stages of clinical development vary among pharmaceutical companies. Overall, metabolite profiling conducted during drug development can be divided into two major categories as outlined in Figure 5.1:

1. studies without the use of radiolabeled material when samples from first-in-human (FIH) studies (plasma and urine) and long-term toxicology studies (steady-state plasma) are collected and profiled to assess metabolic profiles in

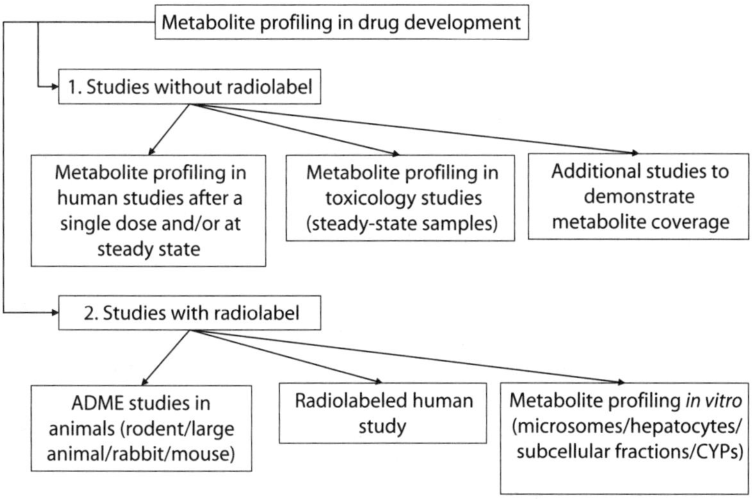

Figure 5.1 An outline of metabolite profiling studies conducted during drug development.

humans and animals and decrease chances of finding human-specific or disproportional metabolites in later stages of the development

2. studies with the use of radiolabeled material when metabolism in humans and animals is investigated according to standard development paradigm as required for registration of a drug for regulatory approval. All qualitative and quantitative information is obtained from absorption, disposition, metabolism, and excretion (ADME) studies after administration of a radiolabeled test compound

Pharmaceutical companies generally utilize a combination of the above-mentioned approaches to answer specific questions depending on the stage of the development program. Standard analytical approaches and mass spectrometric techniques for metabolites profiling in the development setting depending on a sample type (non-radiolabeled vs. radiolabeled) will be discussed in this chapter.

5.2 METABOLITE PROFILING IN STUDIES WITHOUT RADIOLABELED TEST ARTICLE

5.2.1 Use of Samples from FIH and Toxicology Studies

Since the FDA issued the final MIST guidance in February 2008 encouraging sponsors to obtain human metabolism data as early as feasible, safety and tolerance studies in human volunteers, so-called FIH studies, are routinely used by a growing number of pharmaceutical companies to obtain the first read-out on human metabolism. Metabolite profiling in plasma and urine collected in rising-single-dose (RSD) and rising-multiple-dose (RMD) studies represents the greatest challenge. First, the human doses are usually much lower than in animals. Second, there is no definitive information on metabolites and their amounts present as it would be the case after administering a radioactive dose. Therefore, most of the time that an investigator needs to rely on LC-MS data alone to detect drug-derived material and use other MS tools such as the NanoMate (Advion, Ithaca, NY) source to estimate relative abundance of metabolites. The same is true in the case of analyzing steady-state plasma samples from good laboratory practice (GLP)-toxicology studies.

Typically, blood and urine samples are collected during the conduct of the RMD study, although in certain cases (e.g., if RMD study is not planned shortly after the RSD study) it may be beneficial to collect samples from the RSD study to take an early look at what metabolites are formed in humans and if the human *in vivo* data correlate with *in vitro* data generated earlier in the discovery and/or development process using human hepatocytes. However, in order to delineate a profile of human circulating metabolites under steady-state conditions and investigate coverage in preclinical species, samples from the RMD study should be analyzed. Plasma samples collected from GLP-toxicology studies conducted in preclinical species can be banked (provided stability of the parent drug and metabolites is satisfactory) or directly analyzed prior to the initiation of the FIH studies. Alternatively, an additional study (e.g., 14-day study depending on the half-life of a parent drug) in preclinical species can be conducted around the time of FIH studies, so that fresh steady-state plasma samples from rodents and large animals can be collected and analyzed concurrently with human samples (see Section 5.2.2 for more details).

General recommendations for collecting blood and urine samples from RSD or RMD studies are summarized in Table 5.1. Blood samples are typically collected at predose (0 h) and a few time-points postdose (up to 24 h). The blood volume requirements will vary depending on the sample pooling methods discussed in more detail in Section 3.2.3 [12–15]. If samples will be pooled using the time-proportional pooling method to assess metabolite exposure over 0–24 h postdose, then the more time elapses between the individual blood draws, the more volume will be required. In general, the same proportional pooling scheme will be used for day 1 and in the

TABLE 5.1 General Recommendations for Collecting Blood and Urine Samples from RSD or RMD Studies

Number of subjects	Placebo: 2–4 Active: 6–12
Dose	RSD: top dose for easier detection of metabolites RMD: dose intended for administration in large clinical trials
Collection time-points (hours postdose)	RSD—Day 1 RMD—Day 1 and steady state (e.g., Day 14) Blood: predose, 1, 4, 8, 12, 24 (a selection of the same time-points as for robust PK samples analysis) Urine: predose, 0–12, 12–24 (or shorter time blocks if $t_{1/2}$ is short) Time-points will vary depending on the half-life of the parent drug.
Collection volume (mL)	Blood: ~4 mL per time-point (2 for analysis and 2 for back-up) Urine: 100 mL per time block (50 mL for analysis and 50 mL for back-up)

case of RMD at study steady state (e.g., day 14 or 28 depending on the half-life of a drug). Using the following formula, one could estimate the 0–24 h metabolite exposures:

$$\text{Metabolite AUC (0–24 h)} = [\text{\%Extracted Ion Chromatogram Peak Area Ratio of Metabolite to Parent Drug (based on qualitative metabolite profiling analysis)} \times \text{Parent Drug AUC (0–24 h) (as determined by a validated assay}]/100$$

However, if one is interested in understanding how rapidly a metabolite is eliminated and/or how quickly it reaches the steady state, then blood samples should be pooled across subjects at each time-point so that individual time-point samples can be profiled [14]. Furthermore, in order to aid in the metabolite identification and characterization process from nonradioactive samples, it is very useful to collect samples from subjects dosed with placebo. The placebo samples are pooled in an analogous way as the "active" samples, correcting for the fact that typically there are only 2–4 placebo subjects and 6–8 "active" subjects in a given FIH study. It is highly recommended to pool and extract exactly the same volume of the placebo and "active" samples, so that the matrix effects when analyzing samples on an LC-MS platform are comparable. Collection and pooling methods for plasma samples from preclinical species are analogous to the methods described for the collection of human plasma. Instead of placebo and "active" samples, vehicle control and "dosed" samples (from animals dosed with the test article) are collected on day 1 (optional) and at steady state and analyzed [15].

Although it is not absolutely necessary to collect human urine from RSD or RMD studies for the purpose of investigating metabolite safety in the context of the metabolite exposure as dictated by MIST or ICH guidance requirements, the urine samples can be very valuable in gaining an early insight into the route of elimination of the parent drug. If the parent drug is the primary component eliminated in the urine, such data can be used to design an appropriate renal elimination study in the renally impaired subjects to better understand the safety profile of a drug in that population. In cases when renal excretion is the predominant clearance pathway (>90%), a radiolabeled human study may not be necessary for characterization of disposition and metabolism of a drug.

As summarized in Table 5.1, urine samples are collected from the RSD and/or RMD studies on the same days as blood samples. Since it is unknown in nonradiolabeled FIH studies how much dose is actually excreted in a given time period by a human subject, urine samples are typically pooled by a percent of total volume excreted in a given time period (e.g., 2% total volume excreted in 0–24 h) by each subject and also pooled across all active subjects to generate one pooled "active" sample. Urine samples from subjects dosed with a placebo are pooled in an analogous manner to generate one pooled placebo sample. By analyzing the "active" and the placebo samples, one will be able to better distinguish between drug-derived material and matrix, as discussed in Sections 5.2.2 and 5.2.3 [15].

One of the most critical aspects for proper investigation of samples from FIH studies is ensuring adequate sample collection based on a stability profile of the parent drug and metabolites in a biological matrix. Specific techniques for stabilization of biological samples are described in Section 5.3.2.2.1.

5.2.2 Approaches for Metabolite Detection

Detection and identification of drug metabolites has always been a challenging task, in particular when analyzing samples from nonradiolabeled studies such as FIH studies and/or toxicology studies. In the absence of radiotracer, the only practical way to detect drug metabolites masked by background noise from endogenous components is to use HPLC-MS.

Traditionally, a standard approach for metabolite detection and identification in drug development using samples from FIH or GLP-toxicology studies would involve acquiring a full scan HPLC-MS spectrum across a certain *m/z* range (e.g., *m/z* 100–1100). Product ion spectra are obtained for the parent drug and any metabolites that are available as synthetic standards and major metabolites identified in preclinical species. This is done to fully elucidate the fragmentation pattern for the parent drug and its major metabolites so that precursor ion scans (PI) and neutral loss scans (for common conjugates such as glucuronides (*m/z* 176) [16], sulfates (*m/z* 80), or glutathione (*m/z* 129) using a triple quadrupole mass spectrometer can be employed to detect all metabolites circulating in plasma and/or excreted in the urine [3, 17–21]. With the PI and neutral loss (NL) scan data, prospective metabolites are identified and then product ion spectra for a prospective metabolites are required to confirm if the prospective metabolite is indeed drug-derived material or not. The full scan data obtained from the analysis of the placebo sample is used to aid in this metabolite identification process to differentiate between drug-derived material and matrix ions. Such an approach, although sensitive in terms of detecting trace-level metabolites, requires multiple HPLC-MS injections, each of which uses a different acquisition method, and in general is very time-consuming and not suitable for a fast-paced work flow. Furthermore, this approach may not be feasible for compounds that fragment extensively or do not easily fragment. Metabolites that do not generate the expected fragments based on the fragmentation pattern of the parent drug and its major metabolites will not be detected [15, 22].

In addition to using this working paradigm on a QQQ mass spectrometer platform, IT mass spectrometers, including three-dimensional (3D) ITs, linear ITs, and hybrid quadrupole ITs, are often used for qualitative metabolite profiling and identification. A common procedure utilized with quadrupole (3D) IT mass spectrometers involves performing a full scan MS analysis with data-dependent acquisition of MS/MS data for several most intense ions in a given scan. Alternatively, a full scan can be used as a survey scan to search for predicted metabolites (a list-dependent approach) so that once a predicted metabolite is detected the MS/MS acquisition of its product ion spectrum is triggered. The Q-Trap instrument, which combines features of a linear IT and a triple quadrupole mass analyzer, allows for performing PI and NL scan experiments in addition to MS^n fragmentation. Furthermore, because of both QQQ and IT functionalities, this instrument has found its use especially in a discovery setting for a simultaneous acquisition of qualitative metabolite profiling and quantitative metabolite data [9, 23–27]. Improvements in the IT technology in terms of trapping efficiency, ion capacity, sensitivity, scan speed, and MS/MS low-mass cutoff have made two-dimensional (2D) ion traps (LTQs) the instrument of choice for qualitative metabolite profiling in discovery and development. It is mainly due to providing the capability to delineate specific metabolite fragmentation patterns via MS^n experiments. However, with neither approach whether based on a triple quadrupole or an IT platform, an analyst can never be absolutely sure that all common (easily predictable) and uncommon (unconventional, multistep biotransformation) metabolites have been detected.

5.2.3 Use of HRMS

5.2.3.1 Use of HRMS for Metabolite Detection Recent developments in the field of high-resolution mass spectrometry (HRMS) such as the LTQ-FT, LTQ-Orbitrap, and quadrupole time-of-flight (Q-TOF) have had a major impact on the way metabolite profiling of nonradioactive and radioactive samples is conducted today [21, 28, 29]. In the past, HRMS was utilized mostly to determine molecular formulae of metabolites and accurate mass of their fragments to assist in structural characterization [30–32]. Today, HRMS is rapidly becoming the technique of choice for metabolite profiling and identification in both discovery and development setting because high resolution with high mass accuracy greatly improves metabolite detection and structural elucidation [29, 33–39]. The HRMS instruments provide enhanced selectivity against matrix background ions, which increases the reliability of metabolite detection. They also substantially increase sample throughput by empowering metabolite identification software tools. The use of real-time data-dependent acquisition triggered by mass defect filtering (MDF) or an accurate mass NL allows for automated component detection with simultaneous MS/MS spectra acquisition in a single run. Subsequently, post-analysis processing software tools such as MDF, noise reduction algorithm (NoRA), and isotope pattern filtering (IPF), as well as application of biotransformation libraries, enable efficient search for metabolites including the ones with unknown biotransformations [22, 33, 40–45]. If there is good evidence that the detected metabolite might be a major human metabolite, additional work can be carried out to positively elucidate its structure. In this case, approaches would include accurate mass measurement of parent and fragments, hydrogen/deuterium (H/D) exchange experiments, and nuclear magnetic resonance (NMR). *In vitro* incubations or targeted chemical synthesis can be used to synthesize larger amounts of a particular metabolite.

Currently, Q-TOF, LTQ-Orbitrap, and the newly introduced TripleTOF mass spectrometers are commonly used for metabolite profiling studies. Fast scanning and acquisition rates of high-resolution mass analyzers instruments allows their combination with modern fast chromatography systems such as ultra-performance liquid chromatography (UPLC). In addition, all metabolites (predicted and not predicted) can be acquired in a single run, and the use of low and high collision energy data acquisition functions allow for generation of MS/MS or MS^E data in a single run. Together with post-acquisition data filtering and processing software tools such as Metabolynx (Waters, Milford, MA) or MetabolitePilot (AB Sciex, Framingham, MA), metabolite profiling using HRMS instruments can be relatively easy and cost-effective.

LTQ-Orbitrap mass spectrometers are also very attractive because of their capability to generate very high mass accuracy data that are useful for ensuring detection and then filtering of drug-derived material with a limited number of false positives. Moreover, these instruments are capable of generating very high accuracy MS/MS and MS^n data that are very useful for determining sites of biotransformation in a parent molecule.

Figure 5.2 demonstrates the use of different acquisition methods for profiling of microsomal metabolites of the compound LX-0722. The analysis was performed using AB Sciex TripleTOF 5600 mass spectrometer with an acquisition of four

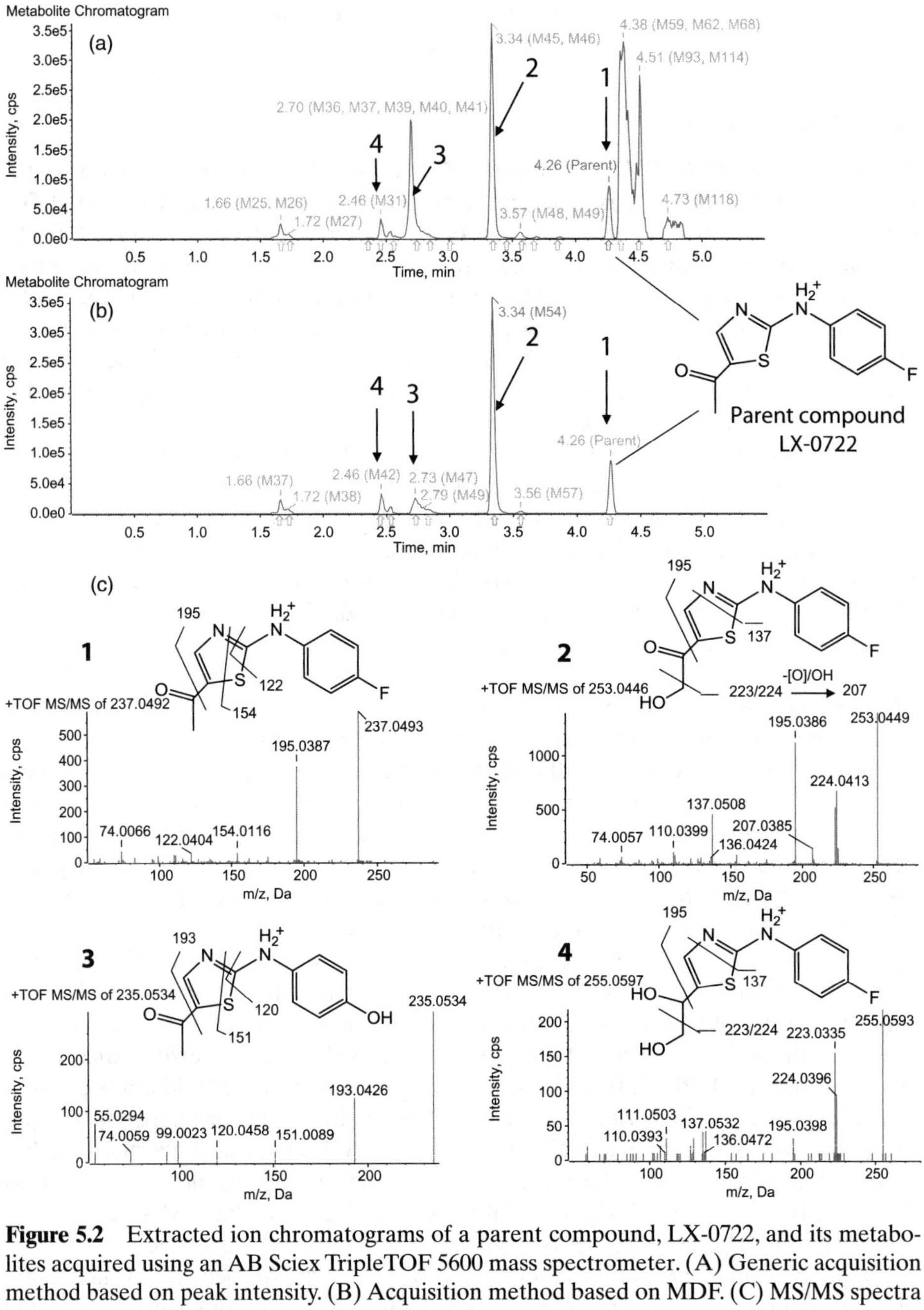

Figure 5.2 Extracted ion chromatograms of a parent compound, LX-0722, and its metabolites acquired using an AB Sciex TripleTOF 5600 mass spectrometer. (A) Generic acquisition method based on peak intensity. (B) Acquisition method based on MDF. (C) MS/MS spectra and metabolites (as marked in A and B).

data-dependent MS/MS spectra based on peak intensity (Fig. 5.2A) and MDF (Fig. 5.2B). In both cases, real-time dynamic background subtraction was utilized to eliminate background ions. Post-run data mining included the application of mass defect filters around the parent compound (50 mDa), phase 1 (50 mDa), phase 2 (50 mDa), and cleavage metabolites. Clearly, both approaches were successful in detecting major metabolites of LX-0722. However, MDF-based acquisition resulted in a clearer metabolite chromatogram devoid of significant background peaks eluting after the parent compound and major interference at 2.73-min retention time (peak 3). One major disadvantage of utilizing the MDF-based acquisition is the additional analyst time needed to set up the MDF-based acquisition methods prior to the analytical run. However, this is acceptable in the drug development setting, if the time invested in the method setup results in the generation of more quality data. Both methods were satisfactory and comparable in acquiring MS/MS spectra of metabolites. Spectra of four major drug-derived components are shown in Figure 5.2. These spectra were sufficient for correct structural elucidation of major microsomal metabolites of LX-0722.

Another advantage of HRMS for metabolite detection is the ability to search for specific fragments or neutral losses in the accurate mass mode. Traditionally, reactive metabolite screening is conducted by searching for 129- and 273-Da losses characteristic of glutathione (GSH) conjugate in the positive ionization mode. However, this approach suffers from both poor sensitivity and a high rate of false positives. The structural information is usually not available from the same analytical run, and in order to elucidate a site of bioactivation, the same sample must be reinjected. Currently, HRMS offers great improvement in the reactive metabolite screening workflow. The use of NL-triggered data-dependent algorithm for the detection of GSH conjugates of clozapine using a TripleTOF™ 5600 mass spectrometer resulted in the detection of 10 different GSH conjugates, including three direct conjugates with *m/z* 632, three conjugates of monohydroxylated clozapine (*m/z* 648), two minor conjugates with addition of a water molecule (*m/z* 650), one conjugate of desmethyl (*m/z* 618), and one of dechlorinated (*m/z* 598) clozapine. The MS/MS spectra of all conjugates in accurate mass mode were also acquired in the same run, which allowed the structural elucidation of the bioactivation site. A similar approach can be used for the detection of other phase 2 metabolites such as glucuronides, sulfates, and amino acid conjugates.

5.2.3.2 Use of HRMS for Structural Elucidation The use of HRMS also allows for improved structural assignment, which is achieved by a combination of the following options available on HRMS instruments:

- assignment of correct molecular formulae for unknown metabolites
- assignment of correct elemental composition for metabolite fragments
- possibility of calculating change in elemental composition due to metabolic modification

In selected cases, the use of HRMS allows the identification of isobaric metabolites and correct assignment of a site of modification, which is not possible with traditional IT and QQQ mass analyzers. For example, if a compound contains aliphatic and fluoro-substituted aromatic rings, metabolites can be formed via desaturation or oxidative defluorination. Both biotransformations result in a decrease jn nominal molecular mass by 2 Da and would not be distinguishable by low-resolution mass spectrometers. With high-resolution high mass accuracy instruments, desaturation (–2.0150 Da) can be clearly distinguished from oxidative defluorination (–1.995 Da), the latter often resulting in defining the exact position of biotransformation.

Occasionally, the formation of isobaric phase 2 metabolites could occur as demonstrated by the metabolism of Compound A containing a phosphate group (Fig. 5.3). MS and MS/MS spectra of Compound A obtained on LTQ-Orbitrap are shown in Figure 5.3A. Compound A produced a protonated molecular ion, $[M + H]^+$ at *m/z* 466. Additionally, a significant in-source fragmentation with the loss of a phosphate moiety (–80 Da) resulted in the presence of an ion with *m/z* 386 in the LC-MS spectrum of Compound A. This ion was also the only major fragment present in the MS/MS spectrum (Fig. 5.3B). After ^{14}C-Compound A was administered to rats and dogs for evaluation of mass balance and metabolism, a metabolite, M10, which produced a protonated molecular ion, $[M + H]^+$ at *m/z* 482, represented a majority of excreted radioactivity. Similar to Compound A, M10 (Fig. 5.3C) exhibited in-source fragmentation with formation of an ion with *m/z* 402 corresponding to the loss of 80 Da. The same ion (*m/z* 402) was the major fragment in the MS/MS spectrum of M10 (Fig. 5.3D). Overall, the spectra of Compound A and M10 were similar, and it appeared that M10 was formed by direct oxidation of the parent compound. However, the mass of M10 measured by HRMS (482.0885 Da) did not match the theoretical mass of hydroxylated Compound A (482.0991 Da). The mass difference of 0.0107 Da clearly indicated that M10 was not formed by direct oxidation of Compound A. Additional analysis of LC-MS spectrum of M10 showed that in-source fragmentation of M10 resulted in the neutral loss of 79.9568 Da, which corresponds to the loss of a sulfate group. Therefore, M10 was identified as a sulfate conjugate of dephosphorylated monohydroxylated Compound A. This example clearly demonstrates that a definitive assignment of biotransformation would not be possible without the use of HRMS instrument in this case.

Overall, modern mass spectrometric techniques such as HRMS combined with post-analysis processing software tools (MDF NoRA and IPF) are successful in detecting metabolites in complex samples. These methods not only allow reliable metabolite detection but also provide improved structural assignment and data quality. HRMS empowers metabolite identification software tools and allows reliable automated metabolite detection, which greatly increases throughput. Therefore, the use of HRMS for metabolite profiling becomes a technique of choice in both discovery and development setting. Mass spectrometric approaches and instruments used in drug metabolism studies for identification and structural elucidation of drug-derived material are listed in Table 5.2.

5.2.4 Quantitation of Metabolites from Nonradiolabeled Studies

5.2.4.1 Use of Nanoelectrospray and Response Normalization Techniques Since analysis of nonradiolabeled samples for metabolites is conducted by HPLC-MS methods, quantitative information about metabolite abundance cannot be

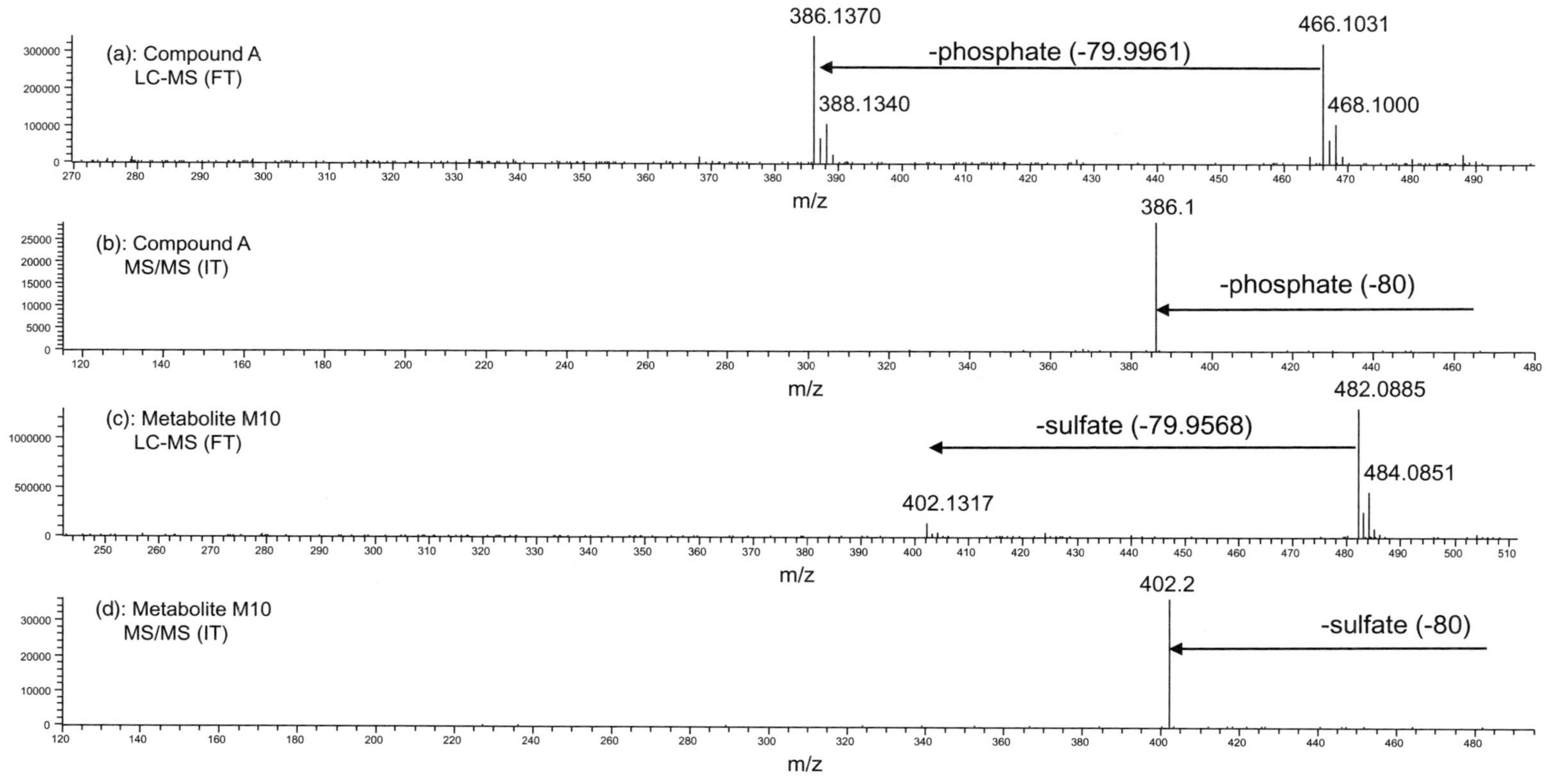

Figure 5.3 LC-MS (FT) and MS-MS (IT) spectra of Compound A and its M10 metabolite.

TABLE 5.2 Mass Spectrometric Approaches and Instruments Used in Drug Metabolism Studies for Identification and Structural Elucidation of Drug-Derived Material

Objective	Target for Monitoring	MS Scan Type	Mass Spectrometer
Metabolic stability	Quantitation of parent (PK)	MRM	QQQ, qtrap, LTQ
	In vitro-MS HTS	DDS	Q-TOF
	PK with simultaneous screening for metabolites (*in vivo*)	Combined, TOFMS	
Reactive metabolite screening	Glutathione conjugate	CNL 129 (positive)	QQQ, qtrap, LTQ
		DDS	
		Stable isotope labeling	
Metabolite profiling	General approach		
	Profiling *in vivo*/*in vitro* samples following administration/incubation of cold or 3H/14C-labeled parent drug		
	LC-MS (with or without radiochromatographic detector) with subsequent MS^2 (MS^n)	LC-MS with targeted MS^n	Any
	More specific approaches		
	1. Glutathione/ Glucuronide/sulfate conjugates	CNL 129 (+)	QQQ, qtrap, LTQ
		DDS	
	Acyl-glucuronide conjugate	CNL 176 (+)	
	Sulfate conjugates	CNL 80 (+)	
	2. Search for metabolites with common fragments	Precursor ion scans (parent scans)	QQQ
	3. Characteristic fragments for N-oxides/S-oxides/N-hydroxylamines (structure-dependent)	Use of different ionization techniques (ESI/APCI/APPI)	Any
	4. Use of DDS to search for metabolites	DDS based on	Any
		a. dynamic exclusion	Any
		b. predicted metabolites	High-resolution MS (Q-TOFs, FT-ICR, FT-Orbitrap [not available yet])
		c. isotopic pattern recognition ($^{14}C/^{12}C$, $^{37}Cl/^{35}Cl$)	
		d. MDF	
	5. Use of mass defect filter to search for drug-related material	LC-MS with high-resolution MS	Q-TOF, FT-ICR, FT-Orbitrap
	Additional procedures for metabolite profiling		
	H/D exchange		
	AMS	$^{14}C/^{12}C$ ratio	AMS
	LC-MS/NMR		
	Fraction collection followed by TopCount/nanospray analysis		Fraction collector, NanoMate or nanospray ion source

directly obtained from the acquired data due to a bias in MS responses for the parent and metabolites. The widely different HPLC-MS response observed for many structurally different compounds limits the use of HPLC-MS in full scan detection mode for quantitative determination of drugs and metabolites without using reference standards. In certain cases, when the flow rate to the ion source is reduced to nanoflow, as in the case of nanospray ionization (NSI) technique, some compounds exhibit comparable MS response. There have been a number of publications exploring nanoflow mass spectrometry and its use for relative metabolite quantitation. A more uniform response between the parent drug and the corresponding metabolites was obtained at flow rates of 10 nL/min or lower. Hop et al. [46] compared the ionization efficiencies of drugs and metabolites in conventional flow (500 uL/min) LC-MS using electrospray ionization with those of a microchip-based infusion NSI-MS and showed that in the latter technique the ionization efficiencies vary to a much lesser extent.

The presence of endogenous components in the biological matrices and constantly changing composition of background ions, which is characteristic of gradient elution, complicate the analysis as well. The solution to diminish the latter was introduced by Ramanathan et al. [47]. To remove the effect of gradient, Ramanathan et al. [47] utilized two HPLC systems: one HPLC unit performed the analytical separation, while the other unit added solvent post-column to make the final composition entering the NSI source isocratic throughout the entire HPLC run. Comparison of compounds of four different structural classes and their metabolites indicated that the influence of the solvent environment on the ionization efficiency was minimized. Although vicriviroc and desloratadine responses obtained using HPLC-NSI-MS were within two- to sixfold from the respective radiochromatographic responses, it is significantly better than the 6- to 20-fold observed with HPLC-electrospray ionization (ESI)-MS. Clearly, response normalization techniques improve quantitative estimates for metabolite levels in exploratory and early clinical drug metabolism studies when radiolabeled drugs are not yet available.

5.2.4.2 Use of Response Factors Determination of response factors for the purpose of obtaining quantitative data from nonradiolabeled metabolite profiling studies has been widely discussed in recent publications. A whole spectrum of methods based on radioactivity, UV, and NMR was evaluated. These approaches help to derive actual amounts of metabolites from data obtained during cold sample analysis. The majority of approaches help to derive actual amounts of metabolites based on HPLC-MS data using the formula below:

$$\%Metabolite = \%metabolite\,(LC\text{-}MS) \times Response\ factor$$

If a synthetic standard of a metabolite is available, the actual amount of drug metabolites in human plasma can be easily estimated by mass spectrometry alone. A known amount of the parent compound and a metabolite standard is spiked into plasma (or other biological matrix for which response factor is evaluated) and the resulting sample is processed with the method identical to the one used for the analysis of human samples. The response factor is then calculated as follows:

$$Response\ factor = \frac{Area(LC\text{-}MS, parent)}{Amount, parent(mol)} \Big/ \frac{Area(LC\text{-}MS, metabolite)}{Amount, metabolite(mol)}$$

The use of reference standards for the estimation of actual amounts of metabolites is limited to metabolites with known structures. Other techniques described below can be used even if the identity of a metabolite is not confirmed.

If human circulating metabolites are formed in appreciable amounts *in vitro* or *in vivo* in preclinical species, radiolabeled samples from those studies can be utilized for the response factor determination. The LC-MS and radiochromatogram peak areas for parent and metabolites are determined, and response factors are calculated according to the following formula:

$$Response\ factor = \frac{Area(LC\text{-}MS, parent)}{Area(radiochromatogram, parent)} \Big/ \frac{Area(LC\text{-}MS, metabolite)}{Area(radiochromatogram, metabolite)}$$

HPLC-MS and radiometric analyses can be conducted simultaneously or separately. A larger amount can be injected to obtain a reliable radiometric signal, while for HPLC-MS analysis response of compounds of interest should fall within linear range; therefore, the injection volume can be smaller to avoid detector saturation. Corresponding HPLC-MS response of metabolites in an unknown sample should be obtained with a similar profile of endogenous compounds; therefore, samples with unknown amounts of metabolites should be mixed with blank samples to diminish a background difference.

$$\begin{matrix}\text{Blank human} \\ \text{plasma}\end{matrix} + \begin{matrix}\text{Radioactive} \\ \text{sample} \\ (\textit{in vitro}\text{ or }\textit{in vivo})\end{matrix} = \begin{matrix}\text{Mixed sample} \\ \text{with known amounts}\end{matrix} \rightarrow \begin{matrix}\text{LC-MS/radiometric} \\ \text{analysis}\end{matrix}$$

$$\begin{matrix}\text{Human} \\ \text{Plasma} \\ \text{(Unknown)}\end{matrix} + \begin{matrix}\text{Blank} \\ \text{sample} \\ (\textit{in vitro}\text{ or }\textit{in vivo})\end{matrix} = \begin{matrix}\text{Mixed sample} \\ \text{with unknown amounts}\end{matrix} \rightarrow \begin{matrix}\text{LC-MS/radiometric} \\ \text{analysis}\end{matrix}$$

The second approach to estimate actual amounts of metabolites with the use of radiolabeled samples was proposed by Cuyckens et al. [44]. In this case unknown sample is mixed with radiolabeled sample which contains the components of interest and the resulting mixture is analyzed by HPLC-MS. The concentration of the unknown component is calculated as follows:

$$C_{unknown} = C_{rad} \times \frac{^{12}C}{^{14}C}$$

where ^{12}C = ^{12}C peak – Y, ^{14}C = ^{14}C peak – X. If the unknown compound does not exhibit prominent isotopic patterns, for example, it does not contain Cl, Br atoms, X mostly comes from the ^{13}C (third isotopic peak) and can be considered negligibly small ($X = 0$). Similarly, *in vitro* samples are the most convenient source of "internal standards" since incubation of a fully enriched ^{14}C compound simplifies the formula

($Y = 0$). In our opinion, the proposed method should be used only when *in vitro* samples are available for calibration. Specific activity of *in vivo* samples is usually relatively low; they contain a high amount of nonlabeled compound, which makes calculations less reliable.

Among the standard-free quantitative techniques, ultraviolet (UV)-based techniques is probably the most widely used to correct for HPLC-MS response differences. The calculation and matrix mixing are conducted as described for radiolabeled samples except that *Area (UV profile)* is used instead of *Area (radiochromatogram)*. For UV-based techniques to be successful, parent NCE and its metabolites must have good UV chromophores, as well as similar maximum UV absorption, which is not always observed.

A new paradigm in understanding metabolite coverage as early as possible is profiling plasma samples from toxicology studies. To minimize matrix-specific effects on ionization of different samples, blank animal plasma was added to the human plasma samples and vice versa to normalize the effect of matrix on HPLC-MS/MS signal [48]. Profiles of circulating metabolites in humans obtained after a nonradiolabeled dose are compared with similar profiles in animals. This approach allows for direct comparison of animal/human XICs without lengthy and often not fully reliable determination of response factors. If for a given metabolite the mass spectrometric peak area ratio is higher than the minimal accepted ratio (in this case it was 2 with 99% confidence), such metabolite is considered covered from the MIST point of view and would not require synthesis, assay development, and validation, or direct administration to animals. In rare cases when animal/human concentration of metabolite is less than 1, further investigation is warranted. One key limitation of this method is that the stability of individual metabolites is not fully established. Reanalysis of selected samples after a certain period of storage could indicate whether an analyte has stability issues as the ratio would change significantly upon assay repetition. If formation of metabolites with known stability issues (e.g., acylglucuronides) is expected, plasma samples should be stabilized immediately following collection.

No matter how much biotransformation information is available in phase 1 from nonradiolabeled studies, a full ADME program is still required for drug registration. Definitive information about mass balance and metabolite amounts can only be obtained after administration of the radiolabeled drug. *In vivo* ADME studies are conducted to investigate the routes of elimination (renal or biliary) of the parent drug and the form in which the parent drug is eliminated (as an intact parent and/or metabolites), which informs on the design of DDI studies; delineate metabolic pathways; compare metabolite profiles across species; and obtain cumulative recovery of total radioactivity as low recovery may indicate compound sequestration into tissues. Additionally, there is always a risk that the major drug-related entities were missed in nonradiolabeled studies for example, because of low ionization in mass spectrometer. A survey from 18 pharmaceutical companies showed that in most cases human radiolabeled ADME studies are still used as the definitive data for triggering MIST-related activities. The majority of sponsors initiated such activities in phase 2 or after proof-of-concept (POC) studies. In rare cases, metabolite characterization in humans is delayed until phase 3.

5.3 METABOLITE PROFILING IN STUDIES WITH RADIOLABELED TEST ARTICLE

5.3.1 *In Vitro* Studies

A common strategy for initial metabolism assessment in the discovery stage is an incubation of a drug candidate with liver microsomes fortified with NADPH to assess phase 1 reactions. The next step is the evaluation of phase 2 reactions using hepatocytes or liver microsomes/S9/cytosol with appropriate cofactors. In the development stage, *in vitro* incubations are usually repeated with the use of radiolabeled compounds. A mixture of nonradiolabeled and ^{14}C- or ^{3}H-labeled compounds (1:1 mixture is optimal for MS detection) is incubated in an appropriate *in vitro* system. Samples are then profiled as described in Sections 5.2.2, 5.2.3 and 5.3.4. If necessary, samples are stabilized and/or extracted to improve sensitivity of detection.

In addition to experiments routinely conducted in a discovery setting, such as microsomal and hepatocyte incubations, experiments to identify CYP450 isoforms responsible for metabolism of the drug can be performed. A careful *in vitro* assessment of the contribution of various CYP isoforms to the total metabolism is important for predicting whether drug–drug interactions (DDIs) might take place. One method of CYP phenotyping involves the use of potent and selective chemical inhibitors in human liver microsomal incubations in the presence of a test compound [49, 50]. The selectivity of such inhibitors plays a critical role in deciphering the involvement of specific CYP isoforms. It is convenient to use radiolabeled compounds for CYP phenotyping as qualitative and quantitative changes in radiochromatographic profiles obtained with and without inhibitors serve as a measure of the role of each particular CYP450 in the metabolism of a drug.

5.3.2 *In Vivo* Studies

Radiolabeled studies are routinely used to characterize excretion patterns, identify routes of clearance and metabolism, and define the rate of elimination by renal or biliary routes. They can also be used to estimate absorption, which is especially important when urinary excretion of the drug and metabolites is low. Although it was not the case a decade ago when human radiolabeled studies were conducted late in drug development, nowadays, the common trend is to conduct radiolabeled human studies as early as phase 1 in order to obtain definitive information about excretion, clearance, and metabolic profiles of all circulating and excretory metabolites. Advancement of the radiolabeled human study to phase 1/2 resulted in moving radiolabeled ADME studies in animals to as early as late discovery, when information about ADME properties of a compound is available for investigational new drug (IND) filing. Full characterization of metabolic profiles in animals provides information about metabolic pathways and circulating radioactivity in preclinical species. This information is later compared with human plasma radioprofiles and used to mitigate any MIST-related questions.

Radioprofiles of human urine and feces provide information on the relative amounts of excretory metabolites and such data are used to determine which enzymes are responsible for the metabolism of a given drug, as well as aid in planning for DDI studies accordingly. More details on the conduct of human AME studies were recently described by Penner et al. [14].

Clearly, preclinical ADME programs in general and human AME studies in particular require substantial investment. First, radiolabeled compound with a label in metabolically stable position should be synthesized, and this synthesis should be conducted according to good manufacturing practices (GMPs). Radiolabeled drug products should be tested to ensure that their quality is acceptable for administration to humans. Second, an appropriate clinical site should be identified, which may be challenging since there are few clinical sites that are capable of conducting mass balance studies in humans. There are even fewer sites that can support administration of a radioactive dose intravenously. Nevertheless, a well-designed radiolabeled study in humans conducted early in development can achieve all the necessary drug metabolism objectives and answer majority of regulatory questions with the minimum use of time and resources.

5.3.2.1 Sample Collection A sample collection schedule should be chosen to characterize fully the routes of elimination of the drug and its metabolic profile, and identify circulating and excreted metabolites. A typical outline of the dose administration and sample collection in a typical AME study is presented in Table 5.3. Duration of the in-life phase of a mass balance/biotransformation study is usually 168 h. In most cases, this time is sufficient to collect ~90% of administered radioactivity. A complete plasma concentration–time profile is collected for up to five half-lives of the parent compound to characterize total radioactivity and unchanged drug in plasma or serum as well as major metabolites. Plasma collection for metabolite profiling depends on the pooling approach used and will be discussed later. Blood is collected into tubes containing an anticoagulant such as heparin or K_2EDTA.

TABLE 5.3 Dose Administration and Sample Collection in a Typical *in vivo* AME Study

Dose route	Determined by the program (typically the intended clinical route for example, PO, IV, dermal)
Number of subjects	Human: 4–6 (males only) Animals: 2–4 for each gender
Dose	Human: dose similar to phase 2, radioactivity as low as feasible, usually ~100 μCi for ^{14}C and ~200 μCi for ^{3}H (Intact and bile-duct cannulated animals for biliary excretion assessment) Animal species: dose is selected based on toxicology studies (low or mid dose from single dose or 28-day studies), specific activity of the dose varies depending on specific activity of radiolabeled material, usually 100–150 μCi/kg for rodents, 20–25 μCi/kg for large animals.
Sampling time-points (hours postdose)	Urine: predose, 0–12, 12–24, 24–48, 48–72, 72–96, 96–120, 120–144, 144–168 h, Feces: predose, 0–24, 24–48, 48–72, 72–96, 96–120, 120–144, 144–168 h Blood for PK: up to 5 $t_{1/2}$ Blood for metabolite profiling: time-points and volumes depend on a pooling strategy Bile: predose, 0–4, 4–8, 8–24, 24–48, 48–72, 72–96 h

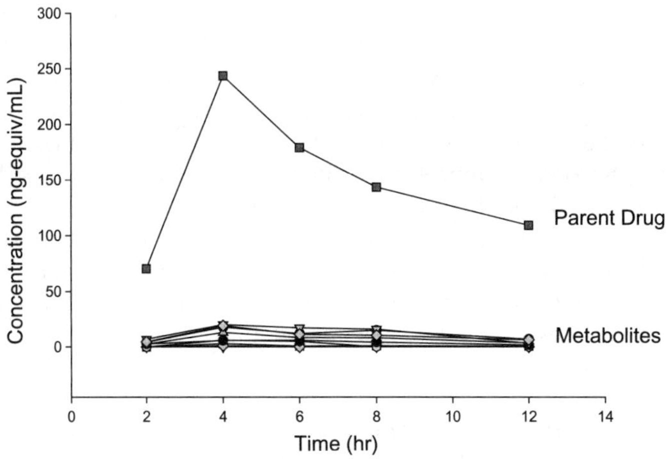

Figure 5.4 Concentration–time plasma profiles for a parent drug and its metabolites obtained by pooling plasma across individual subjects for each time-point (equal volume from each subject).

The comprehensive characterization of circulating human metabolites can be achieved by collecting plasma samples at multiple time-points, pooling plasma across subjects for each time-point (equal volume from each subject), and analyzing them. The amount of parent and metabolites (in ng-equiv/mL) is defined for each time-point, and as a result a pharmacokinetic (PK) profile of the parent and metabolites can be obtained as shown in Figure 5.4. The more time-points are profiled, the more accurate are the data. Six to eight time-points across the profile should be sufficient for a full metabolite characterization. This approach allows for the calculation of PK parameters of the parent drug and metabolites, which helps in simulating their behavior in steady state and eliminates the need for analyzing samples after multiple doses of the drug. As a result, the biggest limitation of human ADME studies—a single dose studies only—would no longer be a problem. Selected time-points can be then used for confirmation of metabolite identity.

It is advisable for ADME studies in animals to be conducted within a short period of time to human AME. It is more convenient to conduct at least one ADME study in animals before human administration to develop profiling and extraction methods and to define requirements for sample stabilization. Plasma sample collection similar to described for humans is also useful for predicting metabolite accumulation. In addition, bile-duct animals (rodent and/or large animal) are used to conduct a biliary excretion study to investigate what percent of the dose is excreted into the bile and what metabolite, if any, is present in the bile.

5.3.2.2 Sample Preparation Metabolite detection in complex mixtures and biological samples is complicated by the presence of a high level of endogenous material. The sensitivity and reliability of detection can be greatly improved by using sample preparation techniques. While dealing with samples for biotransformation studies, there are two main challenges:

- low concentration of drug-derived material in biological samples
- presence of high levels of endogenous components such as proteins, lipids, bile acids, and components of a dosing formulation

The combination of these two factors, which is common for biotransformation work, makes metabolite identification studies challenging and time-consuming. Therefore, three objectives of sample preparation for biotransformation studies are as follows: to bring the level of drug-derived material to a level where analytical instruments can reliably detect it (preconcentration), to remove endogenous interferences, and last but not least, to obtain a sample that is representative of a profile of drug-derived material in a biological sample of interest. The last point means that the stability of a drug and its metabolites should be evaluated and thoroughly considered before starting sample processing. Precautions should be taken when working with unstable analytes.

5.3.2.2.1 Sample Stabilization Techniques Ensuring stability of biological samples collected from radiolabeled or nonradiolabeled studies for biotransformation analysis is a difficult task. Each sample for metabolite profiling is, in fact, a "black box" as there is no information about what is formed before the sample is analyzed. Therefore, it is important to take reasonable precautions when preparing a sample for metabolite identification in order to obtain a representative profile and prevent degradation of unstable metabolites.

There are known metabolic pathways leading to the formation of unstable metabolites. If a parent molecule contains a carboxylic acid group, it is likely that an acylglucuronide conjugate would be formed. Acylglucuronides are unstable at neutral and basic pH, which is common for biological samples, and can degrade with the formation of the parent drug, thus introducing a potentially large error in the evaluation of PK and metabolic profiles. Similarly, acid liable N-glucuronides and N-hydroxyglucuronides require stabilization by adjusting the pH of collected samples.

Instability of a drug and metabolites can also be a desired property, as in the case of prodrugs, which are designed to be unstable in biological conditions. After administration, prodrugs form an active metabolite responsible for biological activity. The rate of conversion of a prodrug to an active component can vary widely between species; therefore, it is important to capture an actual profile to model the pharmacological and toxicological properties of prodrugs intended for humans. Common classes of prodrugs include esters converted by esterases (carboxylesterase [CE], butyrylcholinesterase [BchE], and paraoxonase [PON]) and phosphates converted by phosphatases. Common approaches for the stabilization of unstable metabolites are listed in Table 5.4. In addition, antioxidants such as ascorbic acid, sodium bisulfite, ethylenediaminetetraacetate (EDTA), and mercaptoethanol could be added for the analytes that are chemically oxidized in biological fluids such as GSH adducts.

5.3.2.2.2 Sample Extraction Methods Precipitation of proteins from a biological matrix by adding an organic solvent is the easiest and probably the most widely used procedure for extracting drugs and drug-related material from plasma, bile, and fecal homogenate. An organic solvent, usually acetonitrile, is added to the sample in a ratio of ~3:1; the mixture is well mixed and then centrifuged at

TABLE 5.4 Common Approaches for Sample Stabilization

Metabolite/Type of Pro-drug	Method	Reference
Acylglucuronides	Acidification 4% v/v 2 M acetic acid in water and 1% v/v 0.5 M ascorbic acid 5% phosphoric acid to pH 1–2 Citric acid	[69]
Acid labile *N*-glucuronides or *N*-hydroxyglucuronides	Adjust to pH > 7 and <10	[70]
Esters Enalapril—enalaprilat Valaciclovir—aciclovir	Esterase inhibitors such as 0.1 mM Diisopropylfluorophosphate, and 0.1 mM eserine, sodium fluoride, and phenylmethylsulfony fluoride	[71]
Phosphates Psilocybin—psilocin	1% v/v phosphatase inhibitor such as Calbiochem Phosphatase Inhibitor Cocktail (EMD Millipore, Billerica, MA) added to whole blood before separating plasma	
Amidases/hydrolases	Sodium fluoride and phenylmethylsulfony fluoride	
GSH adducts	Ascorbic acid	[72]

~3500–4000 RPM for 10–15 min. After supernatant is removed, the treatment can be repeated to achieve better recovery. For selected compounds, placing a mixture into a freezer overnight helps to increase extraction recovery. Different solvent mixtures can be used to achieve maximum recovery of drug-derived material. The recommended solvents are acetonitrile and methanol; addition of a small amount of dimethyl sulfoxide (DMSO) can improve recovery.

A disadvantage of protein precipitation is that the resulting sample still contains a high amount of endogenous material, which makes the analysis difficult. Other well-established extraction methodologies, such as solid-phase extraction (SPE) and liquid–liquid extraction (LLE), can be applied to biological samples during metabolite identification studies and usually result in cleaner extracts.

SPE is applicable to liquid samples such as plasma and urine. SPE is the most suitable technique for extracting drug-derived material/radioactivity from large volumes of a sample. A variety of SPE cartridges filled with media of various functionalities are offered by vendors. Reversed-phase or mixed-mode cartridges are the most widely used in metabolite identification studies since the majority of drugs and their metabolites are hydrophobic with weak acidic or basic properties.

An improved sample extraction procedure combining protein precipitation and SPE was used by Prasad and Singh [51]. After precipitation with acetonitrile, samples were centrifuged at 9000 *g* for 10 min, and then a supernatant was separated and placed into a –20°C freezer. An organic layer was decanted from the aqueous frozen layer; the latter was thawed and loaded onto an SPE cartridge. The drug-derived material was eluted with acetonitrile, which was combined with the acetonitrile layer from protein precipitation. An additional step of freeze-liquid separation resulted in a cleaner extract and improved recovery of polar analytes.

After extraction is complete, a supernatant is dried under a stream of nitrogen or in a Speedvac (ThermoFisher Scientific, Waltham, MA), and a residue is then redissolved in solvent(s) compatible with an intended analytical method. Since the majority of separation is performed in the reversed-phase mode, the residue is usually dissolved in a mixture of mobile phases used for separation (various proportions of acetonitrile and a buffer). It is not recommended to use more than 50% of acetonitrile and/or methanol in a final sample injected onto a reversed-phase column, since it can greatly decrease the quality of separation, especially when a large volume of sample is injected. For compounds with low solubility, it is beneficial not to dry extracted samples completely, or reconstitution may result in low recovery. Adding DMSO to a final sample (up to 50%) can help the solubilization of a dried extract. Achieving the highest possible extraction recovery is important for obtaining representative metabolic profiles.

5.3.2.3 Pooling of Biological Samples Pooling is mixing of original samples to obtain a single representative sample. Pooling can be conducted across time-points for one subject, across subjects for one time-point, or both. To obtain a pooled excreta sample, individual samples should be combined in the amounts relative to their total volume/weight collected. Pooling for each subject is decided based on mass balance data. For metabolite profiling of urine, bile, and feces, samples collected at different time intervals can be pooled to account for >90% of the radioactivity excreted in a given matrix. The 90% criterion may be lowered to prevent excessive sample dilution.

For plasma, several pooling approaches can be used depending on the amount of the available sample and on the objective of the study [14]. For quantitative profiling of metabolites in circulation, plasma/serum can be pooled using the time-proportional pooling to obtain a pooled plasma sample reflecting the area under the curve (AUC) of the parent drug and metabolites using the approach of Hamilton and Hop [12, 13] (Table 5.5). This method can be used to minimize the number of profiled plasma samples as only a single sample is obtained for each animal or for each gender.

TABLE 5.5 Example of Volume Calculation for AUC Pooling of Plasma

Time-point, h t_n	Volume to be pooled is proportional to Δt_n	Total sample volume to be pooled, μL V_n
0	0.25	4.5
0.25	0.5	9
0.5	0.75	13.5
1	2.5	45
3	5	90
6	5	90
8	6	108
12	16	288
24	12	216

Calculations are as follows:
$\Delta t_0 = t_1 - t_0$, $\Delta t_1 = t_2 - t_0 \ldots \Delta t_{n-1} = t_n - t_{n-2}$, $\Delta t_n = t_n - t_{n-1}$;
$V_n = K\Delta t_n$, where K is a proportionality constant, in this case $K = 18$.

Comprehensive characterization of circulating human metabolites can be achieved by pooling plasma across subjects for each time-point (equal volume from each subject). The amount of a parent drug and metabolites in ng-equiv/mL is defined for each time-point and as a result a PK profile of the parent and metabolites can be obtained as shown in Figure 5.4. The more time-points are profiled, the more accurate the data are. Six to eight time-points across the profile should be sufficient for a full characterization of the PK of the metabolite. This approach allows for calculation of the PK parameters of the parent drug and metabolites, which helps in simulating their behavior at steady state and eliminates the need for analyzing samples after multiple doses of the drug. The profiles in Figure 5.4 clearly show that no circulating metabolite exceeds 10% of the parent drug exposure or is expected to accumulate upon multiple dosing; therefore, no metabolite-related activities such as safety testing in preclinical species or bioanalytical monitoring were performed.

The time-point pooling method is mostly recommended for human studies to bridge single and multiple dose data. The AUC pooling method is very practical in animal studies where there is no need to know PK parameters of each circulating metabolite or in nonradiolabeled human studies where the objective is to evaluate MIST liability.

5.3.3 Sample Analysis Using HPLC-MS/Radiometry

The studies conducted with a radiolabeled drug usually require long HPLC separation methods to achieve separation of all major metabolites and profile them using radiometric detector. HPLC–MS is coupled with in-line radioactive detector for the simultaneous acquisition of radiochromatographic profile, which is used for quantitative analysis, while HPLC-MS data are used for structural elucidation. A general system setup for biotransformation analysis of radioactive samples is shown in Figure 5.5. The effluent from an HPLC pump after a UV detector (optional, but useful) is split to direct approximately 10–15% of the flow into a mass spectrometer, while the majority of the flow goes into a radiometric detector such as accurate radioisotope counting (ARC) or β-RAM flow scintillation analyzer (FSA), or to a fraction collector. Microplate scintillation counting (MSC) is a viable alternative to online radiometry [52–61]. Fractions of the column effluent are collected into 96- or 384-well plates containing solid scintillant and counted using MSC.

With online radiometry, the radiochromatographic profile is available immediately after an analytical run is completed. Offline methods require longer time for generating a radiochromatogram (due to fraction collection, plate drying, and scintillation counting) but offer several advantages over the online techniques. First, the sensitivity of this method greatly exceeds the sensitivity of classical online radiometry and allows profiling samples with low radioactivity, which is often the case with plasma samples. Second, for low radioactivity samples, if an inadequate profile was obtained from the initial analysis, counting can be repeated with increased counting time to improve peak detection due to increased signal-to-noise (S/N) ratio without need for additional sample [55, 56]. Third, the use of FSA where ~80% of sample is mixed with scintillation cocktail and directed to waste is not optimal for analysis of precious samples. For profiling such samples, a "no sample wasted" approach can be used. The collected fractions are reconstituted post-counting and infused into the

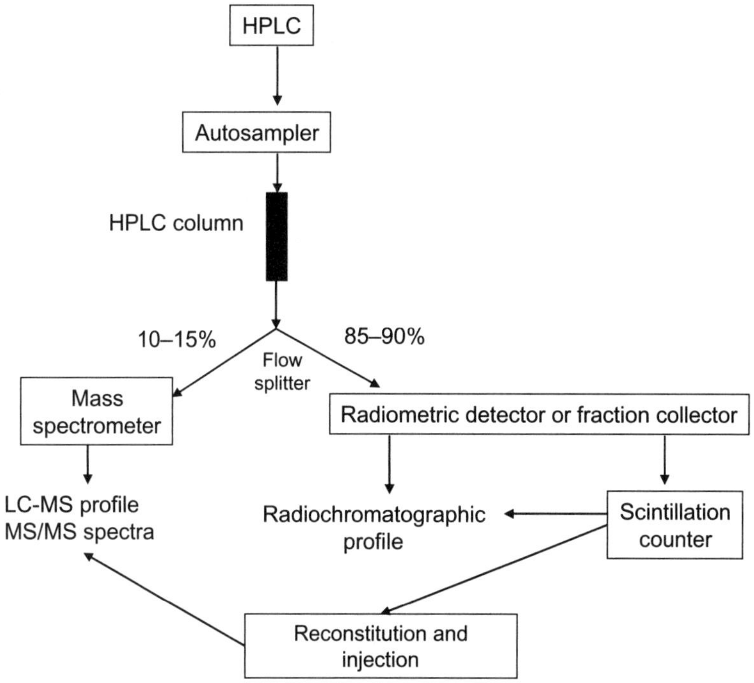

Figure 5.5 A general system setup for biotransformation analysis of radioactive samples.

mass spectrometer via NanoMate to obtain MS and MS/MS spectra necessary for structural elucidation of metabolites.

This approach was successfully utilized for profiling of drug-derived material in urine collected from female rats dosed with ^{14}C-vicriviroc (Fig. 5.6). A representative radiochromatogram of ^{14}C-vicriviroc and its metabolites in 8–24 h rat urine is shown in Figure 5.6A. Separation of all major drug-derived urinary components was achieved in 35 min. Sensitivity of detection is relatively good when ~25,000 disintegrations per minute (DPM) of sample is injected onto the column; however, if the amount of injected sample decreased to 1/100 of the initial injection (Fig. 5.6B), FSA can no longer provide an acceptable profile. This sample with low radioactivity was analyzed on a microplate scintillation counter (TopCount, PerkinElmer, Waltham, MA); fractions of the effluent were collected into 96-well plates containing solid scintillant and then counted. This technique resulted in a well-defined chromatogram. The profile in Figure 5.6C is acceptable for quantitative analysis and similar to the one in Figure 5.6A, which was obtained with the use of FSA after injecting a much higher amount of radioactivity.

Selected fractions were then reconstituted and infused into a Q-TOF mass spectrometer equipped with a NanoMate ion source. Mass spectra were collected for several minutes to acquire MS and MS/MS spectra in accurate mass mode (Fig. 5.6D). The amount of a metabolite at *m/z* 550 contained in a fraction with 257 counts per minute (CPM) of radioactivity is approximately 6 ng. Despite the low amount of injected drug-derived material, this approach allowed spectra sufficient for structural characterization of major metabolites to be obtained. The same was the case for the metabolite at *m/z* 520.

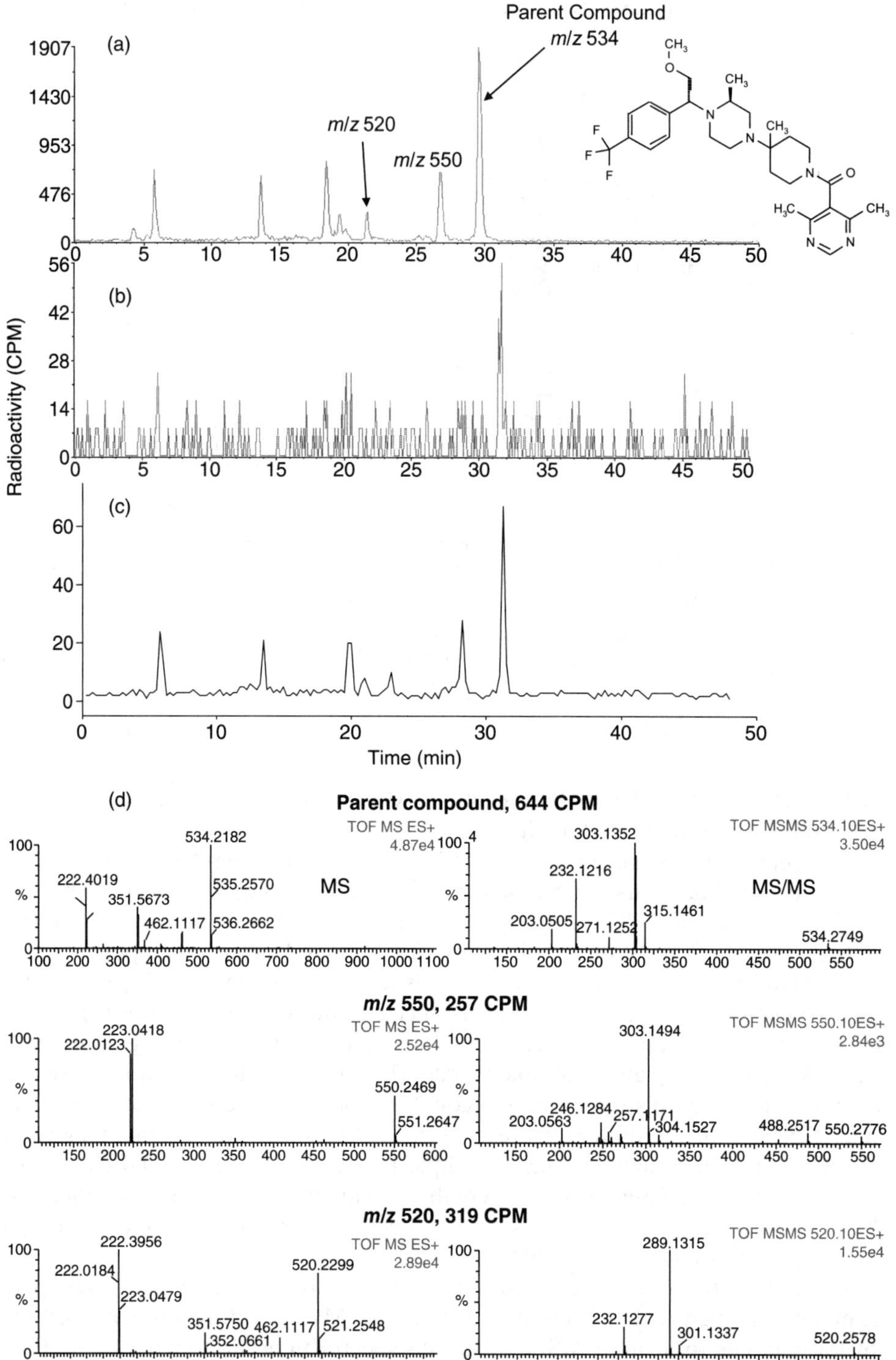

Figure 5.6 Radiochromatograms of ^{14}C-vicriviroc and its metabolites in 8–24 h rat urine. HPLC column: Luna Phenyl-Hexyl 250 × 4.6 mm, 5-μm particle size. (A) Radiochromatogram was acquired using a flow scintillation analyzer. Injection volume of 100 μL. (B) Radiochromatogram was acquired using a flow scintillation analyzer. Injection volume of 1 μL. (C) Radiochromatogram was reconstructed post-microplate scintillation counting. Fraction volume of 250 μL. Injection volume of 1 μL. (D) MS and MS/MS spectra of the parent compound and its two metabolites (*m/z* 550 and *m/z* 520) obtained using a NanoMate ionization source.

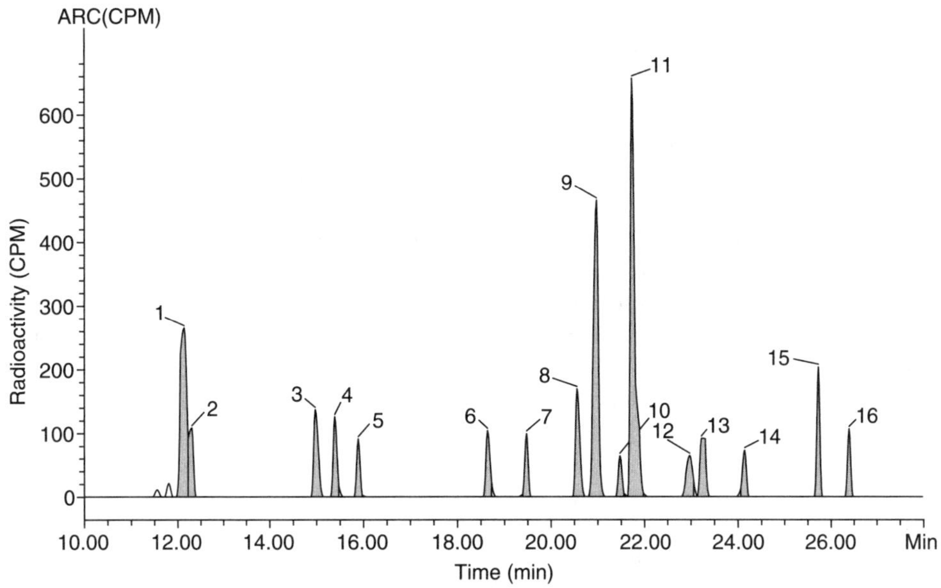

Figure 5.7 Metabolite profile of an *in vivo* sample obtained using an LC-ARC system.

Recent advances in radiometric detectors such as StopFlow LC-ARC (AIM Research Company, Hockessin, DE) with dynamic flow allow using narrow-bore and small particle size columns for profiling radioactive samples. This technology greatly increases resolution of radiochromatographic detection and is capable of detecting and resolving 12- to 15-s wide peaks. Figure 5.7 illustrates the use of LC-ARC for the profiling of an *in vivo* sample. The separation of 16 metabolites was achieved in less than 27 min with complete resolution of all peaks, except for peaks 1 and 2, which eluted 9 s apart. Peaks 3, 4, and 5 were eluted within 1 min and were fully resolved. The use of small particle size columns allows reduction of analysis time and significant savings of mobile phases and scintillation fluid.

5.3.4 Metabolite Detection and Structural Elucidation

The strategies for detection of metabolites and determination of their structures when profiling radiolabeled samples are not different from the strategies applied for profiling nonradiolabeled samples as described in Sections 5.2.2 and 5.2.3. The advantage of using a radiolabeled test article for metabolite profiling is that the retention times of metabolites and their amounts are determined from a radiochromatogram; therefore, there is no need to search for a needle in a haystack as it often happens with nonradiolabeled samples. In addition to automated techniques such as background subtraction, MDF, NL, and fragment-specific analysis, a simple total ion chromatogram search in the area of radiochromatographic peaks can be used for initial analysis or for confirmation of a metabolite presence.

Structural elucidation in radiolabeled studies generally target only major metabolites or metabolites representing more than 1–5% of the administered dose, but the characterization is conducted more thoroughly with the goal of fully defining the structures. Short of using NMR, there are a number of HPLC-MS techniques that can be utilized to assist in definitive structural elucidation, including using alternative ion ionization sources, using chemical derivatization, and hydrogen-deuterium (H/D) exchange. These techniques can become very valuable when isolation of a metabolite of interest is not feasible.

The application of HPLC-MS methods for distinguishing between aliphatic and aromatic C-hydroxylation and between C- and N-oxidation was reported by Ramanathan et al. [62]. HPLC-atmospheric pressure chemical ionization (APCI)-MS was used to distinguish between hydroxylated metabolites and N-oxides of loratadine using IT and QQQ mass spectrometers. HPLC-ESI-MS/MS and HPLC-APCI-MS data showed a predominant loss of water from metabolites with aliphatic hydroxylation, while the loss of water was not favored when hydroxylation was phenolic. N-Oxides showed a minor water loss in the MS/MS spectra but produced distinct $[M + H - O]^+$ ions in the APCI-MS spectra. In a different study [63], HPLC-MS and HPLC-MS/MS methods, combined with H/D exchange, were able to unambiguously differentiate structures of four mono-oxygenated isomeric analogs of desloratadine. Importantly, H/D exchange was required to differentiate between hydroxylamine and N-oxide.

In addition, H/D exchange provided critical structural information for elucidation of structures of two novel metabolites of SCH486757 [32]. In this case, H/D exchange showed no change in *m/z* for both metabolites in the presence of deuterated mobile phase. This finding allowed conclusion that the metabolites contained a quaternary amine group in the structure since they did not have any exchangeable protons and did not require an ionizing proton.

5.3.5 Use of Chemical Derivatization for Structural Elucidation

Chemical derivatization combined with HPLC-MS analysis is an extremely useful approach for structural elucidation of metabolites. Structural elucidation of polar metabolites is often complicated because of low retention time, high mass spectrometric background, and low sensitivity. Chemical derivatization, such as dansylation, is useful for increasing the molecular weight and hydrophobicity of small polar metabolites to obtain better chromatographic separation. An increase in mass and an addition of a dimethylamino group can also improve the S/N ratio of a metabolite in HPLC-MS [3]. Simple chemical reactions can be used to distinguish regioisomers and isobaric compounds, or to confirm the presence of a functional group. A number of reviews are available in the literature [3, 5, 64, 65].

Chemical derivatization, although largely underutilized for metabolite profiling, can be extremely helpful to achieve the following:

- to increase the retention time of hydrophilic metabolites on an HPLC column (methylation, acetylation, dansylation; derivatization with 1,2-dimethylimidazole sulfonyl chloride [DMSIC])
- to increase sensitivity of detection in a mass spectrometer (dansylation; derivatization with DMSIC)
- to improve the stability of unstable metabolites [5]
- to confirm the presence of functional groups

TABLE 5.6 Use of Derivatization Techniques for Distinguishing Sites of Oxidation and Glucuronidation [5, 67]

Functional Group	Acetylation in Nonaqueous Solution in the Presence of Pyridine	Acetylation in Aqueous Solution
Acetylation with acetic anhydride [5, 67]		
Aliphatic alcohols	Yes	No
R-OH		
Ar-OH	Yes	Yes
Aliphatic amines	Yes	Yes
$R\text{-}NH_2$, R_1R_2NH, $R_1R_2R_3N$, $Ar\text{-}NH_2$		
$Ar_2\text{-}NH$	No	No
Methylation with diazomethane [66]		
Aliphatic alcohols	No	
R-OH		
Ar-OH	Yes	
Aliphatic amines	No	
$R\text{-}NH_2$, R_1R_2NH, $R_1R_2R_3N$, $Ar\text{-}NH_2$		
Derivatization with DMSIC [73, 74]		
Aliphatic alcohols	No	
R-OH		
Ar-OH	Yes	
Aliphatic amines	Yes	
$R\text{-}NH_2$, R_1R_2NH, $R_1R_2R_3N$, $Ar\text{-}NH_2$		
R-aliphatic group; Ar—aromatic group		

- to distinguish the position of a modification such as oxidation and glucuronidation [3, 66].

Chemical derivatization was successfully applied to confirm the presence of N-oxide [3, 5] and to distinguish the positions of glucuronidation [5, 67, 68]. The latter is based on selective acetylation of aromatic or aliphatic alcohols, or certain amines, by acetic anhydride in aqueous (protic) or nonaqueous (aprotic) solvents. Nucleophilic groups, such as amines and alcohols/phenols, are readily acetylated in nonaqueous solutions by acetic anhydride in the presence of a base such as pyridine. In aqueous solutions, the more nucleophilic amines are acetylated as well, but not the aliphatic hydroxyls. The groups that are blocked by glucuronides are not accessible to acetylation. Another approach is to use methylation by diazomethane [68]. The same approach can be used for distinguishing sites of hydroxylation. Table 5.6 summarizes the use of derivatization for distinguishing isomeric oxidative metabolites.

5.4 SUMMARY AND FUTURE TRENDS

Metabolite identification has proven to be of great value in drug discovery and development with HPLC-MS as the preferred analytical tool. In the past, it was a labor-intensive and time-consuming activity which highly depended on the availability of skilled scientists. The current advances in HPLC-MS techniques such as the fast acquisition rate, high resolution, accurate mass, and isotope abundance measurement have made the detection and identification of drug metabolites much easier and faster. However, it still remains considerably challenging to obtain quantitative assessment of metabolites without synthetic standards or radiolabeled drugs since metabolism often results in structural modifications and dramatic changes in MS responses. The development of new ionization technologies that minimize the variability in mass spectral responses of metabolites of drugs is highly desirable. Recent studies suggested that the degree of variability in MS responses is much smaller in NSI systems than in conventional HPLC-MS systems. Nanospray methods had shown potential in semi-quantitation of drug metabolites without radioisotopes or authentic reference standards.

Today, with the development of high-resolution mass spectrometers such as the Q-TOF, Orbitrap, or TripleTOF 5600™ instruments that have the ability to operate with relatively high mass resolution and generate accurate mass data (<1 ppm), the data that are produced provide a greater degree of selectivity because of the capability to discriminate between interference ion and drug-derived ions of interest. These instruments can operate at relatively high scanning rates, which makes them ideal for use with UPLC. These instruments have the potential to simultaneously provide both qualitative and quantitative information on multiple analytes, including metabolites. In the future, these instruments are likely to be one of the key instruments for the detection, quantitation, and structural characterization of drug metabolites.

Intelligent data-dependent acquisition processes have been developed to maximize the mass spectral information that can be obtained from a single run. The concomitant acquisition in the full scan mode with additional collection of MS/MS, NL, SRM, and MS^n scans is achievable in today's HRMS systems. In recent years, several post-analysis processing software tools, such as MDF, NoRA, and IPF, as well as application of biotransformation libraries, have been widely used for metabolite identification and integrated into commercial metabolite identification software. Although, the software can facilitate the experimental component of the metabolite identification process, the ultimate interpretation of the spectra remains the rate-limiting step; therefore, tools to automate data interpretation are urgently needed.

There are numerous instances when MS data alone may not be sufficient to determine the metabolite structures. In these cases, multiple analytical and wet chemistry techniques such as HPLC-NMR, chemical derivatization, and H/D exchange combined with MS are required to characterize the structures of metabolites. The flexibility and broad application of MS have allowed its hybridization with itself and other powerful analytical techniques, most notably NMR and radioactivity detectors. Improvements in sample preparation and/or sample introduction technologies is predicted to facilitate higher throughput and more favorable S/N levels concomitantly, enhancing the identification of drug-related MS data. Overall, it results in faster turnaround of impactful metabolite information for the compound of interest. In addition, reliable *in silico* tools capable of predicting metabolite formation are a valuable option. Combining them with the HPLC-MS data acquisition software should provide an even more powerful combination.

REFERENCES

[1] Shaffer CL, Gunduz M, Thornburgh BA, Fate GD. Using a tritiated compound to elucidate its preclinical metabolic and excretory pathways *in vivo*: exploring tritium exchange risk. Drug Metab Dispos 2006;34:1615–23.

[2] Prakash C, Kamel A, Miao Z. Radiochemical tracers: essential tools for the drug metabolism studies. Synthesis and Applications of Isotopically Labelled Compounds, Proceedings of the International Symposium, 8th, 2004, 115–20.

[3] Prakash C, Shaffer CL, Nedderman A. Analytical strategies for identifying drug metabolites. Mass Spectrom Rev 2007;26:340–69.

[4] Gelhaus SL, Blair IA. LC-MS analysis in drug metabolism studies. In Pearson P, Wienkers L, editors. Handbook of Drug Metabolism, 2nd Edition. New York: Informa Healthcare USA, Inc., 2009, pp. 355–72.

[5] Liu DQ, Hop CECA. Strategies for characterization of drug metabolites using liquid chromatography-tandem mass spectrometry in conjunction with chemical derivatization and on-line H/D exchange approaches. J Pharm Biomed Anal 2005;37:1–18.

[6] Hop CECA, Prakash C. Metabolite identification by LC-MS: applications in drug discovery and development. Prog Pharm Biomed Anal 2005;6:123–58.

[7] Ma S, Chowdhury SK, Alton KB. Application of mass spectrometry for metabolite identification. Curr Drug Metab 2006;7:503–23.

[8] Ramanathan R, Josephs JL, Jemal M, Arnold M, Humphreys WG. Novel MS solutions inspired by MIST. Bioanalysis 2010;2:1291–313.

[9] Ramanathan R, Comezoglu SN, Humphreys WG. Metabolite identification strategies and procedures. In Korfmacher WA, editor. Using Mass Spectrometry for Drug Metabolism Studies, 2nd Edition. Boca Raton, FL: CRC Press, 2010, pp. 127–204.

[10] U.S. Department of Health and Human Services FDA, Center for Drug Evaluation and Research, Pharmacology and Toxicology. Safety Testing of Drug Metabolites, Guidance for Industry, 2008.

[11] U.S. Department of Health and Human Services FDA, Center for Drug Evaluation and Research, Pharmacology and Toxicology. M3(R2): guidance on non-clinical safety studies for the conduct of human clinical trials and marketing authorization for pharmaceuticals. Fed Regist 2010;75:3471–2.

[12] Hop CECA, Wang Z, Chen Q, Kwei G. Plasma-pooling methods to increase throughput for *in vivo* pharmacokinetic screening. J Pharm Sci 1998;87:901–3.

[13] Hamilton RA, Garnett WR, Kline BJ. Determination of mean valproic acid serum level by assay of a single pooled sample. Clin Pharmacol Ther 1981;29:408–13.

[14] Penner N, Klunk LJ, Prakash C. Human radiolabeled mass balance studies: objectives, utilities and limitations. Biopharm Drug Dispos 2009;30:185–203.

[15] Penner NA, Zgoda-Pols J, Prakash C. Early assessment of exposure of drug metabolites in humans using mass spectrom in *Handbook of Metabolic Pathways for Xenobiotics*, John Wiley & Sons, 2013, in press.

[16] Ho GD, Anthes J, Bercovici A, Caldwell JP, Cheng K-C, Cui X, et al. The discovery of tropane derivatives as nociceptin receptor ligands for the management of cough and anxiety. Bioorg Med Chem Lett 2009;19:2519–23.

[17] Clarke NJ, Rindgen D, Korfmacher WA, Cox KA. Systematic LC/MS metabolite identification in drug discovery. Anal Chem 2001;73:430A–9A.

[18] Liu DQ, Hop CE. Strategies for characterization of drug metabolites using liquid chromatography-tandem mass spectrometry in conjunction with chemical derivatization and on-line H/D exchange approaches. J Pharm Biomed Anal 2005;37:1–18.

[19] Xia YQ, Miller JD, Bakhtiar R, Franklin RB, Liu DQ. Use of a quadrupole linear ion trap mass spectrometer in metabolite identification and bioanalysis. Rapid Commun Mass Spectrom 2003;17:1137–45.

[20] Jemal M, Ouyang Z, Zhao W, Zhu M, Wu WW. A strategy for metabolite identification using triple-quadrupole mass spectrometry with enhanced resolution and accurate mass capability. Rapid Commun Mass Spectrom 2003;17:2732–40.

[21] Sanders M, Shipkova PA, Zhang H, Warrack BM. Utility of the hybrid LTQ-FTMS for drug metabolism applications. Curr Drug Metab 2006;7:547–55.

[22] Zhu M, Ma L, Zhang D, Ray K, Zhao W, Humphreys WG, et al. Detection and characterization of metabolites in biological matrices using mass defect filtering of liquid chromatography/high resolution mass spectrometry data. Drug Metab Dispos 2006; 34:1722–33.

[23] Hager JW, Le Blanc JC. High-performance liquid chromatography-tandem mass spectrometry with a new quadrupole/linear ion trap instrument. J Chromatogr A 2003;1020: 3–9.

[24] King R, Fernandez-Metzler C. The use of Qtrap technology in drug metabolism. Curr Drug Metab 2006;7:541–5.

[25] Ruan Q, Peterman S, Szewc MA, Ma L, Cui D, Humphreys WG, et al. An integrated method for metabolite detection and identification using a linear ion trap/Orbitrap mass spectrometer and multiple data processing techniques: application to indinavir metabolite detection. J Mass Spectrom 2008;43:251–61.

[26] Yao M, Ma L, Duchoslav E, Zhu M. Rapid screening and characterization of drug metabolites using multiple ion monitoring dependent product ion scan and postacquisition data mining on a hybrid triple quadrupole-linear ion trap mass spectrometer. Rapid Commun Mass Spectrom 2009;23:1683–93.

[27] Rousu T, Herttuainen J, Tolonen A. Comparison of triple quadrupole, hybrid linear ion trap triple quadrupole, time-of-flight and LTQ-Orbitrap mass spectrometers in drug discovery phase metabolite screening and identification *in vitro*—amitriptyline and verapamil as model compounds. Rapid Commun Mass Spectrom 2009;24:939–57.

[28] Lim HK, Chen J, Sensenhauser C, Cook K, Subrahmanyam V. Metabolite identification by data-dependent accurate mass spectrometric analysis at resolving power of 60,000 in external calibration mode using an LTQ/Orbitrap. Rapid Commun Mass Spectrom 2007;21:1821–32.

[29] Zhu M, Zhang H, Humphreys WG. Drug metabolite profiling and identification by high-resolution mass spectrometry. J Biol Chem 2011;286:25419–25.

[30] Sundstrom I, Hedeland M, Bondesson U, Andren PE. Identification of glucuronide conjugates of ketobemidone and its phase I metabolites in human urine utilizing accurate mass and tandem time-of-flight mass spectrometry. J Mass Spectrom 2002;37: 414–20.

[31] Corcoran O, Nicholson JK, Lenz EM, Abou-Shakra F, Castro-Perez J, Sage AB, et al. Directly coupled liquid chromatography with inductively coupled plasma mass spectrometry and orthogonal acceleration time-of-flight mass spectrometry for the identification of drug metabolites in urine: application to diclofenac using chlorine and sulfur detection. Rapid Commun Mass Spectrom 2000;14:2377–84.

[32] Penner NA, Ho G, Bercovici A, Chowdhury SK, Alton KB. Identification of two novel metabolites of SCH 486757, a nociceptin/orphanin FQ peptide receptor agonist, in humans. Drug Metab Dispos 2010;38:2067–74.

[33] Bateman KP, Castro-Perez J, Wrona M, Shockcor JP, Yu K, Oballa R, et al. MSE with mass defect filtering for *in vitro* and *in vivo* metabolite identification. Rapid Commun Mass Spectrom 2007;21:1485–96.

[34] Bushee JL, Argikar UA. An experimental approach to enhance precursor ion fragmentation for metabolite identification studies: application of dual collision cells in an orbital trap. Rapid Commun Mass Spectrom 2011;25:1356–62.

[35] Calbiani F, Careri M, Elviri L, Mangia A, Zagnoni I. Matrix effects on accurate mass measurements of low-molecular weight compounds using liquid chromatography-electrospray-quadrupole time-of-flight mass spectrometry. J Mass Spectrom 2006;41:289–94.

[36] Leblanc A, Shiao TC, Roy R, Sleno L. Improved detection of reactive metabolites with a bromine-containing glutathione analog using mass defect and isotope pattern matching. Rapid Commun Mass Spectrom 2010;24:1241–50.

[37] Maurer HH. Perspectives of liquid chromatography coupled to low- and high-resolution mass spectrometry for screening, identification, and quantification of drugs in clinical and forensic toxicology. Ther Drug Monit 2010;32:324–7.

[38] Ramanathan R, Jemal M, Ramagiri S, Xia Y-Q, Humpreys WG, Olah T, et al. It is time for a paradigm shift in drug discovery bioanalysis: from SRM to HRMS. J Mass Spectrom 2011;46:595–601.

[39] Rojas-Cherto M, Kasper PT, Willighagen EL, Vreeken RJ, Hankemeier T, Reijmers TH. Elemental composition determination based on MS^n. Bioinformatics 2011;27:2376–83.

[40] Zhang H, Zhu M, Ray KL, Ma L, Zhang D. Mass defect profiles of biological matrices and the general applicability of mass defect filtering for metabolite detection. Rapid Commun Mass Spectrom 2008;22:2082–8.

[41] Zhang H, Zhang D, Ray K, Zhu M. Mass defect filter technique and its applications to drug metabolite identification by high-resolution mass spectrometry. J Mass Spectrom 2009;44:999–1016.

[42] Tiller PR, Yu S, Castro-Perez J, Fillgrove KL, Baillie TA. High-throughput, accurate mass liquid chromatography/tandem mass spectrometry on a quadrupole time-of-flight system as a "first-line" approach for metabolite identification studies. Rapid Commun Mass Spectrom 2008;22:1053–61.

[43] Tiller PR, Yu S, Bateman KP, Castro-Perez J, McIntosh IS, Kuo Y, et al. Fractional mass filtering as a means to assess circulating metabolites in early human clinical studies. Rapid Commun Mass Spectrom 2008;22:3510–6.

[44] Cuyckens F, Hurkmans R, Castro-Perez JM, Leclercq L, Mortishire-Smith RJ. Extracting metabolite ions out of a matrix background by combined mass defect, neutral loss and isotope filtration. Rapid Commun Mass Spectrom 2009;23:327–32.

[45] Sleno L. The use of mass defect in modern mass spectrometry. J Mass Spectrom 2012;47: 226–36.

[46] Hop CECA, Chen Y, Yu LJ. Uniformity of ionization response of structurally diverse analytes using a chip-based nanoelectrospray ionization source. Rapid Commun Mass Spectrom 2005;19:3139–42.

[47] Ramanathan R, Zhong R, Blumenkrantz N, Chowdhury SK, Alton KB. Response normalized liquid chromatography nanospray ionization mass spectrometry. J Am Soc Mass Spectrom 2007;18:1891–9.

[48] Gao H, Deng S, Obach RS. A simple liquid chromatography-tandem mass spectrometry method to determine relative plasma exposures of drug metabolites across species for metabolite safety assessments. Drug Metab Dispos 2010;38:2147–56.

[49] Emoto C, Murayama N, Rostami-Hodjegan A, Yamazaki H. Methodologies for investigating drug metabolism at the early drug discovery stage: prediction of hepatic drug clearance and P450 contribution. Curr Drug Metab 2010;11:678–85.

[50] Khojasteh SC, Prabhu S, Kenny JR, Halladay JS, Lu AYH. Chemical inhibitors of cytochrome P450 isoforms in human liver microsomes: a re-evaluation of P450 isoform selectivity. Eur J Drug Metab Pharmacokinet 2011;36:1–16.

[51] Prasad B, Singh S. *In vitro* and *in vivo* investigation of metabolic fate of rifampicin using an optimized sample preparation approach and modern tools of liquid chromatography-mass spectrometry. J Pharm Biomed Anal 2009;50:475–90.

[52] Zhao W, Wang L, Zhang D, Zhu M. Rapid and sensitive characterization of the metabolite formation enzyme kinetics of radiolabeled drugs using stop-flow liquid radiochromatography. Drug Metab Lett 2008;2:41–6.

[53] Kiffe M, Schmid DG, Bruin GJM. Radioactivity detectors for high-performance liquid chromatography in drug metabolism studies. J Liq Chromatogr Relat Technol 2008;31: 1593–619.

[54] Kiffe M, Nufer R, Trunzer M, Graf D. Cytostar-T plates—a valid alternative for microplate scintillation counting of low radioactivity in combination with high-performance liquid chromatography in drug metabolism studies? J Chromatogr A 2007;1157:65–72.

[55] Dear GJ, Patel N, Kelly PJ, Webber L, Yung M. TopCount coupled to ultra-performance liquid chromatography for the profiling of radiolabeled drug metabolites in complex biological samples. J Chromatogr B Analyt Technol Biomed Life Sci 2006;844:96–103.

[56] Zhu M, Zhao W, Vazquez N, Mitroka JG. Analysis of low level radioactive metabolites in biological fluids using high-performance liquid chromatography with microplate scintillation counting: method validation and application. J Pharm Biomed Anal 2005;39: 233–45.

[57] Zhu M, Zhang D, Skiles GL. Quantification and structural elucidation of low quantities of radiolabeled metabolites using microplate scintillation counting techniques in conjunction with LC-MS. Prog Pharm Biomed Anal 2005;6:195–223.

[58] Nassar AEF, Parmentier Y, Martinet M, Lee DY. Liquid chromatography-accurate radioisotope counting and microplate scintillation counter technologies in drug metabolism studies. J Chromatogr Sci 2004;42:348–53.

[59] Kiffe M, Jehle A, Ruembeli R. Combination of high-performance liquid chromatography and microplate scintillation counting for crop and animal metabolism studies: a comparison with classical on-line and thin-layer chromatography radioactivity detection. Anal Chem 2003;75:723–30.

[60] Boernsen KO, Floeckher JM, Bruin GJM. Use of a microplate scintillation counter as a radioactivity detector for miniaturized separation techniques in drug metabolism. Anal Chem 2000;72:3956–9.

[61] Zhu M. Drug ADME analysis, including tissue metabolite profiling, in a few rats using a combination of microplate scintillation counting, capillary LC/MS, and whole-body autoradiography. Abstracts, 37th Middle Atlantic Regional Meeting of the American Chemical Society. New Brunswick, NJ, May 22–25, 2005. 2005:GENE-530.

[62] Ramanathan R, Su AD, Alvarez N, Blumenkrantz N, Chowdhury SK, Alton K, et al. Liquid chromatography/mass spectrometry methods for distinguishing N-oxides from hydroxylated compounds. Anal Chem 2000;72:1352–9.

[63] Penner NA, Zgoda-Pols J, Ramanathan R, Chowdhury SK, Alton KB. LC-MS methods with hydrogen/deuterium exchange for identification of hydroxylamine, N-oxide, and hydroxylated analogs of desloratadine. Mass Spectrom Drug Metab Pharmacokinet 2009:295–310.

[64] Quirke JME, Adams CL, Van Berkel GJ. Chemical derivatization for electrospray ionization mass spectrometry.1. Alkyl halides, alcohols, phenols, thiols, and amines. Anal Chem 1994;66:1302–15.

[65] Lunn G, Hellwig L, Cecchini A. Handbook of Derivatization Reactions of HPLC, 1998.

[66] Johnson K, Shah A, Jaw-Tsai S, Baxter J, Prakash C. Metabolism, pharmacokinetics, and excretion of a highly selective N-methyl-D-aspartate receptor antagonist, traxoprodil,

in human cytochrome P450 2D6 extensive and poor metabolizers. Drug Metab Dispos 2003;31:76–87.

[67] Schaefer WH, Goalwin A, Dixon F, Hwang B, Killmer L, Kuo G. Structural determination of glucuronide conjugates and a carbamoyl glucuronide conjugate of carvedilol: use of acetylation reactions as an aid to determine positions of glucuronidation. Biol Mass Spectrom 1992;21:179–88.

[68] Kondo T, Yoshida K, Yoshimura Y, Motohashi M, Tanayama S. Characterization of conjugated metabolites of a new angiotensin II receptor antagonist, candesartan cilexetil, in rats by liquid chromatograph/electrospray tandem mass spectrometry following chemical derivatization. J Mass Spectrom 1996;31:873–8.

[69] Sparidans RW, Lagas JS, Schinkel AH, Schellens JHM, Beijnen JH. Liquid chromatography-tandem mass spectrometric assay for diclofenac and three primary metabolites in mouse plasma. J Chromatogr B Analyt Tech Biomed Life Sci 2008;872:77–82.

[70] McIntyre IM, Norman TR, Burrows GD. *In vitro* stability of nomifensine in plasma. Clin Chem 1981;27:203–4.

[71] Berry LM, Wollenberg L, Zhao Z. Esterase activities in the blood, liver and intestine of several preclinical species and humans. Drug Metab Lett 2009;3:70–7.

[72] Kassahun K, Davis M, Hu P, Martin B, Baillie T. Biotransformation of the naturally occurring isothiocyanate sulforaphane in the rat: identification of phase I metabolites and glutathione conjugates. Chem Res Toxicol 1997;10:1228–33.

[73] Salomonsson ML, Bondesson U, Hedeland M. Structural evaluation of the glucuronides of morphine and formoterol using chemical derivatization with 1,2-dimethylimidazole-4-sulfonyl chloride and liquid chromatography/ion trap mass spectrometry. Rapid Commun Mass Spectrom 2008;22:2685–97.

[74] Salomonsson ML, Bondesson U, Hedeland M. *In vitro* formation of phase I and II metabolites of propranolol and determination of their structures using chemical derivatization and liquid chromatography-tandem mass spectrometry. J Mass Spectrom 2009;44:742–54.

6

MS ANALYSIS OF BIOLOGICAL DRUGS, PROTEINS, AND PEPTIDES

Yi Du, John Mehl, and Pavlo Pristatsky

6.1 INTRODUCTION

Biological drugs have largely been defined as those substances made from a living organism or its products and are used in the prevention, diagnosis, or treatment of diseases [1]. In this chapter, we will focus on therapeutic monoclonal antibodies (mAbs) and fusion proteins, produced using recombinant DNA technologies. These therapeutic proteins have emerged as one of the most promising drug classes in the biopharmaceutical industry. The past 20 years have witnessed significant development of therapeutic proteins and their addition to the pharmaceutical arsenal to fight various diseases and improve human health [2–5]. Basic research in protein drug discovery requires powerful analytical tools to enable better understanding of these molecules. Protein drug development also requires high quality control standards to be met in order to monitor the complex manufacturing process [6]. With rapid and significant developments in both instrumentation and software, mass spectrometry (MS) has become indispensable for the investigation of the molecular structure of antibodies and large proteins. Several recent reviews have cited many examples of how MS has been applied to facilitate therapeutic protein discovery and development [7–9]. Over the last two decades, MS has evolved as an essential analytical approach that is used widely to support both the selection of appropriate clinical candidate molecules and process development and optimization. This evolution in turn has motivated the continued development of MS techniques.

The application of MS in protein drug discovery and development is shown schematically in the flow chart in Figure 6.1. During the discovery stage, the goal is to ensure sufficient affinity and specificity of the mAb to its target, and to ensure

Mass Spectrometry for Drug Discovery and Drug Development, First Edition. Edited by Walter A. Korfmacher.

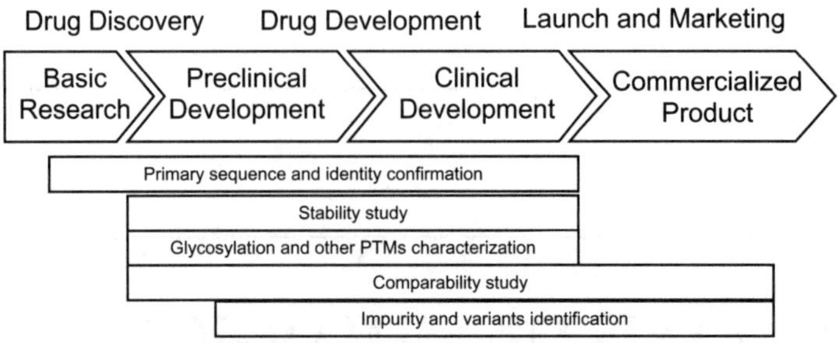

Figure 6.1 Application of mass spectrometry in therapeutic protein discovery and development process.

reliability of animal models for predicting safety and efficacy in humans. Assays used at this stage are focused on determining yields from various cell lines or strains, examining post-translational modifications (PTMs), identifying lead candidates with desirable properties, and screening out molecules that raise red flags regarding potential immunogenicity and half-life issues (e.g., readily oxidized or deamidated molecules). Although immunoassay is used heavily at this stage, MS is often used to confirm the primary sequence, identify protein variants, and provide early structural characterization support [10]. During the discovery stage, these data are used to screen drug candidates and provide valuable guidance on cell line and strain selection, and upstream operating conditions.

Once a protein drug candidate enters the preclinical stage, and beyond, comparability of pivotal lots produced using scaled-up and modified processes needs to be demonstrated. This is especially important when there are differences in the way lots are made. Variation in lots can occur because manufacturing processes for the production of therapeutic proteins generally involve a complex series of steps, each of which can have a dramatic impact on downstream protein quality and activity. The goal of a comparability study is to ensure consistency in the quality, safety, and efficacy of protein drug products. MS is widely used at this stage to evaluate the protein quality at each step in the process that can lead to increased levels of sequence variants, PTMs, or degradation products. The data obtained are also used to support formulation development, define product storage and handling instructions, and determine product specifications [7]. MS analysis of purified protein products can be relatively straightforward; however, the analysis of modified forms or sequence variants, which are typically expressed at very low levels, can become a significant challenge for the sensitivity of MS instrumentation.

In this chapter, we will introduce various MS applications in biological drug discovery and development. Intact mass analysis and peptide mapping are two routine MS assays used to confirm the protein primary sequence. In addition, high-performance liquid chromatography-mass spectrometry (HPLC-MS) and HPLC-tandem mass spectrometry (MS/MS) approaches are commonly used to monitor chemical modifications such as Met oxidation, Asn deamidation, and Asp isomerization. A number of recent examples will be highlighted to illustrate the detailed protein structural characterization for N-glycosylation and disulfide mapping as well as PEGylation analysis. Finally, several new techniques will be discussed as promising future trends that are anticipated to improve the understanding of higher order structure of protein molecules, increase the sequence coverage for peptide mapping, and add tremendous value to the biopharmaceutical industry.

6.2 PRIMARY SEQUENCE CHARACTERIZATION

6.2.1 Intact Mass Analysis

Intact mass analysis using HPLC-MS is one of the most useful analytical techniques for the characterization of therapeutic proteins, and in particular mAbs. Several recent reports have demonstrated the utility of intact mass analysis for studying the heterogeneity of mAbs and other protein drugs [11–18]. Intact mass analysis can be performed in a rapid manner and sample preparation is straightforward, requiring a minimal amount of time. The main goal of intact mass analysis of a protein therapeutic is to confirm that the molecular weight of the expressed and purified protein drug is consistent with the theoretical protein mass. At the same time, modern mass spectrometers, with improved resolution and sensitivity, enable the heterogeneity of the protein to be characterized. In the case of mAbs, the therapeutic protein is a heterogeneous mixture that can be composed of various glycosylation forms and chemical modifications including C-terminal lysine removal, N-terminal pyroglutamic acid formation (cyclization), internal cleavage (clipping), oxidation, and so on. Analysis of the protein heterogeneity by HPLC-MS is commonly applied to evaluate bioprocess development, storage conditions, batch consistency, and so on.

Several types of mass spectrometers have been applied for intact mass analysis of protein therapeutics including matrix-assisted laser desorption ionization-time-of-flight (MALDI-TOF) MS [19, 20], electrospray ionization (ESI)-quadrupole (Q) MS [21], ESI-ion trap MS [22], and ESI-Orbitrap™ MS [23]; however, the instrument of choice has become the ESI-TOF MS (or ESI-QTOF MS) for large proteins, especially mAbs. Zhang et al. [8] provided a thorough explanation of the ideal attributes of ESI-TOF for intact mass analysis of mAbs. These attributes include high resolution and mass accuracy, wide mass range, and high sensitivity.

Coupling of ESI-TOF with HPLC separation results in a very powerful analytical method. Reversed-phase (RP) has found the most utility due to the mobile phase compatibility with ESI, in contrast to size exclusion chromatography (SEC), which has had less applicability [18]. It should be kept in mind that good RP-HPLC separation of large proteins, including mAbs, can require extensive method optimization. Dillon et al. [13] demonstrated the optimization of an RP-HPLC method for the analysis of intact mAbs. RP-HPLC method development led to the choice of using propanol as the organic mobile phase instead of the commonly used acetonitrile, along with high column temperatures (75–80°C) and stable bonded C8 carbon chain stationary phase, which yielded high recovery and good peak shape with minimal tailing. When analyzing smaller proteins, or fragments of mAbs, for example, light and heavy chains, acetonitrile mobile phase is generally sufficient to obtain good chromatography [17]. Polymeric monolithic stationary phase columns are also proving to be successful for RP-HPLC protein separation [24].

Intact mass analysis of mAbs using RP-HPLC-MS can be applied to whole mAb molecules or to fragments of mAbs. Commonly analyzed mAb fragments include

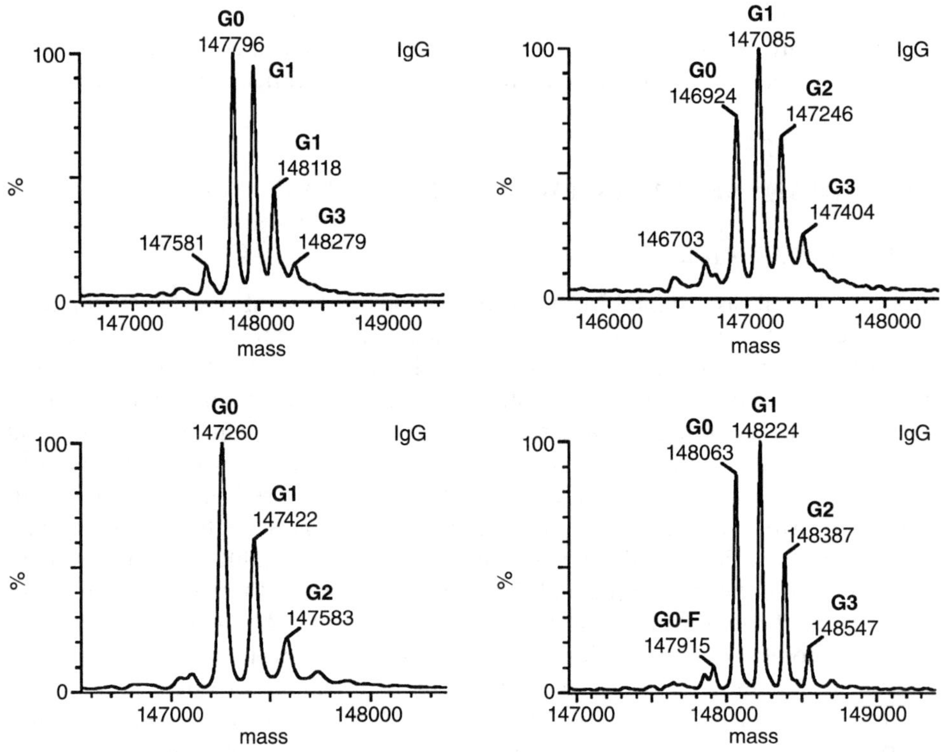

Figure 6.2 Deconvoluted ESI-TOF mass spectra of four different whole molecule mAbs analyzed by HPLC-MS. The intact mass spectra contain multiple species that correlate to the total heterogeneity of galactose (G0, G1, G2, G3) and fucose (F) residues of the carbohydrate moieties. Adapted with permission from Dillon, T.M. et al. *J. Chromatogr. A.* 2006, *1120*(1–2), 112–120.

the heavy chain (HC), light chain (LC), fragment-antibody binding (Fab), and fragment-crystallizable (Fc). HC and LC can be readily released from the mAb by chemical reduction of disulfide bonds. Fab and Fc are released using papain or limited LysC digestion. N-linked glycans can be removed from the mAb by using PNGase F enzyme. By cleaving the mAb into smaller fragments, more information about the protein structure is obtained during the analysis. Figure 6.2 shows the deconvoluted ESI-TOF spectra of four different whole immunoglobulin G (IgG) molecules analyzed by RP-HPLC-MS. The instrument performance of the ESI-TOF enables complete resolution of the various glycoforms attached to the IgG molecules. The mass difference between adjacent peaks ranges from 158 to 163 Da and is due to different amounts of galactose and mannose contained within the glycan chains and is consistent with expression from mammalian cells. The presence of fucose can also be detected in one of the IgG molecules, as shown in Figure 6.2, lower right spectrum. This example illustrates the resolving power and sensitivity of ESI-TOF and the ability to obtain information about the heterogeneity of recombinantly expressed protein drugs [10].

When optimized RP-HPLC is combined with ESI-TOF, it is possible to detect other types of mAb heterogeneity. Dillon et al. [12] chromatographically separated whole mAb molecules differing only by the presence or absence of lysine from the HC C-termini. Cleavage of C-terminal lysine is a common PTM caused by the presence of carboxypeptidase enzyme. The authors observed three chromatographic peaks due to the cleavage of one, two, or zero lysine residues from the C-termini of the HCs. Online ESI-TOF detection showed that the chromatographic peaks differed in mass by 128 Da, consistent with the mass of a lysine residue.

Intact mass analysis of whole molecule IgG and reduced LC and HC subunits can provide valuable insight into the structure of mAb therapeutic proteins. As an example, Dillon et al. [13] performed intact mass analysis on a whole molecule IgG2 by RP-HPLC-MS (using an ESI-TOF) and observed four partially separated species in the chromatogram. Each chromatographic peak contained the same mass and glycoform pattern. The mAb sample was then chemically reduced into HC and LC subunits and reanalyzed by RP-HPLC-MS. The resulting chromatogram contained only two peaks, one for the LC and one for the HC. The observation that the heterogeneity is removed following reduction indicates that the disulfide linkages are the source of the detected heterogeneity. Further characterization of the disulfide heterogeneity could be obtained by analyzing the Fab and Fc subunits.

Another example of the utility of analyzing LC and HC subunits was reported by Yan et al. [17]. The authors applied RP-HPLC-MS to analyze a reduced IgG2 and found that the LC contained the expected mass along with a minor amount of LC with N-terminal pyroglutamate. Cyclization of N-terminal glutamine to pyroglutamic acid is a commonly observed PTM. Other reports have shown that N-terminal glutamic acid can also convert to pyroglutamic acid [14]. Clipping of the LC was also observed. The mass accuracy was sufficient to enable the site of clipping to be determined. Additionally, a mass species of LC was detected that is approximately 1 Da higher in mass than expected. This is evidence of deamidation of the LC. However, it should be noted that a 1-Da mass shift is within the experimental mass accuracy error of the TOF instrument and should only be used to suggest that deamidation has occurred. Further experiments would be required to confirm the presence of deamidation.

Figure 6.3 further illustrates the type of information that can be obtained from the mass analysis of reduced mAbs. The reduced mAb sample was separated into several peaks using RP-HPLC and detected online using ESI-TOF. Figure 6.3 shows deconvoluted spectra of the various HC species separated by RP-HPLC. Figure 6.3A is a spectrum of the expected HC mass containing a glycoform pattern consistent with expression from CHO cells. Figure 6.3B contains the same glycoform pattern as observed in Figure 6.3A, but with a 1- to 2-Da mass shift, suggesting the presence of deamidation. Figure 6.3C,D shows spectra that are due to fragments resulting from internal clipping of the HC. The mass in Figure 6.3C is due to residues 1–275 and does not contain N-linked glycosylation. The masses in Figure 6.3D are due to residues 276–451 and are glycosylated. Figure 6.3E contains peak doublets that are separated by 16 Da possibly due to oxidation. This example illustrates how intact mass analysis provides a tremendous amount of structural information that can be obtained from a single analytical run.

Confirmation of oxidation observed from HPLC-MS analysis of intact mAb molecules or mAb fragments requires additional experimentation to determine the

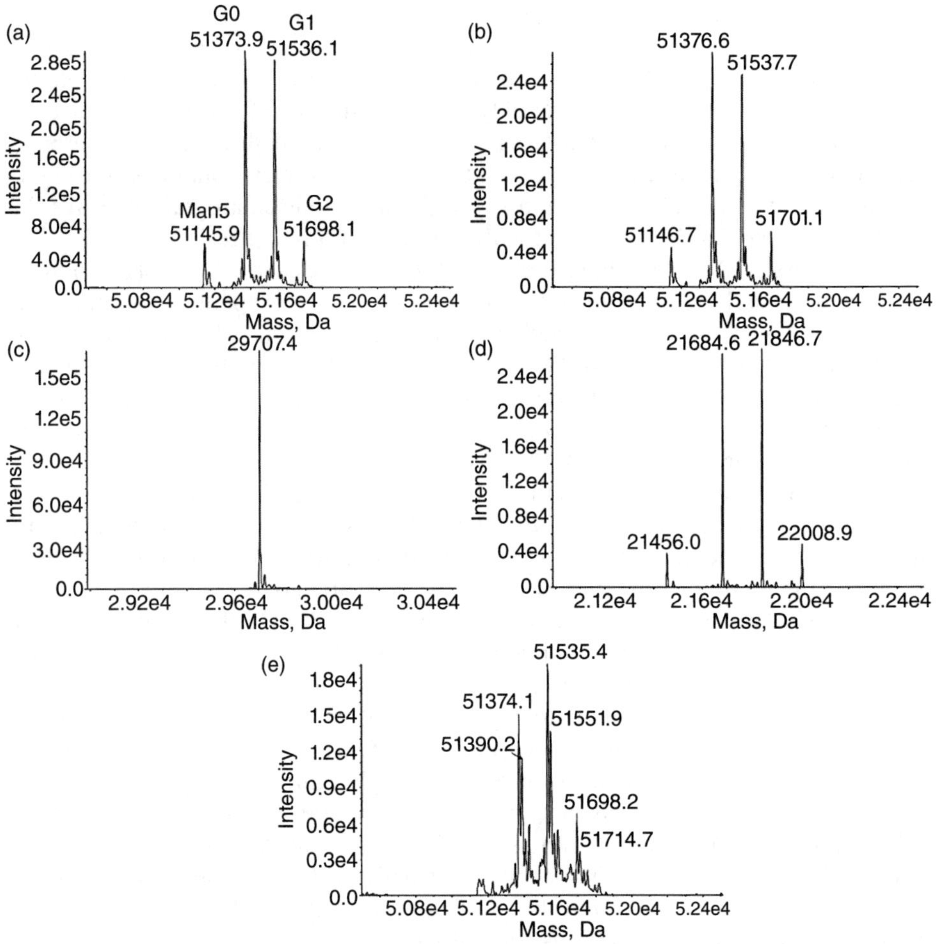

Figure 6.3 Deconvoluted ESI-TOF mass spectra of IgG heavy chain and its variants: (A) heavy chain, residues 1–451; (B) deamidated heavy chain; (C) clipped heavy chain, residues 1–275; (D) clipped heavy chain, residues 276–451; (E) oxidized heavy chain. Adapted with permission from Yan, B. et al. *J. Chromatogr. A.* **2007**, *1164*(1–2), 153–161.

site of oxidation. Yang et al. [15] performed intact mass analysis of deglycosylated (following treatment with PNGase) and reduced HC originating from an IgG2 mAb. The observed mass of the deglycosylated HC was 16 Da higher than the expected value, suggesting oxidation. Using this information, peptide mass mapping analysis was then performed to look for peptides with oxidation (i.e., 16 Da mass shift). A peptide was detected that contained an oxidized tryptophan residue. In this case, the intact mass analysis provided the key information that oxidation may have occurred and then peptide mass mapping was used to determine the site of oxidation.

Chu et al. [16] followed this strategy to identify the origin of a modified form of IgG2 mAb that appeared following storage under stressed conditions. The modified form was first observed by cation exchange (CEX) chromatography. CEX fractions were collected and analyzed by intact mass analysis of the whole mAb molecules. A mass shift of –17 Da was observed. The CEX fractions were then chemically reduced into LC and HC subunits followed by intact mass analysis. This showed that the –17-Da mass shift originated from the LC. A –17-Da mass shift could be due to N-terminal pyroglutamate or conversion of aspartic acid to succinimide. Peptide mass mapping was then used to determine that an aspartic acid residue had converted to succinimide. Additional studies could then be performed to study the kinetics of the succinimide formation using the more convenient CEX method.

6.2.2 Peptide Mapping

As mentioned previously, once a modification for a mAb is observed during intact mass analysis, peptide mapping is usually carried out to further confirm and localize these modifications. Peptide mapping is commonly used in protein drug discovery for primary sequence confirmation and to obtain PTM locations, glycosylation sites, and disulfide linkage information. This approach is also used widely in protein drug development for identity confirmation as well as process monitoring. Product comparability testing by peptide mapping is always conducted following any changes in the manufacturing process, drug formulation, or storage conditions.

In general, peptide mapping requires multiple steps and takes one or two days for sample preparation and MS analysis. The first step normally involves denaturation, reduction, and alkylation to unfold proteins, break the disulfide bonds, and prevent them from scrambling. The sample buffer is then exchanged to the appropriate digestion buffer prior to protease addition. After a certain period of time for digestion, the reaction is quenched and the resulting peptides are ready for MS analysis. Although the HPLC-ultraviolet (UV) trace provides sufficient information for comparability studies and release assays, MS is still needed for characterization in the initial assay development to identify each chromatographic peak. The final objective is to acquire a high-quality, high sequence coverage peptide map with minimum artifacts caused by sample preparation in the shortest amount of time.

Peptide mapping has been applied widely for therapeutic protein characterization to confirm amino acid sequences and to localize PTM sites, especially those that are prone to deamidation and/or isoasparatate formation, because these modifications are critical for mAb stability and efficacy [25, 26]. Application of peptide mapping for glycosylation profiling, disulfide mapping, and other PTMs analysis will be discussed in detail in the following sections. In addition, peptide mapping has also been applied for demonstrating cell line comparability and stability through different drug development stages. This technique provides valuable information to demonstrate comparability of the purified protein product for lot-to-lot consistency. This is particularly important when changes in the bioprocess or the cell bank line are made during clinical development phases or after commercialization [27]. Once the chromatographic peaks are identified by MS, HPLC with a UV detector at 214 nm can be used as the routine assay for demonstrating product consistency, without the need for MS analysis of each lot [28, 29].

Peptide mapping is an information-rich analysis; however, some drawbacks of the peptide mapping approach include labor-intensive sample preparation, limitations in sequence coverage, loss of labile PTMs, and potential artifacts resulting from sample handling. These artifacts will raise concerns when peptide mapping is used in stability studies or for product-related impurities tests. Various efforts have been devoted to improving peptide mapping by optimizing the digestion conditions, reducing the sample processing time, and even evaluating automation of the peptide mapping process. A recent work examines the individual sample preparation steps involved in the peptide mapping procedure for mAbs [30]. A procedure was optimized by combining denaturation and reduction into one step and shortening the overall incubation time, removing urea from the digestion buffer, and changing the pH to near neutral during the tryptic digestion to reduce the artifacts and overall sample preparation time. Some new technologies were applied such as microwave assistance and trypsin spin filter digestion. Commercially available premade solvents were recommended to save time and reduce potential errors during solvent preparation. Overall sample preparation has been reduced from days to hours and the resulting peptide map is nearly free of sample or background artifacts. Another work improved trypsin digestion by removing guanidine completely from the digestion buffer after protein denaturing [31]. In this way, artifacts such as deamidation or N-terminal glutamine cyclization would be largely minimized. Tryptic digestion efficiency was improved and the digestion time can be reduced to 30 min.

Another improvement came from the MS/MS methodology, in which a "data-independent" MS/MS strategy (in contrast to "data-dependent" MS/MS) was developed to achieve higher sequence coverage for peptide mapping. Data-independent acquisition, also called HPLC-MS^E, utilizes MS scanning that alternates between low and elevated collision energy settings [32, 33]. Without isolation of peptide precursor ions, all peptide ions in the first scan (the low energy scan) will be fragmented in the subsequent scan using elevated collision energy. Software tools were developed that correlate the fragment ions with the precursor ions by retention time and identify the peptides. In this way, all peptides are fragmented including those peptides of low abundance, which can be neglected in data-dependent MS/MS methods. Therefore, HPLC-MS^E not only improves the sequence coverage, but also provides an instrument-based solution to characterize low abundant protein variants or product-related impurities. This technique has been demonstrated to achieve 96% sequence coverage for yeast enoloase (Eno1p). Protein impurities and site-specific modifications can also be identified and quantified with peptide mapping [34]. This technique has been applied to compare a biosimilar candidate to an innovator mAb in a rapid manner, as shown in Figure 6.4 [35]. In this work, MS^E was applied to identify a significant difference between the biosimilar and innovator products.

Automated tryptic digestion was also developed for peptide mapping to enable high-throughput analysis [36]. This procedure requires high concentrations of denaturants and employs a custom-designed, 96-well, size-exclusion desalting plate. The completeness and reproducibility of digestion were verified by HPLC-MS/MS analysis. Automation of the digestion process minimizes human intervention during the sample preparation and makes high-throughput peptide mapping feasible.

An alternative to peptide mapping is the "top-down" approach in which intact protein ions are fragmented directly by using high-resolution MS/MS [37, 38]. However, this strategy suffers from limited structural resolution and is generally only successful when used with high-resolution MS instruments. A "middle-down"

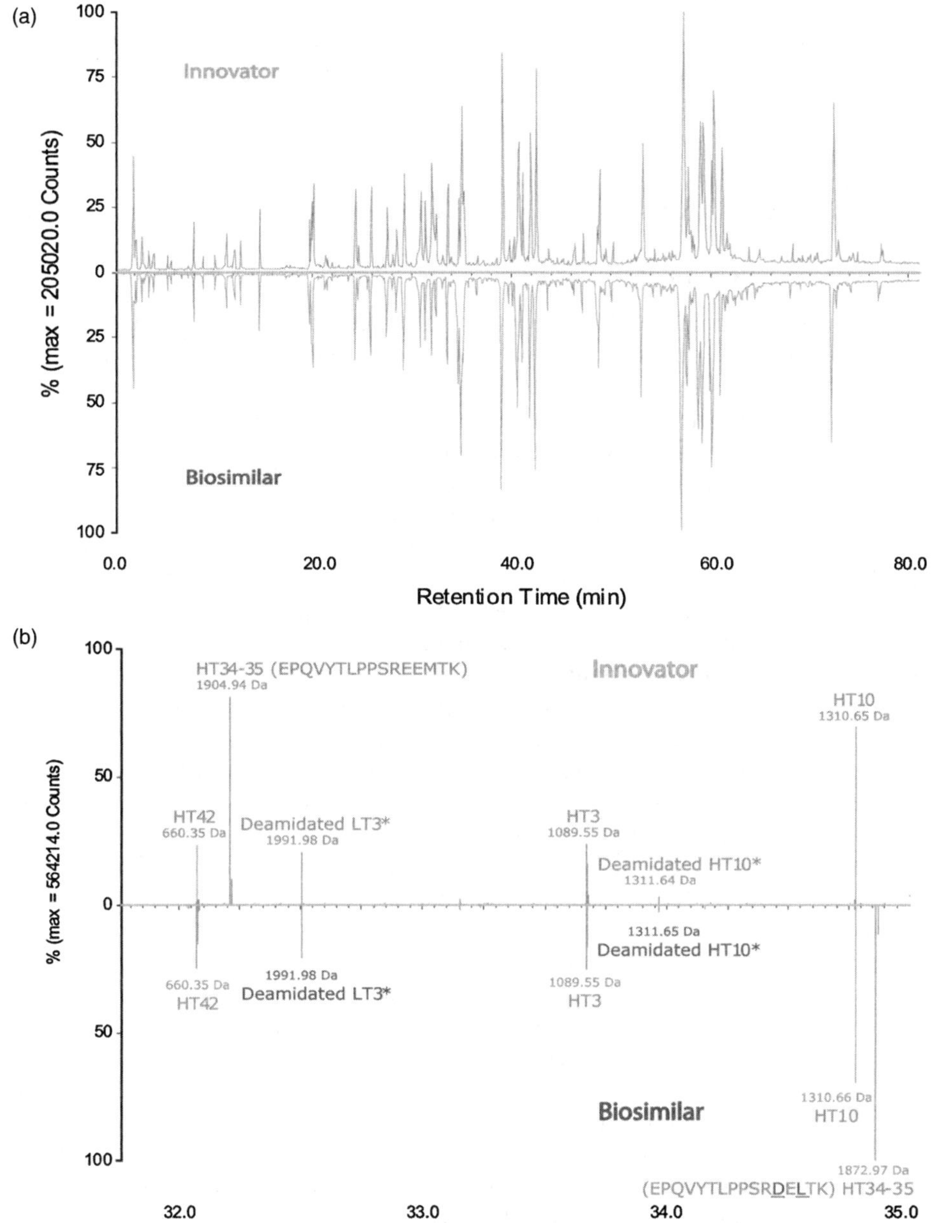

Figure 6.4 Mirror plots of LC-MS peptide maps for the innovator and biosimilar mAbs after 4-h trypsin digestion. (A) LC-MS (TIC); (B) a zoom view of charge-reduced, isotope-deconvoluted LC-MS chromatogram from 32.0 to 35.0 min to show differences between the innovator and biosimilar mAbs. Adapted with permission from Xie, H. et al. *mAbs* **2010**, *2*(4), 1–16.

approach has recently been developed to directly analyze larger peptides generated either by partial digestion or by disulfide bond reduction. This approach could potentially improve the structural resolution with minimal sample preparation. The details of top-down and middle-down MS will be discussed in Section 6.4.

6.3 STRUCTURAL CHARACTERIZATION

6.3.1 Glycosylation Analysis

Glycosylation is a common PTM of proteins. It serves a critical role in the biological and physiochemical properties of a glycoprotein, having an influence on function, activity, stability, and immunogenicity. Proteins are glycosylated by a series of glycosyltransferases in the cell. Biosynthesis of glycans depends on the presence and concentration of specific enzymes, availability of substrates, and the fermentation conditions, but not according to a predetermined template. Therefore, the resultant glycans are very heterogeneous. Also, since nonhuman cells are typically used for production of recombinant therapeutic proteins, the glycans may be different from those produced by human cells and may have a negative impact on the immunogenicity of the glycoprotein. For these reasons it is very important to characterize the glycoforms and to monitor the glycosylation profile of therapeutic glycoproteins for production consistency and quality control. Protein glycosylation and its analysis by MS is extensively reviewed in several articles [8, 39–41]. There are two types of glycans that can be attached to proteins: N- and O-linked. N-linked glycans are attached to the side-chain nitrogen of asparagine residues that are in consensus sequence Asp-X-Ser/Thr, where X represents any amino acid except proline. O-linked glycans can be attached to the side-chain oxygen of any serine or threonine on a protein. Several mass spectrometric approaches are used for the characterization of glycoproteins: intact glycoprotein analysis, released carbohydrate analysis, and glycopeptide analysis.

The analysis of intact glycoproteins was briefly discussed in the intact mass analysis section. This analysis is usually performed using ESI-TOF or ESI-Q-TOF instrumentation. It enables detection of major glycoforms, facilitating a broad overview of the protein glycosylation. As mentioned in the previous section, various glycoforms of an IgG molecule are shown in Figures 6.2 and 6.3. However, due to the heterogeneity and interference from other PTMs, minor glycoforms are not easily detectable using this approach. Also, no site-specific information is obtained. On the other hand, the advantage of this method is that it requires minimal sample handling and small sample amounts, and it is relatively easy to perform.

The released carbohydrate analysis provides information about the composition of the glycan itself. Glycans are released from glycoproteins or glycopeptides either chemically (hydrazinolysis for O- and N-linked, β-elimination for O-linked glycans) or enzymatically (PNGase F for N-linked glycans). Analysis of O-linked glycans is more challenging because an enzyme that would remove intact O-linked glycan is not commercially available at this time. Typically the released glycans are then labeled with a fluorescent tag such as 2-aminobenzoic acid or 2-aminobenzamide and analyzed by normal phase (NP)-HPLC [42] and MALDI-TOF. Relative quantitation of the glycoforms is obtained from the HPLC analysis since each glycan is

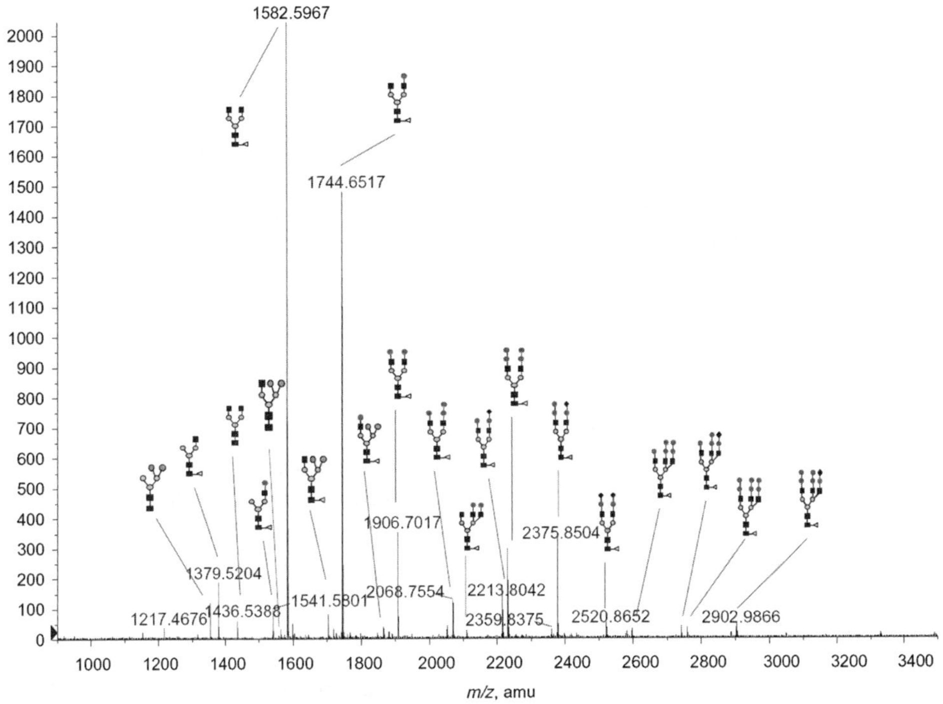

Figure 6.5 Negative ion MALDI Qq–TOF MS spectrum of the 2-AA-labeled N-linked oligosaccharides released from cetuximab by PNGase F. Sugar residues are N-acetylglucosamine (GlcNAc, ■), fucose (Fuc, ◁), mannose (Man, ○), galactose (Gal, ●), and N-glycolyl neuraminic acid (NGNA, ■). Adapted with permission from Qian, J. et al. *Anal. Biochem.* **2007**, *364*(1), 8–18.

stoichiometrically labeled with one fluorescent tag on its reducing end, having the same response factor regardless of the remaining glycan structure. The same labeled glycan pool is also analyzed by MS, usually using MALDI-TOF in negative mode. The MALDI spectrum provides the glycan profile of the glycoprotein. An example of the spectrum is shown in Figure 6.5. An identification of NP chromatographic peaks is performed by fraction collection of individual peaks and subsequent MALDI-TOF MS analysis. A challenge remains to distinguish between isobaric structures. For example, MS is not able to differentiate between various hexoses such as mannose or galactose; also it is not always possible to deduce complete linkage information directly from MS analysis. To overcome this challenge, the glycans can be sequentially digested with highly specific exoglycosidases and analyzed by MS at each step. Thus, Qian et al. used enzymes α(1-3)-galactosidase, Sialidase A, β(1-4)-galactosidase, β-N-acetylhexosaminidase, and α-mannosidase to trim terminal monosaccharides in a stepwise fashion [42]. In addition, the group used MALDI Q-TOF MS instrumentation to perform MS/MS analysis on selected glycans. Mass spectrometric fragmentation of glycans results in three types of cleavages: glycosidic cleavage, cross-ring cleavage, and loss of adduct [43]. Cross-ring fragmentation provided branching and linkage information. Consequently, combination of the data obtained from highly specific enzymatic treatments together with high-resolution MS and MS/MS analyses enabled the researchers to achieve a comprehensive structure characterization of a complex glycan pool released from a therapeutic IgG1 antibody Cetuximab (ImClone Systems, Branchburg, NJ).

Finally, the glycopeptide analysis correlates glycan composition to the attachment site. The sample preparation procedure is basically the same as with the peptide mapping discussed earlier involving digestion of the glycoprotein by a protease and the analysis of the resultant peptides and glycopeptides by RP-HPLC-MS. The peaks representing glycopeptides can be identified by a concomitant deglycosylation by endo- or exo-glycosidases. Peaks that shift following such a treatment represent glycopeptides. Alternatively, an extracted ion chromatogram (EIC) search for glycan-specific diagnostic fragment ions (oxonium ions such as *m/z* 163 for Hex+, *m/z* 204 for HexNAc+, *m/z* 365 for HexNAc-Hex+) may be performed to identify peaks representing glycopeptides [44, 45]. Consequently, both N- and O-linked glycans may be identified. The identity of the glycopeptides is further determined by comparing the measured mass to the theoretical masses of the protein peptides, or by sequencing using MS/MS.

6.3.2 Disulfide Mapping

Formation of disulfide bonds is critical to the structure and function of a protein. The structural stability of a number of proteins and peptides including therapeutic agents depends on the proper formation of disulfide linkages. The disulfides facilitate stabilization of the native structures of proteins by confining mobility to specific regions of the protein. Proteins are folded and the disulfides are formed primarily in the endoplasmic reticulum of a cell [46, 47]. Disulfide bonds form between the thiols (SH groups) of cysteine residues. Complex, multidomain proteins such as antibodies contain multiple inter- and intradomain disulfides. Improper formation of disulfides or misfolding of a protein may lead to heterogeneous product, instability, aggregate formation, and so on. Consequently, regulatory agencies require thorough characterization of cysteine connectivities. For example, the guideline on development, production, characterization, and specification for mAbs and related products published by the European Medicines Agency (EMA) states that the free sulfhydryl groups and disulfide bridges should be determined and the disulfide bridge integrity and mismatches should be analyzed. MS proved to be a very useful tool for structural elucidation of cysteine connectivities including presence and location of free thiols (cysteine residues not involved in disulfide bonds), disulfide, and trisulfide bonds. A historical perspective and some techniques of protein disulfide bond determination by MS are presented in a review paper by Gorman et al. [48]. One of the ways that disulfide bonding is studied involves identification of disulfide-containing peptides by comparing nonreduced and reduced peptide maps. Peaks that disappear after reduction indicate the disulfide bonded structures, and complementarily the new peaks are indicative of peptides that were involved in disulfide bonds prior to reduction. The identity of the disulfide-containing peptide peaks is then determined by matching the measured mass to the calculated masses of all possible combinations of cysteine-containing peptides. The HPLC fractions may also be collected, their disulfide bonds reduced and reanalyzed by HPLC-MS to identify the peptide components. Several examples of the application of MS for cysteine connectivity determination are presented in this section. In addition, the use of the newly developed technique called electron transfer dissociation (ETD) and its application for disulfide determination will be discussed later in the ETD section.

6.3.2.1 Free Thiols Mass spectrometric analysis of free thiols usually involves differential labeling techniques [49–51]. For example, a technique presented by Xiang et al. [51] to study the amount and location of free thiols in mAbs utilized differential labeling with ^{12}C and ^{13}C iodoacetic acid (IAA) followed by LC-MS analysis. First, the free sulfhydryls were selectively labeled with ^{12}C-IAA in Tris buffer at pH 8 in the presence of 6 M guanidine for 30 min. Then, the samples were desalted and reduced using dithiothreitol (DTT). The sulfhydryls generated from the reduction of disulfide bonds were then labeled with ^{13}C-IAA. The differentially alkylated protein was then digested by trypsin. The peptides alkylated by these isotopically labeled agents had identical retention times; however, their mass differed by 2 Da because of the incorporation of ^{13}C isotopes. A typical Q-TOF instrument is capable of resolving these isoforms. The researchers showed that this method was able to detect free sulfhydryls at a level as low as 0.5%. A note must be made that an isotope overlap correction was necessary to correct for the influence of the natural abundance of ^{13}C isotope in the peptides and is detailed in the publication [51].

6.3.2.2 Disulfides of IgG2 An example of a recent important discovery in which MS played an essential role is the structural elucidation of disulfide mediated isoforms of IgG2 antibodies [52, 53]. Researchers at Amgen reported that the widely accepted consensus structure of IgG2 molecules is only a subset of three distinct isoforms termed IgG2-A, IgG2-B, and IgG2-A/B. IgG2-A corresponds to the traditional model with independent Fab and Fc domains connected by the hinge with four disulfide bonds. In the B isoform, two of the four hinge cysteines are bridged to the complementary HC hinge cysteines, the third is bridged to the LC, and the fourth is bridged to a cysteine in the CH1 domain (this CH1 cysteine connects with the LC cysteine in the A isoform). The A/B isoform is an asymmetrical hybrid form containing both A- and B-isoform disulfide linkages.

First, the heterogeneity was observed by various separation techniques (capillary electrophoresis sodium dodecyl sulfate (CE-SDS), CEX-HPLC, and RP-HPLC) under nonreducing conditions. Whole antibody samples or RP-HPLC separated fractions of IgG2 molecules were then subjected to an enzymatic digestion by LysC under nonreducing conditions as well. A portion of the nonreduced digest was analyzed by RP-HPLC with UV and MS detection, and another portion was treated with Tris(2-carboxyethyl) phosphine (TCEP) to create a digest for a reduced peptide map and analyzed in the same manner as the nonreduced digest. The reduced and nonreduced peptide maps were compared to identify peptides connected by disulfide bonds (Fig. 6.6). The peaks that disappeared upon reduction signified disulfide bonded peptides. To determine the identities of each Cys-containing peptide, the disulfide-linked peptides were manually collected from RP-HPLC, reduced with TCEP, and reanalyzed by RP-HPLC-MS/MS. Some of those disulfide-linked structures were quite large (over 25 kDa) or complex (more than one disulfide bond). Such peptides were also treated with GluC after collection from RP-HPLC. This approach enabled identification of a signature region on the chromatogram labeled

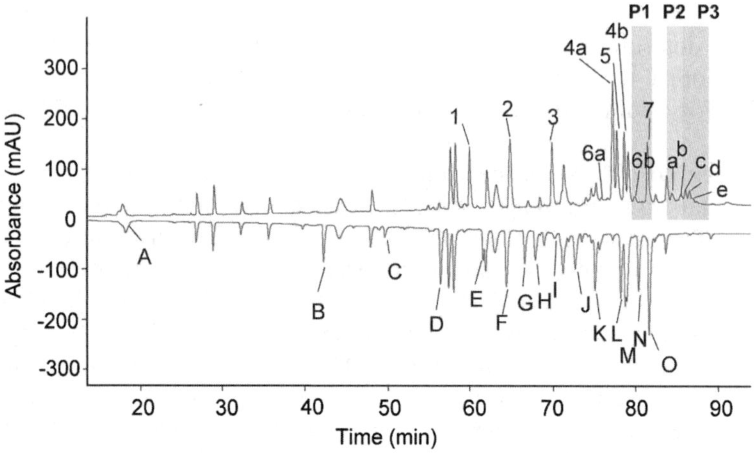

Figure 6.6 UV trace of nonreduced (top) and reduced (bottom) LysC digest of IgG2. Adapted with permission from Wypych, J. et al. *J. Biol. Chem.* **2008**, *283*(23), 16194–16205.

P1, P2, and P3 as shown in Figure 6.6 with peptides specific to each IgG2 isoform. Thereafter, examination of this section of the chromatogram was sufficient for determination of the IgG2 isoform composition of the subsequent samples. Online LC-MS/MS analyses were performed on ion trap mass spectrometer equipped with an ESI ion source.

This work demonstrates how a combination of a well thought out experimental design combined with various analytical techniques including MS resulted in a breakthrough in understanding the structure of IgG2 molecules that are entering the therapeutic protein arena.

6.3.2.3 Trisulfides Recently MS was instrumental in the discovery of another form of cysteine-related modification on therapeutic recombinant mAbs, namely the detection of trisulfide bonds in the hinge of the mAbs fractionated by ion exchange chromatography (IEX) [54]. Previously, trisulfide was believed to be a rather uncommon modification of proteins. It was detected in only a few proteins, specifically *Escherichia coli*-derived hGH [55, 56] interleukin 6 [57], and erythrocyte superoxide dismutase isolated from human erythrocytes [58]. Trisulfide is formed by insertion of a sulfur atom into the disulfide bond, which increases the mass of the protein by 32 Da.

First, the existence of the modification was observed by measuring intact mass of the whole or IEX-fractionated antibody using an ESI-TOF mass spectrometer. Spectrometric resolution of this instrument was enough to suggest that a modification of some kind has occurred on the protein, but not enough to positively characterize the modification at the intact antibody level. The intact mass measurement was able to reveal weakly resolved covalent additions of 25–35 Da to the ~145 kDa protein. Nonreduced LysC peptide mapping was performed to study the modification further. In addition to the expected mass of the hinge peptide, other peaks were detected that corresponded to addition of one or two sulfur atoms. Figure 6.7A

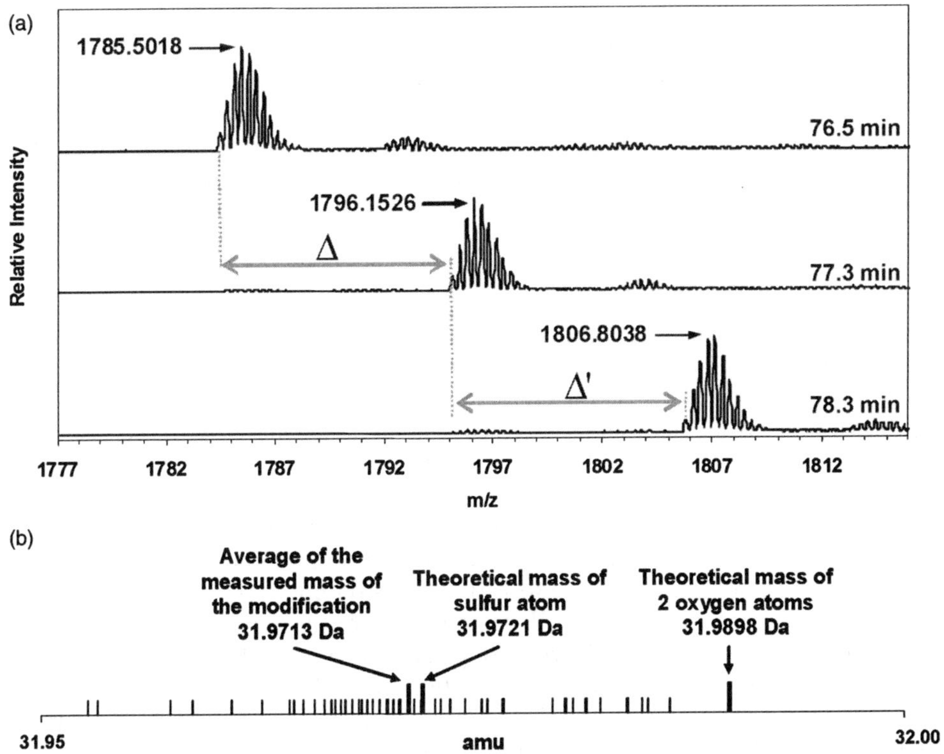

Figure 6.7 Mass spectra of the disulfide-linked hinge dipeptide. (A) A portion of the mass spectra from a LysC digest (performed under nonreducing conditions) of the three RP-HPLC peaks at 76.5, 77.3, and 78.3 min, showing the corresponding isotopic clusters with 3+ charge state of the disulfide-linked hinge dipeptide. (B) Summary of the charge-reduced mass differences measured between the isotope clusters; small vertical bars represent mass differences between corresponding isotope peaks within a pair of isotopic clusters (e.g., Δ and Δ′ represent mass differences between the first isotope peaks from one cluster to the next). The overall average of the measured mass differences is noted and is compared with the theoretical monoisotopic masses of a S atom and two O atoms. Adapted with permission from Pristatsky, P. et al. *Anal. Chem.* **2009**, *81*(15), 6148–6155.

shows the mass spectra of these peaks as isotopic clusters with a charge state of +3. To obtain the most accurate determination of the modification mass, the differences between corresponding pairs of isotope peaks from one cluster to the next were calculated and averaged. Individual mass differences are plotted in Figure 6.7B along with the average value of the differences (31.9713 Da) and the theoretical values for a single S atom (31.9721 Da) and two O atoms (31.9898 Da). This demonstrates the power of the high-resolution mass spectrometer to distinguish between the masses of one sulfur atom versus two oxygen atoms, both of which have a nominal mass of 32 Da. The research was expanded upon by Gu et al. [59] to show the existence of trisulfides not only in the hinge but throughout the IgG molecules as well as in other IgG subtypes.

Analysis of cysteine connectivities using proteolytic peptide mapping approaches poses a challenge because the pKa of sulfhydryl group is close to the optimal buffer pH of most common proteolytic enzymes. Thus, a prolonged incubation with trypsin or LysC under nonreducing conditions might lead to disulfide scrambling. Therefore, care must be taken to minimize the incubation time, as well as to include proper negative controls. Alternatively, digestion with pepsin at acidic pH may be used to prevent disulfide interchange.

6.3.3 PEGylation Analysis

An increasingly popular strategy for improving the pharmacokinetics (PK) of therapeutic peptides and proteins is to conjugate with poly(ethylene glycol) (PEG). The benefits of PEGylation include decreased renal clearance, reduced degradation by proteolytic enzymes, and in some cases, reduced immunogenicity [60]. The first Food and Drug Administration (FDA)-approved PEGylated protein drug was PEG adenosine deaminase (Adagen®, Sigma-Tau Pharmaceuticals, Gaithersburg, MD), in 1997 [61]. Since that time eight more PEGylated drugs have been approved by the FDA and there are a large number of PEGylated proteins under development [62]. PEG molecules can be conjugated to various sites on proteins, depending on what type of chemistry is used [63]. The common sites of attachment are amino groups on lysines and N-termini, sulfhydryl groups on cysteines, and aldehyde groups on carbohydrates of attached glycans. In addition, there is a growing interest in the insertion of non-native amino acids into protein sequences with the goal of enabling highly selective PEG attachment [64].

PEGylated therapeutic proteins present a significant analytical challenge for product characterization. Therapeutic proteins are typically made up of more than one component, for example, protein containing various glycoforms or amino acid modifications such as deamidation, pyroglutamate formation, oxidation, and so on. Couple this heterogeneity with the polydispersity of the PEG reagent, which in itself is a mixture of oligomers, and the resulting product is a complicated mixture that can be further confounded by presence of multiple sites of PEGylation. Characterization of PEGylated therapeutic proteins is required for all steps in the drug development process including patent filing, process development and manufacturing, product release and stability, and clinical studies.

There has been a steady increase in the number of reports describing the application of MS for the characterization of PEGylated therapeutic proteins. One of the earliest examples made use of a dePEGylation reaction to remove the PEG moiety using base hydrolysis leaving a succinyl mass marker at the site where PEG was attached [65]. This technique was applied to a PEGylated super oxide dismutase (SOD) formed by reacting PEG-succinyl succinate with amino groups on the protein. SOD contains 10 lysines that are all possible sites of PEG attachment. Following the de-PEGylation step, the authors went a step further by labeling any unconjugated amino groups with deuterium by reacting with D_4-succinic anhydride. The labeled SOD was then digested into peptides using endoproteinase Asp-N. Mass analysis using fast-atom bombardment (FAB)-MS and ESI-MS of the resulting peptides allowed determination of the extent of PEGylation at each lysine by using the hydrogen/deuterium isotope pattern. Tandem MS of the peptides enabled confirmation of the sites of PEGylation. This approach to characterizing PEGylated proteins provided important information about the sites and extent of PEGylation, but is limited to PEGylation linkages containing an amide bond to the protein and an ester bond to the PEG. There are also several sample manipulation steps involved that make this a cumbersome approach for characterizing multiple samples.

MALDI-TOF MS has been used for a number of years to characterize PEGylated proteins. MALDI-TOF has the advantage of relatively easy sample preparation, high sensitivity, rapid analysis time, and simplified data analysis due to the predominant formation of singly charged ions. An early example was applied for the characterization of PEG-SOD to monitor the extent of PEGylation in a rapid manner [66]. SOD is a 15.6-kDa protein with up to 10 possible sites of PEGylation when conjugation is targeted at amino groups. A spectrum obtained from the intact mass analysis of PEG-SOD using MALDI-TOF is shown in Figure 6.8. The peaks in the spectrum are due to SOD molecules with varying degrees of PEG conjugation (0–7 PEG conjugations). The peak area can be used as a measure of the relative amount of each degree of PEG conjugation in the drug lot. The ease of sample preparation and analytical speed make MALDI-TOF an extremely valuable tool for quickly assessing multiple lots of material using small quantities of sample.

MALDI-TOF has been applied to aid in the optimization of the PEGylation reaction. An example of this is the PEGylation of ricin A-chain (RTA) [67, 68]. RTA has several possible PEG attachment sites, depending on what type of PEG reagent is used. There are two lysine residues and the N-terminus amino group that can be linked to PEG using succinimide chemistry and there are two carbohydrate side chains that can link to PEG using hydrazide chemistry. The number of resulting PEG conjugations is dependent on several factors including pH, molar ratio of reactants, and reaction time. RTA also consists of two molecular variants possessing

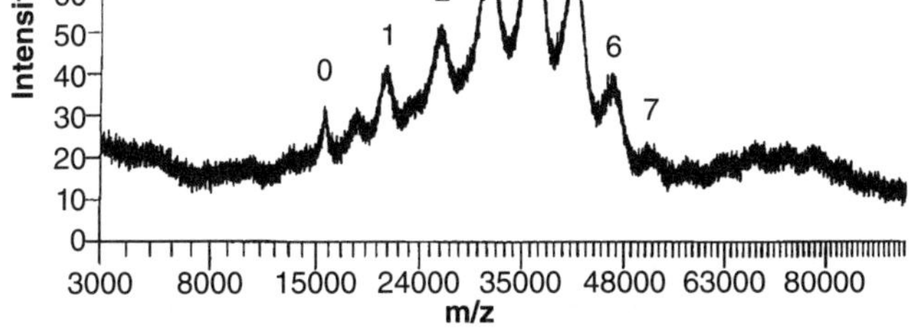

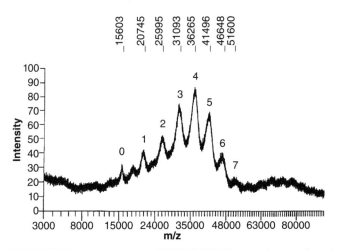

Figure 6.8 MALDI-TOF mass spectrum of PEG-SOD. The numbers assigned to each peak (0–7) indicate the number of PEG molecules attached. The numbers at the top of the spectrum are the *m/z* values of the corresponding peaks. The spectrum is a sum of 100 individual laser shots. 10 pmol of sample was deposited on the probe. Note the square root *m/z* scale. Adapted with permission from Chowdhury, S.K. et al. *J. Am. Soc. Mass Spectrom.* **1995**, *6*(6), 478–487.

identical protein sequences with different glycosylation. The authors used MALDI-TOF to monitor the effect of pH on the PEG conjugation to RTA using either amino-PEGylation or carbohydrate-PEGylation. The MALDI-TOF results indicate that pH 8 produces the highest degree of PEG conjugation for amino-PEGylation and that increasing the pH to 9 decreases the degree of PEGylation, due to hydrolysis of the ester bond on the linker [65]. In the case of carbohydrate PEGylation, the MALDI-TOF results indicate that pH 3 produces the highest degree of PEGylation, with up to five PEG conjugations to RTA.

Determination of the site of PEGylation is important for the full characterization of PEGylated proteins. The traditional method of Edman sequencing can be applied for this; however, large quantities of purified material are required. This can present a challenge if there are multiple sites of PEGylation and separation of proteolytically digested samples is required. Recently, it has been demonstrated that MALDI-TOF, using in-source decay (ISD), enabled the "top-down" determination of the site of PEGylation [69]. ISD [70] is a fragmentation process that occurs on the nanosecond time scale prior to acceleration in the source region of the TOF-MS. During ISD, a gas-phase reaction between the matrix and analyte ions leads to the formation of *c* and *z* fragment ions. ISD cleaves bonds uniformly across the protein backbone and enables large proteins to be sequenced [71].

ISD applied for the determination of the site of PEGylation takes advantage of the fact that in the reflectron mode (reISD), fragment ions conjugated to PEG do not travel through the reflectron. The reISD fragmentation spectrum of a PEGylated protein will be truncated by the missing PEG-conjugated fragments. Yoo et al. [69], applied reISD to the PEGylated peptide Ac-WVTHR*LAGLLSR*SGGVVK*K^{19} NFVPTDVGPX*AF-NH_2, non-natural amino acids R* (citrulline), K* (homoArg), and X* (naphthyl-Ala) have been substituted into the peptide sequence. The lysine at position 19 should be the only site of PEG attachment, as the N-terminus is acetylated and there is a homoArg (K*) substituted for the lysine at position 18. A MALDI-TOF reISD mass spectrum was obtained that contains a series of *c*-ions that covers the sequence from the Arg at position 5 to the homoArg (K*) at position 18. In addition, there are *y*-ions that contain the C-terminus, confirming that the C-terminus is not PEGylated. This example highlights the application of MALDI-TOF reISD MS to a purified PEGylated protein containing a single PEGylation site. MALDI-TOF reISD MS is a potentially powerful method for the identification of PEGylation sites, with the main limitation that a purified sample is required. However, with upfront off-line separation [72, 73], MALDI-TOF reISD MS could be applied to more complicated samples containing multiple PEGylation sites.

Determination of the site-of-PEGylation can also be accomplished using HPLC-MS/MS by combining in-source fragmentation (ISF) with collision-induced fragmentation (CID). Lu et al. [74], used elevated cone voltage to induce ISF of a PEGylated glucagon drug, causing fragmentation along the PEG chain. This resulted in the generation of fragment ions containing the intact peptide with short PEG segments attached. Isolation of the ISF-generated fragment ions followed by CID produced b-ions and y-ions by cleavage along the peptide backbone. The fragment ions were found to be consistent with attachment of PEG at the lysine (K) at position 12. The authors also applied this approach to PEGylated mAb proteins by first digesting the protein using protease enzymes. ESI-ISF-CID is a powerful method because of the well-established coupling of ESI with RP-HPLC separation and it is expected that this type of approach will find continued application for the characterization of PEGylated protein therapeutics.

The application of ESI-MS and HPLC-MS for the analysis of PEGylated proteins is complicated by spectral congestion caused by the heterogeneous polymer distribution and high charge states. In most cases, ESI spectra of PEGylated proteins have little to no analytical utility. To overcome this limitation and to harness the analytical capabilities of HPLC-MS, "charge state reduction" methods have been introduced for the ESI analysis of PEGylated proteins. Bagal et al. [75] accomplished charge reduction by introducing superbase ions by means of a second spray source in the ESI chamber of a commercial Q-TOF instrument. Gas-phase proton transfer from highly charged PEG-protein ions to superbase ions occurs, resulting in charge state reduction. Various superbase compounds can be used; however, 1,8-diazabicycle [0, 4, 5]-undec-7-ene (DBU) was found to be highly useful. The authors showed that by applying charge state reduction to PEGylated proteins, resolved spectra containing valuable structural information can be obtained. When charge state reduction was applied to PEGylated granulocyte colony stimulating factor (PEG-GCSF, average molecular weight [MW] 40.7 kDa), the resulting deconvoluted spectrum had sufficient resolution to separate peaks due to a low abundance oxidized form of the drug. Charge state reduction can also be achieved by post-column addition of amine directly into the chromatographic effluent [76]. Amine adducted ions are formed that are lower in charge state. Figure 6.9 is an ESI-TOF spectrum of PEGylated IgG4 LC and HC. Following spectrum deconvolution, the complete PEGylated protein distribution is obtained. The authors applied this method to study stressed samples of PEG-IgG4 and were able to resolve a water gain (+18-Da mass shift) specie that forms due to maleimide ring opening occurring after storage under stressed conditions.

6.3.4 Other PTMs Analysis

In addition to the characterization of protein modifications discussed above, such as glycosylation and disulfide mapping, MS is well suited for the characterization of other PTMs. A PTM is the chemical modification of a protein after its translation. Examples of PTMs include, but are not limited to deamidation, oxidation, and phosphorylation. Any PTM can have a profound impact on protein structure and function, especially if it is in the active site of the protein.

One of the common modifications or degradations of a protein is deamidation of asparagine (Asn) residues. It is a nonenzymatic reaction [77] where Asn residue converts to a mixture of isoaspartate (iso-Asp) and Asp through a five-membered succinimide ring intermediate at neutral and basic pH. The deamidation rate generally increases at higher pH. Also, the primary sequence and higher order structures have an effect on deamidation susceptibility and rate [77–86]. For example, in a linear sequence, an Asn preceding a Gly is the most susceptible to deamidation. Also, a specific tertiary structure can make an Asn more or less prone to deamidation by imposing structural constraints or by affecting solvent accessibility [79, 81, 82, 84, 86].

A common nonmass spectrometric approach to determine the amount of iso-Asp in proteins is to use a commercially available chemical detection kit. This analysis

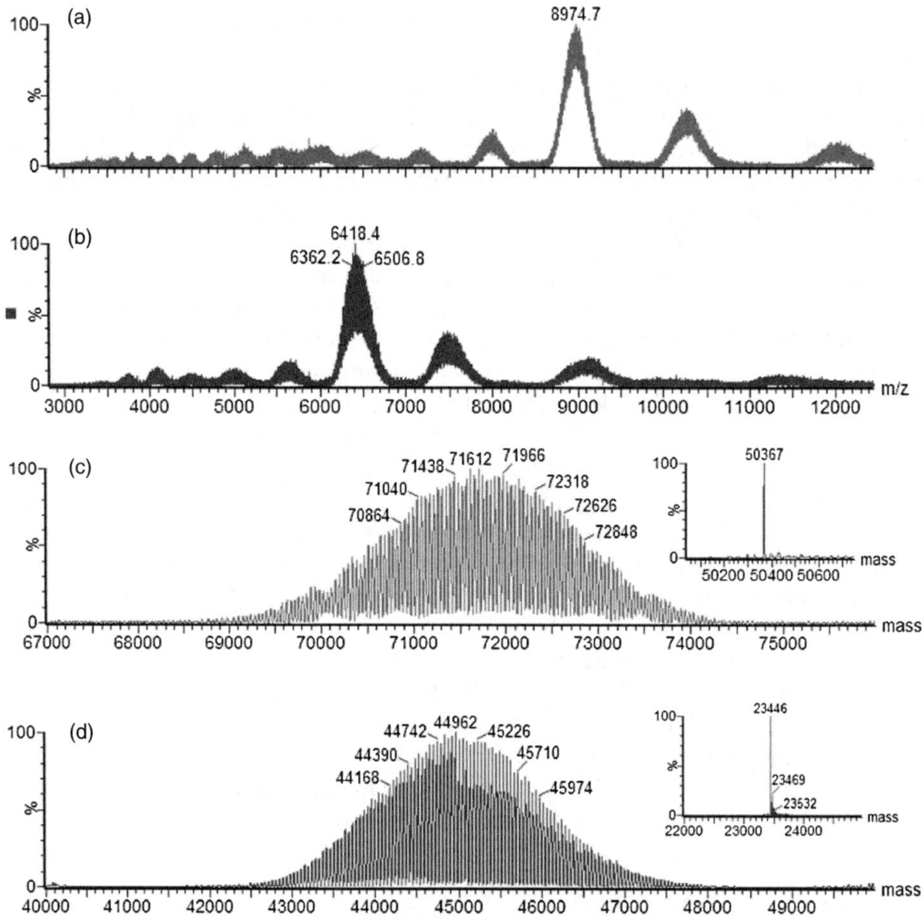

Figure 6.9 ESI-TOF mass spectra of charge reduced PEG-heavy chain (a) and PEG-light chain (b). Deconvoluted mass spectra are also shown, PEG-heavy chain (c), and PEG-light chain (d). Non-PEGylate heavy and light chain were also detected and the deconvoluted spectra are shown in the insets of (c) and (d), respectively. Adapted with permission from Huang, L. et al. *Anal. Chem.* **2009**, *81*(2), 567–577.

provides the total amount of iso-Asp in a protein sample but does not identify where the deamidation has occurred. Also, it only detects iso-Asp, but not Asp that was converted from Asn. Conversion of Asn to iso-Asp or Asp increases the mass of the residue by 0.98 Da and introduces a negative charge by replacing an amide group with a carboxylate group. Both, top-down and bottom-up techniques were used to study deamidation of therapeutic proteins by MS. Vlasak et al. were able to detect a deamidation event on the LC of IgG1 antibody by measuring the mass of the whole LC after reduction by DTT [87]. The analytes were separated on a C8 RP column and analyzed on an ESI-TOF mass spectrometer. The combination of high

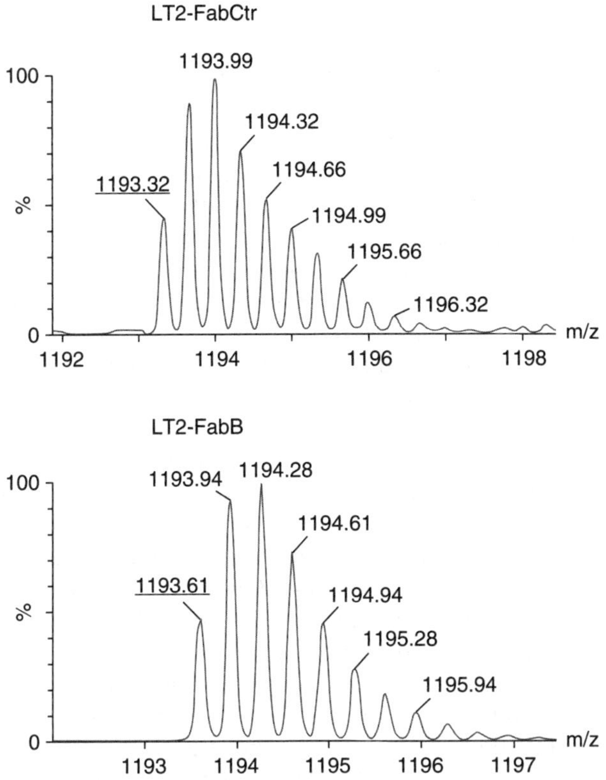

Figure 6.10 ESI-MS isotopic distribution [M + 3H⁺] of a light chain tryptic peptide. Top pane—control, not deamidated; bottom pane—stressed, deamidated. Adapted with permission from Vlasak, J. et al. *Anal. Biochem.* **2009**, *392*(2), 145–154.

mass resolution (10,000) and high mass accuracy (20 ppm) of the mass spectrometer enabled differentiation of the LC with and without deamidation (24,063 vs. 24,064 Da). However, a typical bottom-up approach was necessary to confirm and identify the site of the deamidation. This method involves an enzymatic digestion of a protein to peptides. The peptides are then separated on a C18 RP column and analyzed by MS. The same mass spectrometer was used to localize the deamidation site to the peptide originating from the complementary determining region (CDR1) of the LC. Figure 6.10 shows the mass spectra of the $[M + 3H]^+$ ion of this peptide. The shift in the distribution by approximately +0.3 *m/z* units corresponds to ~+1Da increase in the peptide mass, indicating a deamidation event in the stressed sample compared with the control.

In the case presented above, the samples were obtained by CEX fractionation of stressed material. Therefore, either complete or no deamidation was expected and was consequently observed in the isolated fractions. However, sometimes there is also a need to determine and quantify deamidation on multiple sites in a protein where only partial deamidation has occurred. For example, the extent of deamidation needs to be assessed in a therapeutic protein undergoing a stability study in a particular formulation. In such studies the digestion and LC conditions may need to be optimized on a case by case basis. Ideally, one would strive to achieve a baseline LC separation of the peptides with and without deamidation. If one succeeds, the amount of modified peptide can be quantified either by UV or by ion intensity by integrating the EIC. In case the LC separation is not achieved successfully, it is possible to approximate the amount of deamidation by spectral deconvolution. In any case, if ion signal intensity is used for quantification, one must keep in mind an assumption that the ionization efficiency is similar for the unmodified and modified peptide. This assumption may not necessarily be true for any pair of peptides. Therefore, such information may be used only for estimating the amount of deamidation.

Oxidation of methionine residues to form methionine sulfoxide (Met-ox) is another degradation process that is closely monitored in biopharmaceuticals. Determination of oxidation in therapeutic proteins may be performed for characterization purposes, stability and comparability studies, in-process testing, or as a release assay. Similarly to deamidation analysis, the extent of oxidation is usually determined by digesting a protein with a protease such as trypsin or LysC, and analyzing the digest by RP-HPLC. Both UV and mass spectrometric detection are used at first to identify the chromatographic peaks. Once the identities of Met-containing peptides and their oxidized counterparts are determined, further monitoring of Met oxidation can be done with UV detection only, provided that all peptides of interest are separated chromatographically and there is no interference from other peptides of the protein. However, these ideal conditions are not always achieved. In such cases, MS detection is still necessary for the quantification of the peaks. Houde and coworkers [88] demonstrated that the UV and MS detection are comparable in terms of assay reproducibility, robustness, linearity, accuracy, and precision. In addition, the researchers showed that for quantitative MS detection of Met-containing pairs of peptides, chromatographic separation is not essential. They developed a flow-injection RP-HPLC-MS technique where a protein digest is desalted and eluted by a step gradient. First, the sample was injected onto a peptide macro trap cartridge followed by a step gradient from 5% B to 95% B (mobile phase A: 0.1% trifluoro acetic acid [TFA] in water; mobile phase B: 0.1% TFA in 90% acetonitrile). Thus, all of the peptides eluted in one peak. The extent of oxidation was calculated by integrating the peaks obtained by selected ion monitoring or from EICs of Met-containing pairs of peptides. Note that the possibility of false positives exists in this approach if an isobaric compound is present in the mixture. Therefore, proper controls should be included. The advantage of this technique, however, is rapid analysis time, which is only minutes per sample.

6.4 FUTURE DEVELOPMENT

Significant advancement in MS techniques provides the capability to characterize proteins extensively and enable better understanding of biomolecules under development. Comprehensive characterization of therapeutic proteins, which is required by regulatory agencies, in turn, has motivated continued MS method development. Several recent developments in MS techniques are introduced in this chapter and their potential application in the biopharmaceutical industry is discussed.

6.4.1 Hydrogen Deuterium Exchange (HDX)

The therapeutic properties of protein drugs are uniquely determined by their conformation, since the biological activity of all proteins is directly related to their three-dimensional organization or higher order structure. Partial or complete loss of a native folded structure of a protein drug in solution can dramatically influence its efficacy, or even trigger aggregation [89, 90]. Several analytical methods, such as calorimetry, nuclear magnetic resonance, X-ray crystallography, and advanced fluorescence, are commonly used to understand the higher order structure of protein drugs. Recently, HDX coupled with MS has become a valuable analytical tool for studying the higher order structure of proteins. HDX can provide detailed and complementary information regarding protein dynamics and conformation, and requires far less sample than most other methods [91].

The basic principle of an HDX MS experiment is to measure the exchange of protein backbone amide hydrogen atoms (NHs) with deuterium atoms over a defined time period [92]. In a solution of D_2O, those regions of protein exposed to solvent typically become deuterated faster than regions buried in the core or regions involved in hydrogen bonding. The number of hydrogen atoms exchanged can be determined by using MS to measure the mass difference following exchange with deuterium. In this way, the dynamics and conformation of a protein can be derived from the measured number of exchanged hydrogen atoms as shown in Figure 6.11. The factors that affect the exchange rate include solvent accessibility, hydrogen bonding, pH, and temperature. The HDX experiment is normally carried out at room temperature by adding an excess (10–20-fold) D_2O buffer at pH 7.0. After incubation for a certain time, the exchange process is quenched by adding acidic buffer (pH 2.5) and dropping the temperature to 0°C. The reaction solution can then be analyzed by HPLC-MS directly to measure the molecular weight difference between the exchanged protein and the native protein at different reaction time-points, enabling rapid monitoring of global conformational changes. To obtain localized deuterium exchange information, the deuterium-exchanged protein must be digested into peptide fragments by protease. Since digestion of the labeled protein must occur under quenched (acidic) conditions, acid proteases such as pepsin are preferred. In this case, deuterium exchange information can be obtained down to the peptide level (normally less than 10 amino acid residues). The relationship between the bioactivity and the three-dimensional structure can then be established for the protein under study. For local analysis, to determine the specific location of changes in deuteration, peptides are normally digested online using an immobilized pepsin column. The great challenge for HDX is to reduce the back exchange once the incorporation of the isotope label is complete. Therefore, to minimize back exchange, sample analysis must be carried out rapidly and under 0°C conditions.

HDX MS has been applied widely to analyze all types of proteins and protein systems. It has also been used to analyze whole viruses [93, 94], to investigate antibody epitopes [95], and to study drug-target binding for various diseases [96, 97]. More recently, it has been applied to characterize the higher order structure of protein therapeutics [98–101]. For example, HDX MS has been applied together with IEX chromatography to detect and characterize conformational changes in the biopharmaceutical product interferon β-1a (IFN-β-1a) by inducing alkylation of its

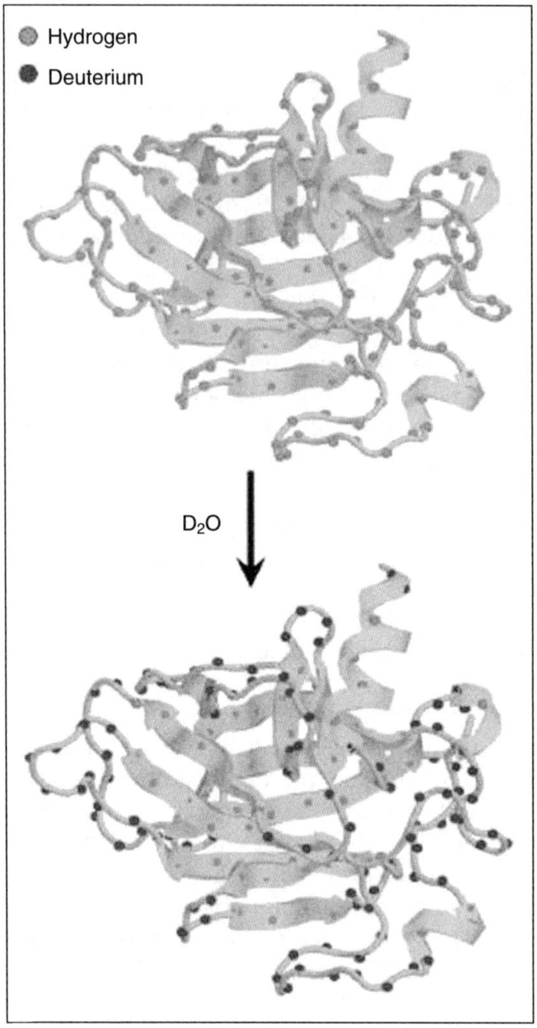

Figure 6.11 HDX in proteins. Light gray dots indicate labile peptide backbone amide hydrogens (NHs) that can be deuterated in a D_2O solution. Dark gray dots indicate those NHs that have been deuterated after a certain reaction time. The rate of deuteration depends on factors including solvent accessibility, hydrogen bonding, pH, and temperature. NHs that are not hydrogen bonded and reside near the surface of the protein can be deuterated rapidly. However, some highly solvent-exposed NHs at the surface are protected by hydrogen bonding and cannot be exchanged. Adapted with permission from Engen, J.R. *Anal. Chem.*, **2009**, *81*(19), 7870–7875.

only free cysteine residue with N-ethylmaleimide [98]. Specific detailed information on the higher order structure was obtained to provide the large-scale dynamic events within the protein that affect its function, which could not be obtained by the standard biophysical techniques commonly used in the biopharmaceutical industry. HDX MS has also be applied to study how glycosylation of the antibody IgG might affect the way it interacts with its receptors by measuring the exchange of protein backbone amide hydrogens in D_2O solution using only picomole quantities of material [99].

Continued efforts, however, are still required to increase the acceptance of HDX MS by the biopharmaceutical industry as a routine analytical approach for protein drug characterization. Challenges include throughput, automation, data analysis, reproducibility, and robustness. As pepsin is a nonspecific protease, the sites of backbone cleavage cannot be predicted from the amino acid sequence. Recently, aspergillopepsin (also known as Factor XIII) has been used as an alternative to pepsin because it can yield higher sequence coverage [102, 103]. Moreover, this enzyme prefers to cleave at the C-terminal side of basic amino acids (Arg, Lys, and His) and thus cleaves with some specificity for peptide identification and yields a higher signal/noise ratio. Further development of HDX MS involving the simultaneous reduction and digestion of proteins containing disulfide bonds was reported with increased sequence coverage [104]. Ion mobility MS was also applied to facilitate HDX studies by helping to reduce spectral complexity [105]. Newly developed fragmentation techniques such as ECD or ETD enabled a top-down approach to provide complementary information to the classical bottom-up approach [106]. The complexity of data analysis is also a challenge preventing HDX MS from being widely applied for therapeutic protein characterization. Some progress has been made toward automating HDX data analysis [107, 108]. A commercialized instrument has recently been developed through a collaboration between Engen's group (Northeastern University, Boston, MA) and Waters Corp. (Milford, MA) and could provide a robust solution to study the dynamics and conformation of biomolecules. As instruments and data analysis continue to develop, advances will make HDX MS more accessible to the biopharmaceutical industry. In the future, this technique could be used as a screening tool to evaluate how protein modifications affect activity.

6.4.2 Ion Mobility Mass Spectrometry

Ion mobility spectrometry (IMS) has been utilized for the analysis of biomolecules since the 1990s following the significant advances in ionization techniques [109–111], primarily ESI [112–114]. However, not until recently, the combination of IMS with MS and HPLC caused it to emerge as a powerful analytical tool for examining complex biomolecular mixtures. Many excellent reviews have been published regarding the details and mechanisms of ion mobility mass spectrometry (IMMS) and its application for protein and protein complex studies [115–119]. In recent years, this technique has drawn significant attention by the biopharmaceutical industry to characterize the higher order structure of protein drugs, and several applications have been reported [120].

Various designs have been developed to enable IMS to be coupled with different MS instruments, including TOF, quadrupole, ion trap, and Fourier transform-ion cyclotron resonance (FT-ICR) [115]. Regardless of design, the basic principles of IMMS are similar: following ionization, the ions are introduced into the IM device. Ion mobility separation takes place under an electric field at either low vacuum or atmospheric pressure. The unique drift velocity of each ion depends on the size, shape, and charge state under a given set of experimental conditions. For example, a tightly folded protein normally travels through the IMS faster and can be separated from less-folded conformers of the same protein. These ions can then be detected by MS following the IM device. Theoretical methods can be developed for elucidating structure information for ions of interest. The main advantages of IMMS are the simplicity and speed of the measurement, the high sensitivity and selectivity, and fast generation of results. What is more, IM can provide an orthogonal separation when coupled with LC to analyze complex biological samples, since multiple dimensions provide enhanced analytical peak capacity [121, 122]. IMMS has shown great promise as an intact protein separation and analysis methodology to characterize higher order structure including the overall size/shape of biopolymers and large macromolecular assemblies [119].

There are two kinds of instruments that have been commercialized for ion mobility. One technique was developed by Ionalytics Inc. (acquired by Thermo Fisher Scientific, San Jose, CA), to position a novel FAIMS (high-field asymmetric waveform ion mobility spectrometry) [123] device between the electrospray and the mass spectrometer. Based on the mobilities of different ions at atmospheric pressure, FAIMS filters out unwanted ions and prevents them from entering the MS; therefore, interfering ions can be minimized and chemical noise can be reduced after liquid chromatographic separation and ESI. Waters Corp. commercialized an IMMS (Synapt HDMS, Waters Corp.) based on traveling waves (T-wave) [124, 125]. In the T-wave implementation of ion mobility, ion separation occurs when a sequence of dc pulses push ions through the mobility cell in the presence of an inert gas at relatively high vacuum pressure. The ability of an ion to "surf" the T-wave depends on its collision cross-section. Ions with compact structures are pushed through the mobility cell faster than ions with more elongated structures. A recent work has demonstrated the application of T-wave IMMS to separate disulfide variants of intact IgG2 antibodies [120]. Two to three gas-phase conformer populations for IgG2s were detected by IMMS, whereas a single gas-phase conformer is observed for both an IgG1 antibody and a Cys-232→Ser mutant IgG2. This indicated that disulfide bond heterogeneity resulted in multiple IgG2 conformers.

Ion mobility MS is ideal for the study of the noncovalent structural features of mAbs, because molecules with the same mass but different conformations might be resolved according to their different collisional cross-sections. IMMS has the potential to be applied for the characterization of intact antibodies and could be used routinely to monitor the higher order structure of protein drugs in the near future.

6.4.3 Top-Down Mass Spectrometry

As described earlier, peptide mapping is a common method used to confirm the primary sequence of therapeutic proteins and a useful means to obtain structural information. However, this so-called bottom-up approach is labor-intensive and often suffers from large sample consumption. In addition, artifacts can be introduced during lengthy sample preparation. Samples with trace amounts, such as degraded products or protein variants, can require repeated fractionation and concentration to obtain enough material for enzymatic digestion. With expanding efforts to produce biosimilar protein drugs [126], it becomes critical to have a rapid and robust analytical method to characterize drug product and drug substance. Top-down approaches have the potential to become a solution to provide useful information in a fast and convenient fashion.

Top-down MS refers to a method where intact protein ions are introduced into the mass spectrometer and fragmented directly using tandem MS. The primary sequence can be obtained, as well as structural information about the protein, without the need to digest the protein by proteases [37, 38]. This analytical strategy usually requires a high-resolution mass spectrometer due to the large number of highly charged fragment ions generated and has been demonstrated to be successful for rapid characterization of small to medium size proteins (below 70 kDa) [127–129]. For large proteins, however, it is still a challenge for the top-down approach, with only limited success reported. Recently, top-down MS has been applied for the characterization of therapeutic proteins undergoing drug development. For example, it has been reported that direct ISF of IgG1 and IgG2 molecules on a linear ion-trap instrument and further analysis by an Orbitrap MS predominantly yielded N-terminal fragments that corresponded to the variable regions of the heavy and light chains

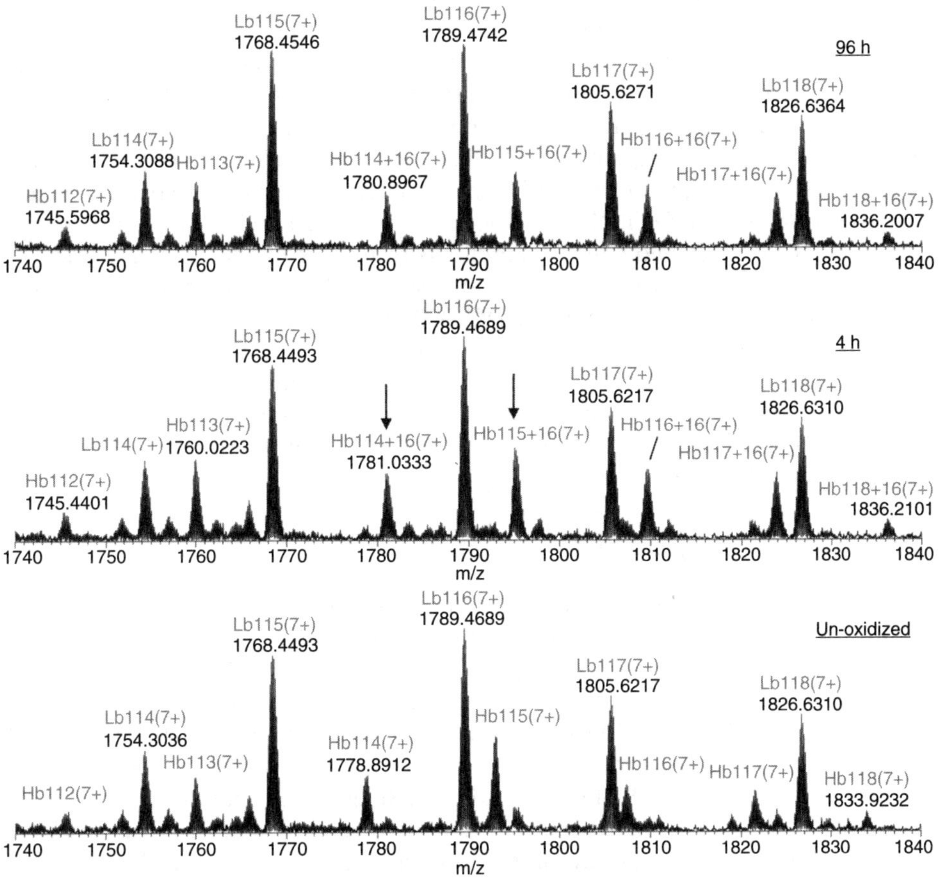

Figure 6.12 Part of in-source fragmentation spectra of the IgG2 samples after oxidation compared with the unoxidized sample. The heavy-chain variable region is oxidized after 4 h as indicated by the key fragment ions and the oxidation site is determined to be Met-114 on the heavy chain. Adapted with permission from Zhang, Z. et al. *Anal. Chem.*, **2007**, *79*(15), 5723–5729.

[23, 130–132]. As shown in Figure 6.12, after 4 h oxidation of the IgG2 sample, the HC variable region, as indicated by Hb114-Hb118 ions, is completely oxidized, while the LC shows no oxidation, even after 96 h. Further analysis determined that the oxidation site is located on Met-114, since the Hb112 and Hb113 ions are not oxidized. This observation makes the top-down MS a convenient method to confirm the sequence or identify modifications within the variable regions of mAbs. Because the variable regions determine the activity and specificity of the mAbs, and the sequences of these regions are unique to each mAb, it is particularly beneficial to have a rapid method to characterize these regions. In addition, top-down MS consumes little material, and can provide useful structural information rapidly.

A drawback with top-down MS for mAbs is its limited structural resolution. If the modification site is located in a region with limited MS/MS data, the modification site cannot be localized to a specific amino acid. In this case, middle-down MS might serve as a valuable alternative approach [8, 127]. In a middle-down experiment, the protein is first cleaved to a few large fragments, either by disulfide bond reduction or by limited proteolysis with a protease such as papain, pepsin, or LysC, before being introduced into the MS for MS/MS analysis. Middle-down experiments could potentially increase the structural resolution, with some limited sample preparations. Some applications have been reported by using middle-down approach to characterize protein therapeutics [133]. Recently, a middle-down LC/MS approach was applied for the rapid quantification and characterization of site-specific methionine oxidation in an IgG1 molecule [134]. LysC was used under limited proteolytic conditions to produce two major components of IgG1: Fab and Fc. These fractions were then reduced to produce three major species: LC, C-terminal region of the HC, and the N-terminal region of the HC. A diphenyl column was used to separate these three fragments by RP-HPLC. The oxidation sites on the LC and HC could be identified by direct fragmentation of these large segments of the protein. Met oxidation is an important modification in protein therapeutics as described previously. This methodology could serve as a rapid assay to monitor the Met oxidation during protein drug development.

6.4.4 ECD and ETD

Tandem MS methods for gas-phase sequencing involving vibrational excitation of peptide bonds, such as CID and infrared multiphoton dissociation (IRMPD), are the most common ways to fragment macromolecular ions to obtain structure information, but are generally limited to small proteins and peptides [135]. Excitation of parent ions via CID or IRMPD normally can lead to elimination of labile PTMs such as glycosylation or phosphorylation prior to backbone fragmentation [136]. Electron capture dissociation (ECD) [137] and ETD [138] and their potential application to top-down MS have been reviewed by several groups [139–143]. The proposed ECD mechanism involves the capture of an electron by a polycation, which leads to an excited radical species that undergoes subsequent bond cleavage. ETD fragments peptides by transferring an electron from a radical anion to a protonated peptide ion to induce peptide backbone cleavage. Nearly exclusive N-alkyl bond cleavage occurs rapidly and forms *c*, and *z*• product ions and labile PTMs are maintained after the fragmentation. This technique is particularly useful for glycan analysis [144, 145] and other PTMs' characterization. ECD is almost exclusively paired with an FT-ICR instrument, while ETD can been installed on an ion trap MS, which is less expensive and easier to maintain than FT-ICR MS. Moreover, implementation of ETD on a high-resolution linear trap-Orbitrap MS makes it possible to characterize much larger peptides, and becomes potentially ideal for middle-down experiments.

Various applications of ECD/ETD in protein modification analysis have been reported. One application is to use ETD for disulfide mapping [146, 147]. As already described, disulfide bonds play a critical role in protein structural stability. Therefore, it is important to characterize disulfide linkages in protein therapeutics, especially for mAbs. Since ETD can break up disulfide bonds before peptide backbone fragmentation, an online HPLC-MS strategy was developed by using CID for MS^3 to fragment product ions generated from ETD for MS^2 to characterize disulfide-containing peptides. In this way, disulfide-containing peptides can be identified by ETD and disulfide linkages can be characterized by MS^3. This approach was applied on a linear ion trap ETD instrument to determine the disulfide linkages of a therapeutic monoclonal antibody. It was demonstrated as a more simplified procedure compared with conventional approaches [48]. Another application is to use ECD/ETD to differentiate isoAsp from Asp products. Deamidation of Asn residue is a common chemical degradation pathway that can occur to mAbs during production and long-term storage. After deamidation, Asn is converted into Asp or isoAsp. In addition, some Asp residues may isomerize to isoAsp residues. All of these modifications can result in loss of protein drug efficacy and even induce immunogenicity. However, it is a great challenge to characterize isomerization of Asp through conventional CID approach since these two isomers share the same MW. Recently, ECD has been applied to differentiate isoAsp from Asp successfully through unique fragment ions [148, 149] and can even provide quantitative information [150]. ETD was also applied to demonstrate a similar result for isoAsp and Asp differentiation [151, 152]. ECD/ETD can provide a unique approach for PTM characterization, which will add great benefit for protein therapeutics development.

6.4.5 Quantitative Analysis of mAbs

Therapeutic mAbs are rapidly becoming one of the most important areas in drug development [153]. This rapid growth is placing increased demands for quantitative methods that are necessary for early drug development studies, PK studies, and drug exposure studies (e.g., toxicity studies). Enzyme-linked immunoassays (ELISA), in particular sandwich format ELISA, is the method of choice for such studies. ELISA methods are highly sensitive and offer high sample throughput. However, method development time can be lengthy due to the need to generate specific antidrug antibodies (5–6 months). In addition, ELISA methods can suffer interference effects and can have limited dynamic range [154]. During the past 15 years, a growing number of reports describing the use of MS for the quantification of proteins have been published [155–157]. MS can provide high analyte selectivity, rapid method development, and high sample throughput. MS used in the selected reaction monitoring mode (SRM) has proven to be a robust approach for protein quantification. SRM methods have high precision and wide dynamic range, and they are readily transferrable between laboratories [158].

When quantifying small protein and peptide therapeutics, SRM methods can be applied without the need for proteolytic digestion [159, 160]. In some instances, researchers found that sample denaturing was required to dissociate the protein drug from interaction with endogenous antidrug antibodies [161]. Such a step is difficult to implement with ELISA techniques and is another advantage that MS offers for the quantification of protein therapeutics. As the MW of the therapeutic protein increases, the need to proteolytically digest the protein into smaller peptides arises because fragmentation of large proteins is generally insufficient to yield a sensitive and selective method. In this case, signature peptides are selected that have unique sequences for the protein of interest. Peptides can be synthetically labeled with ^{13}C and or ^{15}N to be used as internal standards (ISTD) [156]. Signature peptides selected for quantification should be chemically stable and devoid of missed cleavage sites to enable a robust method [162–164].

The first reported example describing the use of LC-MS/MS for the quantification of therapeutic mABs in serum using the signature peptide approach was by Hagman et al. [165]. A tryptic peptide from a complement determining region (CDR) was used for selectivity against the serum background. To reduce the sample complexity, an albumin depletion step was used prior to the digestion step. A limit of quantification (LOQ) of 2 μg/mL was achieved. More recent improvements in using LC-MS/MS for mAb quantification have been published. Heudi et al. [153] demonstrated the advantage of using a full-length isotope-labeled mAb ISTD for quantification. The mAb ISTD was prepared by using $^{13}C/^{15}N$-labeled threonine in the CHO cell growth medium. Use of the full-length mAb ISTD has the advantage of compensating for sample losses occurring on the protein level (prior to the digestion step) and allowing greater flexibility in the selection of the signature peptide. For example, the signature peptide can be readily changed during method development. The sample preparation employed by the authors used digestion of the serum sample directly, without prefractionation, and then used solid-phase extraction (SPE) to clean up the tryptic peptides. The HPLC method used for quantification was 11 min in duration and the LOQ was 5 μg/mL. A direct comparison between the MS method and an ELISA method using serum samples from an animal PK study found that similar drug profiles are obtained. However, the LC-MS/MS method yielded higher drug exposure because in this particular case, the LC-MS/MS measured total drug levels whereas the ELISA method measured free drug levels. This illustrates that in certain cases, LC-MS/MS can be complementary to ELISA and that it is important to understand the fundamentals of each method to properly interpret the results.

Further improvements in mAb quantification methods have been reported. Ji et al. [166] used dimethyl labeling with deuterated formaldehyde on the peptide level to produce labeled ISTD peptides and enable rapid method development time (1–2 weeks). However, this approach does not overcome problems associated with the potential loss of full length mAb prior to the digestion step [167, 168]. The authors applied modern ultra-performance liquid chromatography (UPLC) technology to enable a rapid 6-min gradient method to be used and increase the sample throughput and sensitivity of the method.

Most recently, refinements in sample preparation with a focus on analyte enrichment on the intact protein level have been reported for mAb quantification. Lu et al. [169] compared three different serum sample enrichment strategies: albumin depletion, Protein A purification, and antidrug antibody purification. The authors found that all three approaches could be applied successfully and that the antidrug antibody purification approach produced 100-fold greater sensitivity than the Protein A purification strategy. The Protein A strategy is about 10-fold more sensitive than the albumin depletion method. Albumin depletion is applicable to a broad range of protein therapeutics but offers the poorest sensitivity. Both the Protein A and albumin depletion methods require the least amount of development time compared with the antidrug antibody purification method due to the need to produce a selective antidrug antibody. The authors point out that during early drug development the albumin depletion and Protein A methods are appropriate and that the antidrug antibody offers greater sensitivity that could be applied for clinical studies occurring during late stage drug development.

6.5 SUMMARY

Therapeutic proteins are complex molecules that rely on a variety of analytical methodologies in order to be fully characterized. As therapeutic protein drug candidates progress from the discovery stage, through the development stage, and finally commercialization, analysis of the drug is required. MS, as highlighted in this chapter, plays a critical role in the investigation of protein structure and product quality, and is indispensable in the drug development process. MS applications have advanced to the point where they are now applied to investigate all aspects of protein therapeutic structure, including intact mass analysis, peptide mass mapping, glycosylation analysis, disulfide linkage analysis, PTM analysis, PEGylation site investigation, high order structure, quantification, and so on.

Intact mass analysis is applied to whole molecule protein therapeutics and, in the case of mAbs, can be used to analyze subunits: HC, LC, Fab, and Fc. Intact mass analysis provides a global view of protein structure and is used to confirm that the expressed protein drug is consistent with the theoretical sequence and at the same time enables investigation of protein heterogeneity. Examples of heterogeneity include glycosylation, oxidation, clipping, pyroglutamate formation, deamidation, and disulfide conformers. Peptide mass mapping is used to confirm the primary sequence and to localize the sites of a wide variety of protein modifications. Glycosylation, which influences protein function, activity, stability, and immunogenicity, can be studied globally using intact mass analysis, in a site-specific manner by digesting the protein into glycopeptides, or structurally by releasing the glycans to be analyzed by MS and MS/MS. Free thiols can be investigated by using differential labeling, while disulfides and trisulfides can be investigated using nonreduced proteolytic digestion conditions. PEGylation of protein therapeutics results in a highly heterogeneous product that is challenging to analyze. MALDI-TOF has been used for many years to characterize the degree of PEGylation because the technique is straightforward to apply. Recent advancements in ISD for MALDI-TOF and charge state reduction for ESI-TOF will enable MS to play an increased role in the characterization of PEGylated protein therapeutics. In some cases, qualitative MS methods can be modified to provide quantification analysis if the proper analytical controls are used. For example, peptides generated by digestion of a therapeutic protein can be used as surrogates to allow protein quantification. By using isotopically labeled ISTD peptides or proteins, precise and accurate quantification can be achieved. The approach has a key advantage over ELISA since there is no dependence on immunoreagents and hence method development time can be reduced.

Continued advancement in MS instrumentation and techniques will allow MS to be used to probe protein structure more extensively. HDX, which analyzes proteins exposed to D_2O solvent, allows higher order structure to be studied. Solvent accessible hydrogen atoms exchange deuterium, while hydrophobic regions do not. In this way, the influence of modifications such as glycosylation and the interaction of proteins with receptors can be investigated. Top-down MS, which allows the primary sequence to be determined without proteolytic digestion, can yield very high sequence coverage and provides the ability to study numerous PTMs in a site-specific manner. IMMS combines two powerful ion separation methods and enables highly selective and sensitive analysis of structurally related proteins. IMS separates ions on the basis of size and shape, while MS separates ions on the basis of mass-to-charge. With the introduction of commercial instruments, IMMS will play an increasingly important role in the characterization of protein therapeutics. ETD and ECD provide an alternative ion fragmentation mode to CID and enables protein sequence to be studied more extensively. Detection of labile PTMs like phosphorylation and the ability to distinguish Asp from isoAsp are just two examples of the analytical advantages that ETD and ECD can provide.

As discussed in this chapter, MS is indispensable for analysis of protein therapeutics. Academic and industrial researchers, along with instrument manufactures, have developed a variety of analytical MS applications and instruments that enable protein therapeutics to be evaluated in great detail. As a result, the discovery and development of protein therapeutics have benefited tremendously. MS has, and will continue, to play a critical role in ensuring the quality, safety, and efficacy of protein therapeutics.

ACKNOWLEDGMENTS

We would like to thank Drs. Ray Bakhtiar, Lorenzo Chen, Van Hoang, and Lawrence Dick Jr. for critical reading, helpful feedback, and comments on this chapter.

REFERENCES

[1] http://www.cancer.gov/dictionary/?CdrID=426407.

[2] Reichert JM, Rosensweig CJ, Faden LB, Dewitz MC. Monoclonal antibody successes in the clinic. Nat Biotechnol 2005;23:1073–8.

[3] Leader B, Baca QJ, Golan DE. Protein therapeutics: a summary and pharmacological classification. Nat Rev Drug Discov 2008;7:21–39.

[4] Dimitrov DS, Marks JD. Therapeutic antibodies: current state and future trends—is a paradigm change coming soon? Methods Mol Biol 2009;525:1–27.

[5] Nelson AL, Dhimolea E, Reichert JM. Development trends for human monoclonal antibody therapeutics. Nat Rev Drug Discov 2010;9:767–74.

[6] Crossley L. Protein characterization through the stages. BioPharm Int 2010;(Suppl. S):4–7.

[7] Barnes CAS, Lim A. Applications of mass spectrometry for the structural characterization of recombinant protein pharmaceuticals. Mass Spectrom Rev 2007;26:370–88.

[8] Zhang Z, Pan H, Chen X. Mass spectrometry for structural characterization of therapeutic antibodies. Mass Spectrom Rev 2009;28:147–76.

[9] Chen G, Warrack BM, Goodenough AK, Wei H, Wang-Iverson DB, Tymiak AA. Characterization of protein therapeutics by mass spectrometry: recent developments and future directions. Drug Discov Today 2011;16:58–64.

[10] Kim YJ, Doyle ML. Structural mass spectrometry in protein therapeutics discovery. Anal Chem 2010;82:7083–9.

[11] Adamczyk M, Gebler JC, Harrington CA, Sequeira AF. The use of electrospray ionization mass spectrometry to distinguish the lot-to-lot heterogeneity of an antigen specific monoclonal antibody from a specific cellular clone. Eur. J Mass Spectrom 1999; 5:165–8.

[12] Dillon TM, Bondarenko PV, Ricci MS. Development of an analytical reversed-phase high-performance liquid chromatography-electrospray ionization mass spectrometry method for characterization of recombinant antibodies. J Chromatogr A 2004; 1053:299–305.

[13] Dillon TM, Bondarenko PV, Rehder DS, Pipes GD, Kleemann GR, Ricci MS. Optimization of a reversed-phase high-performance liquid chromatography/mass spectrometry method for characterizing recombinant antibody heterogeneity and stability. J Chromatogr A 2006;1120:112–20.

[14] Chelius D, Jing K, Lueras A, Rehder DS, Dillon TM, Vizel A, et al. Formation of pyroglutamic acid from N-terminal glutamic acid in immunoglobulin gamma antibodies. Anal Chem 2006;78:2370–6.

[15] Yang J, Wang S, Liu J, Raghani A. Determination of tryptophan oxidation of monoclonal antibody by reversed phase high performance liquid chromatography. J Chromatogr A 2007;1156:174–82.

[16] Chu GC, Chelius D, Xiao G, Khor HK, Coulibaly S, Bondarenko PV. Accumulation of succinimide in a recombinant monoclonal antibody in mildly acidic buffers under elevated temperatures. Pharm Res 2007;24:1145–56.

[17] Yan B, Valliere-Douglass J, Brady L, Steen S, Han M, Pace D, et al. Analysis of post-translational modifications in recombinant monoclonal antibody IgG1 by reversed-phase liquid chromatography/mass spectrometry. J Chromatogr A 2007;1164:153–61.

[18] Brady LJ, Valliere-Douglass J, Martinez T, Balland A. Molecular mass analysis of antibodies by on-line SEC-MS. J Am Soc Mass Spectrom 2008;19:502–9.

[19] Hoffstetter-Kuhn S, Alt G, Kuhn R. Profiling of oligosaccharide-mediated microheterogeneity of a monoclonal antibody by capillary electrophoresis. Electrophoresis 1996; 17:418–22.

[20] Beck A, Bussat MC, Zorn N, Robillard V, Klinguer-Hamour C, Chenu S, et al. Characterization by liquid chromatography combined with mass spectrometry of monoclonal anti-IGF-1 receptor antibodies produced in CHO and NS0 cells. J Chromatogr B Analyt Technol Biomed Life Sci 2005;819:203–18.

[21] Feng R, Konishi Y. Analysis of antibodies and other large glycoproteins in the mass range of 150,000–200,000 Da by electrospray ionization mass spectrometry. Anal Chem 1992;64:2090–5.

[22] Le JC, Bondarenko PV. Trap for MAbs: characterization of intact monoclonal antibodies using reversed-phase HPLC on-line with ion-trap mass spectrometry. J Am Soc Mass Spectrom 2005;16:307–11.

[23] Bondarenko PV, Second TP, Zabrouskov V, Makarov AA, Zhang Z. Mass measurement and top-down HPLC/MS analysis of intact monoclonal antibodies on a hybrid linear

quadrupole ion trap-Orbitrap mass spectrometer. J Am Soc Mass Spectrom 2009; 20:1415–24.

[24] Zheng S, Yoo C, Delmotte N, Miller FR, Huber CJ, Lubman DM. Monolithic column HPLC separation of intact proteins analyzed by LC-MALDI using on-plate digestion: an approach to integrate protein separation and identification. Anal Chem 2006;78: 5198–204.

[25] Jenkins N. Modifications of therapeutic proteins: challenges and prospects. Cytotechnology 2007;53:121–5.

[26] Jenkins N, Murphy L, Tyther R. Post-translational modifications of recombinant proteins: significance for biopharmaceuticals. Mol Biotechnol 2008;39:113–8.

[27] Dougherty J, Mhatre R, Moore S. Using peptide maps as identity and purity tests for lot release testing of recombinant therapeutic proteins. BioPharm Int 2003;16:54–8.

[28] Allen D, Baffi R, Bausch J, Bongers J, Costello M, Dougherty J, Jr., et al. Validation of peptide mapping for protein identity and genetic stability. Biologics and biotechnology section, pharmaceutical research and manufacturers of America. Biologicals 1996; 24:255–75.

[29] Bongers J, Cummings JJ, Ebert MB, Federici MM, Gledhill L, Gulati D, et al. Validation of a peptide mapping method for a therapeutic monoclonal antibody: what could we possibly learn about a method we have run 100 times? J Pharm Biomed Anal 2000; 21:1099–128.

[30] Dick LW, Jr., Mahon D, Qiu D, Cheng KC. Peptide mapping of therapeutic monoclonal antibodies: improvements for increased speed and fewer artifacts. J Chromatogr B Analyt Technol Biomed Life Sci 2009;877:230–6.

[31] Ren D, Pipes GD, Liu D, Shih LY, Nichols AC, Treuheit MJ, et al. An improved trypsin digestion method minimizes digestion-induced modifications on proteins. Anal Biochem 2009;392:12–21.

[32] Silva JC, Gorenstein MV, Li GZ, Vissers JPC, Geromanos SJ. Absolute quantification of proteins by LCMSE: a virtue of parallel MS acquisition. Mol Cell Proteomics 2006;5:144–56.

[33] Chakraborty AB, Berger SJ, Gebler JC. Use of an integrated MS-multiplexed MS/MS data acquisition strategy for high-coverage peptide mapping studies. Rapid Commun Mass Spectrom 2007;21:730–44.

[34] Xie H, Gilar M, Gebler JC. Characterization of protein impurities and site-specific modifications using peptide mapping with liquid chromatography and data independent acquisition mass spectrometry. Anal Chem 2009;81:5699–708.

[35] Xie H, Chakraborty A, Ahn J, Yu YQ, Dakshinamoorthy DP, Gilar M, et al. Rapid comparison of a candidate biosimilar to an innovator monoclonal antibody with advanced liquid chromatography and mass spectrometry technologies. MAbs 2010; 2:1–16.

[36] Chelius D, Xiao G, Nichols AC, Vizel A, He B, Dillon TM, et al. Automated tryptic digestion procedure for HPLC/MS/MS peptide mapping of immunoglobulin gamma antibodies in pharmaceutics. J Pharm Biomed Anal 2008;47:285–94.

[37] Kelleher NL, Lin HY, Valaskovic GA, Aaserud DJ, Fridriksson EK, McLafferty FW. Top down versus bottom up protein characterization by tandem high-resolution mass spectrometry. J Am Chem Soc 1999;121:806–12.

[38] Du Y, Parks BA, Sohn S, Kwast KE, Kelleher NL. Top-down approaches for measuring expression ratios of intact yeast proteins using Fourier transform mass spectrometry. Anal Chem 2006;78:686–94.

[39] Dalpathado DS, Desaire H. Glycopeptide analysis by mass spectrometry. Analyst 2008; 133:731–8.

[40] An HJ, Froehlich JW, Lebrilla CB. Determination of glycosylation sites and site-specific heterogeneity in glycoproteins. Curr Opin Chem Biol 2009;13:421–6.

[41] Brooks SA. Strategies for analysis of the glycosylation of proteins: current status and future perspectives. Mol Biotechnol 2009;43:76–88.

[42] Qian J, Liu T, Yang L, Daus A, Crowley R, Zhou Q. Structural characterization of N-linked oligosaccharides on monoclonal antibody cetuximab by the combination of orthogonal matrix-assisted laser desorption/ionization hybrid quadrupole–quadrupole time-of-flight tandem mass spectrometry and sequential enzymatic digestion. Anal Biochem 2007;364:8–18.

[43] Harvey DJ. Identification of protein-bound carbohydrates by mass spectrometry. Proteomics 2001;1:311–28.

[44] Conboy JJ, Henion JD. The determination of glycopeptides by liquid chromatography/mass spectrometry with collision-induced dissociation. J Am Soc Mass Spectrom 1992; 3:804–14.

[45] Huddleston MJ, Bean MF, Carr SA. Collisional fragmentation of glycopeptides by electrospray ionization LC/MS and LC/MS/MS: methods for selective detection of glycopeptides in protein digests. Anal Chem 1993;65:877–84.

[46] Benham AM. Protein folding and disulfide bond formation in the eukaryotic cell: meeting report based on the presentations at the European Network Meeting on Protein Folding and Disulfide Bond Formation 2009 (Elsinore, Denmark). FEBS J 2009;276:6905–11.

[47] Riemer J, Bulleid N, Herrmann JM. Disulfide formation in the ER and mitochondria: two solutions to a common process. Science 2009;324:1284–7.

[48] Gorman JJ, Wallis TP, Pitt JJ. Protein disulfide bond determination by mass spectrometry. Mass Spectrom Rev 2002;21:183–216.

[49] Schilling B, Yoo CB, Collins CJ, Gibson BW. Determining cysteine oxidation status using differential alkylation. Int J Mass Spectrom 2004;236:117–27.

[50] Chumsae C, Gaza-Bulseco G, Liu H. Identification and localization of unpaired cysteine residues in monoclonal antibodies by fluorescence labeling and mass spectrometry. Anal Chem 2009;81:6449–57.

[51] Xiang T, Chumsae C, Liu H. Localization and quantitation of free sulfhydryl in recombinant monoclonal antibodies by differential labeling with 12C and 13C iodoacetic acid and LC-MS analysis. Anal Chem 2009;81:8101–8.

[52] Wypych J, Li M, Guo A, Zhang Z, Martinez T, Allen MJ, et al. Human IgG2 antibodies display disulfide-mediated structural isoforms. J Biol Chem 2008;283:16194–205.

[53] Dillon TM, Ricci MS, Vezina C, Flynn GC, Liu YD, Rehder DS, et al. Structural and functional characterization of disulfide isoforms of the human IgG2 subclass. J Biol Chem 2008;283:16206–15.

[54] Pristatsky P, Cohen SL, Krantz D, Acevedo J, Ionescu R, Vlasak J. Evidence for trisulfide bonds in a recombinant variant of a human IgG2 monoclonal antibody. Anal Chem 2009;81:6148–55.

[55] Jespersen AM, Christensen T, Klausen NK, Nielsen F, Sørensen HH. Characterisation of a trisulphide derivative of biosynthetic human growth hormone produced in *Escherichia coli*. Eur J Biochem 1994;219:365–73.

[56] Andersson C, Edlund PO, Gellerfors P, Hansson Y, Holmberg E, Hult C, et al. Isolation and characterization of a trisulfide variant of recombinant human growth hormone formed during expression in *Escherichia coli*. Int J Pept Protein Res 1996;47:311–21.

[57] Breton J, La Fiura A, Bertolero F, Orsini G, Valsasina B, Ziliotto R, et al. Structure, stability and biological properties of a N-terminally truncated form of recombinant

human interleukin-6 containing a single disulfide bond. Eur J Biochem 1995; 227:573–81.

[58] Okado-Matsumoto A, Guan Z, Fridovich I. Modification of cysteine 111 in human Cu,Zn-superoxide dismutase. Free Radic Biol Med 2006;41:1837–46.

[59] Gu S, Wen D, Weinreb PH, Sun Y, Zhang L, Foley SF, et al. Characterization of trisulfide modification in antibodies. Anal Biochem 2010;400:89–98.

[60] Gaberc-Porekar V, Zore I, Podobnik B, Menart V. Obstacles and pitfalls in the PEGylation of therapeutic proteins. Curr Opin Drug Discov Devel 2008;11:242–50.

[61] Parveen S, Sahoo SK. Nanomedicine: clinical applications of polyethylene glycol conjugated proteins and drugs. Clin Pharmacokinet 2006;45:965–88.

[62] Levy Y, Hershfield MS, Fernandez-Mejia C, Polmar SH, Scudiery D, Berger M, et al. Adenosine deaminase deficiency with late onset of recurrent infections: response to treatment with polyethylene glycol-modified adenosine deaminase. J Pediatr 1988; 113:312–7.

[63] Roberts MJ, Bentley MD, Harris JM. Chemistry for peptide and protein PEGylation. Adv Drug Deliv Rev 2002;54:459–76.

[64] Schultz P, Wang L, Anderson JC, Chin J, Liu DR, Magliery TJ, et al. Methods and compositions for the production of orthogonal tRNA-aminoacyl tRNA synthetase pairs. US Patent Number: 7083970, 2006.

[65] Vestling MM, Murphy CM, Keller DA, Fenselau C, Dedinas J, Ladd DL, et al. A strategy for characterization of polyethylene glycol-derivatized proteins. A mass spectrometric analysis of the attachment sites in polyethylene glycol-derivatized superoxide dismutase. Drug Metab Dispos 1993;21:911–7.

[66] Chowdhury SK, Doleman M, Johnston D. Fingerprinting proteins coupled with polymers by mass spectrometry: investigation of polyethylene glycol-conjugated superoxide dismutase. J Am Soc Mass Spectrom 1995;6:478–87.

[67] Na DH, Youn YS, Lee KC. Optimization of the PEGylation process of a peptide by monitoring with matrix-assisted laser desorption/ionization time-of-flight mass spectrometry. Rapid Commun Mass Spectrom 2003;17:2241–4.

[68] Na DH, Youn YS, Lee KC. Matrix-assisted laser desorption/ionization time-of-flight mass spectrometry for monitoring and optimization of site-specific PEGylation of ricin A-chain. Rapid Commun Mass Spectrom 2004;18:2185–9.

[69] Yoo C, Suckau D, Sauerland V, Ronk M, Ma M. Toward top-down determination of PEGylation site using MALDI in-source decay MS analysis. J Am Soc Mass Spectrom 2009;20:326–33.

[70] Lennon JJ, Walsh KA. Direct sequence analysis of proteins by in-source fragmentation during delayed ion extraction. Protein Sci 1997;6:2446–53.

[71] Reiber DC, Grover TA, Brown RS. Identifying proteins using matrix-assisted laser desorption/ionization in-source fragmentation data combined with database searching. Anal Chem 1998;70:673–83.

[72] Lee KC, Moon SC, Park MO, Lee JT, Na DH, Yoo SD, et al. Isolation, characterization, and stability of positional isomers of mono-PEGylated salmon calcitonins. Pharm Res 1999;16:813–8.

[73] Na DH, Park MO, Choi SY, Kim YS, Lee SS, Yoo SD, et al. Identification of the modifying sites of mono-PEGylated salmon calcitonins by capillary electrophoresis and MALDI-TOF mass spectrometry. J Chromatogr B Biomed Sci Appl 2001;754:259–63.

[74] Lu X, Gough PC, DeFelippis MR, Huang L. Elucidation of PEGylation site with a combined approach of in-source fragmentation and CID MS/MS. J Am Soc Mass Spectrom 2010;21:810–8.

[75] Bagal D, Zhang H, Schnier PD. Gas-phase proton-transfer chemistry coupled with TOF mass spectrometry and ion mobility-MS for the facile analysis of poly(ethylene glycols) and PEGylated polypeptide conjugates. Anal Chem 2008;80:2408–18.

[76] Huang L, Gough PC, DeFelippis MR. Characterization of poly(ethylene glycol) and PEGylated products by LC/MS with postcolumn addition of amines. Anal Chem 2009;81:567–77.

[77] Geiger T, Clarke S. Deamidation, isomerization, and racemization at asparaginyl and aspartyl residues in peptides. Succinimide-linked reactions that contribute to protein degradation. J Biol Chem 1987;262:785–94.

[78] Stephenson RC, Clarke S. Succinimide formation from aspartyl and asparaginyl peptides as a model for the spontaneous degradation of proteins. J Biol Chem 1989;264: 6164–70.

[79] Wearne SJ, Creighton TE. Effect of protein conformation on rate of deamidation: ribonuclease A. Proteins 1989;5:8–12.

[80] Patel K, Borchardt RT. Chemical pathways of peptide degradation. III. Effect of primary sequence on the pathways of deamidation of asparaginyl residues in hexapeptides. Pharm Res 1990;7:787–93.

[81] Xie M, Schowen RL. Secondary structure and protein deamidation. J Pharm Sci 1999;88:8–13.

[82] Capasso S, Salvadori S. Effect of the three-dimensional structure on the deamidation reaction of ribonuclease A. J Pept Res 1999;54:377–82.

[83] Robinson NE, Robinson AB. Molecular clocks. Proc Natl Acad Sci U S A 2001;98: 944–9.

[84] Athmer L, Kindrachuk J, Georges F, Napper S. The influence of protein structure on the products emerging from succinimide hydrolysis. J Biol Chem 2002;277:30502–7.

[85] Robinson NE, Robinson ZW, Robinson BR, Robinson AL, Robinson JA, Robinson ML, et al. Structure-dependent nonenzymatic deamidation of glutaminyl and asparaginyl pentapeptides. J Pept Res 2004;63:426–36.

[86] Chelius D, Rehder DS, Bondarenko PV. Identification and characterization of deamidation sites in the conserved regions of human immunoglobulin gamma antibodies. Anal Chem 2005;77:6004–11.

[87] Vlasak J, Bussat MC, Wang S, Wagner-Rousset E, Schaefer M, Klinguer-Hamour C, et al. Identification and characterization of asparagine deamidation in the light chain CDR1 of a humanized IgG1 antibody. Anal Biochem 2009;392:145–54.

[88] Houde D, Kauppinen P, Mhatre R, Lyubarskaya Y. Determination of protein oxidation by mass spectrometry and method transfer to quality control. J Chromatogr A 2006;1123:189–98.

[89] Wurm FM. Production of recombinant protein therapeutics in cultivated mammalian cells. Nat Biotechnol 2004;22:1393–8.

[90] Grillberger L, Kreil TR, Nasr S, Reiter M. Emerging trends in plasma-free manufacturing of recombinant protein therapeutics expressed in mammalian cells. Biotechnol J 2009;4:186–201.

[91] Engen JR. Analysis of protein conformation and dynamics by hydrogen/deuterium exchange MS. Anal Chem 2009;81:7870–5.

[92] Wales TE, Engen JR. Hydrogen exchange mass spectrometry for the analysis of protein dynamics. Mass Spectrom Rev 2006;25:158–70.

[93] Tuma R, Coward LU, Kirk MC, Barnes S, Prevelige PE, Jr. Hydrogen-deuterium exchange as a probe of folding and assembly in viral capsids. J Mol Biol 2001; 306:389–96.

[94] Wang L, Smith DL. Capsid structure and dynamics of a human rhinovirus probed by hydrogen exchange mass spectrometry. Protein Sci 2005;14:1661–72.

[95] Coales SJ, Tuske SJ, Tomasso JC, Hamuro Y. Epitope mapping by amide hydrogen/deuterium exchange coupled with immobilization of antibody, on-line proteolysis, liquid chromatography and mass spectrometry. Rapid Commun Mass Spectrom 2009;23:639–47.

[96] Gajiwala KS, Wu JC, Christensen J, Deshmukh GD, Diehl W, DiNitto JP, et al. KIT kinase mutants show unique mechanisms of drug resistance to imatinib and sunitinib in gastrointestinal stromal tumor patients. Proc Natl Acad Sci U S A 2009;106:1542–7.

[97] Zhang HM, Yu X, Greig MJ, Gajiwala KS, Wu JC, Diehl W, et al. Drug binding and resistance mechanism of KIT tyrosine kinase revealed by hydrogen/deuterium exchange FTICR mass spectrometry. Protein Sci 2010;19:703–15.

[98] Bobst CE, Abzalimov RR, Houde D, Kloczewiak M, Mhatre R, Berkowitz SA, et al. Detection and characterization of altered conformations of protein pharmaceuticals using complementary mass spectrometry-based approaches. Anal Chem 2008;80:7473–81.

[99] Houde D, Arndt J, Domeier W, Berkowitz S, Engen JR. Characterization of IgG1 conformation and conformational dynamics by hydrogen/deuterium exchange mass spectrometry. Anal Chem 2009;81:2644–51.

[100] Kaltashov IA, Bobst CE, Abzalimov RR, Berkowitz SA, Houde D. Confirmation and dynamics of biopharmaceuticals: transition of mass spectrometry-based tools from academe to industry. J Am Soc Mass Spectrom 2010;21:323–37.

[101] Houde D, Peng Y, Berkowitz SA, Engen JR. Post-translational modifications differentially affect IgG1 conformation and receptor binding. Mol Cell Proteomics 2010;9:1716–28.

[102] Cravello L, Lascoux D, Forest E. Use of different proteases working in acidic conditions to improve sequence coverage and resolution in hydrogen/deuterium exchange of large proteins. Rapid Commun Mass Spectrom 2003;17:2387–93.

[103] Zhang HM, Kazazic S, Schaub TM, Tipton JD, Emmett MR, Marshall AG. Enhanced digestion efficiency, peptide ionization efficiency, and sequence resolution for protein hydrogen/deuterium exchange monitored by Fourier transform ion cyclotron resonance mass spectrometry. Anal Chem 2008;80:9034–41.

[104] Zhang HM, McLoughlin SM, Frausto SD, Tang H, Emmett MR, Marshall AG. Simultaneous reduction and digestion of proteins with disulfide bonds for hydrogen/deuterium exchange monitored by mass spectrometry. Anal Chem 2010;82:1450–4.

[105] Iacob RE, Murphy JP, Engen JR. Ion mobility adds an additional dimension to mass spectrometric analysis of solution-phase hydrogen/deuterium exchange. Rapid Commun Mass Spectrom 2008;22:2898–904.

[106] Kaltashov IA, Bobst CE, Abzalimov RR. H/D exchange and mass spectrometry in the studies of protein conformation and dynamics: is there a need for a top-down approach? Anal Chem 2009;81:7892–9.

[107] Althaus E, Canzar S, Ehrler C, Emmett MR, Karrenbauer A, Marshall AG, et al. Computing H/D-exchange rates of single residues from data of proteolytic fragments. BMC Bioinformatics 2010;11:1–12.

[108] Kazazic S, Zhang HM, Schaub TM, Emmett MR, Hendrickson CL, Blakney GT, et al. Automated data reduction for hydrogen/deuterium exchange experiments, enabled by high-resolution Fourier transform ion cyclotron resonance mass spectrometry. J Am Soc Mass Spectrom 2010;21:550–8.

[109] Karas M, Hillenkamp F. Laser desorption ionization of proteins with molecular masses exceeding 10,000 daltons. Anal Chem 1988;60:2299–301.

[110] Tanaka K, Waki H, Ido Y, Akita S, Yoshida Y, Yoshida T. Protein and polymer analyses up to *m/z* 100,000 by laser ionization time-of-flight mass spectrometry. Rapid Commun Mass Spectrom 1988;2:151–3.

[111] Fenn JB, Mann M, Meng CK, Wong SF, Whitehouse CM. Electrospray ionization for mass spectrometry of large biomolecules. Science 1989;246:64–71.

[112] Wittmer D, Chen YH, Luckenbill BK, Hill HH, Jr. Electrospray ionization ion mobility spectrometry. Anal Chem 1994;66:2348–55.

[113] Clemmer DE, Hudgins RR, Jarrold MF. Naked protein conformations: cytochrome c in the gas phase. J Am Chem Soc 1995;117:10141–2.

[114] von Helden G, Wyttenbach T, Bowers MT. Conformation of macromolecules in the gas phase: use of matrix-assisted laser desorption methods in ion chromatography. Science 1995;267:1483–5.

[115] Kanu AB, Dwivedi P, Tam M, Matz L, Hill HH, Jr. Ion mobility–mass spectrometry. J Mass Spectrom 2008;43:1–22.

[116] Ruotolo BT, Benesch JLP, Sandercock AM, Hyung SJ, Robinson CV. Ion mobility-mass spectrometry analysis of large protein complexes. Nat Protoc 2008;3:1139–52.

[117] Bohrer BC, Merenbloom SI, Koeniger SL, Hilderbrand AE, Clemmer DE. Biomolecule analysis by ion mobility spectrometry. Annu Rev Anal Chem 2008;1:293–327.

[118] Fenn LS, McLean JA. Biomolecular structural separations by ion mobility-mass spectrometry. Anal Bioanal Chem 2008;391:905–9.

[119] Uetrecht C, Rose RJ, van Duijn E, Lorenzen K, Heck AJR. Ion mobility mass spectrometry of proteins and protein assemblies. Chem Soc Rev 2010;39:1633–55.

[120] Bagal D, Valliere-Douglass JF, Balland A, Schnier PD. Resolving disulfide structural isoforms of IgG2 monoclonal antibodies by ion mobility mass spectrometry. Anal Chem 2010;82:6751–5.

[121] Valentine SJ, Plasencia MD, Liu X, Krishnan M, Naylor S, Udseth HR, et al. Toward plasma proteome profiling with ion mobility-mass spectrometry. J Proteome Res 2006;5:2977–84.

[122] Liu X, Valentine SJ, Plasencia MD, Trimpin S, Naylor S, Clemmer DE. Mapping the human plasma proteome by SCX-LC-IMS-MS. J Am Soc Mass Spectrom 2007;18:1249–64.

[123] Guevremont R, Purves RW. Atmospheric pressure ion focusing in a high-field asymmetric waveform ion mobility spectrometer. Rev Sci Instrum 1999;70:1370–83.

[124] Pringle SD, Giles K, Wildgoose JL, Williams JP, Slade SE, Thalassinos K, et al. An investigation of the mobility separation of some peptide and protein ions using a new hybrid quadrupole/travelling wave IMS/oa-ToF instrument. Int J Mass Spectrom 2007;261:1–12.

[125] Shvartsburg AA, Smith RD. Fundamentals of traveling wave ion mobility spectrometry. Anal Chem 2008;80:9686–99.

[126] Kresse GB. Biosimilars—science, status, and strategic perspective. Eur J Pharm Biopharm 2009;72:479–86.

[127] Breuker K, Jin M, Han X, Jiang H, McLafferty FW. Top-down identification and characterization of biomolecules by mass spectrometry. J Am Soc Mass Spectrom 2008;19:1045–53.

[128] Kellie JF, Tran JC, Lee JE, Ahlf DR, Thomas HM, Ntai I, et al. The emerging process of Top Down mass spectrometry for protein analysis: biomarkers, protein-therapeutics, and achieving high throughput. Mol Biosyst 2010;6:1532–9.

[129] Garcia BA. What does the future hold for Top Down mass spectrometry? J Am Soc Mass Spectrom 2010;21:193–202.

[130] Zhang Z, Shah B. Characterization of variable regions of monoclonal antibodies by top-down mass spectrometry. Anal Chem 2007;79:5723–9.

[131] Ren D, Pipes GD, Hambly D, Bondarenko PV, Treuheit MJ, Gadgil HS. Top-down N-terminal sequencing of immunoglobulin subunits with electrospray ionization time of flight mass spectrometry. Anal Biochem 2009;384:42–8.

[132] Zhang J, Liu H, Katta V. Structural characterization of intact antibodies by high-resolution LTQ Orbitrap mass spectrometry. J Mass Spectrom 2010;45:112–20.

[133] Kleemann GR, Beierle J, Nichols AC, Dillon TM, Pipes GD, Bondarenko PV. Characterization of IgG1 immunoglobulins and peptide-Fc fusion proteins by limited proteolysis in conjunction with LC-MS. Anal Chem 2008;80:2001–9.

[134] Pipes GD, Campbell P, Bondarenko PV, Kerwin BA, Treuheit MJ, Gadgil HS. Middle-down fragmentation for the identification and quantitation of site-specific methionine oxidation in an IgG1 molecule. J Pharm Sci 2010;99:4469–76.

[135] Domon B, Aebersold R. Mass spectrometry and protein analysis. Science 2006; 312:212–7.

[136] Chi A, Huttenhower C, Geer LY, Coon JJ, Syka JEP, Bai DL, et al. Analysis of phosphorylation sites on proteins from *Saccharomyces cerevisiae* by electron transfer dissociation (ETD) mass spectrometry. Proc Natl Acad Sci U S A 2007;104:2193–8.

[137] Zubarev RA, Kelleher NL, McLafferty FW. Electron capture dissociation of multiply charged protein cations. A nonergodic process. J Am Chem Soc 1998;120:3265–6.

[138] Syka JEP, Coon JJ, Schroeder MJ, Shabanowitz J, Hunt DF. Peptide and protein sequence analysis by electron transfer dissociation mass spectrometry. Proc Natl Acad Sci U S A 2004;101:9528–33.

[139] Bakhtiar R, Guan Z. Electron capture dissociation mass spectrometry in characterization of post-translational modifications. Biochem Biophys Res Commun 2005; 334:1–8.

[140] Bakhtiar R, Guan Z. Electron capture dissociation mass spectrometry in characterization of peptides and proteins. Biotechnol Lett 2006;28:1047–59.

[141] Mikesh LM, Ueberheide B, Chi A, Coon JJ, Syka JEP, Shabanowitz J, et al. The utility of ETD mass spectrometry in proteomic analysis. Biochim Biophys Acta 2006; 1764:1811–22.

[142] Udeshi ND, Shabanowitz J, Hunt DF, Rose KL. Analysis of proteins and peptides on a chromatographic timescale by electron-transfer dissociation MS. FEBS J 2007; 274:6269–76.

[143] Wiesner J, Premsler T, Sickmann A. Application of electron transfer dissociation (ETD) for the analysis of posttranslational modifications. Proteomics 2008;8:4466–83.

[144] Catalina MI, Koeleman CA, Deelder AM, Wuhrer M. Electron transfer dissociation of N-glycopeptides: loss of the entire N-glycosylated asparagine side chain. Rapid Commun Mass Spectrom 2007;21:1053–61.

[145] Perdivara I, Petrovich R, Allinquant B, Deterding LJ, Tomer KB, Przybylski M. Elucidation of O-glycosylation structures of the beta-amyloid precursor protein by liquid chromatography-mass spectrometry using electron transfer dissociation and collision induced dissociation. J Proteome Res 2009;8:631–42.

[146] Wu SL, Jiang H, Lu Q, Dai S, Hancock WS, Karger BL. Mass spectrometric determination of disulfide linkages in recombinant therapeutic proteins using online LC-MS with electron-transfer dissociation. Anal Chem 2009;81:112–22.

[147] Wu SL, Jiang H, Hancock WS, Karger BL. Identification of the unpaired cysteine status and complete mapping of the 17 disulfides of recombinant tissue plasminogen activator

using LC-MS with electron transfer dissociation/collision induced dissociation. Anal Chem 2010;82:5296–303.

[148] Cournoyer JJ, Pittman JL, Ivleva VB, Fallows E, Waskell L, Costello CE, et al. Deamidation: differentiation of aspartyl from isoaspartyl products in peptides by electron capture dissociation. Protein Sci 2005;14:452–63.

[149] Cournoyer JJ, Lin C, O'Connor PB. Detecting deamidation products in proteins by electron capture dissociation. Anal Chem 2006;78:1264–71.

[150] Cournoyer JJ, Lin C, Bowman MJ, O'Connor PB. Quantitating the relative abundance of isoaspartyl residues in deamidated proteins by electron capture dissociation. J Am Soc Mass Spectrom 2007;18:48–56.

[151] O'Connor PB, Cournoyer JJ, Pitteri SJ, Chrisman PA, McLuckey SA. Differentiation of aspartic and isoaspartic acids using electron transfer dissociation. J Am Soc Mass Spectrom 2006;17:15–9.

[152] Chan WYK, Chan TWD, O'Connor PB. Electron transfer dissociation with supplemental activation to differentiate aspartic and isoaspartic residues in doubly charged peptide cations. J Am Soc Mass Spectrom 2010;21:1012–5.

[153] Heudi O, Barteau S, Zimmer D, Schmidt J, Bill K, Lehmann N, et al. Towards absolute quantification of therapeutic monoclonal antibody in serum by LC-MS/MS using isotope-labeled antibody standard and protein cleavage isotope dilution mass spectrometry. Anal Chem 2008;80:4200–7.

[154] Hoofnagle AN, Wener MH. The fundamental flaws of immunoassays and potential solutions using tandem mass spectrometry. J Immunol Methods 2009;347:3–11.

[155] Barr JR, Maggio VL, Patterson DG, Jr., Cooper GR, Henderson LO, Turner WE, et al. Isotope dilution—mass spectrometric quantification of specific proteins: model application with apolipoprotein A-I. Clin Chem 1996;42:1676–82.

[156] Gerber SA, Rush J, Stemman O, Kirschner MW, Gygi SP. Absolute quantification of proteins and phosphoproteins from cell lysates by tandem MS. Proc Natl Acad Sci U S A 2003;100:6940–5.

[157] Barnidge DR, Goodmanson MK, Klee GG, Muddiman DC. Absolute quantification of the model biomarker prostate-specific antigen in serum by LC-Ms/MS using protein cleavage and isotope dilution mass spectrometry. J Proteome Res 2004;3:644–52.

[158] Addona TA, Abbatiello SE, Schilling B, Skates SJ, Mani DR, Bunk DM, et al. Multi-site assessment of the precision and reproducibility of multiple reaction monitoring-based measurements of proteins in plasma. Nat Biotechnol 2009;27:633–41.

[159] Yamaguchi K, Takashima M, Uchimura T, Kobayashi S. Development of a sensitive liquid chromatography-electrospray ionization mass spectrometry method for the measurement of KW-5139 in rat plasma. Biomed Chromatogr 2000;14:77–81.

[160] Dai S, Song H, Dou G, Qian X, Zhang Y, Cai Y, et al. Quantification of sifuvirtide in monkey plasma by an on-line solid-phase extraction procedure combined with liquid chromatography/electrospray ionization tandem mass spectrometry. Rapid Commun Mass Spectrom 2005;19:1273–82.

[161] Becher F, Pruvost A, Clement G, Tabet JC, Ezan E. Quantification of small therapeutic proteins in plasma by liquid chromatography-tandem mass spectrometry: application to an elastase inhibitor EPI-hNE4. Anal Chem 2006;78:2306–13.

[162] Kuhn E, Wu J, Karl J, Liao H, Zolg W, Guild B. Quantification of C-reactive protein in the serum of patients with rheumatoid arthritis using multiple reaction monitoring mass spectrometry and 13C-labeled peptide standards. Proteomics 2004;4:1175–86.

[163] Zhang F, Bartels MJ, Stott WT. Quantitation of human glutathione S-transferases in complex matrices by liquid chromatography/tandem mass spectrometry with signature peptides. Rapid Commun Mass Spectrom 2004;18:491–8.

[164] Lin S, Shaler TA, Becker CH. Quantification of intermediate-abundance proteins in serum by multiple reaction monitoring mass spectrometry in a single-quadrupole ion trap. Anal Chem 2006;78:5762–7.

[165] Hagman C, Ricke D, Ewert S, Bek S, Falchetto R, Bitsch F. Absolute quantification of monoclonal antibodies in biofluids by liquid chromatography-tandem mass spectrometry. Anal Chem 2008;80:1290–6.

[166] Ji C, Sadagopan N, Zhang Y, Lepsy C. A universal strategy for development of a method for absolute quantification of therapeutic monoclonal antibodies in biological matrices using differential dimethyl labeling coupled with ultra performance liquid chromatography-tandem mass spectrometry. Anal Chem 2009;81:9321–8.

[167] Bystrom CE, Salameh W, Reitz R, Clarke NJ. Plasma renin activity by LC-MS/MS: development of a prototypical clinical assay reveals a subpopulation of human plasma samples with substantial peptidase activity. Clin Chem 2010;56:1561–9.

[168] Hoofnagle AN. Peptide lost and found: internal standards and the mass spectrometric quantification of peptides. Clin Chem 2010;56:1515–7.

[169] Lu Q, Zheng X, McIntosh T, Davis H, Nemeth JF, Pendley C, et al. Development of different analysis platforms with LC-MS for pharmacokinetic studies of protein drugs. Anal Chem 2009;81:8715–23.

7

CHARACTERIZATION OF IMPURITIES AND DEGRADATION PRODUCTS IN SMALL MOLECULE PHARMACEUTICALS AND BIOLOGICS

HUI WEI, GUODONG CHEN, AND ADRIENNE A. TYMIAK

7.1 INTRODUCTION

Modern drug discovery generally involves disease target identification, lead compound generation and optimization, preclinical studies, scale-up production, formulation development, and clinical testing [1]. Drug safety is another important consideration in drug development. This includes identifying impurities and degradation products in pharmaceuticals. Small molecule impurities present in the active pharmaceutical ingredient (API) need to be characterized to ensure the absence of mutagenic or toxic substances. Drug product degradation profiles are often established to provide guidance on formulation development and drug shelf life assignment. For large molecule therapeutics or biologics, special considerations are given for full characterization of biologics due in part to their production process (recombinant DNA technology) and inherent micro-heterogeneity. Various modifications are common for biologics as a result of expression systems, purification steps, and storage conditions, leading to the formation of impurities and degradants. Characterization of impurities and degradation products in drug products is a critical task in pharmaceutical product development. Furthermore, the analysis of these low-level unknown impurities and degradants, particularly for biotherapeutics, is analytically challenging [2].

As one of the most highly utilized analytical techniques, mass spectrometry (MS) has been widely used for analysis of impurities and degradants due to its analytical

Mass Spectrometry for Drug Discovery and Drug Development, First Edition. Edited by Walter A. Korfmacher.

sensitivity, selectivity and specificity [3]. A general MS-based strategy has been developed to analyze small molecule impurities and degradants [4]. The first step is to determine the molecular weight (MW) of the unknown by suitable ionization methods, such as electrospray ionization (ESI) [5, 6] for polar compounds and atmospheric pressure chemical ionization (APCI) [7] for nonpolar compounds. A second step is to measure elemental compositions of unknown molecular ions by high-resolution (HR) MS experiments. This type of experiment can be performed on different instrument configurations, including magnetic sector, time-of-flight (TOF) [8, 9], Fourier transform-ion cyclotron resonance (FT-ICR) [10, 11] and Orbitrap [12–15] instruments. Structural information is often obtained by tandem MS experiments (MS/MS) on the unknown and parent molecule [16, 17]. For the analysis of complex mixtures, hyphenated analytical techniques such as high-performance liquid chromatography (HPLC)-MS (HPLC-MS) and gas chromatography-MS (GC-MS) are used for analytical experiments. Additional chemical derivatization experiments can be performed to support structural assignments, including hydrogen/deuterium (H/D) exchange experiments [18–27]. This method measures the difference in MW of a compound before and after deuterium exchange with heavy water to determine the solvent-exposed, exchangeable hydrogen atoms in a molecule and assist in structure elucidation.

For characterization of biologics, the first step is MW determination. This can be achieved by either ESI or matrix-assisted laser desorption/ionization (MALDI) [28] MS. With the formation of multiply charged ions for proteins in ESI, a deconvoluted mass spectrum can be generated to give an average MW of the protein by calculating from successive multiply charged ions. This procedure enhances the mass measurement precision and accuracy because all of the observed multiply charged ions are summed (typically better than 0.01% for masses up to 100 kDa) [29]. The MALDI-MS technique has a higher tolerance of impurities (salts) and often displays singly charged protonated molecules for simple data interpretation. The protein characterization or sequence determination can be obtained using two alternative and complementary approaches: "top-down" [30, 31] and "bottom-up" [32]. A top-down experiment involves HR-MS measurements of intact protein ions and their direct fragmentation [33]. The fragment ions obtained allow protein identification by database retrieval, quick positioning of the N- and C-termini, and precise localization of modifications. Challenges associated with this method include accessibility of expensive MS instrumentation, development of suitable MS/MS instrumentation for data acquisition, and appropriate database searching algorithms. In contrast to the top-down approach, the bottom-up method starts with enzymatic digestion of proteins into smaller peptides with subsequent analysis of the digested peptides. These peptides can be separated / detected, and either directly searched against a genome or protein database for protein identification (peptide mass mapping), or further dissociated in MS/MS experiments to produce fragment ions for sequence confirmation and database searching to support protein identification (sequence tagging) [34, 35]. Comparisons of peptide mass mapping for unmodified /modified proteins can be used to locate the site of modifications, followed by MS/MS experiments to examine the nature of modifications. The bottom-up approach is the current gold standard in protein characterization due to its high-throughput format, mature instrumentation, and excellent software development [36, 37].

7.2 CHARACTERIZATION OF SMALL MOLECULE IMPURITIES

Small molecule impurities in pharmaceutical products are largely generated during the synthetic process from starting materials, intermediates, and by-products. For example, in the course of large-scale production of mometasone furoate (a highly potent chlorinated corticosteroid) [1, 2, 38], several very low-level impurities were observed (Fig. 7.1) including two co-eluting components with protonated molecular ions at *m/z* 535 and 581 (peak B). The data suggest that the MWs of these two components were 534 Dalton (Da) and 580 Da, respectively. Their isotopic patterns also indicate two-chlorine patterns for *m/z* 535 and one-chlorine pattern for *m/z* 581.

High-resolution tandem MS experiments were performed on mometasone furoate molecular ions and impurity ions to obtain structural information using an LTQ-Orbitrap mass spectrometer. The advantages of using the LTQ-Orbitrap include enhanced capability of performing HR-HPLC-MS and HPLC-MS^n experiments with high resolving power (up to 200,000), good mass accuracy (<3 ppm with external calibration), and large dynamic range (over 10^3) [14, 15]. Common fragment ions observed for mometasone furoate include the loss of water, furoate, HCl, and the cleavages of the steroid rings (Fig. 7.2). The product ion at *m/z* 503 is due to the loss of water from the molecular ion at *m/z* 521. Further loss of the furoate from *m/z* 503 results in the fragment ion at *m/z* 391. Two product ions at *m/z* 279 and 319 are likely produced by the loss of the ClCHCO moiety and HCl from *m/z* 355,

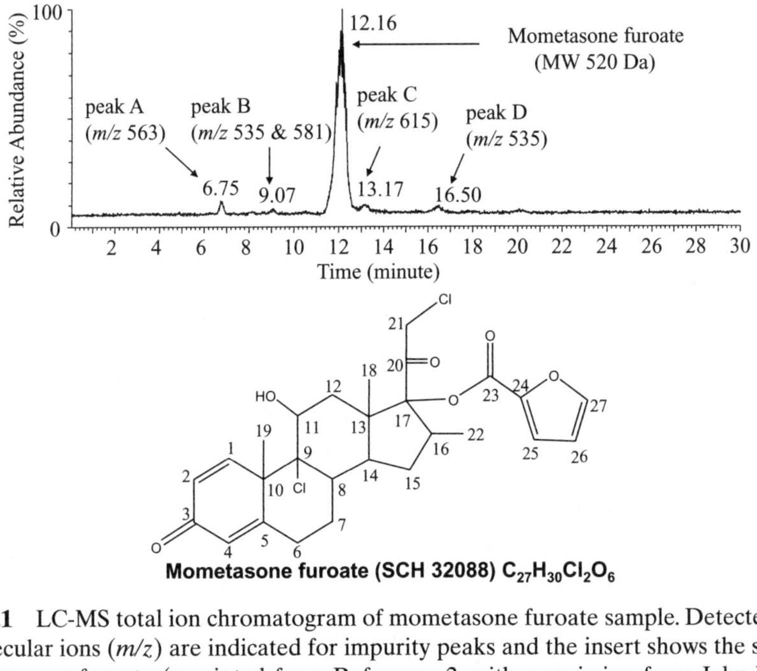

Figure 7.1 LC-MS total ion chromatogram of mometasone furoate sample. Detected impurity molecular ions (*m/z*) are indicated for impurity peaks and the insert shows the structure of mometasone furoate (reprinted from Reference 2, with permission from John Wiley & Sons, Ltd.).

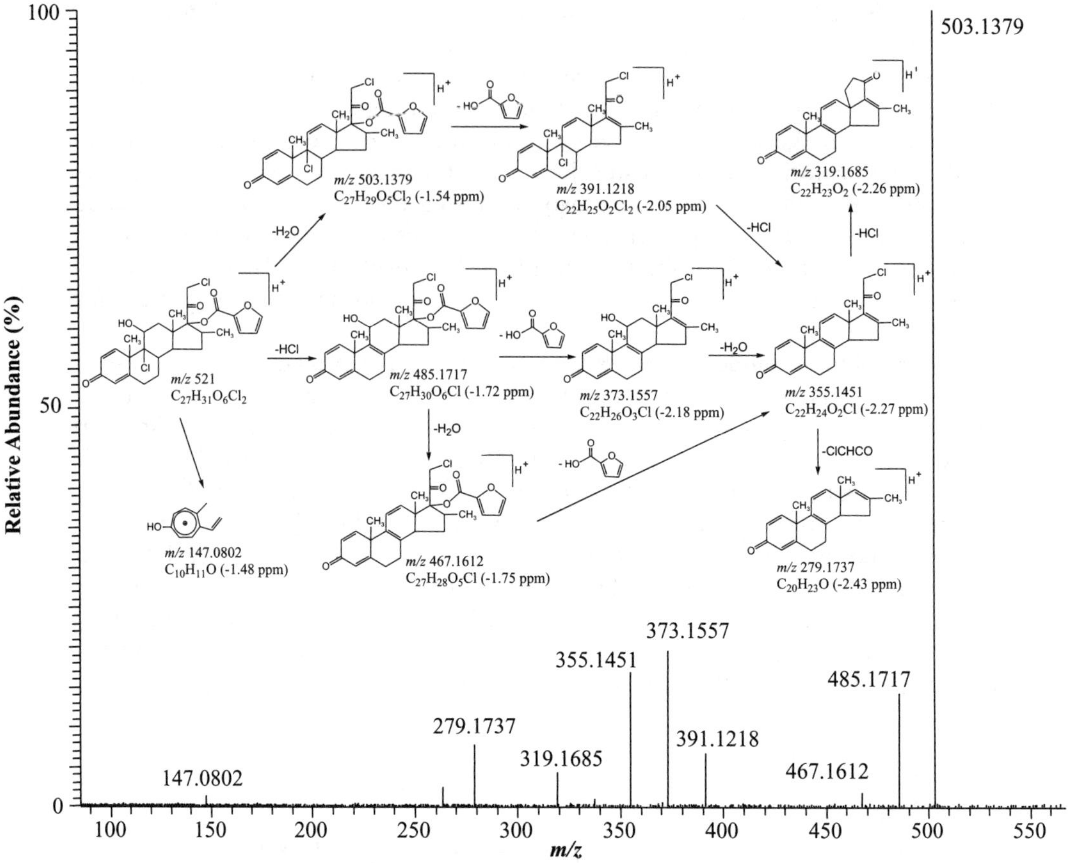

Figure 7.2 Product ion mass spectrum of mometasone furoate molecular ion at *m/z* 521. The insert exhibits the fragment ions with accurate mass data (reprinted from Reference 3, with permission from Wiley).

respectively. Another fragmentation pathway includes *m/z* 521 → 485 (loss of HCl from *m/z* 521) → 373 (loss of furoate from *m/z* 485) → 355 (loss of water from *m/z* 373). The fragment ion at *m/z* 485 could also lose water to yield *m/z* 467, which subsequently loses furoate to form *m/z* 355. For the impurity ion at *m/z* 535 (peak B), its accurate mass data give the elemental composition of $C_{27}H_{29}O_7Cl_2$ (−0.41 ppm), suggesting the addition of one oxygen atom and loss of two hydrogen atoms compared with the parent molecule. The most abundant fragment ion at *m/z* 423 corresponds to the loss of furoate from *m/z* 535 (Fig. 7.3). A highly abundant fragment ion at *m/z* 135 indicates a stabilized product ion. This ion is not observed in the product ion mass spectrum of *m/z* 521 under similar activation conditions, suggesting a 6-keto structure for this impurity. Ultimately, the 6-keto structure was confirmed by comparison with a synthetic standard [1].

Characterization of the impurity ion at *m/z* 581 was initially carried out using a low-resolution triple quadrupole mass spectrometer. Its low-resolution product ion

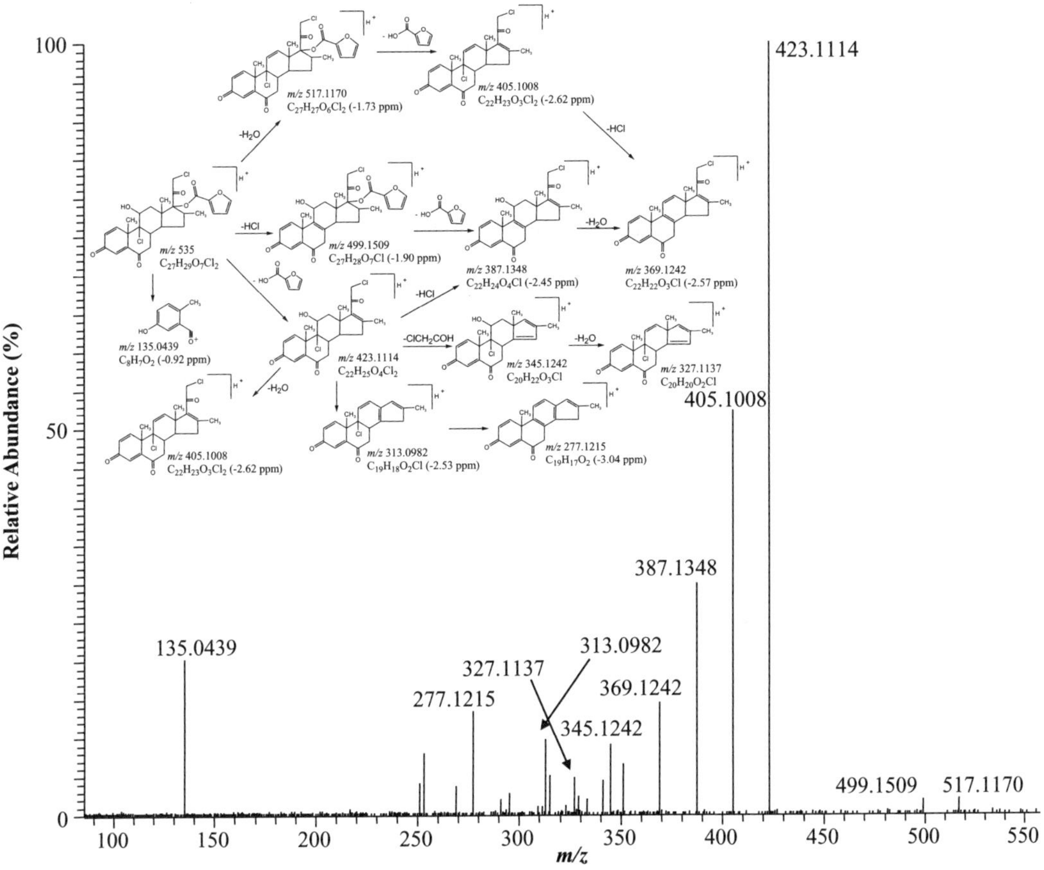

Figure 7.3 Product ion mass spectrum of impurity molecular ion at *m/z* 535 in peak B. The insert displays fragment ions with accurate mass data (reprinted from Reference 3, with permission from Wiley).

mass spectrum appeared to fit with the originally proposed structure where the 21-Cl was replaced by furoate. However, the accurate mass measurement did not support this assignment with an error of over 10 ppm. Based on detected isotopic patterns and most likely element combinations, the best possible formula match for *m/z* 581 was $C_{28}H_{34}O_9ClS$ with a mass accuracy of 0.18 ppm. Compared with the elemental composition of mometasone furoate ($C_{27}H_{31}O_6Cl_2$, $[M + H]^+$), this suggests a net addition of CH_3O_3S with reduction of one chlorine atom. There are two possible structures: an open structure involving replacement of 21-Cl with $-O-S(O)_2-CH_3$ moiety and a related structure involving cyclization at the 20-keto position to form a free –OH group. These two structures are isobaric but contain a different number of exchangeable hydrogen atoms, that is, one exchangeable hydrogen atom for the simple adduct and two exchangeable hydrogen atoms for the cyclized version. Online H/D HPLC-MS exchange experiments were further performed and measured molecular ions were found to be shifted to *m/z* 583.1731 ($C_{28}H_{32}D_2O_9ClS$,

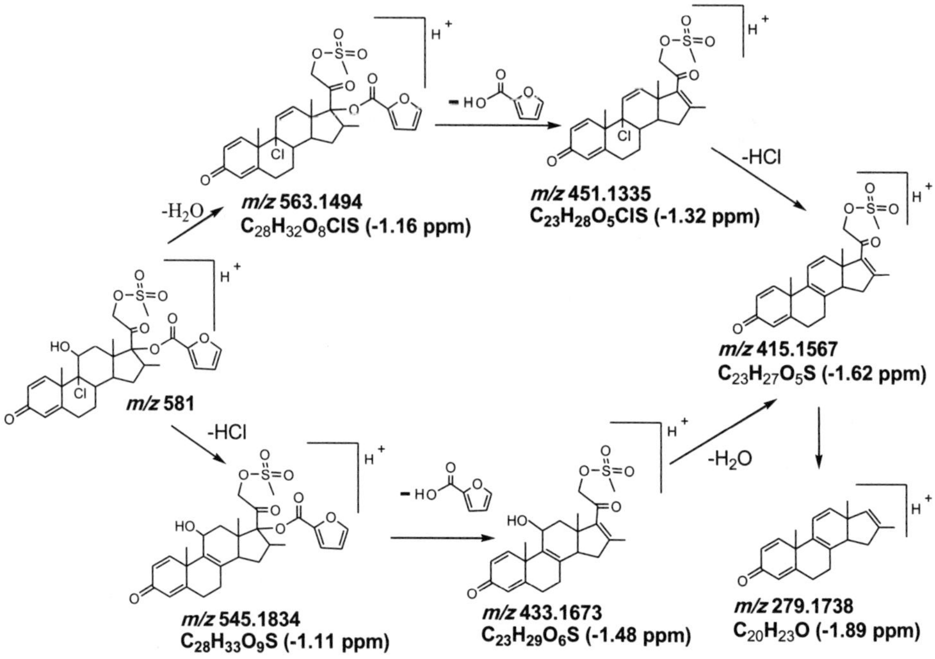

Scheme 7.1 Fragmentation patterns for impurity molecular ion at *m/z* 581 (peak B) (reprinted from Reference 3, with permission from Wiley).

$[M + D]^+$, –0.27 ppm), consistent with one exchangeable hydrogen atom in the molecule having no additional cyclization. It is likely that this impurity (*m/z* 581) is formed through a reaction between mometasone furoate and CH_3SO_2Cl, a reagent used in the synthesis. Further HR-LC/MS/MS experiments on *m/z* 581 support its structural assignment as the replacement of 21-Cl with the sulfur-moiety, as evidenced by accurate mass measurements of product ions (Scheme 7.1).

The impurity ion at *m/z* 535 (peak D) has the same nominal MW of 534 Da as the 6-keto impurity in peak B. However, the accurate mass data is consistent with an elemental composition of $C_{28}H_{33}O_6Cl_2$ ($[M + H]^+$, –0.41 ppm), and a net addition of CH_2 to mometasone furoate. HR-HPLC/MS/MS experiments on the impurity ion at *m/z* 535 also display a different fragmentation pattern (Fig. 7.4) from those of impurity ion at *m/z* 535 (peak B) (Fig. 7.3). The base peak in the product ion mass spectrum of *m/z* 535 (peak D) is due to the loss of water from the molecular ion at *m/z* 535. A very low abundant fragment ion at *m/z* 161corresponding to $C_{11}H_{13}O$ is also observed, suggesting a 6-methyl structure. Complete analysis of fragmentation patterns of the impurity ion at *m/z* 535 (peak D) further supports the 6-methyl assignment.

The structures of all five impurities detected in the sample are illustrated in Scheme 7.2. Mass accuracy data for measured elemental compositions of impurities and drug substance are less than 1 ppm even with large differences in ion abundances among impurities and drug substance. The ability to obtain low ppm mass accuracy data enables ready differentiation of isobaric impurities with the same

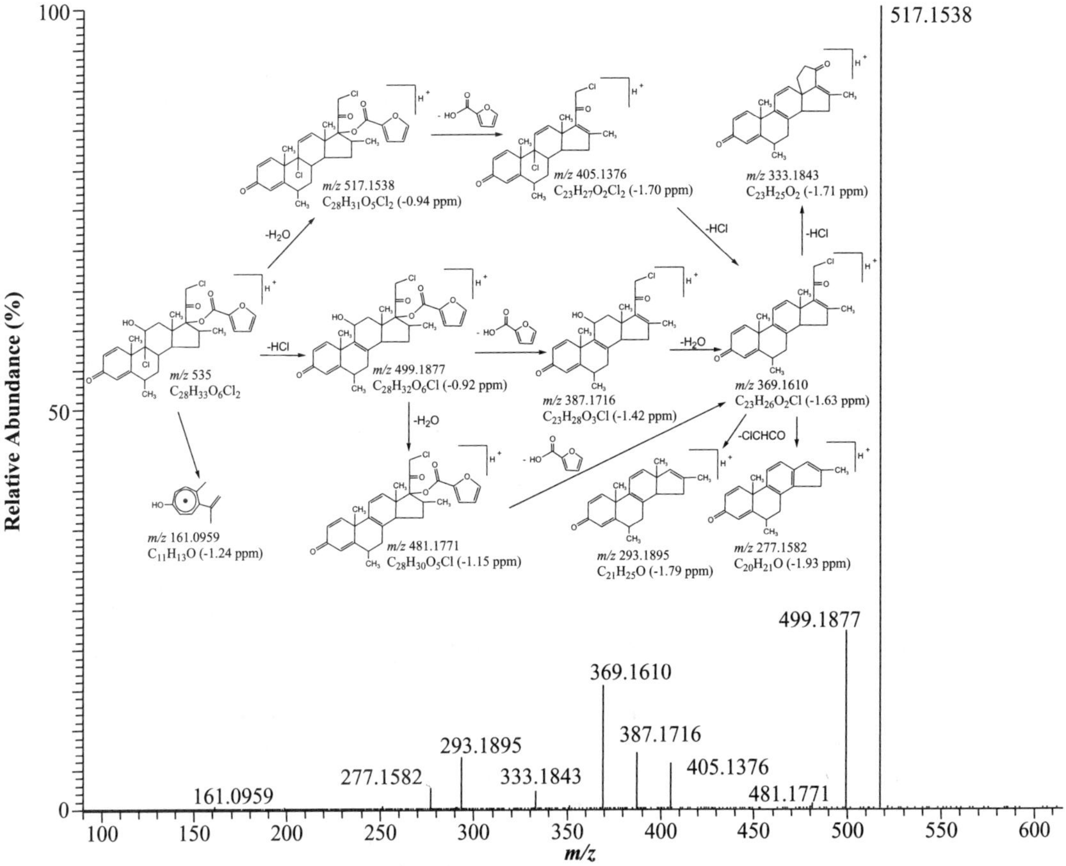

Figure 7.4 Product ion mass spectrum of impurity molecular ion at *m/z* 535 in peak D. The insert shows the fragment ions with accurate mass data (reprinted from Reference 3, with permission from Wiley).

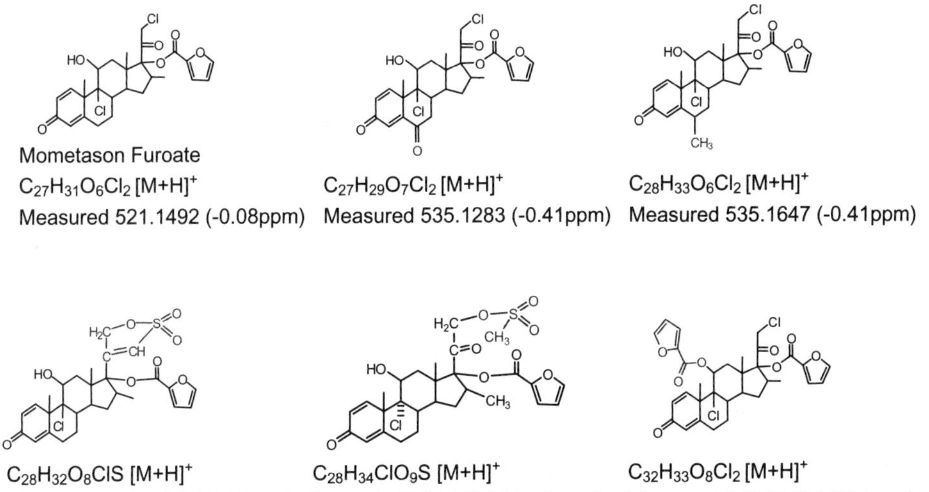

Scheme 7.2 Proposed structures of impurities related to mometasone furoate, including accurate mass data on elemental compositions (reprinted from Reference 2, with permission from John Wiley & Sons, Ltd.).

nominal mass, as shown in the case of 6-keto and 6-methyl impurities. Accurate mass data on fragment ions provide additional evidence to support structural assignments.

7.3 CHARACTERIZATION OF SMALL MOLECULE DEGRADATION PRODUCTS

Degradation profiling of pharmaceuticals is an integral part of safety and potency assessments of drug candidates in development. Various stress-testing methods have been designed to simulate stresses that a compound might experience during production processes and storage. These methods expose drug candidates to forced degradation conditions such as acid, base, heat, oxidation, and exposure to light. A rapid and accurate identification of degradation products can enhance our understandings of the degradation pathways of drug candidates and improve formulation development.

Posaconazole (SCH 56592) is a novel triazole antifungal agent [39]. It has been shown to exhibit higher potency against a broad range of fungal pathogens including *Asperigillus*, *Candida*, and *Cryptococcus* [40–44]. The drug substance is stable at room temperature, but the compound starts to form degradation products under stress conditions. As a part of stability studies, degradation products of SCH 56592 were characterized in order to understand its degradation pathway [45].

Four major degradation products were observed under various stress conditions (Fig. 7.5). The presence of these four elution peaks in all three stressed samples suggests a generalized degradation pathway for SCH 56592. Structures of degradants A, B, C, and D were completely elucidated by HPLC-NMR, HPLC-MS, and HPLC-MS/MS, as illustrated in Scheme 7.3.

The initial HPLC-MS data provided the MW information for these four degradants: A (714 Da), B (730 Da), C (702 Da), and D (414 Da). Further HPLC-MS/MS experiments were performed using a triple quadrupole mass spectrometer for structural information.

Compared with the MW (700 Da) of SCH 56592, the degradant B has additional 30 mass units and might be the oxidized product of SCH 56592 via the addition of two oxygen atoms and the reduction of two hydrogen atoms in the molecule. The major fragment ions from tandem MS experiments include *m/z* 685, 441, 372, 317, and 299 (Fig. 7.6a). Loss of a CH_3CHOH-moiety from *m/z* 731 can form the fragment ion at *m/z* 685. Two product ions at *m/z* 441 and 317 are the result of the breakdown of the central piperazine ring. Loss of the triazole group ($-C_2H_2N_3$) from the ion at *m/z* 441 leads to the fragment ion at *m/z* 372. Further loss of water from the ion at *m/z* 317 leads to the formation of the fragment ion at *m/z* 299. The MS/MS data suggest that SCH 56592 is cleaved into *NN′*-formyl diamine at the piperazine ring in the core of the molecule. Further HR-MS/MS experiments on an LTQ-Orbitrap provide additional accurate mass data to support these structural assignments (Fig. 7.7).

The degradant C (MW 702 Da) represents a possible loss of a carbonyl group (–CO) from the degradant B (MW 731 Da). The most abundant product ion at *m/z* 441 corresponds to the left portion of the molecule via the cleavage of C–N bond in the open form of the central piperazine ring (Fig. 7.6b). Further loss of the triazole

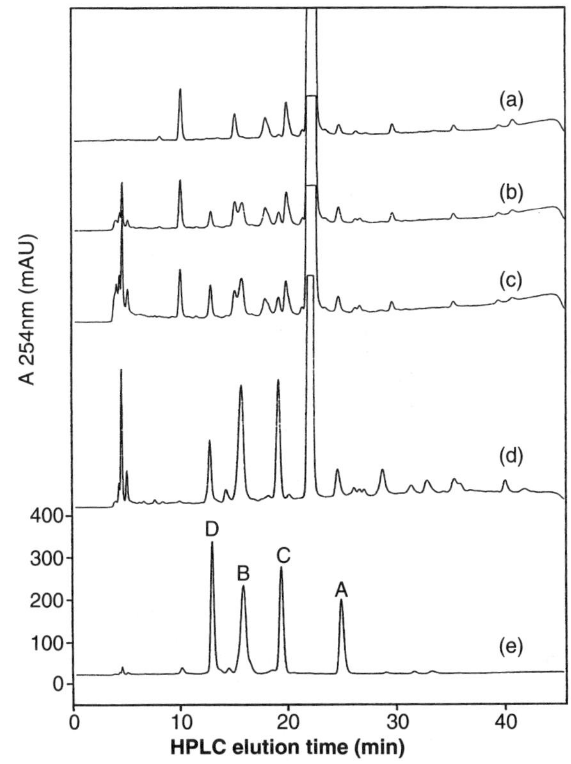

Figure 7.5 LC/UV chromatograms of SCH 56592 (a), stressed SCH 56592 sample by visible light (b), UV light (c), heat (d), as well as the partially purified mixture containing the major heat-stressed degradation products of SCH 56592 (e) (reprinted from Reference 45, with permission from Elsevier Science B.V.).

group ($-C_2H_2N_3$) from *m/z* 441 yields the fragment ion at *m/z* 372. The consecutive loss of water from the protonated molecule at *m/z* 703 results in fragment ions at *m/z* 685 and 667. Degradant D has a MW of 414 Da, indicating a possible breakdown of SCH 56592 molecule. Its product ion mass spectrum shows two abundant fragment ions at *m/z* 127 and 150 (Fig. 7.6c). The fragment ion at *m/z* 127 is likely to be a stable di-fluoro benzonium ion. Accurate mass data were obtained on fragment ions for degradant C and D, and were found to be consistent with the structural assignments. Degradant A has an MW of 714 Da, suggesting a possible oxidative product of SCH 56592 (MW 700 Da) with the reduction of two hydrogen atoms in the molecule. HR-MS/MS experiments provide insightful structural information with accurate mass data on fragment ions (Fig. 7.8).

Based on the determined degradant structures, an oxidative degradation pathway of SCH 56592 was proposed. The air oxidation of SCH 56592 would initially yield a mixture of oxidation products at C10 or C11 positions (degradant A). Further oxidation at both C10 and C11 leads to the formation of *N N′*-diformyl structure in degradant B. Deformylation of degradant B results in the formation of degradant C with one *N*-formyl group remaining and one secondary amine. Subsequent

(a) SCH 56592
MW 700
$C_{37}N_8F_2O_4H_{42}$

(b) Component A
MW 714
$C_{37}N_8F_2O_5H_{40}$

(c) Component B
MW 730
$C_{37}N_8F_2O_6H_{40}$

(d) Component C
MW 702
$C_{36}N_8F_2O_5H_{40}$

(e) Component D
MW 414
$C_{21}N_4F_2O_3H_{20}$

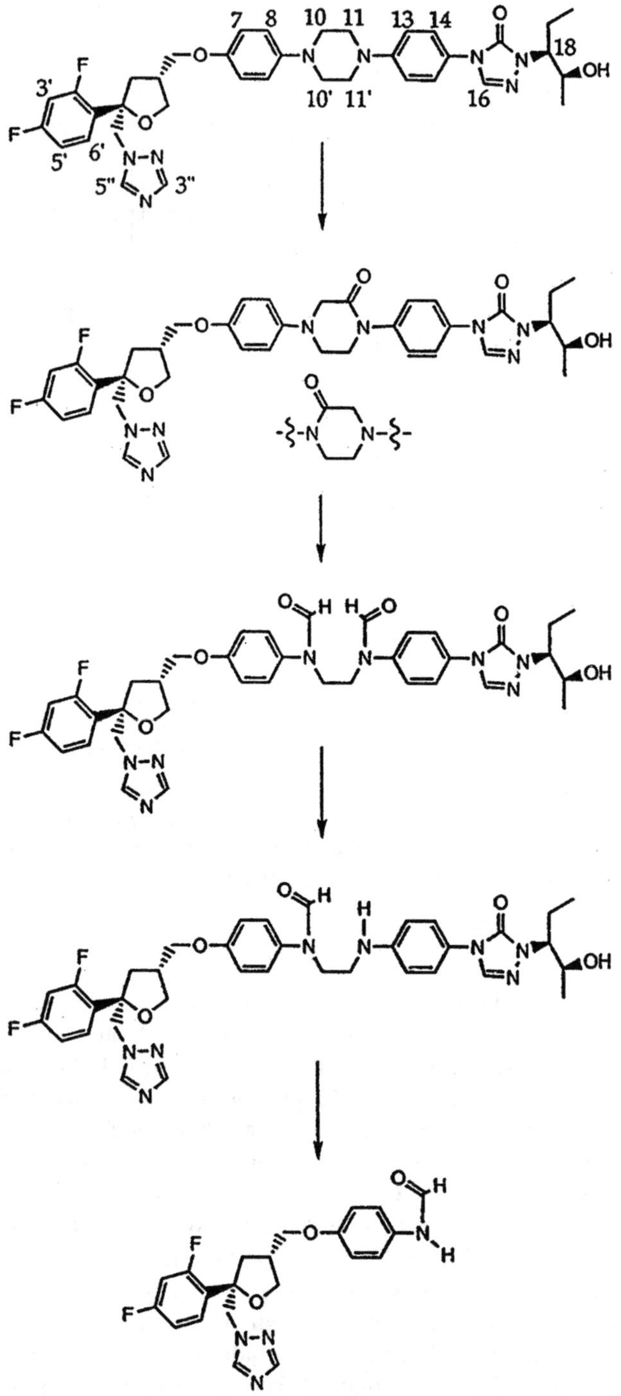

Scheme 7.3 Structures of SCH 56592 (a) and four major degradants A–D (reprinted from Reference 45, with permission from Elsevier Science B.V.).

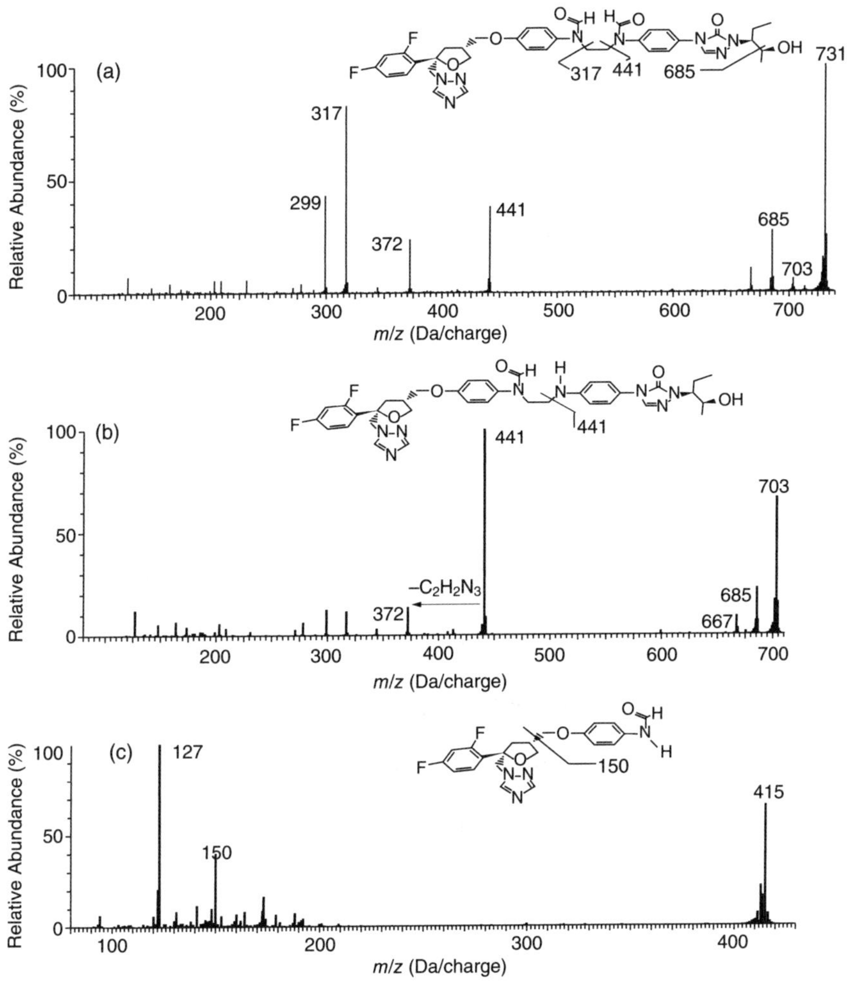

Figure 7.6 Product ion mass spectra of protonated molecular ions at *m/z* 731 (degradant B) (a), 703 (degradant C) (b), and 415 (degradant D) (c) (reprinted from Reference 45, with permission of Elsevier Science B.V.).

oxidative cleavage of degradant C causes the breakdown of the structure into degradant D. This oxidative degradation pathway is believed to be a dominant one under the stress conditions in the studies.

7.4 CHARACTERIZATION OF DEGRADATIONS IN BIOLOGICS

In comparison to small molecule drugs, differentiating characteristics of biologic pharmaceuticals typically include higher specificity, lower toxicity, potential clinical immunogenicity, higher success rates during development, and more complex functions that cannot be mimicked by simple chemical compounds [46]. The US market

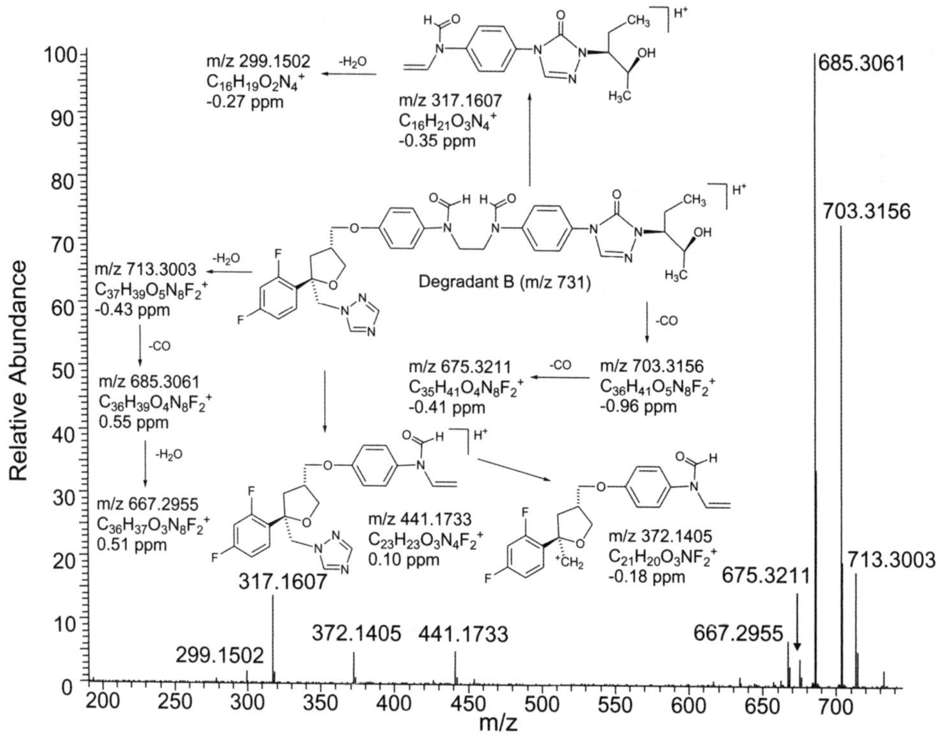

Figure 7.7 Product ion mass spectrum of degradant B of SCH 56592 degradation products at *m/z* 731 obtained from an LTQ-Orbitrap mass spectrometer. The insert shows the fragment ions with accurate mass data (reprinted from Reference 3, with permission from Wiley).

for biologics drugs has maintained a double digital growth rate in recently years, with the total sales in the US biotech market increased to $46.5 billion in 2008 [46, 47]. Therapeutic proteins produced using recombinant DNA technologies are generally complex, heterogeneous, and subject to a variety of enzymatic, chemical, or physical modifications/degradations [48, 49]. There are over 300 known protein modifications/degradations, among which the commonly observed ones are summarized in Table 7.1. Some modifications, such as glycosylation, phosphorylation, and disulfide linkage formation, are required for correct protein structure and function [50]. Other modifications such as oxidation, deamidation, and glycation that occur during protein production, processing, and storage are considered to be impurities/degradants, with the potential to affect the activity, stability, and safety of protein drugs [51]. Therefore, these modifications are closely monitored to ensure consistency in product quality. Since the introduction of ESI and MALDI, the two "soft" techniques for ionization of large, polar, and nonvolatile molecules [6, 28], MS has been widely used for structural analysis of proteins and peptides. Such characterization can lead to detailed information regarding amino acid modifications and sequence alterations as most of the degradations result in changes in the molecular mass. Here we described MS based methodologies for the

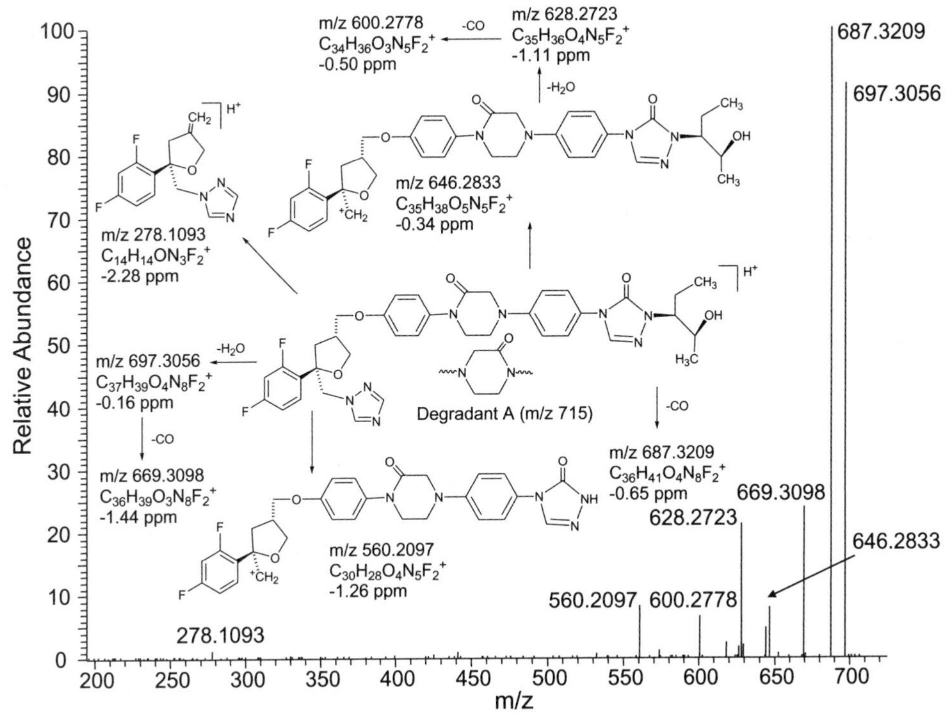

Figure 7.8 Product ion mass spectrum of degradant A of SCH 56592 degradation products at *m/z* 715 obtained from an LTQ-Orbitrap mass spectrometer. The insert shows the fragment ions with accurate mass data (reprinted from Reference 3, with permission of Wiley).

TABLE 7.1 Some Common Post-Translational Modifications in Proteins/Peptides and Mass Changes

Modification	Monoisotopic Mass Change (Da)	Average Mass Change (Da)
C-terminal Lys	−128.0950	−128.17
Pyroglutamic acid formed from Gln	−17.0265	−17.03
Disulfide bond formation (Cys)	−2.0157	−2.02
Isomerization (Asp, Glu)	0.00	0.00
Deamidation (Asn, Gln)	0.9840	0.98
Methylation (Asp, Glu)	14.0157	14.03
Oxidation	15.9949	16.00
Acetylation (N-terminus, Lys)	42.0106	42.04
Phosphorylation (Ser, Thr, Tyr)	79.9663	79.98
Sulfation (Ser, Thr, Tyr)	79.9568	80.06
Hexosamines (GalN, GlcN)	161.0688	161.16
Glycation	162.0528	162.14
Hexoses (Fru, Gal, Glc, Man)	162.0528	162.14
N-acetylhexosamines (GalNAc, GlcNAc)	203.0794	203.20
N-acetylneuraminic acid (Sialic acid, NeuAc, NANA, SA)	291.0954	291.26

characterization of chemical degradations including deamidation, isomerization, oxidation, disulfide variants, and fragmentation, as well as some physical degradations such as aggregation.

7.4.1 Chemical Degradation

Deamidation of Asn is one of the most commonly observed degradations in therapeutic proteins produced using recombinant DNA technology. Many proteins such as recombinant human DNase [52] and recombinant soluble CD4 [53] have been reported to show *in vitro* or *in vivo* biological activity changes due to deamidation, while proteins such as recombinant human growth hormone (hGH) [54] appear to be unaffected by the modification. Thus, it is important to establish methods to identify the sites of deamidation and to evaluate the effect on biological activity and antigenicity. Deamidation most frequently occurs at Asn residues. The rate of deamidation depends on pH, on neighboring amino acid residues, and on the higher order structure of the folded protein [55]. Under neutral or basic pH conditions, the deamidation of Asn residues occurs primarily through intramolecular rearrangement involving cyclic succinimide intermediates, which hydrolyze to isoAsp and Asp residues (Fig. 7.9a). In simple peptide studies, the branching ratio of this experiment is usually 3:1 in favor of the isoAsp residue [56]; however, this ratio is unlikely to hold in whole proteins because of conformational constraints. Under acidic conditions, deamidation occurs by direct hydrolysis of the Asn residue to yield only Asp residues [49]. The highest frequency of deamidation occurs at Asn residues with a C-terminal [57]. Deamidation is also observed at intermediate frequency at Asn residues followed by small side-chain residues (i.e., Ser, Ala). The most stable Asn residues are typically followed by a hydrophobic amino acid with a bulky side chain, presumably due to the protective effects of steric hindrance. Susceptible Asp residues and flanking regions also need to reside within a conformationally flexible region of the protein as solvent accessibility is necessary for the reaction. Deamidation also occurs at Gln residues but at a much lower rate. Due to entropic factors, Gln rarely forms a succinimide within a polypeptide chain. However, when Gln is located at the N-terminus, it readily cyclizes with the N-terminal primary amine, forming pyro-Glu [58] with a loss of 17 Da in the molecular mass (Table 7.1).

Detection of deamidated peptides and proteins can be facilitated by cation exchange chromatography since the degradants should elute earlier as a result of their additional negative charge [59, 60]. Isoelectric focusing (IEF) can also be used to resolve and detect deamidated products [61]. The method of choice to identify and quantify deamidation at specific positions in proteins, for example, mAbs, with multiple potential deamidation sites, is though enzymatic digestion of the protein into peptides followed by HPLC-MS/MS analysis [62, 63]. Each deamidation generates a mass shift of 0.984 Da (Table 7.1). Figure 7.9b is a typical chromatogram for the separation of a tryptic peptide and its deamidated products. A recombinant mAb was reduced, alkylated, and digested with trypsin followed by a comprehensive HPLC/MS/MS peptide mapping analysis. Peptide GFYPSDIAVEWESNGQPENNYK has three Asp residues in the backbone sequence, with the first Asp positioned within the preferred sequence motif (NG) for deamidation. The main peak in the chromatogram is the peptide without deamidiation. Peak 1 and peak 2 correspond to the degradants with the first Asp residue converted to isoAsp and Asp,

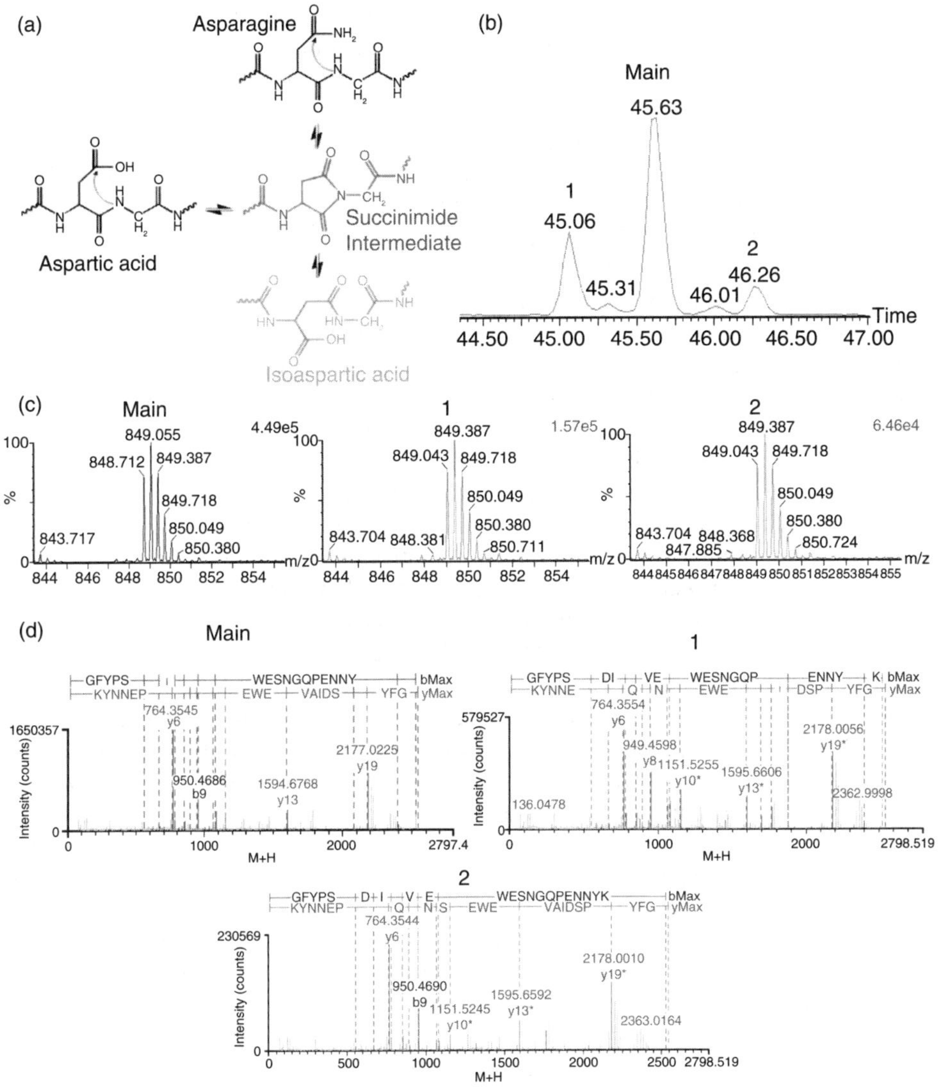

Figure 7.9 Asn deamidation/isomerization scheme (a); elution pattern of a tryptic peptide (GFYPSDIAVEWESNGQPENNYK) of a mAb and its deamidation products in HPLC-MS analysis (b); precursor ion spectra of the unmodified peptide (main) and deamidated products (1 and 2) (c); product ion spectra of the unmodified peptide and deamidated products (d). y-series ions (marked in bold) starting from y9 clearly prove the deamidation. y^n* indicates y ions with modification (deamidation) in sequence.

respectively. The MW of the peptide and its degradants can be obtained from the precursor mass in MS spectra as shown in Figure 7.9c. A mass difference of 0.33 Da was observed for the triply charged peaks of unmodified (main peak) and deamidated species (peak 1, 2), indicating a +1 Da mass difference. Furthermore, the peptides were fragmented in collision-induced dissociation (CID) analysis. The product ion spectra, as shown in Figure 7.9d, clearly indicated that the y and b ions increased by 1 Da starting from y8 and b9 ions, respectively, in the MS/MS spectra of peak 1 and 2. This set of data unambiguously identified the first Asn in the NG site as the deamidation site. Although the two deamidated products (peak 1 and peak 2) display the same MW in MS spectra and similar fragmentation patterns in CID product ion spectra, isoAsp and Asp species can usually be distinguished by retention time in HPLC, with isoAsp eluting earlier than Asp, as well as by their relative abundance (1:3) (Fig. 7.9b). Nonconventional fragmentation methods can also be used to distinguish these isomeric degradants, as described below.

Asp and Glu can also cyclize and hydrolyze to form isoAsp and isoGlu residues, respectively, with the isomerization of Asp residues occurring more frequently. This reaction happens in a manner identical to the deamidation reaction described above (Fig. 7.9a). The rate of isomerization is highly dependent on pH and the presence of a small residue at the N + 1 position. Therefore, Asp-Gly is the sequence that is most susceptible to the formation of isoAsp residues [64]. The formation of isoAsp lengthens the protein backbone by one methylene unit, potentially resulting in disruption of the folded protein structure; therefore, it is believed to be partly responsible for the inactivation, aggregation, and aging of proteins in tissue [64–66]. In order to unambiguously differentiate Asp from isoAsp products, nonconventional fragmentation techniques have been used such as electron transfer dissociation (ETD) [67], electron capture dissociation (ECD) [68], and more recently, electron ionization dissociation (EID) [69]. Unlike CID, ECD involves a gas-phase reaction of low-energy electrons with multiply charged peptide/protein ions that primarily cleave the $n—c_\alpha$ bond. Because of the absence of an $n—c_\alpha$ bond in isoAsp, the ECD spectra of Asp-containing peptides versus isoAsp-containing peptides are substantially different, allowing differentiation of these two residues. ECD experiments performed on an FT-ICR MS using model peptides with Asp residues and their isoAsp analogs demonstrated the ability to distinguish these two isomers using unique diagnostic fragment ions: the backbone cleavages at $c_n + 58$ and z_{l-n}-57 (n = position of Asp; l = total number of amino acids in the peptide) for the isoAsp form and a side-chain cleavage at M-60 Da for the Asp form [68]. Similarly, the ETD method employs transfer of an electron to multiply charged peptide/protein ions using singly charged anions, generating odd-electron fragment ions via radical-based rearrangements. The unique fragment ions for isoAsp and Asp can be observed in ETD as well. For example, ETD experiments carried out on an ion trap MS allowed differentiation of isoAsp and Asp residues using backbone cleavage ions ($c_n + 57$ and z_m-57), although the low resolution of the ion trap instrument made detection of the diagnostic peak of the Asp residue difficult due to interfering side-chain fragment ions from Arg residues [67].

Oxidation is another covalent modification that is frequently seen in proteins. It is often induced by reactive oxygen as a result of stress or contamination [70]. Protein oxidation can occur on many amino acids and is stimulated by different mechanisms [71, 72]. Met residues are often the most susceptible to oxidation and can be oxidized to Met sulfoxide or even Met sulfone. Met oxidation is of distinct importance in the biopharmaceutical industry as it has been shown to occur in a wide variety of proteins and lead to significant changes such as reduction of biological activity, aggregation, increased immunogenicity, and proteolysis [73–75]. For example, the oxidation of two susceptible Met residues of a recombinant fully human monoclonal antibody resulted in decreased stability of the CH_2 domain [76] and the binding of this antibody to protein A and protein G [77]. The exposure of susceptible residues on the cell surface can determine the extent of Met oxidation, and cell culture media components and formulation excipients can be used to protect the protein from oxidative damage [78]. Identification of Met oxidation in proteins is problematic for conventional amino acid analysis, since Met sulfoxide is converted to Met during acid hydrolysis. However, oxidation of Met can result in a mass increase of +16 (sulfoxide) or +32 Da (sulfone) and is readily detectable by MS even at the intact protein level or with limited digestion for mAbs [79]. Peptides containing oxidized Met can also be separated from natural fragments on a weak ion-exchange column [80] or by hydrophobic interaction chromatography (HIC) [81]. With more comprehensive peptide mapping HPLC-MS/MS analysis, the extent and the location of the oxidized Met residues can be determined. Typically a Met containing tryptic peptide from a therapeutic protein elutes a few minutes later than the oxidized peptide containing Met sulfoxide in reversed-phase chromatography. The identity and relative quantity of the degradant can be obtained from the MS spectra as well as the peak area of each component. Good linearity and reproducibility ($R2 > 0.99$; relative standard deviations: 4–9%) has been achieved in assessing the extent of Met oxidation by the peptide mapping MS method [82].

Degradants or impurities related to Cys residues are of key importance in therapeutic proteins. Most Cys residues in recombinant proteins are involved in the formation of disulfide bonds that maintain the protein three-dimensional folded structure. However, free Cys residues may still be present in freshly prepared material, which could cause degradation via disulfide bond rearrangement, protein misfolding, or covalent aggregation upon processing or storage. Significant efforts continue to be devoted to the control of disulfide scrambling and to the mapping of disulfide bonds to ensure drug quality. One effective way to reduce the free thiol content and increase disulfide bond formation of recombinant mAbs produced in Chinese hamster ovary (CHO) cells is to add low amounts (up to 100 μM) of the oxidizing agent copper sulfate. This can result in a 10-fold reduction of free thiols, and a significant (3-fold) reduction has been observed with as little as 5 μM copper sulfate [83]. Total free thiol content can be rapidly assessed spectrophotometrically using Ellman's reagent or a fluorescent thiol-sensitive maleimide dye, or by HIC [83]. Mapping of intra- or intermolecular disulfide bonds can be achieved using diagonal electrophoresis [84] or peptide mapping analysis using MALDI [85] or HPLC-MS [86]. Tandem MS is required to generate fragment ions so that the linkage patterns within some disulfide-linked peptides can be defined. This requirement applies where disulfide bonds are intertwined or in the case of intrachain disulfide bonds. While fragmentation in CID occurs preferentially in the peptide backbone, disulfide bond cleavages are preferred to peptide backbone fragmentation in ETD and ECD. Recently, ETD performed on a linear ion trap MS was successfully applied to determine disulfide linkages in proteins in a single online HPLC-MS experiment involving multiple stage fragmentation [87]. Disulfide-linked peptide ions were identified by CID and ETD fragmentation, and the disulfide-dissociated (or partially dissociated) peptide ions were characterized in the subsequent MS^3 step.

One of the less documented modification related to Cys disulfide bonds is the formation of trisulfide linkages. The number of proteins in which a trisulfide has been unambiguously identified is small. A trisulfide bond can be generated by the post-translational insertion of a sulfur atom into a disulfide bond. It has been found in *Escherichia coli (E. coli)*-derived recombinant hGH and recombinant truncated interleukin-6, as well as in Cu, Zn-superoxide dismutase isolated from human erythrocytes. It has also been observed in the disulfide bond between heavy chains of a recombinant human immunoglobulin G2 (IgG2) mAb variant. Although no evidence has been observed to show any changes in protein function or receptor binding affinities due to trisulfide modification [88, 89], its prevalence may be underestimated, particularly with the increasing evidence for significant pools of sulfides in living tissues and their possible roles in cellular metabolism [90]. The location and type of trisulfide modifications in mAbs were recently studied by peptide mapping of nonreduced proteins using HPLC-MS and tandem MS [89]. The paper also showed that, by adding reducing agent such as Cys, trisulfide bonds can be nearly completely converted back to native disulfides without disulfide scrambling or a resulting increase in free sulfhydryls. An efficient on-column conversion of IgG1 trisulfide linkages to native disulfides via a Cys wash step incorporated into Protein A affinity column purification has also been reported [91].

The C-terminal lysine variation is commonly observed in biopharmaceutical mAbs. The CHO cells that are commonly used to express antibodies contain carboxy peptidase that removes the C-terminal Lys residue leaving Gly as the terminal residue [92]. Monitoring this modification can be important since it is found to be sensitive to the production process. It has also been reported that Gly, now at the C-terminus, is further enzymatically removed in a humanized IgG1 antibody expressed in CHO cells, leaving an amidated Pro at the C-terminus [93]. As both C-terminal variants have different ionic charges from the unmodified protein, the methods commonly used to probe this charge variation include IEF, capillary IEF, ion-exchange chromatography, and HPLC-MS. It has been demonstrated that peptide mapping by LC-MS after trypsin digestion provides reproducible relative percentage information on C-terminal variants and has significant advantages over other methods [92].

Antibody fragmentation is one of the major degradation pathways of therapeutic proteins including recombinant mAbs in liquid formulation [94]. In mAbs, the hinge region is the least structured, thus most susceptible for enzymatic and chemical cleavage, resulting in fragments such as the Fab region, Fc region, and antibody lacking one Fab arm (one-armed antibody). Antibodies are routinely incubated at stressed conditions such as high temperature with the proper formulation conditions to generate fragments within a short period of time to help assess the stability of the molecules. The fragments are readily separated by size-exclusion chromatography (SEC) or HPLC and identified by MS or directly by MALDI-TOF MS. Fragment identification begins with matching the observed versus calculated MWs of presumed fragments, and is further evaluated based on additional information such as the known post-translational modifications, peak clusters, the likelihood of such fragment, and other modifications such as oxidation of Met residues to further confirm the assignment. Figure 7.10 exhibits a deconvoluted spectrum of a mAb and its fragments generated after incubation at 40°C under different pH conditions for 10 weeks [95]. Data were obtained from HPLC followed by ion detection on a quadrupole time-of-flight (Q-TOF) MS. Incubation at pH 6 led to the generation of the fewest fragments, whereas the number of fragments and their concentration increased at pH values either lower or higher than 6. Other truncation sites in mAbs

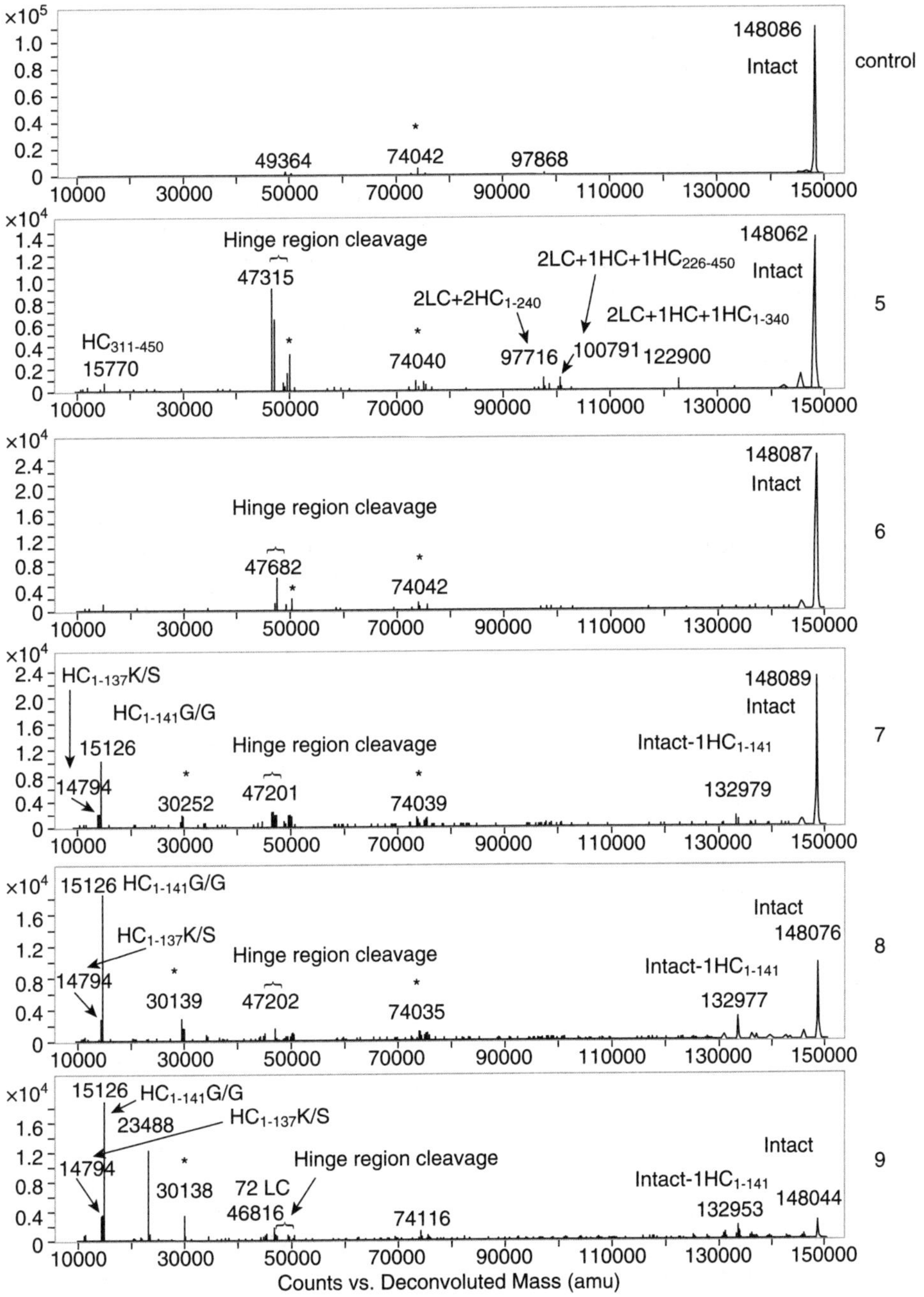

Figure 7.10 Deconvoluted mass spectra of the control and samples incubated at 40°C for 10 weeks at pH 5 (5), 6 (6), 7 (7), 8 (8), and 9 (9). Fragments corresponding to the peaks are labeled in the figures. Peaks labeled with * are deconvolution artifacts (reprinted from Reference 95, with permission from Elsevier BV).

include, but are not limited to, peptide bond cleavage at the carboxy sides of Asn and Gln [56]. Peptide bond cleavage at these residues is more likely to take place after deamidation to Asp or Glu [96].

Many other chemical degradations are known to exist in biologics, such as carbohydrate variants, amino acid insertion/mutations, phosphorylation, and glycation variants. Some of these are discussed in Chapter 6 of this book. As HPLC-MS-based methodology has been improved with enhanced separation power, instrument sensitivity, resolution, and mass accuracy, many chemical degradants can be detected in a single comprehensive analysis. It has been reported that target proteins were identified with site-specific modifications, such as Met oxidation, Asn deamidation, and N-terminal acetylation, and impurity proteins were identified and quantified using a peptide mapping approach with HPLC-MS operated in a data-independent acquisition mode (LC-MS^E) [97]. However, in many cases, detection of these modifications/degradations are challenging due to the low abundance of the modified species or a low signal response in the analytical method used. Affinity-based enrichment of post-translationally modified proteins and peptides can serve to increase the relative abundance of a select class of modified polypeptide species in a sample before the subsequent analysis. Such an enrichment strategy can be applied at the intact protein level or at the peptide level, that is, after enzymatic or chemical degradation of proteins, and has been discussed in the literature [98, 99].

7.4.2 Physical Degradation

Proteins, unlike small molecule drugs, have secondary, tertiary, and quaternary structures, and exist as ensembles of different conformations in equilibrium instead of a single rigid structure. Although chemical bonds remain unchanged, proteins can have profoundly different functions depending on higher order structures; for instance, conformational heterogeneity may lead to differences in ligand binding affinity [100]. Protein functions in the cell require native configurations, and defects in protein folding characterize a number of human genetic disorders. Among all "defects," protein denaturation and aggregation are two of the major changes that can affect the efficacy and safety of therapeutic proteins.

Protein aggregation is widely considered a nonspecific coalescence of misfolded proteins, driven by interactions between solvent-exposed hydrophobic surfaces that are normally buried within a protein's interior [101]. Aggregation has been associated with disease states such as Alzheimer's and the transmissible spongiform encephalopathies [102]. Any conditions that induce protein unfolding and promote protein–protein interactions, such as incubation at high temperature, contact with hydrophobic surfaces, exposure to extreme pH, and exposure to shear forces, potentially will increase the amount of aggregates [94, 101]. SEC has become one of the standard techniques used to detect aggregates. Traditionally a high salt mobile phase is used for the SEC separations and hence SEC is incompatible with ESI-MS. In recent years a novel method has been developed that uses MS compatible mobile phases for the coupling of SEC to ESI-MS, allowing the characterization of arsenic interactions with peptides and proteins [103] as individual components can be directly characterized by MS. Some artifacts may be introduced under the high pressure on column in SEC, and the large aggregates may not enter the column. Due to these limitations, orthogonal analytical techniques, such as using analytical ultra-centrifugation (AUC) [104] or laser light scattering [105], are typically needed to confirm SEC results. Although MS approaches have not been used for high-throughput analysis of noncovalent aggregates, H/D exchange combined with MALDI-MS has been applied for the conformational characterization of aggregates of polyvaline and a polyleucine-Helix. Infusion of proteins in their native state for ESI MS analysis has also been used to characterize insulin aggregation and amyloid fibril formation [106]. It is important to note that experimental conditions for these studies need to be carefully controlled to avoid nonspecific binding, and sometimes reference proteins or deconvolution methods need to be applied [107, 108].

7.5 ARTIFICIAL DEGRADATION IN PEPTIDE MAPPING

Peptide mapping using HPLC-MS/MS is like a fingerprint of a protein, and is the key step in determining the primary sequence and any degradations/modifications. There are several steps involved in peptide mapping, starting from protein denaturation to unfold the protein, reduction and carboxymethylation to break the disulfide bonds and prevent the reformation of sulfer linkages, sample cleanup if any harsh buffers are used, then digestion with enzymes or chemicals, followed by quenching of the reaction prior to HPLC-MS or HPLC-MS/MS analysis. The cleavage of proteins can be accomplished enzymatically (e.g., trypsin, chymotrypsin, LysC endoprotease, Asp-N, Glu-C, pepsin, etc.) or by chemical digestion (e.g., cyanogen bromide, o-iodosobenzoic acid, 2-nitro-5-thiocyanobenzoic acid). Trypsin, which cleaves proteins on the C-terminal sides of Lys and Arg residues, is the most widely used protease for peptide mapping [109]. The trypsin digestion step can be performed in many different ways. If not carefully controlled, undesired artificial covalent degradation of the protein can be induced in the process [110, 111].

Urea is commonly used to denature proteins and help solubilize proteins in peptide mapping methods. However, it contains varying amounts of cyanate that can react with free amine groups in Lys residues or at the N-terminus, resulting in carbamylation of the peptide resulting in a 43 mass units higher species with a longer retention time [112, 113]. Urea also reduces the trypsin activity. It has been shown that significantly higher levels of missed cleavages were seen in samples digested with urea than with urea removed [114]. Several other denaturants can be used as alternatives to urea. Guanidine hydrochloride (GdnHCl) is an ionic and inert substance that is recommended in many digestion protocols. However, a high concentration of GdnHCl is required and the guanidine group can form a hydrogen bond/salt bridge with the carboxyl group of Asp189 in trypsin, and prevents the interaction of trypsin and target protein(s) [115]. Rapigest surfactant helps solubilize proteins to make them more susceptible to enzymatic cleavage without inhibiting enzyme activity or inducing protein modifications. It is also hydrolyzed at low pH and is therefore easy to remove at the end of the digestion. In general, regardless of the reducing agent is used, upon denaturation, the steric arrangement of the native protein is lost, with Asn residues being the most susceptible to deamidation [116]. Therefore, the time for denaturing proteins should be carefully controlled to minimize deamidation during sample handling.

The most commonly used reducing agent to cleave protein disulfide bonds is dithiothreitol (DTT). It works best at high pH under conditions, which also increase the risk for deamidation. One way to minimize the degree of artificial deamidation is to reduce the pH to near neutral, but then it lowers the concentration of DTT in the anionic form in solution. Tris (2-carboxyethyl) phosphine (TCEP) is an alternative reducing agent and is also widely used in peptide mapping. It has the advantages of being odorless, more powerful as a reducing agent, irreversible, and more resistant to air oxidation. TCEP also works in a wide pH range and can therefore be used in neutral or slightly acidic solutions, where the risk of deamidation is small. However, one risk of using a reducing agent is the reduction of oxidized Met residues in the protein, generating false negative data for monitoring Met oxidation in the quality control process. It has been reported that TCEP reduces oxidized Met while DTT does not induce significant reduction of Met sulfoxide [117]. It was also observed that TCEP cleaved the polypeptide backbone of Cys-containing proteins at both the N- and C-terminal sides of Cys in mild conditions [118]. In this work, a synthetic peptide containing 16 residues, SGTASVVCLLDNFYPR (from human Ig κ chain C region), was incubated at 0.04 mg/mL (~23 μM) with 10 mM TCEP for 24 h at 37°C in buffers with different pHs: 2.2, 4.0, 6.0, 8.0, and 10.0. Samples were analyzed by HPLC-MS/MS on an LTQ-FT mass spectrometer. The intact peptide (P16) has a monoisotopic mass of 1741.85 Da. After incubation, six peptides as degradation products were identified with monoisotopic mass of 618.34 (F618), 619.40 (F619), 646.32 (F646), 1036.56 (F1036), 1064.57 (F1064), and 1108.55 Da (F1108). Each cleaved product was identified with amino and carboxy terminal structures by MS/MS fragmentation. Three of them are found to be cleaved at the N-terminal side of Cys and three are cleaved at the C-terminal side of Cys. This TCEP side reaction on Cys-containing proteins is also pH dependent, where the yields of the six cleaved products and the amount of P16 vary at each different pH. At pH 8.0, all six products had the highest concentration while the intact peptide had the lowest, indicating that the peptide is least stable at this pH [118].

Besides the choice of denaturant and reducing agent, other factors such as temperature, pH of the buffer and length of digestion time, can also affect the extent of induced degradation. Trypsin is most efficient at alkaline pH and elevated temperature (usually 37°C) and the digestion is usually performed for 3–24 h. The risk of deamidation during digestion increases with increasing temperature and alkaline pH, and degradants accumulate as incubation times increase. For example, one study has shown that at pH 8.5 and 37°C the rate of deamidation in a protein containing the sequence –Asn-Gly- is six times higher than at pH 7.5 and 25°C [117]. It has also been observed that 70% to 80% of protocol-induced deamidation on peptides with -Asn-Gly- sequence take place after a 12-h tryptic digestion at 37°C [110]. Cyclization of the N-terminal Gln residue on tryptic peptides is another degradation product produced during digestion [119]. It has been reported that trypsin digestion performed for 24 h can cause approximately 55% cyclization of Gln residues when they are on the N terminus of tryptic peptides [111]. Therefore, to minimize these artificial degradation products, near neutral pH, lower temperature, and shorter digestion time should be used in the digestion protocol. For proteins of rigid structure and large size such as mAbs, digestion at a lower temperature may result in incomplete digestion; therefore a higher temperature is still needed, but the digestion time can be reduced to minimize induced degradation. Trypsin digestion is typically performed overnight, but many studies have shown that 3–4 h of digestion at 37°C should be sufficient for mAbs [114]. It has also been demonstrated using many mAbs that peptides produced in trypsin digestion at 55°C for 1 h and at 37°C for 3 h are comparable, with no obvious differences [114]. The shortened digestion time counterbalanced the negative impact induced by elevated temperature. Recent studies showed that when trypsin digestions were performed in a microwave [120] or high-pressure [121] devices, the digestion time can be reduced dramatically. These methods have been used successfully in the field of proteomics research. Denaturants such as urea and GdnHCl reduce the trypsin activity as already mentioned; therefore, they should be removed from the buffer before the addition of trypsin to maximize the proteolytic activity. With such reagents removed, an improved trypsin digestion of IgG molecule was achieved at a 1:25 enzyme/substrate ratio at 37°C for 30 min with little protocol-induced Asn deamidation or N-terminal Gln cyclization product observed [115].

With the above-mentioned criteria in mind, all steps of the sample preparation procedure need to be scrutinized to minimize artificial degradation. With well-controlled conditions, peptide mapping HPLC-MS/MS is still the most powerful way to characterize and quantify modifications in biologics.

7.6 CONCLUSIONS

The use of HPLC-MS and HPLC-MS/MS methodologies has become the method of choice in structural analysis of unknown impurities and degradation products in small molecule pharmaceuticals and biologics, as described in this chapter. For small molecule applications, accurate mass measurements on molecular ions provide critical information to deduce elemental compositions of unknowns. Obtaining accurate mass data on fragment ions is important to structural assignments. Additional online H/D exchange HR-HPLC-MS approach further facilitates structural assignments for unknowns. For biologics applications, MS techniques are uniquely positioned to provide detailed modification information on proteins for structural characterization. It is important to note that problem solving is largely dependent on the scientific knowledge and creativity of scientists and their implementation of new analytical strategies and technologies. With technological advances in MS and separation science, HPLC-MS-based techniques will continue to play important roles in the characterization of impurities and degradation products in the future.

REFERENCES

[1] Pramanik BN, Bartner PL, Chen G. The role of mass spectrometry in the drug discovery process. Curr Opin Drug Discov Devel 1999;2:401–17.

[2] Chen G, Pramanik BN, Liu YH, Mirza UA. Applications of LC/MS in structural identifications of small molecules and proteins in drug discovery. J Mass Spectrom 2007;42:279–87.

[3] Pramanik BN, Lee MS, Chen G, editors. Characterization of Impurities and Degradants Using Mass Spectrometry. Hoboken, NJ: Wiley, 2011.

[4] Chen G, Zhang LK, Pramanik BN. LC/MS: theory, instrumentation and applications to small molecules. In Kazakevich Y, LoBrutto R, editors. HPLC for Pharmaceutical Scientists. New York: Wiley, Inc., 2007, pp. 281–346.

[5] Yamashita M, Fenn JB. Electrospray ion source. Another variation on the free-jet theme. J Phys Chem 1984;8:4451–9.

[6] Fenn JB, Mann M, Meng CK, Wong SF, Whitehouse CM. Electrospray ionization for mass spectrometry of large biomolecules. Science 1989;246:64–71.

[7] Horning EC, Horning MG, Carroll DI, Dzidic I, Stillwell RN. New picogram detection system based on a mass spectrometer with an external ionization source at atmospheric pressure. Anal Chem 1973;45:936–43.

[8] Mamyrin BA, Karataev VI, Shmikk DV, Zagulin VA. Mass reflectron. New nonmagnetic time-of-flight high-resolution mass spectrometer. Zh Eksp Teor Fiz 1973;64:82–9.

[9] Guilhaus M. Principles and instrumentation in time-of-flight mass spectrometry. Physical and instrumental concepts. J Mass Spectrom 1995;30:1519–32.

[10] Lawrence EO, Edlefsen NE. On the production of high speed protons, Science 1930;72:376.

[11] Comisarow MB, Marshall AG. Fourier transform ion cyclotron resonance spectroscopy. Chem Phys Lett 1974;25:282–3.

[12] Hardman M, Makarov AA. Interfacing the orbitrap mass analyzer to an electrospray ion source. Anal Chem 2003;75:1699–705.

[13] Hu Q, Noll RJ, Li H, Makarov A, Hardman M, Cooks RG. The Orbitrap: a new mass spectrometer. J Mass Spectrom 2005;40:430–43.

[14] Makarov A, Denisov E, Kholomeev A, Balschun W, Lange O, Strupat K, Horning S. Performance evaluation of a hybrid linear ion trap/Orbitrap mass spectrometer. Anal Chem 2006;78:2113–20.

[15] Makarov A, Denisov E, Lange O, Horning S. Dynamic range of mass accuracy in LTQ orbitrap hybrid mass spectrometer. J Am Soc Mass Spectrom 2006;17:977–82.

[16] Cooks RG, Beynon JH, Caprioli RM, Lester GR. Metastable Ions. Amsterdam, The Netherlands: Elsevier, 1973.

[17] Busch K, Glish G, Mcluckey S. Mass Spectrometry/Mass Spectrometry: Techniques and Applications of Tandem Mass Spectrometry. New York: VCH publishing, 1988.

[18] Ohashi N, Furuuchi S, Yoshikawa M. Usefulness of the hydrogen-deuterium exchange method in the study of drug metabolism using liquid chromatography-tandem mass spectrometry. J Pharm Biomed Anal 1998;18:325–34.

[19] Olsen MA, Cummings PG, Kennedy-Gabb S, Wagner BM, Nicol GR, Munson B. The use of deuterium oxide as a mobile phase for structural elucidation by HPLC/UV/ESI/MS. Anal Chem 2000;72:5070–8.

[20] Liu DQ, Hop CE, Beconi MG, Mao A, Chiu SH. Use of on-line hydrogen/deuterium exchange to facilitate metabolite identification. Rapid Commun Mass Spectrom 2001;15:1832–9.

[21] Lam W, Ramanathan R. In electrospray ionization source hydrogen/deuterium exchange LC-MS and LC-MS/MS for characterization of metabolites. J Am Soc Mass Spectrom 2002;13:345–53.

[22] Tolonen A, Turpeinen M, Uusitalo J, Pelkonen O. A simple method for differentiation of monoisotopic drug metabolites with hydrogen-deuterium exchange liquid chromatography/electrospray mass spectrometry. Eur J Pharm Sci 2005;25:155–62.

[23] Wolff JC, Laures AM. "On-the-fly" hydrogen/deuterium exchange liquid chromatography/mass spectrometry using a dual-sprayer atmospheric pressure ionisation source. Rapid Commun Mass Spectrom 2006;20:3769–79.

[24] Novak TJ, Helmy R, Santos I. Liquid chromatography-mass spectrometry using the hydrogen/deuterium exchange reactions as a tool for impurity identification in pharmaceutical process development. J Chromatogr B Biomed Sci Appl 2005;825:161–8.

[25] Liu DQ, Hop CECA. Strategies for characterization of drug metabolites using liquid chromatography-tandem mass spectrometry in conjunction with chemical derivatization and on-Line H/D exchange approaches. J Pharm Biomed Anal 2005;37:1–18.

[26] Chen G, Khusid A, Daaro I, Irish P, Pramanik BN. Structural identification of trace level enol tautomer impurity by on-line hydrogen/deuterium exchange HR-LC/MS in a LTQ-Orbitrap hybrid mass spectrometer. J Mass Spectrom 2007;42:967–70.

[27] Chen G, Daaro I, Pramanik BN, Piwinski JJ. Structural characterization of *in vitro* rat liver microsomal metabolites of antihistamine desloratadine using LTQ-Orbitrap hybrid mass spectrometer in combination with online hydrogen/deuterium exchange HR-LC/MS. J Mass Spectrom 2009;44:203–13.

[28] Hillenkamp F, Karas M, Beavis RC, Chait BT. Matrix-assisted laser desorption / ionization mass spectrometry of biopolymers. Anal Chem 1991;63:1193A–203A.

[29] Smith RD, Loo JA, Edmonds CG, Barinaga CJ, Udseth HR. New developments in biochemical mass spectrometry: electrospray ionization. Anal Chem 1990;62:882–99.

[30] McLafferty FW. Tandem mass spectrometry. Science 1981;214:280–7.

[31] Hirayama K, Takahashi R, Akashi S, Fukuhara K, Oouchi N, Murai A, Arai M, Murao S, Tanaka K, Nojima I. Primary structure of paim I, an alpha-amylase inhibitor from *Streptomyces corchorushii*, determined by the combination of Edman degradation and fast atom bombardment mass spectrometry. Biochemistry 1987;26:6483–8.

[32] Henzel WJ, Billeci TM, Stults JT, Wong SC, Grimley C, Watanable C. Identifying proteins from two-dimensional gels by molecular mass searching of peptide fragments in protein sequence databases. Proc Natl Acad Sci U S A 1993;90:5011–5.

[33] Kelleher NL. Top-down proteomics. Anal Chem 2004;76:197A–203A.

[34] Hunt DF, Yates JR, III, Shabanowitz J, Winston S, Hauer CR. Protein sequencing by tandem mass spectrometry. Proc Natl Acad Sci U S A 1986;83:6233–7.

[35] Yates JR, III. Mass spectrometry and the age of the proteome. J Mass Spectrom 1998; 33:1–19.

[36] Nelson RW. The use of bioreactive probes in protein characterization. Mass Spectrom Rev 1997;16:353–76.

[37] Chen G, Warrack BM, Goodenough AK, Wei H, Wang-Iverson DB, Tymiak AA. Characterization of protein therapeutics by mass spectrometry: recent developments and future directions. Drug Discov Today 2011;16:58–64.

[38] Chen G, Daaro I, Pramanik BN. Structural identifications of mometasone furoate steroid related impurities in a LTQ-Orbitrap hybrid mass spectrometer. Proceedings of the 55th ASMS Conference on Mass Spectrometry and Allied Topics. Abstract # A074176, Indianapolis, IN, June 3–7, 2007.

[39] Nomeir AA, Pramanik BN, Heimark L, Bennett F, Veals J, Bartner P, Hilbert M, Saksena A, McNamara P, Girijavallabhan V, Ganguly AK, Lovey R, Pike R, Wang H, Liu YT, Kumari P, Korfmacher W, Lin CC, Cacciapuoti A, Loebenberg D, Hare R, Miller G, Pickett C. Posaconazole (Noxafil, SCH 56592), a new azole antifungal drug, was a discovery based on the isolation and mass spectral characterization of a circulating metabolite of an earlier lead (SCH 51048). J Mass Spectrom 2008;43:509–17.

[40] Perfect JR, Cox GM, Dodge RK, Schell WA. *In vitro* and *in vivo* efficacies of the azole SCH56592 against *Cryptococcus neoformans*. Antimicrob Agents Chemother 1996; 40:1910–3.

[41] Galgiani JN, Lewis ML. *In vitro* studies of activities of the antifungal triazoles SCH56592 and itraconazole against *Candida albicans*, *Cryptococcus neoformans*, and other pathogenic yeasts. Antimicrob Agents Chemother 1997;41:180–3.

[42] Graybill JR, Bocanegra R, Najvar LK, Luther MF, Loebenberg D. SCH56592 treatment of murine invasive aspergillosis. J Antimicrobial. Chemotherapy 1998;42:539–42.

[43] Uchida K, Yokota N, Yamaguchi H. *In vitro* antifungal activity of posaconazole against various pathogenic fungi. Int J Antimicrob Agents 2001;18:167–72.

[44] Frampton JE, Scott LJ. Posaconazole: a review of its use in the prophylaxis of invasive fungal infections. Drugs 2008;68:993–1016.

[45] Feng W, Liu H, Chen G, Malchow R, Bennett F, Lin E, Pramanik B, Chan TM. Structural characterization of the oxidative degradation products of an antifungal agent SCH 56592 by LC-NMR and LC-MS. J Pharm Biomed Anal 2001;25:545–57.

[46] Leader B, Baca QJ, Golan DE. Protein therapeutics: a summary and pharmacological classification. Nat Rev Drug Discov 2008;7:21–39.

[47] Aggarwal S. What's fueling the biotech engine—2008. Nat Biotechnol 2009;27:987–93.

[48] Mimura Y, Nakamura K, Tanaka T, Fujimoto M. Evidence of intra- and extracellular modifications of monoclonal IgG polypeptide chains generating charge heterogeneity. Electrophoresis 1998;19:767–75.

[49] Kroon DJ, Baldwin-Ferro A, Lalan P. Identification of sites of degradation in a therapeutic monoclonal antibody by peptide mapping. Pharm Res 1992;9:1386–93.

[50] Walsh CT, Garneau-Tsodikova S, Gatto GJ, Jr. Protein posttranslational modifications: the chemistry of proteome diversifications. Angew Chem Int Ed Engl 2005;44: 7342–72.

[51] Jenkins N, Murphy L, Tyther R. Post-translational modifications of recombinant proteins: significance for biopharmaceuticals. Mol Biotechnol 2008;39:113–8.

[52] Cacia J, Quan CP, Vasser M, Sliwkowski MB, Frenz J. Protein sorting by high-performance liquid chromatography. I. Biomimetic interaction chromatography of recombinant human deoxyribonuclease I on polyionic stationary phases. J Chromatogr 1993;634:229–39.

[53] Teshima G, Porter J, Yim K, Ling V, Guzzetta A. Deamidation of soluble CD4 at asparagine-52 results in reduced binding capacity for the HIV-1 envelope glycoprotein gp120. Biochemistry 1991;30:3916–22.

[54] Becker GW, Tackitt PM, Bromer WW, Lefeber DS, Riggin RM. Isolation and characterization of a sulfoxide and a desamido derivative of biosynthetic human growth hormone. Biotechnol Appl Biochem 1988;10:326–37.

[55] Robinson NE, Robinson AB. Prediction of protein deamidation rates from primary and three-dimensional structure. Proc Natl Acad Sci U S A 2001;98:4367–72.

[56] Geiger T, Clarke S. Deamidation, isomerization, and racemization at asparaginyl and aspartyl residues in peptides. Succinimide-linked reactions that contribute to protein degradation. J Biol Chem 1987;262:785–94.

[57] Wright HT. Nonenzymatic deamidation of asparaginyl and glutaminyl residues in proteins. Crit Rev Biochem Mol Biol 1991;26:1–52.

[58] Chelius D, Jing K, Lueras A, Rehder DS, Dillon TM, Vizel A, et al. Formation of pyroglutamic acid from N-terminal glutamic acid in immunoglobulin gamma antibodies. Anal Chem 2006;78:2370–6.

[59] Weitzhandler M, Farnan D, Horvath J, Rohrer JS, Slingsby RW, Avdalovic N, et al. Protein variant separations by cation-exchange chromatography on tentacle-type polymeric stationary phases. J Chromatogr A 1998;828:365–72.

[60] Zhang W, Czupryn MJ. Analysis of isoaspartate in a recombinant monoclonal antibody and its charge isoforms. J Pharm Biomed Anal 2003;30:1479–90.

[61] Volkin DB, Verticelli AM, Bruner MW, Marfia KE, Tsai PK, Sardana MK, et al. Deamidation of polyanion-stabilized acidic fibroblast growth factor. J Pharm Sci 1995; 84:7–11.

[62] Huang L, Lu J, Wroblewski VJ, Beals JM, Riggin RM. *In vivo* deamidation characterization of monoclonal antibody by LC/MS/MS. Anal Chem 2005;77:1432–9.

[63] Liu YD, van Enk JZ, Flynn GC. Human antibody Fc deamidation *in vivo*. Biologicals 2009;37:313–22.

[64] Robinson NE, Robinson AB. Molecular clocks. Proc Natl Acad Sci U S A 2001; 98:944–9.

[65] Roher AE, Lowenson JD, Clarke S, Wolkow C, Wang R, Cotter RJ, et al. Structural alterations in the peptide backbone of beta-amyloid core protein may account for its deposition and stability in Alzheimer's disease. J Biol Chem 1993;268:3072–83.

[66] Reissner KJ, Aswad DW. Deamidation and isoaspartate formation in proteins: unwanted alterations or surreptitious signals? Cell Mol Life Sci 2003;60:1281–95.

[67] O'Connor PB, Cournoyer JJ, Pitteri SJ, Chrisman PA, McLuckey SA. Differentiation of aspartic and isoaspartic acids using electron transfer dissociation. J Am Soc Mass Spectrom 2006;17:15–9.

[68] Cournoyer JJ, Pittman JL, Ivleva VB, Fallows E, Waskell L, Costello CE, et al. Deamidation: differentiation of aspartyl from isoaspartyl products in peptides by electron capture dissociation. Protein Sci 2005;14:452–63.

[69] Sargaeva NP, Lin C, O'Connor PB. Identification of aspartic and isoaspartic acid residues in amyloid beta peptides, including Abeta1-42, using electron-ion reactions. Anal Chem 2009;81:9778–86.

[70] Shacter E, Williams JA, Hinson RM, Senturker S, Lee YJ. Oxidative stress interferes with cancer chemotherapy: inhibition of lymphoma cell apoptosis and phagocytosis. Blood 2000;96:307–13.

[71] Hovorka S, Schoneich C. Oxidative degradation of pharmaceuticals: theory, mechanisms and inhibition. J Pharm Sci 2001;90:253–69.

[72] Scislowski PW, Foster AR, Fuller MF. Regulation of oxidative degradation of L-lysine in rat liver mitochondria. Biochem J 1994;300(Pt 3):887–91.

[73] Taggart C, Cervantes-Laurean D, Kim G, McElvaney NG, Wehr N, Moss J, et al. Oxidation of either methionine 351 or methionine 358 in alpha 1-antitrypsin causes loss of anti-neutrophil elastase activity. J Biol Chem 2000;275:27258–65.

[74] Stadtman ER. Cyclic oxidation and reduction of methionine residues of proteins in antioxidant defense and cellular regulation. Arch Biochem Biophys 2004;423:2–5.

[75] Vogt W. Oxidation of methionyl residues in proteins: tools, targets, and reversal. Free Radic Biol Med 1995;18:93–105.

[76] Liu H, Gaza-Bulseco G, Xiang T, Chumsae C. Structural effect of deglycosylation and methionine oxidation on a recombinant monoclonal antibody. Mol Immunol 2008; 45:701–8.

[77] Gaza-Bulseco G, Faldu S, Hurkmans K, Chumsae C, Liu H. Effect of methionine oxidation of a recombinant monoclonal antibody on the binding affinity to protein A and protein G. J Chromatogr B Analyt Technol Biomed Life Sci 2008;870:55–62.

[78] Soenderkaer S, Carpenter JF, van de Weert M, Hansen LL, Flink J, Frokjaer S. Effects of sucrose on rFVIIa aggregation and methionine oxidation. Eur J Pharm Sci 2004;21:597–606.

[79] Liu H, Gaza-Bulseco G, Zhou L. Mass spectrometry analysis of photo-induced methionine oxidation of a recombinant human monoclonal antibody. J Am Soc Mass Spectrom 2009;20:525–8.

[80] Chumsae C, Gaza-Bulseco G, Sun J, Liu H. Comparison of methionine oxidation in thermal stability and chemically stressed samples of a fully human monoclonal antibody. J Chromatogr B Analyt Technol Biomed Life Sci 2007;850:285–94.

[81] Lam XM, Yang JY, Cleland JL. Antioxidants for prevention of methionine oxidation in recombinant monoclonal antibody HER2. J Pharm Sci 1997;86:1250–5.

[82] Houde D, Kauppinen P, Mhatre R, Lyubarskaya Y. Determination of protein oxidation by mass spectrometry and method transfer to quality control. J Chromatogr A 2006;1123:189–98.

[83] Chaderjian WB, Chin ET, Harris RJ, Etcheverry TM. Effect of copper sulfate on performance of a serum-free CHO cell culture process and the level of free thiol in the recombinant antibody expressed. Biotechnol Prog 2005;21:550–3.

[84] Lin HJ, Lin CH, Tseng HC, Chen YH. Detecting disulfide crosslinks of high-molecular weight complexes in mouse SVS proteins by diagonal electrophoresis. Anal Biochem 2006;352:296–8.

[85] Wu J, Watson JT. A novel methodology for assignment of disulfide bond pairings in proteins. Protein Sci 1997;6:391–8.

[86] Gorman JJ, Wallis TP, Pitt JJ. Protein disulfide bond determination by mass spectrometry. Mass Spectrom Rev 2002;21:183–216.

[87] Wu SL, Jiang H, Lu Q, Dai S, Hancock WS, Karger BL. Mass spectrometric determination of disulfide linkages in recombinant therapeutic proteins using online LC-MS with electron-transfer dissociation. Anal Chem 2009;81:112–22.

[88] Thomsen MK, Hansen BS, Nilsson P, Nowak J, Johansen PB, Thomsen PD, et al. Pharmacological characterization of a biosynthetic trisulfide-containing hydrophobic derivative of human growth hormone: comparison with standard 22 K growth hormone. Pharmacol Toxicol 1994;74:351–8.

[89] Gu S, Wen D, Weinreb PH, Sun Y, Zhang L, Foley SF, et al. Characterization of trisulfide modification in antibodies. Anal Biochem 2010;400:89–98.

[90] Nielsen RW, Tachibana C, Hansen NE, Winther JR. Trisulfides in proteins. Antioxid Redox Signal 2011;15:67–75.

[91] Aono H, Wen D, Zang L, Houde D, Pepinsky RB, Evans DR. Efficient on-column conversion of IgG1 trisulfide linkages to native disulfides in tandem with Protein A affinity chromatography. J Chromatogr A 2010;1217:5225–32.

[92] Dick LW, Jr., Qiu D, Mahon D, Adamo M, Cheng KC. C-terminal lysine variants in fully human monoclonal antibodies: investigation of test methods and possible causes. Biotechnol Bioeng 2008;100:1132–43.

[93] Johnson KA, Paisley-Flango K, Tangarone BS, Porter TJ, Rouse JC. Cation exchange-HPLC and mass spectrometry reveal C-terminal amidation of an IgG1 heavy chain. Anal Biochem 2007;360:75–83.

[94] Wang W, Singh S, Zeng DL, King K, Nema S. Antibody structure, instability, and formulation. J Pharm Sci 2007;96:1–26.

[95] Liu H, Gaza-Bulseco G, Lundell E. Assessment of antibody fragmentation by reversed-phase liquid chromatography and mass spectrometry. J Chromatogr B Analyt Technol Biomed Life Sci 2008;876:13–23.

[96] Joshi AB, Sawai M, Kearney WR, Kirsch LE. Studies on the mechanism of aspartic acid cleavage and glutamine deamidation in the acidic degradation of glucagon. J Pharm Sci 2005;94:1912–27.

[97] Xie H, Gilar M, Gebler JC. Characterization of protein impurities and site-specific modifications using peptide mapping with liquid chromatography and data independent acquisition mass spectrometry. Anal Chem 2009;81:5699–708.

[98] Jensen ON. Modification-specific proteomics: characterization of post-translational modifications by mass spectrometry. Curr Opin Chem Biol 2004;8:33–41.

[99] Salzano AM, Crescenzi M. Mass spectrometry for protein identification and the study of post translational modifications. Ann Ist Super Sanita 2005;41:443–50.

[100] Austin RH, Beeson KW, Eisenstein L, Frauenfelder H, Gunsalus IC. Dynamics of ligand binding to myoglobin. Biochemistry 1975;14:5355–73.

[101] Carpenter JF, Kendrick BS, Chang BS, Manning MC, Randolph TW. Inhibition of stress-induced aggregation of protein therapeutics. Methods Enzymol 1999;309: 236–55.

[102] Ross CA, Poirier MA. Protein aggregation and neurodegenerative disease. Nat Med 2004;10(Suppl):S10–S17.

[103] Schmidt AC, Fahlbusch B, Otto M. Size exclusion chromatography coupled to electrospray ionization mass spectrometry for analysis and quantitative characterization of arsenic interactions with peptides and proteins. J Mass Spectrom 2009;44:898–910.

[104] Pekar A, Sukumar M. Quantitation of aggregates in therapeutic proteins using sedimentation velocity analytical ultracentrifugation: practical considerations that affect precision and accuracy. Anal Biochem 2007;367:225–37.

[105] Ye H. Simultaneous determination of protein aggregation, degradation, and absolute molecular weight by size exclusion chromatography-multiangle laser light scattering. Anal Biochem 2006;356:76–85.

[106] Nettleton EJ, Tito P, Sunde M, Bouchard M, Dobson CM, Robinson CV. Characterization of the oligomeric states of insulin in self-assembly and amyloid fibril formation by mass spectrometry. Biophys J 2000;79:1053–65.

[107] Sun J, Kitova EN, Wang W, Klassen JS. Method for distinguishing specific from nonspecific protein-ligand complexes in nanoelectrospray ionization mass spectrometry. Anal Chem 2006;78:3010–8.

[108] Daubenfeld T, Bouin AP, van der Rest GA. Deconvolution method for the separation of specific versus nonspecific interactions in noncovalent protein-ligand complexes analyzed by ESI-FT-ICR mass spectrometry. J Am Soc Mass Spectrom 2006;17: 1239–48.

[109] Dong MW. Tryptic mapping by reversed phase liquid chromatography. Adv Chromatogr 1992;32:21–51.

[110] Krokhin OV, Antonovici M, Ens W, Wilkins JA, Standing KG. Deamidation of -Asn-Gly- sequences during sample preparation for proteomics: consequences for MALDI and HPLC-MALDI analysis. Anal Chem 2006;78:6645–50.

[111] Dick LW, Jr., Kim C, Qiu D, Cheng KC. Determination of the origin of the N-terminal pyro-glutamate variation in monoclonal antibodies using model peptides. Biotechnol Bioeng 2007;97:544–53.

[112] Stark GR. Reactions of cyanate with functional groups of proteins. 3. Reactions with amino and carboxyl groups. Biochemistry 1965;4:1030–6.

[113] Lippincott J, Apostol I. Carbamylation of cysteine: a potential artifact in peptide mapping of hemoglobins in the presence of urea. Anal Biochem 1999;267:57–64.

[114] Dick LW, Jr., Mahon D, Qiu D, Cheng KC. Peptide mapping of therapeutic monoclonal antibodies: improvements for increased speed and fewer artifacts. J Chromatogr B Analyt Technol Biomed Life Sci 2009;877:230–6.

[115] Ren D, Pipes GD, Liu D, Shih LY, Nichols AC, Treuheit MJ, et al. An improved trypsin digestion method minimizes digestion-induced modifications on proteins. Anal Biochem 2009;392:12–21.

[116] Lewis UJ, Cheever EV, Hopkins WC. Kinetic study of the deamidation of growth hormone and prolactin. Biochim Biophys Acta 1970;214:498–508.

[117] Lundell N, Schreitmuller T. Sample preparation for peptide mapping—a pharmaceutical quality-control perspective. Anal Biochem 1999;266:31–47.

[118] Liu P, O'Mara BW, Warrack BM, Wu W, Huang Y, Zhang Y, et al. A tris (2-carboxyethyl) phosphine (TCEP) related cleavage on cysteine-containing proteins. J Am Soc Mass Spectrom 2010;21:837–44.

[119] Bongers J, Cummings JJ, Ebert MB, Federici MM, Gledhill L, Gulati D, et al. Validation of a peptide mapping method for a therapeutic monoclonal antibody: what could we possibly learn about a method we have run 100 times? J Pharm Biomed Anal 2000;21:1099–128.

[120] Pramanik BN, Mirza UA, Ing YH, Liu YH, Bartner PL, Weber PC, et al. Microwave-enhanced enzyme reaction for protein mapping by mass spectrometry: a new approach to protein digestion in minutes. Protein Sci 2002;11:2676–87.

[121] Lopez-Ferrer D, Petritis K, Hixson KK, Heibeck TH, Moore RJ, Belov ME, et al. Application of pressurized solvents for ultrafast trypsin hydrolysis in proteomics: proteomics on the fly. J Proteome Res 2008;7:3276–81.

8

LIQUID EXTRACTION SURFACE ANALYSIS (LESA): A NEW MASS SPECTROMETRY-BASED TECHNIQUE FOR AMBIENT SURFACE PROFILING

DANIEL EIKEL AND JACK D. HENION

8.1 INTRODUCTION

Studying the distribution of new chemical entities in the whole body of laboratory animals (or single organs/tissues of interest) is a major research focus in drug discovery and drug development in the pharmaceutical industry. Currently, the most utilized technique for this type of study is quantitative whole-body autoradiography (QWBA) [1–3], an approach that is both very sensitive and can provide a high spatial resolution of up to ca. 30 μm. Also, QWBA provides valuable quantitative data and is required by regulatory agencies. However, QWBA has some limitations: autoradiography requires a radiolabeled compound whose synthesis might be expensive and time-consuming, and it cannot distinguish between the radiolabeled parent (dosed compound) and its metabolites. Also, metabolites not carrying the radiolabel would not be detected, making the interpretation of QWBA data challenging and limiting its ability to study metabolites that could be pharmacologically active or contribute to toxicity.

Any surface analysis technology that utilizes mass spectrometry (MS) as the detector could potentially overcome the above limitations of QWBA. Mass spectrometry is capable of detecting the parent drug and its metabolites in the same experiment and therefore allows drug distribution and biotransformation analysis in one step without the cost of employing radiolabeled drug substances. The use of MS for the detection and distribution of drugs and their metabolites in tissue slices has been referred to as mass spectrometry imaging (MSI) [4–10]. MSI includes, but

Mass Spectrometry for Drug Discovery and Drug Development, First Edition. Edited by Walter A. Korfmacher.

is not limited to, techniques such as matrix-assisted laser desorption ionization mass spectrometry (MALDI MS) ([11–14], also Chapters 10 and 11 in this book), desorption electrospray ionization mass spectrometry (DESI-MS) [15], secondary ion mass spectrometry imaging (SIMS-Imaging) [16], laser ablation electrospray ionization (LAESI) [17], and electrospray-assisted laser desorption/ionization MS [18] or live single cell mass spectrometry (nano-manipulation nESI-MS) [19–21].

More recent developments in the context of ambient surface analysis and ionization are approaches utilizing a liquid junction and extraction from the surface [22], for example, the liquid micro junction surface sampling probe (LMJ-SSP) [23] and the liquid extraction surface analysis mass spectrometry (LESA-MS) [24].

Not only does the LESA-MS approach has potential advantages over QWBA as already mentioned, but it also has distinct characteristics compared with other MSI approaches in that no additional sample preparation is required (which compared with MALDI could also eliminate potential low molecular weight MALDI matrix mass interferences). LESA-MS uses nESI, which is a soft ionization process, potentially providing qualitative and possibly quantitative information on fragile drugs or metabolites. Furthermore, LESA-MS allows simple optimization of the extraction/spray solvent to reflect optimal analyte extraction and ionization conditions in both positive and negative ion modes.

In this chapter we will give an overview of the LESA-MS technique and discuss examples of its application in drug discovery and drug development. We will also briefly describe LESA applications in other areas and give an outlook of interesting developments that might shape the use of LESA in the future.

8.2 LESA: HOW DOES IT WORK?

Liquid extraction surface analysis is an ambient air surface analysis and ionization technique. The only commercially available system that can analyze samples via LESA is currently the Advion (Ithaca, NY) TriVersa-NanoMate robotic ion source (http://www.advion.com). In the LESA experiment, the robot picks up a sample pipette tip, aspirates a programmable volume of extraction solvent from a reservoir (usually 1–2 μL), and moves the tip close to the surface location of interest (programmable location within 50 μm, with a distance to the surface usually within 200 μm). The robot then dispenses the extraction solvent onto the target, forming a micro liquid junction of 1–2 μL, and a static micro extraction process commences. Extraction time on the surface, repeated dispense/aspirate steps, and wait times between the extraction cycle are programmable and can help optimize the extraction process.

Following sample surface extraction, the solvent mixture is delivered to a single nanoelectrospray (nESI) emitter located within an array of 400 emitter nozzles of a micro fabricated silicon chip whereupon a voltage of 1–1.5 kV is applied to generate ions by the electrospray process (Figs. 8.1 and 8.2). Samples may be analyzed by a wide variety of atmospheric pressure ionization (API) mass spectrometer types and models as long as they are compatible with the TriVersa NanoMate. This list includes most of today's commercially available MS systems. The enabling part of the technology is the chip-based nESI emitter array that allows the use of a very small volume of extraction solvent. The volume of solvent should be maintained as

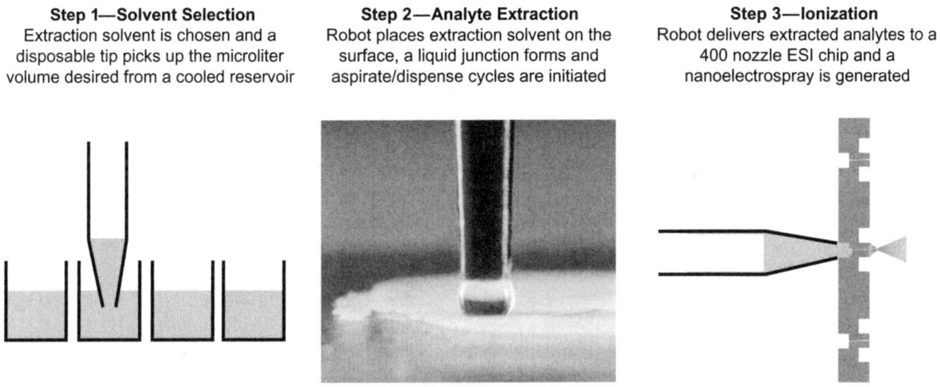
Step 1—Solvent Selection
Extraction solvent is chosen and a disposable tip picks up the microliter volume desired from a cooled reservoir
Step 2—Analyte Extraction
Robot places extraction solvent on the surface, a liquid junction forms and aspirate/dispense cycles are initiated
Step 3—Ionization
Robot delivers extracted analytes to a 400 nozzle ESI chip and a nanoelectrospray is generated

Figure 8.1 Schematic representation of the liquid extraction surface analysis (LESA) workflow. Step 1—Extraction solvent is robotically aspirated from the reservoir of choice. Step 2—Solvent is placed on the surface, a liquid junction forms and the static extraction process starts. Step 3—The extract is delivered to the nESI chip, a voltage applied and electrospray ionization commences with analysis via mass spectrometry. Reprinted from Reference 28 with permission from John Wiley & Sons, Ltd.

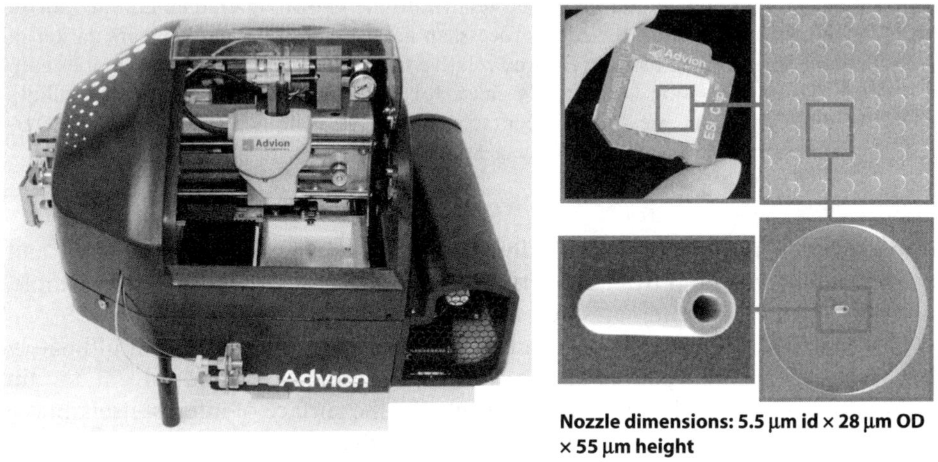
Advion
Nozzle dimensions: 5.5 μm id × 28 μm OD × 55 μm height

Figure 8.2 The Advion TriVersa-NanoMate is currently the only commercially available system that can perform LESA analysis. The enabling technology is the 400 nozzle nESI chip that allows nanonelectrospray from only 1–2 μL of solvent for an extended period of time (2–4 min), hence allowing complex mass spectrometry experiments.

low as possible in order to increase overall sensitivity (ESI-MS behaves as a concentration-dependent analysis technique [25]) and an optimal area on surface to drop volume ratio is found between 1 and 2 μL. At the same time, such an infusion nanoelectrospray experiment can generate ions for an extended period of time consuming only minimal volume (usually ca. 2–4 min for an extract of 1–2 μL).

Such time can be used for information-dependent analysis (IDA) MS experiments, complex MS^n tree experiments, high-resolution MS, or any combination of selected reaction monitoring (SRM) and full scan MS/MS experiments, of which the

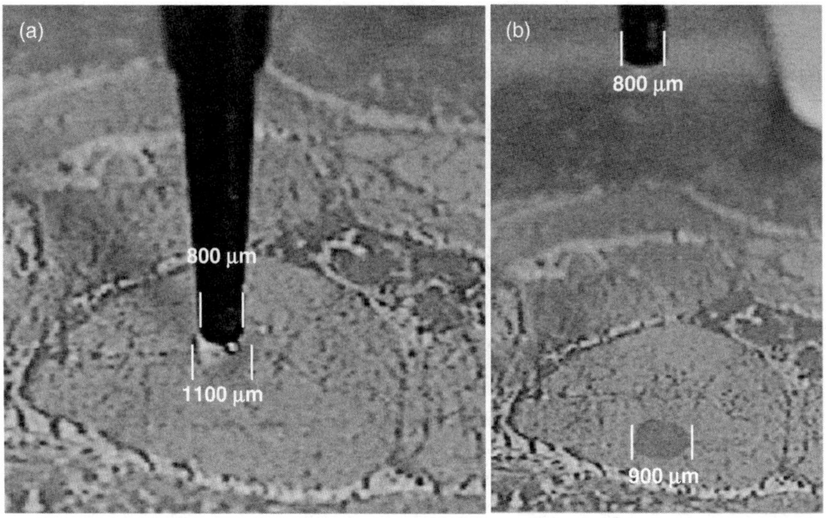

Figure 8.3 Example of a liquid junction formed on a whole-body thin section of mouse brain during LESA extraction. Area sampled is about 1.0 mm using an extraction solvent of 80/20 methanol/water 0.1 vol% formic acid and a solvent volume of 1.1 μL. (A) The liquid junction shows about 1.1 mm in diameter measured relative to the known tapered end of the pipette tip of 0.8 mm. (B) After the extraction process, an area of about 0.9 mm appears darker in color than the surrounding tissue measured relative to the known diameter of the pipette tip of 0.8 mm. Tissue hydration is the likely cause for the observed discoloration and allows measurement of the tissue previously in contact with extraction liquid. Reprinted from Reference 28 with permission from John Wiley & Sons, Ltd.

respective MS is capable. This flexibility greatly enhances the information and quality of data acquired from the surface of interest. Also, the one tip, one sample, one nozzle approach eliminates potential cross-contamination.

The spatial resolution of LESA and the area sampled on the surface is ultimately dependent on the interplay of surface tension from the extraction solvent, the tip used to accommodate the extraction solvent, and the surface of interest itself. In the case of a single organ or whole-body cryo-section, the minimal area sampled is typically 1.0 mm in diameter using an extraction solvent of 80/20 methanol/water 0.1 vol% formic acid and a solvent volume of 1.1 μL (Fig. 8.3, [24]). As such, the spatial resolution on the surface is currently lower than the resolution for QWBA or MSI techniques such as MALDI. This is one reason why LESA is typically referred to as a profiling technique rather than an imaging approach.

8.3 EXAMPLES OF LESA APPLICATIONS IN DRUG DEVELOPMENT AND DRUG DISCOVERY

One of the strengths of LESA is the ability to optimize the extraction solvent for both the extraction process and the analyte ionization. Current knowledge suggests that the extraction of drugs and their metabolites from tissue section surfaces seems to be mostly dependent on the vol% organic present in the solvent and the choice of the organic solvent. For example, for small molecule drug compounds, an 80/20 methanol/water solvent gives good extraction results and is paired with a reasonable surface spread after forming the micro liquid junction (Fig. 8.3). Solvents can be varied from acetonitrile, methanol, tetrahydrofuran (THF) to isopropanol, with some special cases such as lipid extractions where the extraction/spray solvent also contains chloroform or methyl-tert-butyl-ether. However, it is the addition of a modifier such as formic acid or ammonium hydroxide in 0.1 vol% that can make a significant difference in the ionization efficiency of the analyte and the overall sensitivity of the assay. A good example for this effect is fluticasone, a topical corticosteroid used in the treatment of asthma. ESI MS/MS fragmentation shows a rich fragmentation pattern suitable for identification and quantification in both positive and negative nESI-MS/MS using 0.2 μM analytical standard in 80/20 methanol/water with either 0.1 vol% formic acid or ammonium hydroxide (Fig. 8.4).

In one study, fluticasone was dosed *ex vivo* into a mechanically vented and perfusated guinea pig lung model to show its distribution pattern within the lung [26]. The lung was sectioned and analyzed using LESA-MS in both positive and negative ionization modes, and the negative ion showed higher sensitivity, resulting in a 30-fold increase in absolute signal and an approximately sixfold increase of the signal-to-noise (S/N) ratio compared with undosed tissue. A LESA-MS profiling of the complete lung sections (Fig. 8.5) showed one area of highest signal intensity in the longitudinal section A and a double signal peak in the vertical section B. This is consistent with a drug distribution throughout the trachea and into the bronchial part of the lung and is the expected result for an antiasthma drug. Such a model system could be used to both screen for drugs and optimize applicators for inhalants to optimize for drug localization, particle size, and distribution within the lung.

Other recent publications show the successful application of LESA to study the skin penetration of glucocorticoid receptor agonists [27]. Marshall et al. compared the depth penetration of three test compounds into pig ears with both LESA profiling and MALDI imaging. Both MS-based approaches correlated well with each other and gave an explanation for prior unusual findings with regard to a positive skin blanching test for all three compounds and conflicting *in vitro* agonist potency test results. In this interesting example, the information gained from LESA experiments was equivalent to information gained from MALDI; however, based on a truly orthogonal analytical approach, the researchers could verify the model assumption that skin penetration is an important factor in the consideration for positive glucocortocoid receptor agonists.

A more common application for LESA is the whole-body distribution of drugs in dosed rodents. As already discussed, LESA currently provides limited spatial resolution. However, in the case of whole-body drug distribution, the tissues of interest are comparably large (kidney, liver, stomach content, etc.), and it is quite sufficient to sample four locations each on three section depths to gain a rapid understanding of the drug's distribution in the animal model. Figure 8.6 shows a graphical representation of a typical LESA analysis after 50 mg/kg PO terfenadine dosage to a mouse [28]. The drug is extensively metabolized to fexofenadine, and both drug and metabolite are distributed throughout the body with a distribution pattern supporting the assumption of hepatic clearance as well as a smaller renal component to the clearance. Terfenadine and fexofenadine were detected at trace levels in the brain after 1 and 4 h, and considering the fairly high initial dose level,

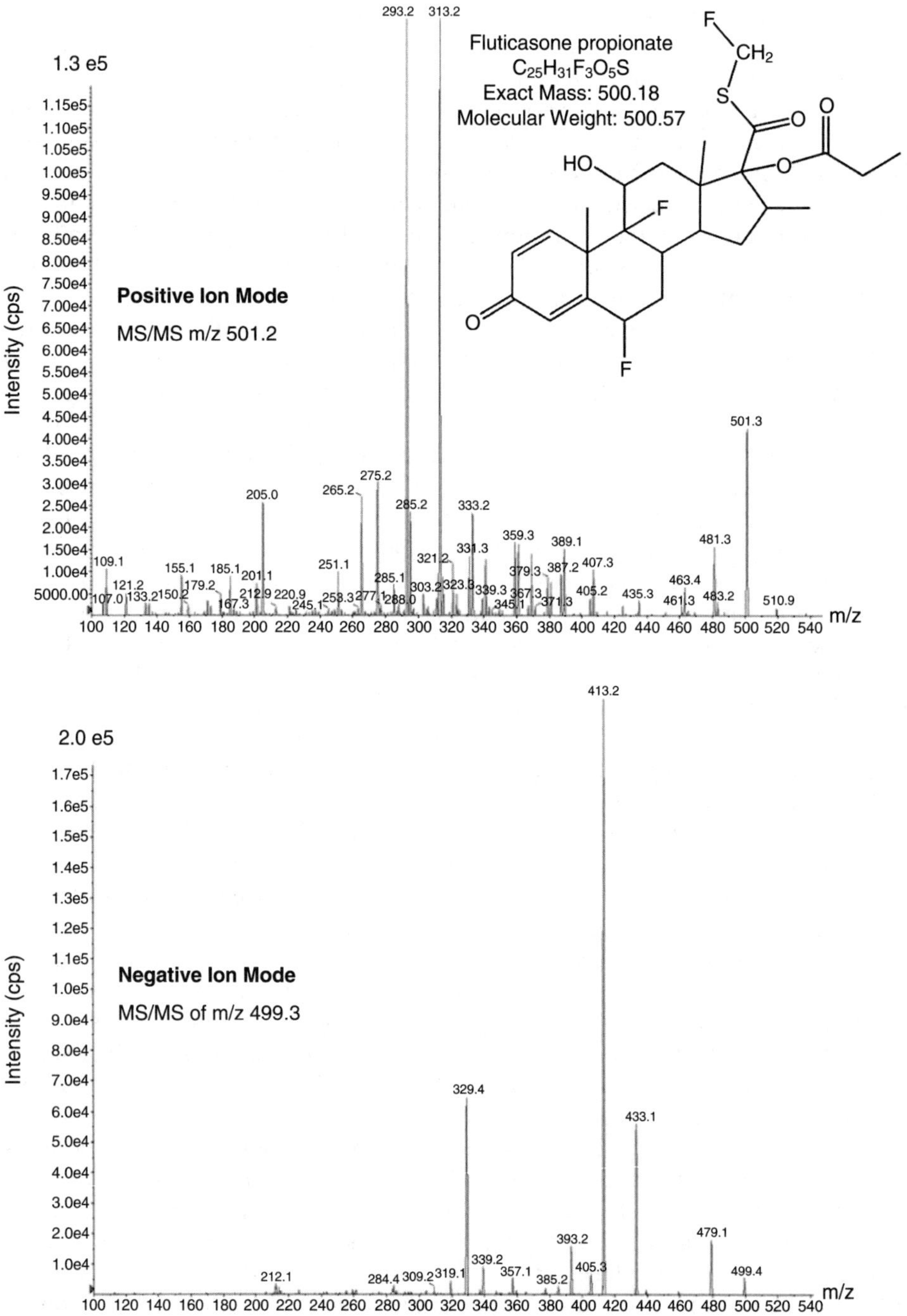

Figure 8.4 Fluticasone MS/MS fragmentation shows a rich fragmentation spectrum for both positive and negative ion mode MS using 200 nM analytical standard in 80/20 methanol/water using either 0.1 vol% formic acid or 0.1 vol% ammoniumhydroxide. SRM tissue analysis later on showed an improved LESA sensitivity based on signal-to-noise ratio against untreated tissue when using negative ion mode SRM versus positive ion mode SRM.

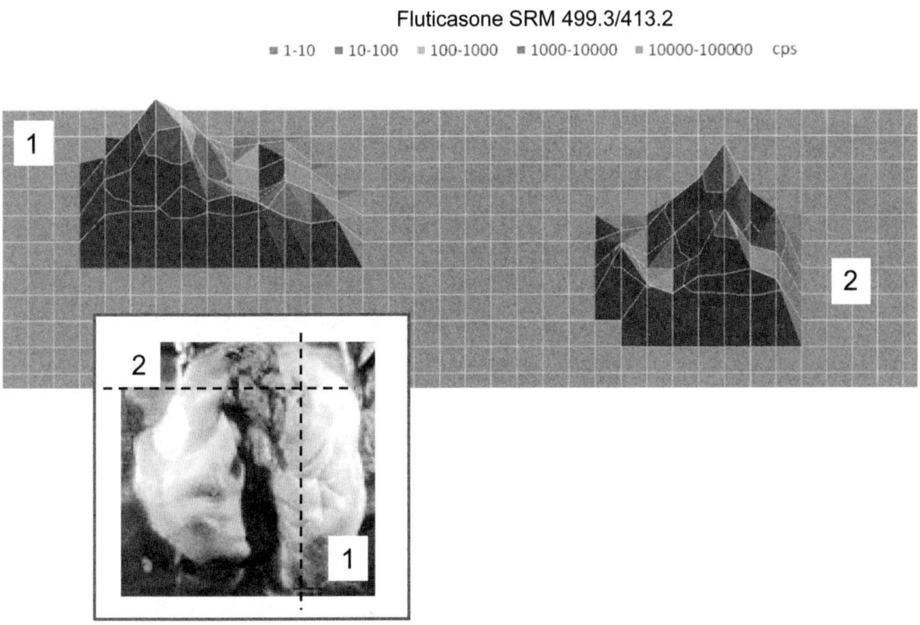

Figure 8.5 See color insert. LESA-MS-SRM in negative ion mode nESI profiling of fluticasone (SRM *m/z* 499.3 > *m/z* 413.2) in lung tissue after aerosol dosage into a mechanically vented and perfusated guinea pig lung model. Distribution pattern shows a single area with intense analyte response along the longitudinal section A and two areas with strong analyte signal along the vertical section B. This is consistent with a drug distribution through the trachea and into the bronchial part of the lung only, the desired distribution to the site of action for this topical corticosteroid drug used in asthma treatment.

the LESA-MS findings suggest limited, if any, ability of these compounds to cross the blood–brain barrier. These findings are in excellent agreement with known literature properties of both drugs. Whereas terfenadine was discontinued, fexofenadine is currently used as an antihistamine antiallergen drug. It is noteworthy that a MALDI imaging experiment at the same dosage level showed less information about the drug distribution in the body [29], and the higher spatial resolution of MALDI typically used for such experiments (around 100 μm) is rather unnecessary and can result in some loss of sensitivity. Compared with liquid chromatography (LC)–MS/MS experiments it was estimated that LESA-MS was able to detect terfenadine and fexofenadine in the low pmol/g concentration in various tissues; however this sensitivity will be compound and tissue dependent [28].

It is worth noting that the LESA sampling process on the surface is currently not well understood and hence can only be optimized empirically. To date, it is unknown how far into the tissue the extraction reaches or how long it takes to reach extraction equilibrium in this essentially static extraction process. For example, a 10th extraction sequence from the same kidney location 1 h post PO dose of 50 mg/kg terfenadine still results in ca. 50 % of the original signal intensity for both drug and metabolite, suggesting either a very limited extraction within each step or an insufficient time to equilibrate the extraction given the high initial concentration of analyte present in this tissue [28]. Despite this limited extraction efficiency, the

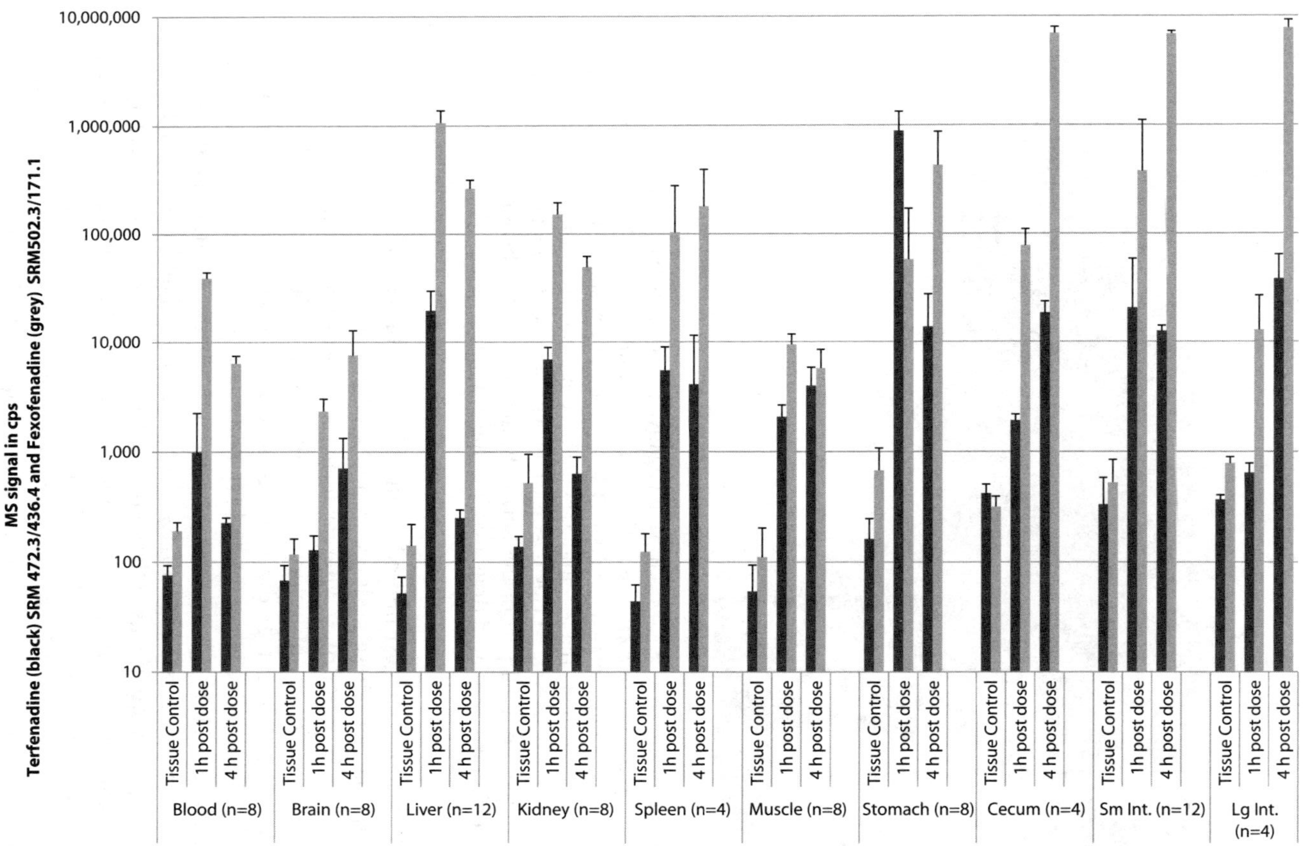

Figure 8.6 LESA-MS-SRM analysis of terfenadine (black columns, SRM *m/z* 472.3 > *m/z* 436.4) and fexofenadine (gray columns, SRM *m/z* 502.3 > *m/z* 171.1) in the positive ion mode at different tissues/organs of mouse whole-body cryo-sections and at different times after PO dose of 50 mg/kg terfenadine. Average cps signal of 4 to 12 locations from up to three different section depths of each organ and the respective standard deviation is shown. Drug distribution pattern is in excellent agreement with known pharmacokinetic properties of the drug and its metabolite. Reprinted from Reference 28 with permission from John Wiley & Sons, Ltd.

overall sensitivity of the method was very good as discussed earlier and compared favorably with MALDI imaging and QWBA. A better understanding of the underlying extraction mechanism could lead to a faster optimization process and an improved overall sensitivity for LESA-MS. However, others have found, for example, that extraction from solid-phase extraction (SPE) cards was exhausted after four extractions, suggesting that the extraction is surface and/or compound dependent.

The original LESA publication from Kertez and van Berkel [24] has demonstrated that LESA-MS is also able to detect glutathione (GSH)-metabolites of propanolol from thin tissue sections. Additional successful examples of metabolite detection and distribution analysis have been shown, including phase 1 hydroxylation and oxidation as stated above for terfenadine [28], figopitant [30], hydroxylation of diclofenac [26, 31, 32], and phase 2 conjugation metabolites such as glucuronidation of diclofenac [32]. Figure 8.7 shows an example for the diclofenac glucuronide detected in mouse liver 15 min after PO dose of 10 mg/kg diclofenac using a high-resolution MS approach, supporting the suggestion that the softer ionization process of ESI might enable the detection of drug metabolites, especially of phase 2-type metabolites, which might otherwise be challenging to ionize and detect by MS employing alternative ionization techniques.

LESA has also been used for shotgun lipidomic approaches to the detection and characterization of artherosclerotic plaques [33]. Here, the LESA technique was compared with a standard liquid–liquid tissue extraction approach for shotgun lipidomics, and the authors found equivalent data quality between the two approaches and could present comprehensive lipid profiling of artherosclerotic plaques and surrounding tissue. Lipid analysis via LESA was further shown to be useful for profiling of different brain regions of rats. Again, a direct LESA comparison with MALDI imaging and a third approach utilizing laser micro dissection followed by liquid extraction appears to favor the LESA approach [34]. Understanding of the lipid composition of target structures, its variance in healthy and diseased state as well as under drug influence is a major focus of current research.

8.4 LESA APPLICATIONS BEYOND DRUG DEVELOPMENT AND DRUG DISCOVERY

LESA can also be used in areas that might not reveal immediate relevance to drug development and discovery. However, such uses are interesting to consider since they might encourage the use of this new analysis tool in ways currently not envisioned. For example, LESA was used to identify pesticides from fruits and vegetables [35]. It is interesting to note that surfaces ranging from hard to smooth such as apples, kiwi, and lettuce were successfully analyzed. Also, these surfaces were not planar like an organ or whole-body section but rather convex. The overall method sensitivity was sufficient to detect a variety of pesticides in the 100 ppb range in fruit pulp and 20-fold below the US Environmental Protection Agency (EPA) tolerance level from the skin surface.

A direct analysis of tablets is also possible using LESA-MS [36]. Again, the surface was not necessarily planar and often porous, making the solvent delivery to the surface and formation of the liquid junction challenging. The authors found LESA to be favorable compared with direct analysis in real time (DART) applied

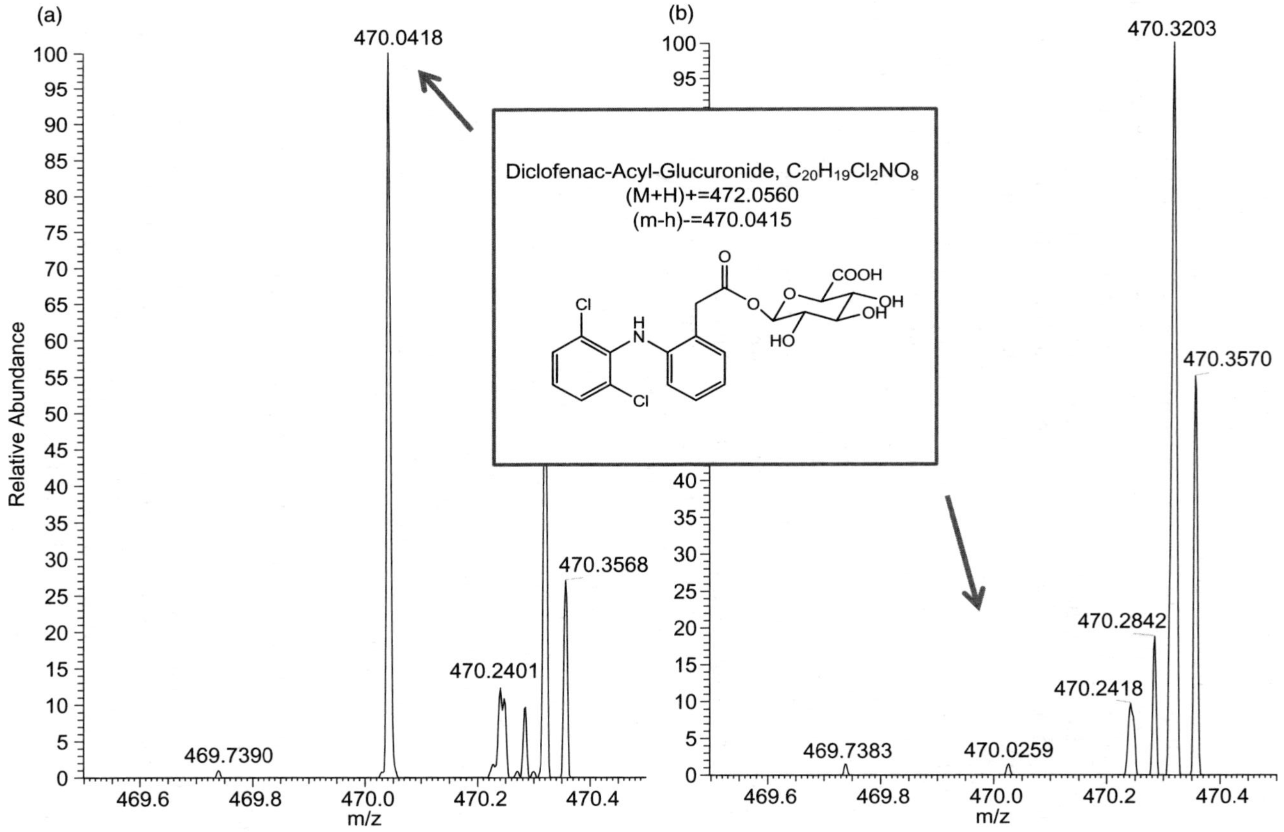

Figure 8.7 Example for LESA detection of phase 2 metabolites of diclofenac. Negative ion mode HRMS of mouse liver tissue 15 min after 10 mg/kg PO diclofenac showing a glucuronide metabolite within 0.6 ppm deviation from the theoretical value in dosed tissue A. However, the signal is absent in control tissue B (nearest *m/z* signal is 33 ppm away from theoretical value).

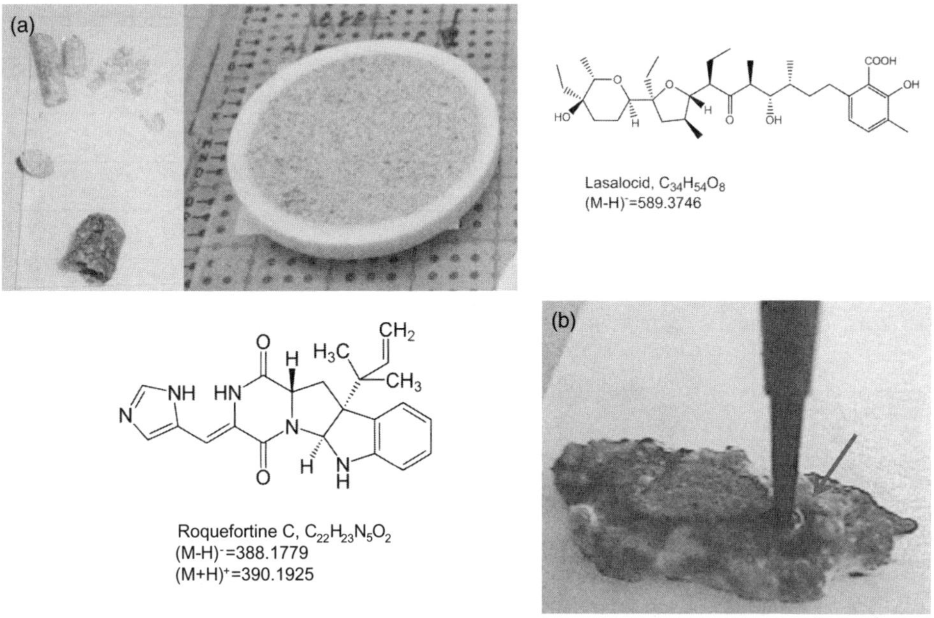

Figure 8.8 LESA-MS applied to a toxicological sample. Analysis of porous surfaces such as feed pellets (A) require either a silicone pretreatment or grinding and pressing into a feed waver. From pretreated samples, lasalocid could be readily detected in alpaca feed at 10 ppm. Also, dog vomitus could be analyzed via LESA-HRMS to detect Roquefortin C as one of the two suspected toxins causing acute tremor in dog (B), picture showing the liquid junction formed on a small piece of wood.

to the same samples, and point out the no carry-over strategy using one tip-one sample-one nozzle as well as the method simplicity as strong points. Porosity of the sample surface was also shown to be challenging for toxicology applications of LESA testing the surface of feed pellets [37]. Feed had to be ground up and compressed into a wafer or the feed pellets had to be pretreated with silicone to enable the formation of a liquid junction and LESA analysis of, for example, 10 ppm lasalocid in alpaca feed. A small wood piece from dog stomach vomitus, on the other hand, was successfully analyzed without any prior treatment and Roquefortin C was confirmed as one of the two suspected toxins causing acute tremor in dogs (Fig. 8.8). Another example of how to analyze nonplanar surfaces via LESA was shown by investigating benzoxazinoids from crown and primary roots of maize plants as a response to insect herbivores [38]. Here, it was shown that benzoxazinoids can be directly quantified on roots mounted in the LESA analysis system.

Another recent publication showed the use of LESA to automatically extract analytes from SPE cards in a 96-well format [39]. Both qualitative and quantitative analysis of herbicides, peptides, and small molecule drugs allowing parallel offline sample preparation was demonstrated. In the original LESA publication from the same group [24] it had been shown that LESA could also be utilized to analyze sitamaquine via dried blood spots (DBS) from punched cellulose-based DBS cards. In order to directly sample from the DBS card, however, a card treatment with silicone spray was necessary to facilitate the formation of a liquid junction [40]. Such

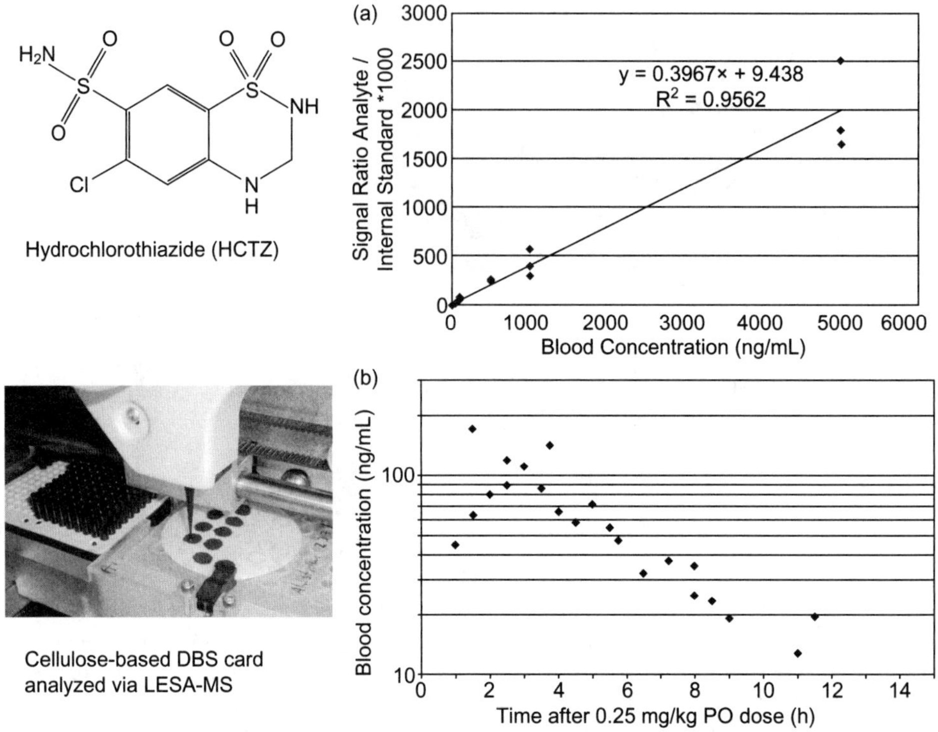

Figure 8.9 Proof-of-concept LESA-MS analysis of hydrochlorothiazide from dried blood spot cards after surface modification using a silicone spray approach. (A) Standard calibration curve from nine fortified blood calibrants covering the therapeutical range from 5 to 5000 ng/mL blood. (B) Pharmacokinetic profiling of two volunteers, samples taken on three consecutive days for 12 h after 0.25 mg/kg PO hydrochlorothiazide (HCTZ) at time zero.

an approach was also applied for the proof-of-concept to quantify hydrochlorothiazide (HCTZ) from DBS later on [41] as shown in Figure 8.9.

LESA analysis from DBS cards for the analysis of hemoglobin variants from blood [42] was recently demonstrated without card treatment. Here, the authors showed the successful identification of a variety of hemoglobin variants, which points to a DBS-LESA-MS approach as a potential screening tool for sickle cell and thalassemia programs. Another LESA related approach to analyze cellulose-based DBS cards, without sample pretreatment, might be the mechanical enclosure of a small portion of the card via a clamping/holding device [41, 43].

The studies discussed in this chapter may indicate a future for LESA applications in the screening and quantification of drugs from animals or humans provided that a suitable surface system can be developed that combines the current advantages of DBS cards (small format, rapid method development, simplicity, minimal cost in shipping and storing, patient acceptance) with sufficient hydrophobicity and a surface porosity allowing formation of a micro liquid junction. Also, a surface chemistry might be included that could combine steps of the classical wet lab sample preparation with the collection card.

Other areas of current interest for LESA-MS analysis are pigments and leachables from packaging materials, medical devices, plastic containers, or technical surfaces. A first publication in this field investigated the degradation of plastic additives during aging of the product [44].

8.5 OUTLOOK AND FUTURE DEVELOPMENT OF LESA

Aside from new application areas for LESA (which might affect its use in drug distribution and discovery), there are other developments that will likely influence how this new tool can be used in the future. Three of these are particularly interesting. As already mentioned, the spatial resolution of LESA is currently limited to about 1.0 mm. However, the resolution is dependent on three surface characteristics: the surface tension of the solvent used, the sample surface investigated, and the cross-sectional area of the tip used to handle and form the liquid junction with the solvent extract. In a recent publication, it was shown that using a large inner diameter (ID) fused silica capillary instead of a plastic tip one can achieve a resolution of ca. 200 μm on target [45]. This would bring LESA spatial resolution closer to the currently routinely used spatial resolution for MALDI and at least into the range of current QWBA resolution of 30 μm.

In this LESA setup, the extract was also injected onto a nano-LC column and a chromatographic separation was coupled with LESA, allowing separation of different lipids as well as separation of lipids from matrix [45]. As expected, the LESA-LC-MS approach can increase the sensitivity of LESA since it reduces matrix ion suppression and concentrates the analyte into a narrow chromatographic peak. However, both developments come at the cost of analysis time given the extra time needed for chromatographic separation or to handle the increased number of locations per sample. The latter is the classical conundrum of all MSI methods striving for higher spatial resolution; for example, MALDI imaging is technically capable of scanning at sub-1 μm resolution [11]; however, due to restrictions to instrumentation, sensitivity, analysis time, and data storage/handling, a spot size of ca. 200 μm is often preferred and sufficient for many applications.

One could envision a LESA system though that would allow different modes of operation using low-resolution drug profiling with an emphasis on whole-body sections and speed of analysis and a higher resolution analysis for single organs or structures of interest such as tumor tissues. In other instances, a LESA-LC approach would be utilized for added sensitivity or cases where a chromatographic separation is necessary such as isobaric hydroxyl metabolites that also fragment similarly or may not separate in ion mobility MS.

A third interesting development is the absolute quantitation from tissue samples via MSI methods. Although QWBA can give absolute quantitative results, MSI approaches could not easily provide this information until fairly recently—a clear disadvantage of MSI methods. However, there are currently two interesting paths toward MSI absolute quantitative data, that is, by either placing a reference standard on control tissue that was processed concurrent with the analytical sample or—more promising—using homogeneous and fortified control tissue [46] or fortified surrogate matrix such as liquid egg [47] in order to analyze these samples together with the tissue of interest (both standard fortified control tissue and sample frozen and sectioned on the same target). This allows the generation of a calibration curve and consideration of matrix effects and may mimic the real conditions on analytical samples. Both ideas toward absolute quantification could, of course, be applied to LESA-MS as well, and a first successful attempt was shown recently [48].

8.6 CONCLUSIONS

LESA is a relatively new surface analysis technique in the realm of mass spectrometric imaging that was first described by Kertez and Van Berkel at the 57th Conference of the American Society of Mass Spectrometry 2009 and published shortly thereafter [24]. It is characterized as an ambient surface analysis and ionization technique and as such has attracted considerable interest in the last two years. Despite its novelty, a commercial realization is available, and LESA-MS is used in a variety of applications ranging from technical to biological surfaces, with the majority of publications in the area of drug development and discovery as shown in this chapter.

Compared with QWBA and whole tissue extract coupled with LC-MS, the two gold standards in drug distribution and drug metabolism analysis, MSI methods in general have advantages such as no need for a radiolabeled compound, analysis of the drug and its metabolites in one analysis step, and the provision of a more detailed view of drug distribution due to the potentially higher spatial resolution possible. Within the toolbox of MSI approaches, LESA-MS has a unique characteristic and nicely complements the analytical capabilities available to researchers as reviewed recently [49]. Its ease-of-use and fit-for-purpose approach make it readily adoptable by many labs and the significant increase of contributions to both conferences and publications in peer-reviewed journals shows the interest of the scientific community. One of the strong points of LESA certainly is the fact that analyte extraction and, even more importantly, ionization can be optimized by choosing different solvent compositions and modifiers. In combination with the soft ionization nESI provides, LESA profiling appears suitable for a wide range of drugs and metabolites.

The currently available instrumentation is compatible with most commercially available MS systems and its nESI chip is the enabling technology for LESA since nESI not only reduces ion matrix suppression effects [50], but also generates an extended electrospray time from a very small volume of solvent (1–2 μL sufficient to generate electrospray ions for 2–4 min). Such time can be used for extensive MS experiments allowing in-depth analysis of the surface of interest.

The limitations of LESA-MS lay in the 1-mm spatial resolution, thus limiting the ability to investigate smaller regions within biological structures of interest—a reason why LESA should be considered a profiling approach. This spatial resolution can be improved as demonstrated by a spatial resolution of ca. 200 μm [45]. However, as with other MSI approaches, this raises the question of analysis time, data file size generated, and sensitivity when sampling from smaller areas of a surface.

Another crucial point for LESA is the static extraction process from the tissue. This process is currently not well characterized or understood, making the optimization an empirical process. Even under optimized conditions, the extraction efficiency appears limited, and hence the overall method sensitivity is lower than might be possible. Although first sensitivity comparisons to QWBA, MALDI imaging, or DART show favorable results for LESA-MS analysis, a better understanding of the extraction process is still needed. Also, LESA depends on the formation of the liquid junction itself, which requires a surface that is sufficiently hydrophobic. Thus, an empirical process is also necessary to determine which sample type is suitable for LESA-MS analysis, and although surface treatments with silicone have been demonstrated, such added sample preparation should ideally be avoided and is not necessary for thin cryo-sections of whole animal or organs.

In summary, we believe LESA MS is an exciting novel approach in the field of MSI that complements the tools available to scientists working in the field of drug development and drug distribution.

REFERENCES

[1] Coe RA. Quantitative whole-body autoradiography. Regul Toxicol Pharmacol 2000;31:1–3.

[2] Shigematsu A, Motoji N, Hatori A, Satoh T. Progressive application of autoradiography in pharmacokinetic and metabolic studies for the development of new drugs. Regul Toxicol Pharmacol 1995;22:122–42.

[3] Bénard P, Burgat V, Rico AG. Application of whole-body autoradiography in toxicology. Crit Rev Toxicol 1985;15:181–215.

[4] Rubakhin SS, Sweedler JV. Mass spectrometry imaging: principles and protocols. In Walker JM, editor. Methods in Molecular Biology. New York: Springer, 2010, p. 487.

[5] McDonnell LA, Heeren RM. Imaging mass spectrometry. Mass Spectrom Rev 2007;26:606–43.

[6] Reyzer ML, Caprioli R. MS imaging: new technology provides new opportunities. In Korfmacher, W, editor. Using Mass Spectrometry for Drug Metabolism Studies, 1st Edition. Boca Raton, FL: CRC Press, 2005, pp. 305–24.

[7] Li F. Imaging mass spectrometry for small molecules. In Korfmacher, W, editor. Using Mass Spectrometry for Drug Metabolism Studies, 2nd Edition. Boca Raton, FL: CRC Press, 2009, pp. 333–59.

[8] Caldwell RL, Caprioli RM. Tissue profiling by mass spectrometry: a review of methodology and applications. Mol Cell Proteomics 2005;4:394–401.

[9] Chaurand P, Schwartz SA, Reyzer ML, Caprioli RM. Imaging mass spectrometry: principles and potentials. Toxicol Pathol 2005;33:92–101.

[10] Chaurand P, Schwartz SA, Caprioli RM. Imaging mass spectrometry: a new tool to investigate the spatial organization of peptides and proteins in mammalian tissue sections. Curr Opin Chem Biol 2002;6:676–81.

[11] Bouschen W, Schulz O, Eikel D, Spengler B. Matrix vapor deposition/recrystallization and dedicated spray preparation for high-resolution scanning microprobe matrix-assisted laser desorption/ionization imaging mass spectrometry (SMALDI-MS) of tissue and single cells. Rapid Commun Mass Spectrom 2010;24:355–64.

[12] Hopfgartner G, Varesio E, Stoeckli M. Matrix-assisted laser desorption/ionization mass spectrometric imaging of complete rat sections using a triple quadrupole linear ion trap. Rapid Commun Mass Spectrom 2009;23:733–36.

[13] Hsieh Y, Chen J, Korfmacher WA. Mapping pharmaceuticals in tissues using MALDI imaging mass spectrometry. J Pharmacol Toxicol Methods 2007;55:193–200.

[14] Reyzer ML, Hsieh Y, Ng K, Korfmacher WA, Caprioli RM. Direct analysis of drug candidates in tissue by matrix-assisted laser desorption/ionization mass spectrometry. J Mass Spectrom 2003;38:1081–92.

[15] Wiseman JM, Ifa DR, Zhu Y, Kissinger CB, Manicke NE, Kissinger PT, Cooks RG. Desorption electrospray ionization mass spectrometry: imaging drugs and metabolites in tissues. PNAS 2008;105:18120–5.

[16] Burns MS. Applications of secondary ion mass spectrometry (SIMS) in biological research: a review. J Microscopy 1982;127:237–58.

[17] Nemes P, Vertes A. Laser ablation electrospray ionization for atmospheric pressure, *in vivo*, and imaging mass spectrometry. Anal Chem 2007;21:8098–106.

[18] Shiea J, Huang MZ, Hsu HJ, Lee CY, Yuan CH, Beech I, Sunner J. Electrospray-assisted laser desorption/ionization mass spectrometry for direct ambient analysis of solids. Rapid Commun Mass Spectrom 2005;19:3701–4.

[19] Tsuyama N, Mizuno H, Tokunaga E, Masujima T. Live single-cell molecular analysis by video-mass spectrometry. Anal Sci 2008;5:559–61.

[20] Masujima T. Live single-cell mass spectrometry. Anal Sci 2009;8:953–60.

[21] Brown JM, Hoffmann WD, Alvey CM, Wood AR, Verbeck GF, Petros RA. One-bead, one-compound peptide library sequencing via high-pressure ammonia cleavage coupled to nanomanipulation/nanoelectrospray ionization mass spectrometry. Anal Biochem 2010;1:7–14.

[22] Wachs T, Henion J. Electrospray device for coupling microscale separations and other miniaturized devices with electrospray mass spectrometry. Anal Chem 2001;73:632–8.

[23] Van Berkel GJ, Kertesz V, Koeplinger KA, Vavrek M, Kong AN. Liquid microjunction surface sampling probe electrospray mass spectrometry for detection of drugs and metabolites in thin tissue sections. J Mass Spectrom 2008;4:500–8.

[24] Kertesz V, Van Berkel GJ. Fully automated liquid extraction-based surface sampling and ionization using a chip-based robotic nano-electrospray platform. J Mass Spectrom 2010;45:252–60.

[25] Hopfgartner G, Bean K, Henry R, Henion JD. Ion spray mass spectrometry detection for liquid chromatography: a concentration or a mass-flow device? J Chromatogr 1993;647:51–61.

[26] Eikel D, Alpha C, Rule GS, Prosser SJ, Henion JD. Liquid extraction surface analysis (LESA) combined with nESI-MS as a novel tool in early ADME studies of drug candidates. 58th Conference of the American Society for Mass Spectrometry 2010—oral contribution. 2010.

[27] Marshall P, Toteu-Djomte V, Bareille P, Perry H, Brown G, Baumert M, Biggadike K. Correlation of skin blanching and percutaneous absorption for glucocorticoid receptor agonists by matrix-assisted laser desorption ionization mass spectrometry imaging and liquid extraction surface analysis with nanoelectrospray ionization mass spectrometry. Anal Chem 2010;18:7787–94.

[28] Eikel D, Vavrek M, Korfmacher WA, Henion JD. Liquid extraction surface analysis mass spectrometry (LESA-MS) as a novel profiling tool for drug distribution and metabolism analysis: the Terfenadine example. Rapid Commun Mass Spectrom 2011;25:3587–96.

[29] Chen J, Hsieh Y, Knemeyer I, Crossman L, Korfmacher WA. Visualization of first-pass drug metabolism of terfenadine by MALDI-imaging mass spectrometry. Drug Metab Lett 2008;2:1–4.

[30] Schadt S, Kalbach S, Almeida R, Sandel J. Investigation of figopitant and its metabolites in rat tissue by combining whole-body autoradiography with liquid extraction surface analysis mass spectrometry. Drug Metab Dispos 2012;40(3):419–25.

[31] Cox D, Covey T, Leblanc JCY, Schnieder B, VanBerkel GJ, Kertesz V, Moench P, Flarakos J. Analysis of tissue samples by liquid extraction surface analysis (LESA) coupled to differential ion mobility (DMS) high resolution MS/MS. 59th Conference of the American Society for Mass Spectrometry 2011—poster contribution. 2011.

[32] Prosser SJ, Eikel D, Linehan ST, Murphy K, Heller D, Rudewicz PJ, Henion JD. Liquid extraction surface analysis mass spectrometry (LESA MS): drug distribution and metabolism of diclofenac in mouse. 59th Conference of the American Society for Mass Spectrometry 2011—oral contribution. 2011.

[33] Stegemann C, Drozdov I, Shalhoub J, Humphries J, Ladroue C, Didangelos A et al. Comparative lipidomics profiling of human atherosclerotic plaques. Circ Cardiovasc Genet 2011;3:232–42.

[34] Taguchi R, Ikeda K, Tajima Y. Several different approaches for the analysis in localization profile of lipid molecular species by mass spectrometry. 59th Conference of the American Society of Mass Spectrometry 2011—poster contribution. 2011.

[35] Eikel D, Henion JD. Liquid extraction surface analysis (LESA) of food surfaces employing chip-based nano-electrospray mass spectrometry. Rapid Commun Mass Spectrom 2011;16:2345–54.

[36] Ray AD, Weston DJ, Blatherwick EQ, Snelling JR, Scrivens JH. Investigating liquid extraction surface analysis (LESA) for rapid analysis of pharmaceutical tablets. Annual meeting of the British Mass Spectrometry Society 2010—poster contribution. 2010.

[37] Henion JD, Eikel D, Li Y, Ebel J. Automated surface sampling of veterinary toxicological samples with analysis by nano electrospray mass spectrometry. 59th conference of the American Society of Mass Spectrometry 2011—poster contribution. 2011.

[38] Robert CAM, Veyrat N, Glauser G, Marti G, Doyen GR, Villard N, Gaillard MDP, Koellner TG, Giron D, Body M, Babst BA, Ferrieri RA, Turlings TCJ, Erb M. A specialist root herbivore exploits defensive metabolites to locate nutritious tissues. Ecol Lett 2012;15:55–64.

[39] Walworth MJ, Elnaggar MS, Stankovich JJ, Witkowski C, Norris JL, Van Berkel GJ. Direct sampling and analysis from solid-phase extraction cards using an automated liquid extraction surface analysis nanoelectrospray mass spectrometry system. Rapid Commun Mass Spectrom 2011;17:2389–96.

[40] Stankovich JJ, Walworth MJ, Kertesz V, King R, VanBerkel GJ. Liquid microjunction surface sampling probe analysis of dried blood spots using an automated chip-based nano-ESI infusion device. 58th Conference of the American Society of Mass Spectrometry 2010—oral contribution. 2010.

[41] Henion J, Eikel D, Rule G, Vega J, Prosser S, Jones J. Liquid extraction surface analysis (LESA) of dried blood spot cards via chip-based nanoelectrospray for drug and drug metabolite monitoring studies. 58th Conference of the American Society of Mass Spectrometry 2010—oral contribution. 2010.

[42] Edwards RL, Creese AJ, Baumert M, Griffiths P, Bunch J, Cooper HJ. Hemoglobin variant analysis via direct surface sampling of dried blood spots coupled with high-resolution mass spectrometry. Anal Chem 2011;6:2265–70.

[43] Eikel D, Henion J, Alpha C, Vega J. Mechanical Holder for surface analysis. United States Patent Application 13/100383 Filed May 4, 2011.

[44] Paine MR, Barker PJ, Maclauglin SA, Mitchell TW, Blanksby SJ. Direct detection of additives and degradation products from polymers by liquid extraction surface analysis employing chip-based nanospray mass spectrometry. Rapid Commun Mass Spectrom 2012;4:412–8.

[45] Almeida R. Liquid extraction surface analysis (LESA) combined with nESI-MS for direct sampling of planar tissues. 1st AB SCIEX European Conference on MS/MS 2011, Noordwijkerhout, The Netherlands—oral contribution.

[46] Hamm G, Bonnel D, LeGouffe R, Pamelard F. Could mass spectrometry imaging be a drug quantification technique? 59th conference of the American Society of Mass Spectrometry 2011—oral contribution. 2011.

[47] Wagner DS, Groseclose MR, Richards-Peterson L, Gorycki P, Castellino S. Integration of imaging mass spectrometry into drug development and the importance of distribution and quantitation. 59th Conference of the American Society of Mass Spectrometry 2011—oral contribution. 2011.

[48] Blatherwick EQ, Pickup KJ, Sarda S, Schultz-Utermoehl T, Wilson ID, Weston DJ, Scrivens JH. Imaging mass spectrometry based approaches to the localization of drugs and drug metabolites in animal tissue. 59th Conference of the American Society of Mass Spectrometry 2011—poster contribution. 2011.

[49] Blatherwick EQ, Van Berkel GJ, Pickup K, Johansson MK, Beaudoin ME, Cole RO, et al. Utility of spatially-resolved atmospheric pressure surface sampling and ionization techniques as alternatives to mass spectrometric imaging (MSI) in drug metabolism. Xenobiotica 2011;8:720–34.

[50] Karas M, Bahr U, Dulcks T. Nano-electrospray ionization mass spectrometry: addressing analytical problems beyond routine. Fresenius J Anal Chem 2000;6–7:669–76.

9

MS APPLICATIONS IN SUPPORT OF MEDICINAL CHEMISTRY SCIENCES

MAARTEN HONING, BENNO INGELSE, AND BIRENDRA N. PRAMANIK

9.1 INTRODUCTION

In the drug discovery process, medicinal chemistry plays a crucial role in the optimal design and synthesis (production) of new chemical entities (NCEs) or new biological entities (NBEs); this is true for a single entity as well as a complete compound library. A key requirement in this field of science is the optimization of both the chemical and the biological properties of these potential drug-like compounds. Important criteria are the design of efficient synthesis routes, the balancing of the physical chemical properties with optimal *in vitro* and *in vivo* drug metabolism and pharmacokinetic (DMPK) processes, and a profound understanding of the pharmacodynamics in the appropriate pharmacological models. In the lead optimization stage, an important step is scaling up the synthesis route as needed for the execution of toxicology studies and the screening of suitable drug formulations. In summary, medicinal chemistry support includes participating in a large variety of activities and decision making processes that play a key role in the search for either a "first in class" new drug or a "best in class" new drug [1].

These "drug discovery" processes are continuously under pressure, as the research and development (R&D) costs continue to increase and more data are requested before new drug candidates are taken into (pre-)clinical development. Hence, the choice of "targeted" high-throughput screening (HTS) libraries, the process of hit appraisal, and the understanding of target–ligand interactions can be considered the starting points for a new drug discovery program [2]. Next, the main focus is on the discovery of a lead compound; typically, this is a compound showing promising data regarding target affinity, reasonable efficacy in cellular models, good physical chemical parameters (i.e., a "drug-like" compound), and acceptable oral bioavailability in

Mass Spectrometry for Drug Discovery and Drug Development, First Edition. Edited by Walter A. Korfmacher.

a rat model together with a good metabolic stability in an *in vitro* assay [3, 4]. In the subsequent lead optimization process, these aspects are further fine-tuned before entering the *in vivo* phase.

Initiatives like six sigma and lean thinking focus on the improvement of the efficiency of the complete discovery process while maintaining the quality [5, 6]. As a result, high-throughput synthesis (combinatorial chemistry), efficient purification procedures or (open access) purity analysis, fast screening of various physicochemical properties and metabolic stability, high-throughput screening technologies for target–ligand screening (IC50 assays) and functional assays have all received tremendous attention [7]. Thus, each step of the lead identification process has been optimized in recent years in order to make this stage of new drug discovery as efficient as possible.

Mass spectrometry (MS) has played an important role in these enhancements. The introduction of the various atmospheric pressure ionization (API) techniques such as electrospray ionization (ESI), atmospheric pressure chemical ionization (APCI), and more recently, atmospheric pressure photoionization (APPI) has dramatically changed the utility of MS and allowed it to become the premier analytical technique for many new drug discovery assays [8–13]. In addition to being the "gold standard" for bioanalytical (BA) applications supporting various *in vitro* and *in vivo* DMPK studies, high-performance liquid chromatography-tandem mass spectrometry (HPLC-MS/MS) has become the most important analytical detection technique for various open-access services, the purification of compound libraries, the assessment of physicochemical properties including chemical stability, and the growing need to better understand the target–ligand interactions [14, 15]. For example, over the past five years, the determination of metabolic hotspots in a compound series has become a standard new drug discovery expectation. The current methods typically provide for the identification of the areas of the molecular scaffold that are metabolized rather than resolving the absolute chemical structure of the specific metabolites.

This chapter will provide an overview of all the medicinal chemistry activities that rely on various mass spectrometric applications. In addition to a discussion of current analytical strategies, the chapter will include a limited outlook into the nearby future. Recent developments in biological drug discovery or development, for example, lead optimization of chemically modified biologics, such as PEGylated insulin or interferon, are outside the scope of this chapter.

9.2 SYNTHESIS AND IDENTIFICATION OF NEW CHEMICAL AND BIOLOGICAL ENTITIES

9.2.1 Open-Access Analysis

Along with the introduction and further development of the various API interfaces, the robustness, and even more important the stability of the ion optics and detectors, mass analyzers rapidly matured throughout the 1990s [16]. In addition, the coupling of MS with stable HPLC pumps, robust and reproducible analytical HPLC columns, and the introduction of improved autosamplers that are able to handle rather "dirty" samples such as crude reaction products together with the improvements in easy-to-use software, all led to the successful implementation of so-called open-access HPLC-MS [17].

These new methodologies rapidly found their way into medicinal chemistry departments, mainly replacing the "low-resolution" thin layer chromatography (TLC) and the "troublesome" HPLC-ultraviolet (UV) systems [18, 19]. In many cases, UV detection lacks sufficient sensitivity for these applications; in addition, UV-visible (UV-VIS) spectra are often not helpful in proving structural changes. Besides these constraints, the implementation of new synthesis approaches such as combinatorial chemistry resulted in a far larger number of new drug-like compounds being produced on a weekly basis. Thus, the required higher throughput was easily provided by HPLC-MS systems, but would have been a problem if these departments had not made the switch to using HPLC-MS for these assays.

Besides the increased number of new test compounds, the ever-growing variety in chemical structures, including rather apolar steroids and heteronuclear bases and acids, being submitted to HPLC-MS analysis, generated a new analytical challenge [17]. In general, electrospray (the most commonly applied ionization interface) is an excellent ionization technique for polar molecules while APCI can be used for many nonpolar compounds. Still, the need for an even more universal and normalized (ionization efficiency is the same for all molecules) ionization technique still exists. Various attempts to develop "dual ionization" sources as well as the introduction of APPI have been reported as potential solutions to this need [20–25]. In a comparison using 241 proprietary compounds, colleagues at Wyeth showed the potential value of APPI; their report showed $M^{+\cdot}$ ions for compounds with low polarity and low proton affinity and (as is seen for ESI or APCI) protonated molecules, $[M+H]^+$, for structures with high polarity and/or proton affinity. The potential for APPI to handle nearly all types of chemical structures could result in HPLC-MS becoming a truly generic analytical technique.

The diversity in molecular structure, the type of samples, and the information needed led to a further differentiation of the open-access tools. As an example, to support reaction monitoring or optimization, fast flow injection analysis (FIA), with typical run times of 30 s, was considered sufficient. Nonetheless, the frequent occurrence of ion suppression due to improper sample preparation led to the introduction of stable fast gradient methods utilizing a 5-cm analytical column with typical analysis cycle times of 2 min [17, 26]. The recent introduction of ultra-performance liquid chromatography (UPLC) and hydrophilic interaction liquid chromatography (HILIC) column technologies and the further optimization of the LC-MS interfaces, offered increased resolution, chromatographic stability, and a further reduction of the analysis time [25, 27–30]. Consequently, in our laboratories, one UPLC-MS system per 20 medicinal chemists serves for all purity analysis and largely replaced the FIA-MS approach.

In addition to these new developments in separation technologies, the MS vendors commercialized robust, high-resolution (HR) time-of-flight (TOF) mass spectrometers [31]. These technologies with a mass resolution well in excess of 10,000 and mass accuracies of approximately 2–3 mDa allow the accurate assessment of the empirical formula of all masses detected; in addition, improved software tools also facilitate in the efficient calculation and reporting of the data. In combination with nuclear magnetic resonance (NMR), the accurate assessment of the

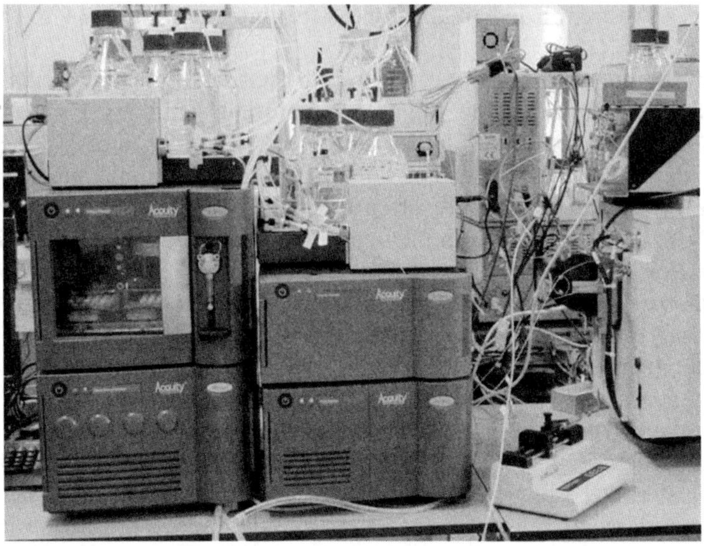

Figure 9.1 Example of a multicolumn UPLC-UV-Q-TOF system (including post-run calibration tool) for the purity screening of compound libraries with medicinal chemistry departments.

empirical formula of all masses has been shown to be a powerful tool for structural assignments and is an important tool for the correct interpretation of structure activity and property relationships.

In conclusion, after the introduction of low-resolution LC-MS in the 1900s, this technology has rapidly developed to being a key benchtop tool for the medicinal chemist, offering both accurate and high-quality purity data and structural information of a large variety of samples. The recent developments in analytical supercritical fluid chromatography (SFC)-MS as a promising and robust approach for small molecule analysis or purification may provide solutions for some of these issues [32–35]. Still, further improvements are needed, especially for the analysis of enantiomeric purity and molecular identity. It is possible that LC-Q-TOF MS/MS in combination with multicolumn (in parallel) HPLC systems will evolve to a generic analytical technology for handling samples originating from all R&D specialists (Fig. 9.1).

9.2.2 MS Coupled to (Semi)-Preparative LC and SFC

While the number of test compounds synthesized per medicinal chemist has significantly increased over the last two decades, and, at the same time, the throughput of analytical LC-MS has increased so that it was not on the critical time path; the time-limiting factor became the purification of crude compounds step. The result was (in an analogy with the open-access LC-MS methodologies) the introduction of MS-directed purification of chemical libraries evolved as the standard approach. Its main advantage of selecting the "correct compound" with a unique *m/z* value resulted in a major improvement in the overall quality (purity) and recovery (yield) of purified test compounds [36, 37].

At first, rather complex and highly automated purification LC-MS platforms were developed for the purification of large HTS screening libraries containing over 100,000 compounds in small amounts, typically sub-milligram to milligram quantities [38, 39]. At the same time, the application of (semi)-preparative LC-MS became an important tool in the later lead optimization stages of discovery. Here, the purification of approximately 10 mg to sub-gram quantities is required for the *in vitro* and *in vivo* PK and efficacy studies. This need required a completely different setup of the MS platform as the column dimension and the character of the solid phases changed significantly. In addition, while the purity requirements for many of the early screening libraries was> 95%, for late stage lead optimization (LO) compounds (used in large species PK/pharmacodynamics [PD] and toxicology testing) the purity requirements were >99%. All these developments resulted in an extensive number of manuscripts describing different tactics mainly focused on the LC technology reducing the time necessary to reach these high-purity requirements [38–40]. In all cases, MS was applied as the detector, typically equipped with ESI or sometimes with the APCI interface.

Interestingly, the area of automated compound purification systems has been changing rather rapidly with the introduction of high-quality SFC-UV and SFC-MS purification platforms [41–43]. This development, mainly triggered by the fact that single enantiomers (as compared with racemates) provide better and safer toxicology profiles, led to the need to generate LO compounds with a well-defined enantiomeric purity, preferably containing less than 1% of the undesired stereoisomer(s). Classical chiral LC platforms seem to be outdated due to the high amounts of "hazardous" organic solvent and the rather expensive (semi)-preparative columns. As a result, supercritical fluid chromatography (SFC) was reborn and put to use as a way to provide single enantiomers for drug testing.

The better understood processes underlying and the application of SFC with UV and/or mass spectrometric detection have proven its added value, in general providing superior chiral resolution, improved load ability (Fig. 9.2), and a higher success rate than chiral LC-MS [35, 44–47]. Recently, robust (semi)-preparative SFC-UV and SFC-MS systems were commercially introduced. This revival of SFC-MS has led to new mass spectrometric challenges [48–50]. While in the 1980s and 1990s many research groups developed SFC-MS interfaces utilizing the "hard" electron and chemical ionization techniques [51, 52], the renewed focus on using SFC in hyphenated MS system opens a new analytical tool for the structural assignment of unknowns. Together with the ongoing improvements with respect to selectivity and sensitivity, newly developed software tools will guarantee the continuing pivotal role of MS in (semi)-preparative platforms.

9.2.3 Purity Analysis and Impurity Profiling

Definitely, analytical scientists working in the field of impurity profiling and determination of the overall purity of both drug substance and product will benefit from the previously described developments. The still ongoing instrumental improvements in selectivity (resolution), specificity, and sensitivity go hand in hand with the continuous improvements of the "analytical" requirements for a variety of "impurities" such as unwanted stereoisomers [53], synthesis by-products leading to possible genotoxic compounds [54], residual solvents, and degradation products [55]. As described by

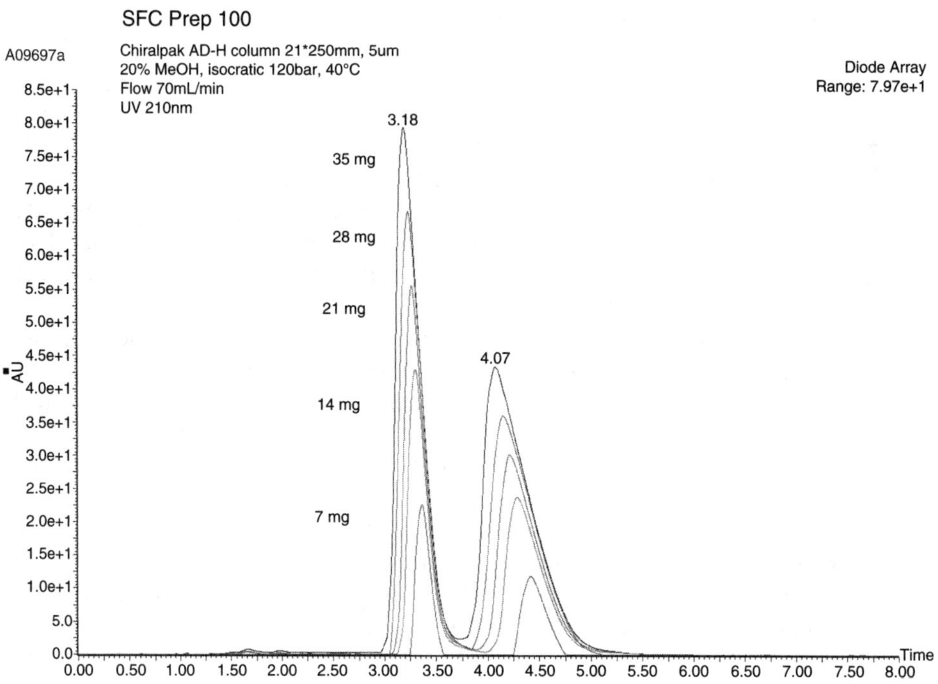

Figure 9.2 A preparative SFC-UV (210 nm) chromatogram of a racemic mixture with various loadings at milligram scale.

one of the first and most experienced pioneers in this field, HPLC in combination with UV and MS found its way in the late 1980s to become the most utilized analytical tool in impurity profiling [56]. Their levels of detection, qualification, and identification as described in the International Conference on Harmonization (ICH) Q3 and Q6 guidelines have not changed dramatically over the last 20 years [53, 55]; various strategies for applying HPLC-UV-MS have been reported [57, 58]. Hence, the quantification of overall purity still relies on HPLC-UV data with MS detection used as the "confirmation" tool.

It is important to note that the accuracy and day-to-day stability needed in fully validated assay as used in drug development is of less importance in the discovery stage. While in the development stage a limited number of samples are analyzed in significant detail, scientists in the discovery stage are faced with the need to provide purity assessment for thousands of compounds on a monthly, if not weekly, basis [59, 60]. In addition, impurity profiling in drug development is mainly a tool to understand and above all gain control over the production process and in addition to prove the safety of the total product. In discovery, on the other hand, the purity requirements are mostly defined by the need to have reliable test data. Binding affinity data can be altered by poorly defined combined libraries. The solubility of test compounds in both lead finding and LO are sensitive for the presence of impurities, and abundant impurities can complicate the metabolic profile from *in vitro* assays. In addition, the read out from *in vitro* and *in vivo* non-good laboratory practice (GLP) safety pharmacology studies can differ.

As a rule of thumb, the purity for test compounds in the *in vitro* test phase should be >95%, and no impurity should be >1%. When emerging from early "rapid rat"

PK studies toward full nonrodent PK/PD and toxicology studies, the purity should be slowly upgraded to 98–99%. An exact reproduction of the impurity profile of the various "discovery batches" is of lesser importance; in case of unexplained findings, the identity of impurities can be requested. In contrast to the drug development stages, the large diversity in chemical structures and structural properties, in combination with the required short analysis times, has led to the introduction of more "generic" high-throughput HPLC-UV-MS platforms for this purpose [60].

While HPLC-UV-MS has become very useful for drug purity screening, there are some practical considerations that should be understood. One issue is the rather large difference in mass spectrometric response for various compounds; some of the variation in response is due to the fact that the ionization process in ESI is dependent on the solution phase conditions and the desorption process for preformed ions going into the gas phase. In some cases, subsequent gas-phase reactions or adduct formation can further complicate the slope of the response curves. This response variation, although to a much lower extent, is also true for APCI. These variations may hamper the correct interpretation of HPLC-MS data for impurity profiling experiments. As an example, at Astrazeneca, using ESI in a standard approach resulted in a satisfactory MS response for not more than 80% of the test compounds [23]. The analytical scientists also tested an interface combining both ESI and APCI (referred to as ESCi) in a so-called dual source and showed that with the ESCi source, 10% more samples could be correctly analyzed compared with each of the other (ESI and APCI) interfaces [23]. The observed differences between the APCI and the ESCi sources were the most significant in the positive ion mode. In contrast, with the ESCi source in the negative ion mode, abundant $[M-H]^-$ ions were encountered.

Other proposed improvements in the ionization technique were the sonic spray devices and the introduction of APPI [20, 61–65]. The latter interface showed its usefulness in the analysis of rather nonpolar steroidal molecules. Despite these developments, HPLC-MS platforms supporting purity analysis or impurity profiling are typically hyphenated with UV-VIS (DAD), evaporative light scattering detection (ELSD), or chemiluminescent nitrogen detector (CLND) type of detectors that provide a normalized detector response for an extreme broad range of molecular structures [60].

Other solutions to the "limited" throughput issue and the goal to have a universal detector for a large variety of molecules in terms of size and polarity included the so-called multichannel electrospray inlets (Niessen) and the nano-ESI interfaces and even chip-based MS [59, 66–70]. Improved mass resolution when using TOF, Orbitrap™, or Fourier transform-ion cyclotron resonance(FT-ICR)-MS systems reduces the need to have ultimate chromatographic peak capacity, especially when analyzing complex mixtures of peptides in, for example, combinatorial libraries. The co-eluting peptides could be easily distinguished by their accurate mass using the high-resolution measurements. The ability to obtain the elemental composition for larger molecules is another strong advantage of using these powerful MS systems.

9.2.4 The Role of MS in the Elucidation of Chemical Structures

Despite the tremendous improvements in mass resolution and mass accuracy with the introduction of TOF, Orbitrap, and FT-ICR-MS systems hyphenated to HPLC or chromatographic technologies, tandem mass spectrometry (MS/MS) or MS^n is still the method of choice for the elucidation of chemical structures [71]. These multistage MS systems became a necessity because ESI and APCI typically produce protonated (or deprotonated) molecules or adduct ions, such as $[M+Na]^+$ or $[M+K]^+$. These mass spectra lack any compound-specific fragment ions as one would get in classical electron ionization mass spectra. On the other hand, polarity switching and a good sense for proton transfer reactions in "buffered" aqueous solutions or the gas phase can be used as a first stage in identifying specific functionalities in molecules [72, 73]. Abundant response in the negative ion mode may indicate the presence of carboxyl groups, while, in general, aromatic hydroxyls tend to be acidic. Likewise, in the positive ion mode, abundant alkali metal adducts could indicate neighboring aliphatic hydroxyl groups, which in the negative mode lead to acetic or formic acid adduct ions.

Second, it must be emphasized that based on accurate mass measurements and isotopic distribution analysis, a good estimation of the elemental composition can be made which allows only the confirmation of a possible structural identity to be given. The number of possible elemental compositions rapidly increases for molecules with masses above 400 Da; even with a limited set of atoms (C, H, N, O, S) and including the nitrogen rule, over 70 elemental compositions are feasible at a nominal mass of 500 [74]. Even when following the Lipinski "rule of five" [3], the molecular masses of pharmaceutically interesting molecules can range from 300 to 1000 Da (antibiotics), and determining the chemical structure of an unknown compound based solely on a high-resolution mass spectrum is challenging.

In the new drug discovery process, especially for the lead compound or drug development candidate, an in-depth knowledge of the absolute chemical structure of the new compound is crucial for understanding the structure–activity relationships (SAR), or structure–property relationships (SPR); in addition, it is important to have the correct description of the molecule in a patent application. The three-dimensional representation of the molecule allows the medicinal chemist to better understand the biological resemblance between a variety of structures and to be able to correlate biological properties to conformers or even stereoisomers. This, in some cases, also accounts for the pharmacologically or toxicologically relevant metabolites that are formed during *in vivo* PK/PD studies, as well as synthesis by-products or degradants in a drug substance and product. As an example, metabolites originating from the hydroxylation of aliphatic groups showed improved target selectivity without loss of potency and efficacy in a drug discovery program. As a definition, "absolute chemical structure" encompasses the correct structural configuration with special emphasis on stereoisomers and the various conformers (energy minima); in Figure 9.3, the various terms and the structural representation is given. Traditionally, its assessment relies on a combination of spectroscopy techniques such as two-dimensional (2D)-NMR [75–78], optical rotation [79], X-ray, and vibrational circular dichroism [80].

A diverse set of mass spectrometric approaches to elucidate molecular structure, with multistage MS techniques being the most prominent, have been reported. Of great importance is the well-thought design of MS^n experiments as well as a profound knowledge of the fragmentation pathways, an essential component for maximizing the structural information derived from sometimes rather complex MS/MS data. As an example, the ammoniated and sodiated molecules of monosaccharide tend to fragment in a different manner, allowing the mass spectrometrist to have a

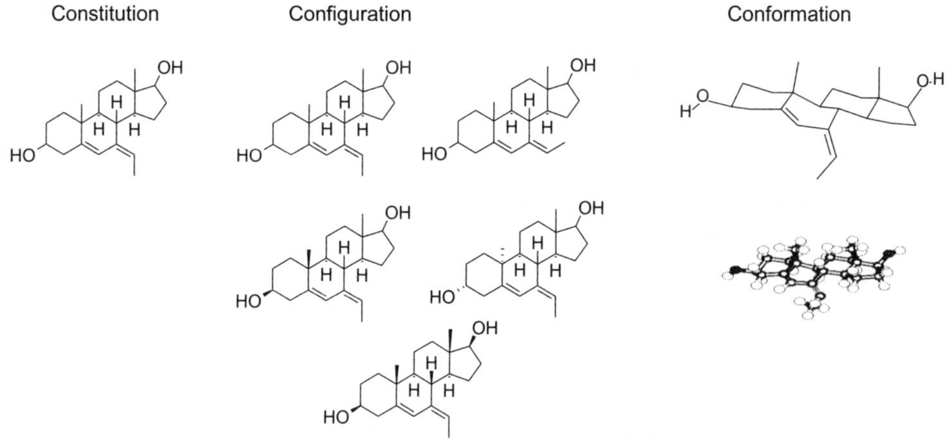

Figure 9.3 Schematic representation of the terms constitution, configuration, and conformation.

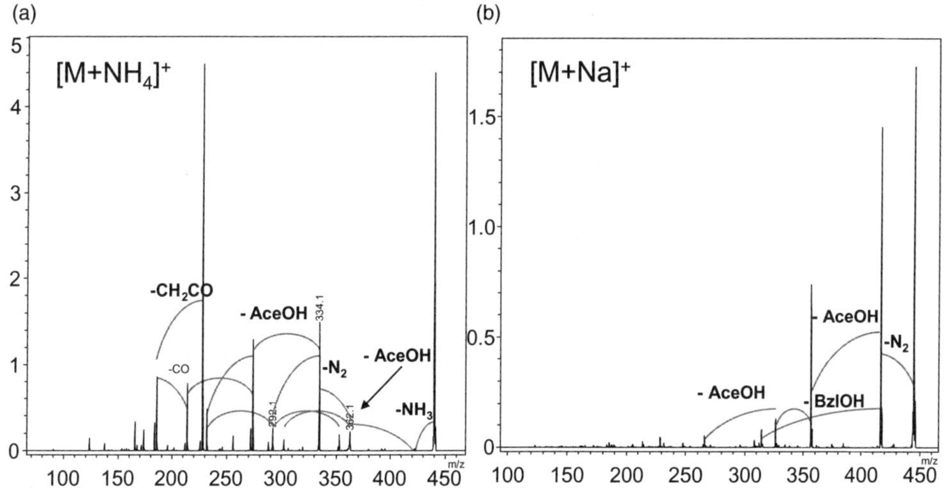

Figure 9.4 MS/MS spectra of a monosaccharide originating from the ammoniated (A) and sodiated (B) ions.

detailed knowledge of the structure and the available functional groups. Loss of N_2 is closely related to the presence of N_3 groups (Fig. 9.4). In an elaborate review, the fragmentation behavior of some 570 pharmaceutical relevant molecules was investigated in great detail [81]. In this study, an AB Sciex (Framingham, MA) API 365 triple quadrupole mass spectrometer was used and spectra at three collision energies (20, 35, and 50 eV) were acquired. The discussed regular and structure-dependent fragmentation pathways, such as how the loss of methoxy radicals ($CH_3O^{\cdot}$) or even Cl/Br are evidence of their substitution on an aromatic ring [82, 83], can serve as a start for more specific MS/MS research projects that are focused on the interpretation of structural look-a-likes from defined compound libraries.

In our experience, and likely to be true for other analytical departments in pharmaceutical R&D, using such an approach, together with various software packages, facilitates the mapping of fragmentation pathways and thus the assessment of the chemical structures for metabolites or impurities. As a matter of fact, together with this knowledge, and before starting profiling studies, a good understanding of the synthesis route and reaction conditions reduces the overall sample analysis time and guarantees high-quality structural proposals. As an example, knowledge of the formation of N-acetylglucosamine conjugation of metabolites and the specific fragmentation thereof led to the development of highly sensitive and robust LC-MS/MS screening and identification assays for these phase 2 metabolites [84]. As another example, Plumb and coworkers reported MS/MS methodologies including low and high collision energies as a tool for structure elucidation [85]. It is not unusual to lose information on the molecular mass at high collision energies while missing specific fragment ions in low energy collision spectra. As mentioned before, in ESI and APCI, gas phase reactivity can alter the mass spectra and complicate the interpretation thereof. On the other hand, controlled gas-phase synthesis at atmospheric pressure can result in improved sensitivity [86].The fast hydrogen/deuterium exchange process can be used to facilitate the identification of specific functional groups in some molecules; these exchange reactions take place when "labile" hydrogen atoms in groups such as –OH, –SH, NH, $-NH_2$, and –COOH are present in the target molecule [87]. In conclusion, many creative approaches have been reported that maximize the amount of pivotal information that can be obtained from MS and MS/MS experiments.

The potential for MS methodologies to detect the structural configuration of stereoisomers has been explored by using gas-phase reactivity of the target molecule with chiral selective agents in ion trap mass spectrometers. For example, MS/MS experiments on "chiral" complexes that measure the specific dissociation energies with the "kinetic method" [88] was previously discussed by Djerassi and coworkers [89]. In these communications from the 1960s, a thorough understanding of electron ionization mass spectra allowed one to resolve structural and stereochemical problems for steroidal molecules [90]. The relative lack of 3D structural information from MS spectra is compensated when highly automated LC-UV-MS/MS-NMR platforms are utilized, as has been shown in several examples [91–94]. As previously stated, hyphenated SFC-MS also plays an important role in these studies [95]. As a final statement, it is our belief that reintroduction of significantly improved interfaces such as the particle beam would place MS into an even more important position for the assessment of the absolute chemical structure for a very large variety of pharmaceutically interesting compounds [96–99].

9.3 ASSESSMENT OF PHYSICOCHEMICAL AND BIOLOGICAL PROPERTIES

In the modern drug discovery process, parallel testing of physicochemical properties, metabolic fate, and biological efficacy is the first stage in understanding the structure and activity relationships of a compound series in both the lead finding and optimization stages [100–102]. Balancing of all these parameters in the search for the optimal exposure and drug efficacy on the basis of physicochemical properties (rule of five) and the biopharmaceutical classification system and "optimal" dosing vehicles are all essential parts of the drug discovery process [3, 103, 104]. "Just a little more time on improving the integrity of a formulation in discovery could significantly improve the chances for success in lead identification and optimization, and save significant time and money" [105]. Moreover, for medicinal chemists, this can be a game within narrow boundaries; in many cases very interesting compounds with low nM target–ligand binding (IC50) values tend to be rather lipophilic in character (log $P > 4$), consequently showing limited oral exposure (bioavailability) [106].

Time and money are crucial parameters in the rapid implementation of high-throughput assay methods for the determination of molecular properties such as lipophilicity, pKa, solubility, chemical stability, and processes such as permeability [100, 102, 107]. Typically, such *in vitro* screening assays should be fast enough to support "testing time cycles" of 1 to 2 weeks for hundreds of compounds. As was discussed in the previous paragraphs, hyphenated LC-MS/MS has gained a strong position as the assay method of choice for assessing physicochemical parameters such as solubility, lipophilicity (LogP/LogD), pKa, and permeability, and for the identification of so-called metabolic hotspots [102, 108–110].

On the other hand, for those "drug candidates" close to the preclinical development stage, a search for the optimal pharmaceutical formulation is initiated and the solid state characterization of the drug substance is implemented. These initiatives require a more in-depth assessment of the chemical stability, solubility, and "crystalline" character of the active pharmaceutical ingredient (API) [104]. In one approach, the coupling of thermogravimetric analysis (TGA) with a gas chromatography (GC)-MS system allows for the identification of solvents captured from the amorphous or crystalline material. In the following paragraphs the main attention is for the description of mass spectrometric tools in high-throughput HPLC-MS/MS-based assay methods.

9.3.1 Lipophilicity, Permeability, pKa, and Solubility Testing

HPLC-MS and HPLC-MS/MS-based assay methods offer improved sensitivity, selectivity, and dynamic ranges for the quantitative analysis of target compounds when compared with "older" approaches such as UV, fluorescence, or ELSD in combination with plate reader systems, or HPLC methods utilizing UV, ELSD, or CLND detection.

An excellent example where the overall sensitivity places a crucial role is the assessment of lipophilicity, defined as the logarithm of the partition coefficient (LogP) or distribution coefficient at a well-defined pH (LogD). That is, the standardized and traditional shake flask methodologies typically require amounts of tens of mg per compound. This is not a practical amount of compound in the early stages of new drug discovery. In our laboratories, between 2 and 10 mg of compound is submitted and 96-well plates with μL quantities from a 1 μM solution are distributed to the various *in vitro* testing groups. Moreover, the molar extinction coefficient can vary tremendously between compound series for a variety of biological targets. One solution is to use liquid chromatography with well-end-capped C_{18} type of stationary phase allowing a rather good correlation between the retention time index (logk) and LogP to be made, the general setup being the calculation of the LogP by comparing the retention time of the test compounds with the retention time of well-known standards. A variety of semi-quantitative isocratic and even faster gradient methods using different commercially available HPLC columns, and ultimately the new UPLC technique, coupled to both MS and MS/MS detection, have been reported and are based on using very small amounts of the compound [101, 108, 111]. An interesting HPLC-MS-based method able to simultaneously assess the lipophilic in terms of the LogD and the purity/compound integrity was described in 2003 by Kerns et al. [110]. One remaining challenge is to solve the differences in response utilizing traditional ESI interfaces, by applying either the so-called dual ion sources (ESI and APCI in one setup) or APPI, before this method can become a generic analytical platform in large discovery organizations.

Furthermore, various HPLC-MS, HPLC-MS/MS, and nanoelectrospray-based assay methods have been reported for the determination of permeability of test compounds using artificial membranes (PAMPA) or Caco-2 cell lines (gastrointestinal transport) [67, 68, 112–118]. In addition, and partly an issue for the LogP analysis, is the possibility of compounds to form aggregates in aqueous solutions, especially for lipophilic compounds. Low test concentrations are needed to prevent these interfering processes; for example, in PAMPA, the diffusion of a compound over a phospholipid membrane from a donor to an acceptor compartment is measured. The donor sample is directly analyzed from a 96-well plate format. In many cases and due to limited solubility, the so-called co-solvent method was used for the assay. The presence of impurities, for example, solubilizers like Tween 80, Cremophor EL, or extractables, and especially solvents like dimethylsulfoxide (DMSO) or acetonitrile can hamper UV detection. Fast and sensitive UPLC-MS/MS methods have proven to serve as a generic platform for samples originating from PAMPA incubations [114].

Reported attempts to implement HPLC-MS methods for the assessment of the solution dissociation coefficient, pKa, are rather limited. While MS permitted increased throughput, the correlation of the retention time dependence on the solution pH does not work well for compounds with multifunctional (basic and acidic) groups [108].

Solubility is another important parameter. Solubility can be defined as the concentration of a test compound at equilibrium with an excess of solid material with the most stable (relevant) crystal form (not amorphous) under specific chemical and physical conditions. Solubility is highly dependent on conditions such as ionic strength and polarity of the solvent [119–122]. As an example, it plays a crucial role in the adequate design of *in vitro* assays (solubility in the test medium) as aggregation can seriously hamper the reliability of the assay. In order to reach optimal exposure, the solubility in the test formulation for *in vivo* dosing experiments has an enormous effect influencing, for example, the absorption of the drug from the intestines. The accurate measurement of this parameter is often performed toward the end of the discovery stage, but may be required at the start of the lead finding stage, where the optimization and inter-comparison of multiple new test compounds is under way. As a result, large varieties of solubility assays have been reported; various solubility prediction tools have also been developed but have not yet been found to generate reliable data.

The various solubility assay methods can be roughly divided into the "kinetic" and the "thermodynamic" assays; the latter has been considered the more accurate and preferred method and is usually employed at the end of the discovery stage. The "kinetic" methods are based on the addition of a very small volume of a (typically) 1 mM DMSO sample to a buffer solution (resulting in a saturated aqueous solution), followed by the determination of the final concentration after 24 h incubation [122]. Compared with plate reader assays, or HPLC-UV methods, HPLC-MS offers increased sensitivity and specificity (especially when solubilizers are used) while not jeopardizing the high-throughput character of the assay [119]. On the other hand, for compounds with a high solubility, HPLC-UV may be the best option (no standard curve samples). Alelyunas et al. [119] described a dual approach based on both HPLC-UV and HPLC-MS/MS quantitative analysis. In our laboratories, we used an UPLC-UV assay based on different diluted samples and after a 24 h incubation period. As a result, a larger dynamic range together with considerable throughput could be achieved (400 samples per week). In the case of any degradation of the test compound, the samples were submitted to the generic HPLC-MS/MS system used for purity screening.

In conclusion, except for pKa assays, (semi)-quantitative HPLC-MS/MS is, from an analytical point of few, by far the "method of choice" for the determination of the physicochemical properties in the *in vitro* and the *in vivo* phases of new drug discovery. In many cases, financial reasons underlay the choice to use the older (non-MS) methods.

9.3.2 Chemical and Photostability Testing

The profiling and identification of impurities in both the drug substance and drug product is one of the most extensively and best regulated areas (ICH Q3A, B, C guidance's) in the pharmaceutical R&D process. These compounds, not being the active entity or excipients are classified as: residual solvents, inorganic and organic impurities, with the latter including the synthesis byproducts, as well as degradation and decomposition products [77, 123]. Hence, the assessment of the complete chemical structure and the mechanism of formation of impurities originating from the chemical instability or from reactions of the API with excipients is first of all a production quality issue. In addition, they are expected to be fully qualified through toxicological evaluation in order to address possible safety concerns. As a consequence, clear thresholds for their levels in both drug substance and products are defined; below or above 2 g/day intake these values are 0.1% (or 1 mg whichever is lower) or 0.05%, respectively [77].

Historically, the investigation of chemical stability was not considered a high priority by medicinal chemists, as their main focus was on potency, selectivity, and eventually novelty of compound classes. Nonetheless, along with the growing awareness of physicochemical properties, absorption, distribution, metabolism, and excretion (ADME) profiling and solid state characterization studies in discovery, the assessment of chemical stability has received growing attention [124]. In the development phase, the stability of the solid state (changes in polymorphy) and the drug product are of major concern, while in discovery, instability of the NCEs in solution or decomposition due to exposure to light can severely influence the experimental results [125–127]. As an example, the integrity of the data originating from *in vitro* testing using buffered solutions (potency, physicochemical properties, ADME) can be questioned, storage of the "HTS compound libraries" at mM levels in DMSO is crucial in safeguarding one of the most precious assets of a pharmaceutical company, and instability due to strong differences in pH (stomach) can explain reduced exposure.

Another important aspect is the development of "stability indicating" and in most cases; HPLC-UV-based purity assays, commonly at the start of the preclinical development phase. Key is the search for "optimal" orthogonal assay methods, for example, HPLC-UV-MS/MS; these are able to provide, on the basis of relative retention time, identification and quantitation of all the degradation products. In order to cover "all" degradation products, so-called forced degradation studies, under light, heat, and humidity stressed conditions, are performed according to ICH proposed guidance [128]. Hence, for completeness, assay methods on the basis of ion chromatography or capillary electrophoresis are frequently applied for biologicals or more complex pharmaceutical ingredients. MS and MS/MS for these types of molecules are of great importance as traditional spectroscopy fails as result of their complexity. Logically, all analytical strategies for the structural identification of the degradants start with the hyphenation of HPLC with UV (unique absorption profiles for chemical reduction of a compound), infrared (IR), optical rotation detection, MS, or even NMR. The applicability of all these techniques is highly related to the degradation processes [77]; thus, stereoselective conversion, hydrolysis, hydration, or oxidative stress, the analytical technique may vary depending on the process. Still, MS and especially high-resolution MS and MS/MS are still considered the best approach in most cases.

Compared with UV, the variety in ionization techniques and mass analyzers allows the detection of most, if not all, possible chemical structures, especially for molecules containing a considerable number of hetero atoms, specific functional groups, and molecules lacking protons, making NMR more challenging. As ESI, APCI, and APPI are expected to generate protonated molecules, $[M+H]^+$, and radical cations or molecular ions, $[M^{+\cdot}]$, and with some mass analyzers the accurate mass and isotopic distribution can be assessed, which can lead to the elemental composition of the compound, so that the number of N atoms and double bonds can be determined. This first stage in evaluating the MS data, especially in combination with UV, IR, and NMR data, and knowledge of the structure, already provides a good idea on the structural properties of the impurities, or the degradants or decomposition products. More likely in ESI, is the presence of "adduct ions," such as $[M+K]^+$, $[M+Na]^+$, $[M+NH_4]^+$, or $[M+CH_3COO]^-$. An in-depth knowledge of the physicochemical properties of molecules allows the tentative identification of some functional groups. As an example, intense adducts with alkali-metal ions are related to cis-dihydroxylated entities, while adducts with acetic acids would suggest the presence of proton-donating hydroxyl groups, as has been shown for an number of steroidal drug candidates [72]. Switching from the most commonly applied positive ion mode to the detection of negative ions can help to determine the presence of such hydroxyl group or carboxylic entities. More sophisticated approaches such as hydrogen/deuterium (H/D) exchange MS has also proven their value in the effective resolving of known chemical structures [129, 130]. Therefore, the detailed investigation of mass spectra derived from different ionization techniques both in the positive and negative ion modes can often give rise to a high level of structural information.

In combination with tandem MS, the molecular constitution, defined as a two-dimensional (2D) representation of a molecular structure, can be assigned. However, a complete assessment of the absolute structure configuration(s) remains the sole property of NMR, X-ray, and VCD. A fundamental understanding of specific MS/MS fragmentation pathways of odd and even electron ions from compound series can be used to construct the unknown structure [81]. As an example, a methoxy group on an aromatic ring leads to the loss of formaldehyde, a neutral with a low heat of formation (ΔHf). Throughout the last decades, much information on specific functional groups or compound series has been published; however a well-documented handbook on MS/MS fragmentation mechanism is not yet available. One reason for the lack of such a document is the large differences in the MS/MS spectra as a function of the mass analyzer types and, in some cases, even between the same types of MS instruments. Ion source geometries, focusing lens potentials, the local gas pressure in the collision cell, or even the type of collision gas, commonly nitrogen or argon, are the well-known causes of these differences.

The choice of mass analyzer is a rather difficult one since the analytical group tends to investigate molecules ranging from small nonpolar steroids to large chemically modified peptides, oligosaccharides, or even humanized monoclonal antibodies. In our experience, reliable, fast scanning Q-TOF MS/MS systems with a mass resolution of approximately 18,000, together with accurate mass and isotopic distribution analysis, serve well as workhorse systems. For a large variety of molecules, combining a Q-TOF with a UPLC-UV system exhibiting three different reversed-phase columns (C_{18}, C_8, and phenyl), using multiple gradient/solvent methods, and monitoring the UV trace at 210, 245, and 280 nm is of great value for impurity and degradation profiling in a drug discovery setting.

9.3.3 Metabolic Hotspot Screening

In drug discovery, newly synthesized compounds are first screened on potency, and subsequently, active compounds undergo a number of screening tests to determine physicochemical properties (pKa, log D, PAMPA) and ADME properties, such as plasma protein binding and metabolic stability [10, 101]. The *in vitro* determination of the metabolic stability of drug compounds is one of the most important components in the lead optimization stage [10]. Traditionally, these stability studies would focus only on the quantitation of the newly synthesized drug compound during incubation with either liver microsomes or hepatocytes. The resulting information on half-life or intrinsic clearance will then be used to (de)select compounds. The actual metabolic pathway was usually not determined until much later in the lead optimization stage and was performed by specialized analytical chemists utilizing top-end MS instruments and NMR. Medicinal chemists, however, require information on the metabolic pathway to direct the design of new compounds as early as possible. Early knowledge of the metabolic fate of a compound can provide vital guidance toward improving the metabolic stability of a chemical series or to assess the potential role of active metabolites.

The most common metabolic stability studies are directed toward phase 1 metabolism and utilize human and rat liver microsomes. In larger pharmaceutical companies, the throughput of these assays is in the hundreds of compounds per week, and (liver) microsomes of all relevant preclinical species are incubated in parallel. BA support of these studies therefore generally involves HPLC-MS/MS with automated compound tuning using triple quadrupole instruments and fast generic LC gradients.

The data obtained are targeted quantitation of the parent drug. Several approaches can be followed to obtain qualitative information on metabolites, and these will be discussed in this section.

9.3.3.1 Design The design of metabolite screening studies depends much on the type of MS instrument that is being utilized. Targeted MS techniques, specifically tandem MS, require an upfront selection of potential analytes. The instrument only "sees" what an operator asks it to see. Therefore, masses of potential metabolites have to be selected before starting the analyses. When using a scanning instrument, specifically a TOF or Orbitrap, no prior selection of potential metabolites is needed. The instrument potentially "sees" every compound that is present in the sample. With appropriate control samples in the study design, the metabolite profile is the result of an analyte-control comparison. Assignment of potential metabolite structures for the metabolites found can be based on mechanistic metabolic rules and/or *in silico* predictions after the analytical run. The main advantage of applying scanning instruments is that information regarding potential metabolites should be present in the data at all times. This allows the analytical chemist to answer questions regarding putative metabolites generally arising from lead optimization (LO) project teams long after running the metabolic stability samples. Otherwise, this would have required reincubation and reanalysis.

9.3.3.2 Metabolite Screening Using Targeted MS When using tandem MS for metabolite screening, an upfront selection of potential metabolites is required. To do so, two different strategies can be applied. First, without any prior knowledge of the investigated drug compound, standard biotransformations can be proposed [131]. As most metabolic reactions involve oxidation, the resulting mass shifts are typically M+16 (hydroxylation), M+14 (keto-formation), and M+32 (dihydroxylation). The tandem mass spectrometric transitions of these anticipated metabolites can then be monitored. Alternatively, metabolites can be predicted based on knowledge of metabolic rules, generally in cooperation with DMPK or medicinal chemistry scientists. Preferably, *in silico* tools are used to generate a list of potential metabolites.

The calculation of the specific selected reaction monitoring (SRM) information should be based on the measured optimal SRM information of the parent compound and the predicted mass changes of the metabolites. This results in two SRM functions per predicted metabolite (with or without the mass difference in the product ion).

This approach has several drawbacks. First, uncommon metabolites will not be detected, even not if they are not major metabolites. Second, the fragmentation of the metabolites has to be similar to that of the parent compound. If the expected product ion is not formed, then obviously the metabolite will not be detected. Third, quantitation of the metabolites, based on these derivatives, SRM transitions should be avoided at all times. Since SRM parameters are not optimized, metabolites showing small peaks could easily be the major metabolite and vice versa. An advantage, on the other hand, of using SRM transitions to screen for metabolites is that some insight into the site of metabolism can be obtained based on the presence or absence of a mass shift in the product ion.

Tandem mass spectrometry can also be used in neutral loss mode or in product ion scan mode [132]. These scanning modes facilitate the search for drug-related compounds (i.e., metabolites) in a complex sample by specifically screening for metabolite fragments that are similar to the parent drug fragment [133]; this approach is also applicable to accurate tandem MS systems, such as the Q-TOF.

More recent developments in MS techniques have increased metabolite screening capacities of quadrupole-type instruments. The hybrid triple quadrupole linear ion trap allows SRM-triggered information-dependent acquisition of potential metabolites [134]. However, prerun selection of possible metabolites remains a requirement. A novel design of the collision cell in the latest generation of tandem quadrupole mass spectrometers allows fast and sensitive simultaneous collection of multiple SRM and full scan MS data. This allows detection of metabolites without SRM preselection, although the post-acquisition data filtering can be hampered by the limited resolution of the instrument compared with TOF and Orbitrap MS systems [135].

9.3.3.3 Metabolite Screening Using Scanning MS Mass spectrometers that have high sensitivity over a wide mass range offer clear advantages for metabolite screening. Upfront selection of potential metabolites is no longer required. In principle, all metabolites that are present in a sample will be detected and can be found during data analysis using the right tools. Ion trap or Orbitrap mass spectrometers have been applied for this purpose [136, 137]. Ion trap only instruments lack mass accuracy and, in a hybrid configuration with an Orbitrap, suffer from a slow data acquisition rate compared with TOF instruments and are therefore not suitable for fast chromatographic applications, especially when simultaneously quantifying parent drugs and screening for metabolites. Although a faster scanning benchtop Orbitrap was introduced recently, the most commonly used mass spectrometer for fast metabolite screening, especially in combination with parent quantification, is the TOF or Q-TOF instrument. Modern TOF instruments have good linear response (3–4 orders of magnitude) and good sensitivity. The main advantage of the TOF over the Orbitrap instrument is its sensitivity at a fast acquisition rate. Metabolic stability assays are typically performed using ultra-performance liquid chromatography (UPLC), with 1–2 min total run time, requiring high scan rates (10 scans/s). This approach was already proposed by Zhang and colleagues in 2000 when TOF analyzers still lacked the required sensitivity [138]. Recent developments in TOF technology have significantly improved sensitivity. Detection limits of 1–5 nM can be easily obtained for most drug compounds. In a standard metabolic stability assay, drugs are incubated at a concentration of 1 μM, and after sample processing (typically protein crash using 3 volumes of acetonitrile and a subsequent 1:1 dilution with a buffer), the concentration in the t_0 sample will be 125 nM. The sensitivity of modern TOF analyzers therefore allows monitoring a two-order decrease in drug concentration and is thus sufficient for an accurate half-life calculation. At the same time, the major metabolites will be detected. To obtain structural information in the same analytical run, the Q-TOF instrument can also be used in the "MS^E" mode. In the MS^E mode, high and low collision energy settings are alternated and, consequently, both molecular ion and nonselective MS/MS fragment ion data are obtained for all detected compounds

[139]. The quality of these MS/MS spectra is compromised by the fast chromatography. It is therefore recommended to perform one extra run in the MS^E mode using a 5-min gradient time, at a preselected incubation time (at 5 or 10 min). This extra injection would then be on top of the (typically) 10 fast runs that are required for half-life calculations.

9.3.3.4 Data Mining Tools for the Detection and Identification of Metabolites

Several tools are offered to help the mass spectrometrist to analyze the large data sets and come up with a list of detected metabolites. The most basic approach is to look for mass differences resulting from the most common metabolic reactions. This is similar, as discussed earlier, for the preselection of SRM information. As an example, a mass filter of M+ 15.994 would be applied to look for mono-oxidation reactions in a TOF data set. Most metabolic data acquisition software packages offer tools for selecting a range of common metabolic transitions to help find common metabolites.

An increasingly popular software tool is the so-called mass defect filter. In this process, both common and uncommon metabolites can be detected based on definable narrow ranges in the mass defect dimensions of high-resolution data regardless of the molecular masses or isotope patterns of the metabolites [140]. The mass defect filter has recently been improved with a dealkylation tool [141]. With this tool, potential metabolic cleavages of the parent drug are identified. These cleavages include single, exocyclic, C-heteroatom, and heteroatom–heteroatom. With this tool the mass defect becomes a function of the mass of the predicted metabolites, and it allows the mass spectrometrist to narrow down the filtering range and thus improve the removal of the matrix background. However, the user of the mass defect filter should be aware that narrowing down the filtering range can, at the same time, result in filtering out true metabolites [141]. The filter works particularly well if the parent drug contains (heavy) heteroatoms such as Br. The presence of these structural elements will push the mass defect of the drug molecule away from that of endogenous compounds containing only C, H, O, and N. The mass defect filtering capacity can also be challenged by difficult matrices, such as urine.

In silico prediction of metabolites can also be used as an intelligent data mining tool, both for targeted (preacquisition) and for scanning MS approaches (post-acquisition). Examples of software packages that can be used are Meta [142], MetaSite [143], Meteor [144], and SyGMa [145]. SyGMa has proven particularly suitable as it not only predicts potential metabolites efficiently and based on a broad set of reaction rules, but also assigns an empirical probability score to each predicted metabolite. This score can be used to rank the predicted metabolites and select a list of most relevant masses to either set up a targeted MS experiment or to apply as filter in the analysis of full scan high-resolution MS data. The best results, as always, are obtained by combining the best of both worlds, that is, by applying the mass defect filter and SyGMa in an orthogonal approach. In addition, the input from medicinal chemists and DMPK experts on the expected metabolic pathways will also be very useful.

This use of data mining tools is demonstrated in Figure 9.5a. SyGMa was used to predict phase 1 metabolites of diltiazem and *m/z* values of predicted structures were imported into MetaboLynx and used for an "expected metabolite" search.

(a)

Home **SyGMa Predicted Metabolites**

	Struct.	LogP	Reaction Sequence	Prob.Score Human vitro	MIM	Rel. Mol. Formula
1 ☑	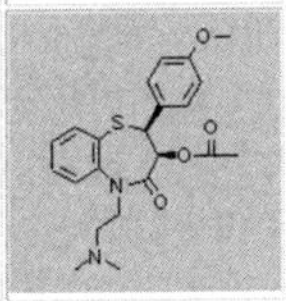	3.65	1. PARENT: diltiazem	1.0000	414.1613	C22H26N2O4S
2 ☑	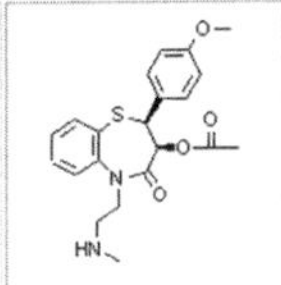	3.06	1. N-demethylation (R-N (CH3)2)	0.6269	400.1456	-CH2
3 ☑	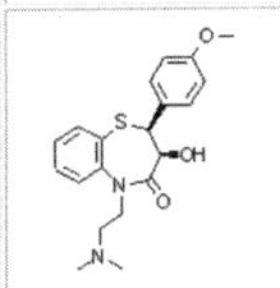	2.96	1. O-deacetylation	0.3636	372.1507	-C2H2O
4 ☑	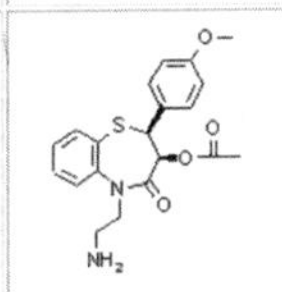	2.77	1. N-demethylation (R-N (CH3)2) 2. N-demethylation (R-NHCH3)	0.3319	386.1300	-C2H4

(b)

S...	M...	Mass	Metabolite Name	Formula	Mass Difference	m/z Found	mDa	PPM	Label	Peak ID	Time	Trace	Area Abs	Area %
	YES	386.1300	-C2H4	C20H22N2O4S	-28.0283	387.1408	3.0	7.7		2	1.84	1: TOF MS ES+ :387.138 0.05Da	1.00	0.67
	YES	386.1300	-C2H4	C20H22N2O4S	-28.0294	387.1397	1.9	4.8		7	2.30	1: TOF MS ES+ :387.138 0.05Da	1.70	1.14
	YES	400.1457	-CH2	C21H24N2O4S	-14.0138	401.1554	1.9	4.8		4	1.88	1: TOF MS ES+ :401.154 0.05Da	1.40	0.94
	YES	400.1457	-CH2	C21H24N2O4S	-14.0164	401.1528	-0.7	-1.7		11	2.40	1: TOF MS ES+ :401.154 0.05Da	35.30	23.64
	YES	414.1613	PARENT	C22H26N2O4S	-0.0009	415.1682	-0.9	-2.2		15	2.42	1: TOF MS ES+ :415.169 0.05Da	108.80	72.87
	YES	430.1562	+O	C22H26N2O5S	15.9939	431.1630	-1.0	-2.4		17	2.45	1: TOF MS ES+ :431.164 0.05Da	1.10	0.74

(c)

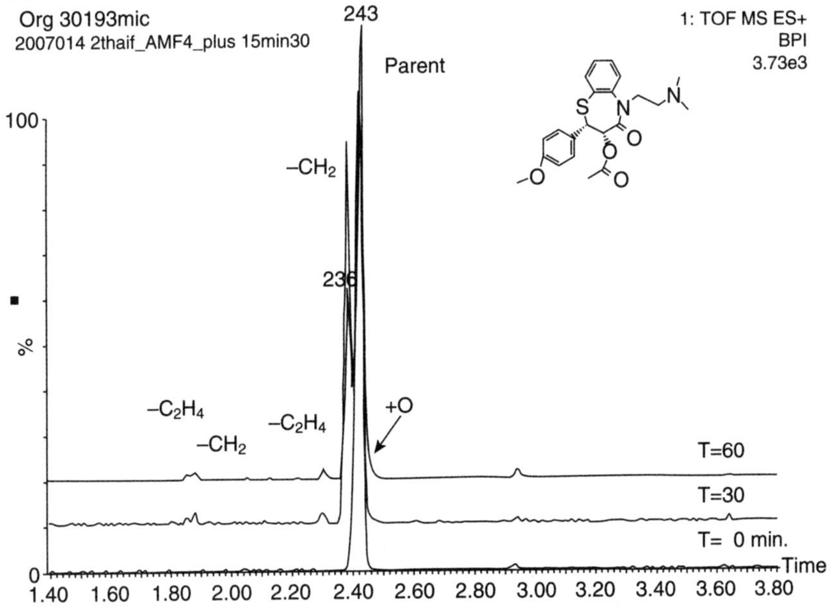

Figure 9.5 (A) Parent structure and top ranked SyGMa predicted phase 1 human metabolites of diltiazem. (B) From the 464 metabolites predicted by SyGMa, 88 different molecular formulae were obtained and imported into MetaboLynx for an expected metabolite search. Besides the remaining parent, five potential metabolites were observed in the $T = 60$ min sample. (C) Mass defect filtered base peak chromatogram. The parent mass defect was used with a filter of +15 mDa and –30 mDa in the mass range of 330–500 Da to check for the presence of additional components.

Accurate masses that passed the search criteria were exported together with their corresponding formulae to show only predicted metabolites (probability ranked) that matched these Q-TOF results (Fig. 9.5b). The most abundant metabolite (at 23.64%) was also predicted with the highest probability (Fig. 9.5a). Next to that, the mass defect filter was applied to check for the presence of additional components. In sum, the search result of five potential metabolites and the parent was confirmed (Fig. 9.5c) [146].

9.3.4 Reactivity and Covalent Binding: Trapping Assays

Metabolites from *in vitro* incubations of nonradiolabeled (cold) compounds are identified by HPLC-MS/MS methods to provide basic metabolism information to assist compound optimization goals and to investigate metabolism related issues. The overall goal is to accelerate the pace of drug discovery and to support medicinal chemistry in producing safe and high-quality recommendation candidates [10, 147]. Typically, the *in vitro* incubations in liver enzyme preparations are analyzed by HPLC with both UV and MS detection. UV detection provides a semi-quantitative measure of the relative amounts of metabolites, while the structure determination is conducted by MS/MS fragmentation studies. When necessary, incubations are scaled up to provide sufficient materials for isolation and NMR studies for further confirmation of the structures.

The most commonly encountered PK issue is rapid metabolic clearance. Identification of the sites of metabolism is the first step in modifying the structures to improve PK properties. One advantage of using *in vitro* systems is that a large number of new compounds can be quickly evaluated. The second common issue investigated is reactive metabolites [148]. Small organic molecules can undergo extensive bioactivation *in vivo* and depletion of detoxifying pathways (glutathione [GSH] trapping), and the reactive electrophilic species generated by metabolism can form covalent adducts with biological macromolecules and subsequently elicit organ toxicity. Thus, it is important to minimize reactive intermediate formation to the extent possible by appropriate structure modification during the lead optimization stage. The first step of this effort is to identify reactive electrophilic species by using chemical trapping agents, such as reduced glutathione (GSH) or cyanide [149]. They form stable adducts with reactive intermediates that can be readily characterized by HPLC-MS/MS and/or NMR methods [150].

GSH, an abundant physiological nucleophile representing cysteine sulfhydryl group, captures reactive metabolites to form S-substituted adducts. Typically, *in vitro* glutathione experiments are conducted at 0.2–5 mM GSH concentration [151]. GSH adducts usually elute early in the chromatogram and display characteristic losses of 129 and 75 Da in the MS/MS spectra. All GSH-related species can be detected using either the full scan MS mode to search for anticipated molecular weights of conjugates or constant neutral loss scan for 129 Da (gamma-glutamyl moiety). It should be noted that some GSH adducts are subject to chemical or enzymatic degradation, thereby escaping detection. N-acetyl cysteine, which is less efficient than GSH, can be used in place of glutathione. The cyanide anion has been used to trap certain electrophilic drug metabolites (an example is the trapping of iminium in the intermediate). Rather recently, eports describing methodologies and analytical strategies to investigate the covalent binding to large biomolecules have been published [152].

9.4 TARGET CHARACTERIZATION AND TARGET–LIGAND INTERACTIONS

The drug discovery process starts with the identification of drug targets and the validation of binding or functional assays able to determine the affinity of ligands or even the kinetics of the commonly noncovalent target–ligand (T-L) interactions. These "drugable targets" can be categorized as cellular surface receptors (ion-channel-linked, G-protein coupled receptors (GPCRs), enzyme-linked) and intracellular receptors (nuclear receptors and kinases), while ligands range from small steroidal molecules to peptides, highly glycosylated proteins, or even monoclonal antibodies. The effort in finding new targets, known as proteomics, encompasses the "structural characterization" of the targets applying hyphenated MS and MS/MS methodologies.

9.4.1 Target Characterization

One might expect the full characterization of the clinically validated drug targets produced by recombinant or other biotechnologies that are applied in (ultra)HTS and other screening assays. In practice, the most stable form highly resembling the activity of the human drug-target, sometimes only the ligand binding domain of the complete target, is engineered followed by well-defined quality assessments. Such an evaluation includes activity assays, Western blotting, and a variety of hyphenated (CE, μLC, or capLC) ESI or MALDI MS/MS characterization methodologies. At this point, there should be good confidence that besides the amino acid sequence and the folding, the shape of the binding pocket is highly comparable between the different batches. This is not trivial; the variation in the position and percentage of phosphorylation of a kinases can cause large fluctuations in the data from screening assays, making the comparison of the binding affinities, reported as IC50, in large screening libraries rather troublesome [153].

9.4.2 Target-Ligand Screening Assays

These (ultra)high-throughput affinity-based or functional assays have a biochemical or cellular character, respectively, and commonly rely on fluorescence, chemiluminesence, or radioactivity detection in 364- or 1536-plate formats. Lower throughput analytical technologies such as surface plasmon resonance, microcalorimetry, NMR, and circular dichroism lack sensitivity, require labeling, are system dependent, or are unable to assess the stoichiometry of the interaction. The potential of MS-based technologies to overcome these drawbacks has led to the development of various screenings approaches and their routine application within the pharmaceutical industry together with more than 300 scientific manuscripts [153–157].

For completeness and interest of the reader, it must be mentioned that the sensitivity and resolution of more recent high-field NMR technologies is rapidly improving and, as part of the fragment or structure-based drug design cycles, has

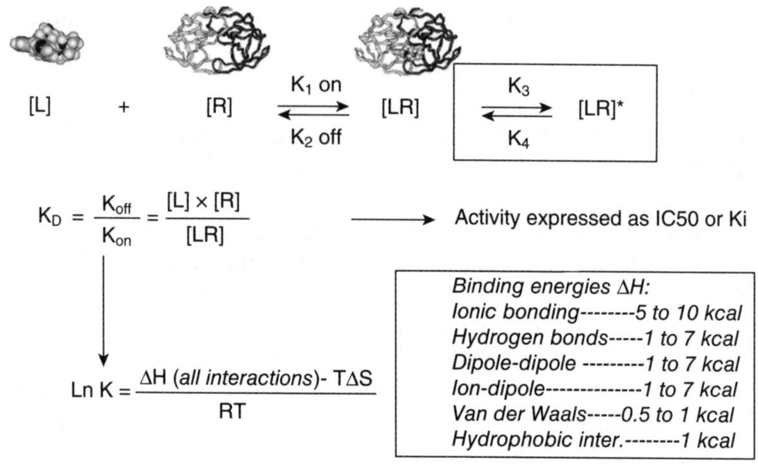

Figure 9.6 Schematic representation of the kinetics and energies involved in target–ligand interaction.

proven its great potential in combining structural information (binding pocket) with detailed knowledge of the reaction kinetics. That is, more and more the thermodynamic parameters such as the enthalpy (ΔH) and entropy (ΔS) of the "reaction" and especially the so-called residence time being directly related with the Koff are of interest to medicinal chemists [158–160]. In Figure 9.6a, schematic representation of the kinetics and energies involved is given. The fact that electrospray mass spectra can be related to the processes or interactions of molecules in solution opened a new window for direct target–ligand interaction studies.

With the direct ESI methodologies, the incubated sample is (nano)sprayed into the mass spectrometer and analyzed after the laborious and tedious optimization of the (nano or capillary) spray-MS interface settings (ion optics). Logically, an important factor such as the transfer of the noncovalent complex from the liquid into the gas phase can prevent any correlation to be determined [161]. In a critical review, the limitations, the failures, and the potentials of this approach were extensively discussed [162]. In this still rather experimental field of science, already several successes have been reported showing good correlation with well-known and validated "standard assays" for the P38α MAP kinase, the Src-family lymphocyte-specific protein tyrosine kinase LCK [163, 164], or the soluble domain of the human growth hormone receptor [165]. Interestingly, nonspecific and allosteric binding kinetics could be distinguished from first-order kinetics. Different setups such as the kinetic method or titration method were chosen to obtain information of the IC50 and the quantitative analysis of the Kd.

Likewise, MALDI mass spectrometry in combination with traditional "low-throughput" H/D exchange experiments originating from the beginning of the 1950s is another promising methodology for the assessment of binding kinetics and Kd [166]. Here, in a method called SUPREX (stability and unpurified proteins from rates of H/D exchange), the ability to localize the ligand binding site, monitor conformational changes as a result of the target–ligand interaction, and the affinity of the ligand can be determined [167–169]. In essence, after incubation of the target in the presence/absence of ligand in deuterated solvent, mass shifts of the peptides from the digested protein are measured. The resolving power of the mass spectrometer is highly dependent on the complexity/mass of the target and the stability of the deuterium-labeled peptides as well as the resolution and analysis time of the chromatographic step used to separate all the peptides. By combining fast chromatographic steps and Fourier transform mass spectrometers, the ultimate performance can be obtained [170].

Such high-resolution mass spectrometers, which are still not routine in general R&D laboratories, are not essential when utilizing all "high-throughput indirect mass spectrometric screening assays." Among these, the affinity selection–MS (ALIS, ultrafiltration-ESI-MS), the MS hyphenated online post-column affinity assay, frontal affinity chromatography–MS, and the rapidly emerging methodologies combining surface plasmon resonance–MS in one analytical setup are well described. Within Schering-Plough (now Merck) over the last decade, the "industrially validated" ALIS method has proven to be a robust, reliable, and truly HT methodology [171]. Moreover, it has been supportive in finding new hits and even new leads, with considerable success [171–174].

With this ALIS methodology, not only water soluble targets, such as the estrogen nuclear receptor or the P38 MAP kinases, but also rather difficult-to-handle/express GPCRs, such as the muscarinic M2 acetylcholine receptor, could be tested. Basically, after incubation of the target (T) and ligand (L), the T-L complex is separated from the target and ligand-free fractions by size exclusion chromatography. Upon UV detection and an automated valving system, the T-L complex is directed toward a reversed-phase analytical column for dissociation, desalting, and elution of any ligands into the ESI-MS. As ESI-TOF instruments facilitate both the mass measurement over a large m/z range combined with a good sensitivity, cocktails of compounds can be analyzed in this setup. In this competitive screening assay, the affinity compared with a "standard" compound allows the ranking of the individual compounds. The titration of their concentration will generate vulnerable Kd values. Interestingly, the concentration of the compounds in a combinatorial mixture did not influence the Kd value. This approach is of great value for medicinal chemists, as they can compare or even make a limited SAR of compounds from one experiment.

The same advantage holds for those technologies that are based on the inline or online coupling of MS to homogeneous or heterogeneous reaction environments. In the latter technology either the target or the test compound is immobilized on a surface, and the interaction takes place in a flow-based or static measurement. A well-described example of the former methodology is frontal affinity chromatography or "dynamic protein affinity selection mass spectrometry" [175]. In the most common setup, the target protein is immobilized on the HPLC column material [176], while ligands flow through and bind to the target on the basis of their affinity. The corrected breakthrough volume is a characteristic of the ligand. A detailed explanation of this technology is given elsewhere [177, 178]. Again, mixtures of compounds including references can be utilized to screen GPCR, nicotine acetylcholine, or opioid receptors, when coupled with MS detection. The rather recent efforts in the offline and online coupling of surface plasmon resonance with MS has proven its potential to combine the "multiresidue" analysis with so-called multichannel chip to provide the possibility of generating valuable data on the binding characteristics of a ligand [179, 180]. In this more static experimental setup, besides the Kd, the Koff and consequently an estimation of the residence time can all be determined. Moreover, with the SPR technologies, nonspecific binding of ligands can be detected and corrected.

The tackling of this problem has been addressed in the hyphenated methodologies based on the coupling of liquid chromatography with post-column biochemical assays and MS [181–185]. In some examples, high-resolution mass spectrometry (HRMS), tandem mass spectrometry, or even inductively coupled plasma (ICP)-MS were utilized to assess the chemical empirical formula and structure of the "unknown" ligand. In most applications, fluorescence enhancement and detection form the basis of the biochemical assays, and therefore throughput is determined by the separation power of the HPLC system. A large advantage here is the possibility to couple various biochemical assays online and to include inline solid-phase extraction (SPE)-NMR along with the MS/MS detection. In such an experimental design, besides the IC50 of the ligands, the selectivity toward two different receptors (ERα and ERβ) and their absolute chemical structure can be determined on the basis of MS (Fig. 9.7A) and NMR data (Fig. 9.7B).

In general, in the last decade many developments in the hyphenation of HPLC, biochemical reactors, SPR, or other techniques with MS have been reported, each with their own successes and added value. In case of the ALIS technology, MS has gained its position in a truly high-throughput screening process of new ligand libraries. In addition, other technologies have shown their potential and need further "industrial validation" in order to become recognized as an alternative to current methods. Overall, the role of MS in target–ligand screening has helped medicinal chemists in finding new ways of assessing the binding kinetics together with an SAR, allowing the best compounds to be pushed forward in the drug discovery process.

9.5 FINAL REMARKS

From a general analytical position, MS is considered the most generic platform for the identification and quantitation of nearly all model compounds. Logically, it can be postulated that the same would account for its role in the support of medicinal chemistry, ranging from the determination of physical chemical properties, identification and quantitation of impurities, degradation products, and metabolites in biological, to even target–-ligand interaction studies. The applications and potentials of this technology described in this chapter as a whole clearly underline these conclusions.

Nonetheless, it does not mean that the specific methodologies have reached their final stage of development. On the contrary, new developments are expected to continue; these will focus on improving selectivity, resolution, speed, and accuracy. Coupling of MS with SFC is expected to mature over the following years, initiating the revival of old ionization techniques. New and more fundamental experience with ionization techniques would hopefully lead to interfaces able to volatize and ionize large chemical space, and in combination with fast scanning ultrahigh mass resolution (~250,000), facilitate the accurate analysis of small and large molecular complexes.

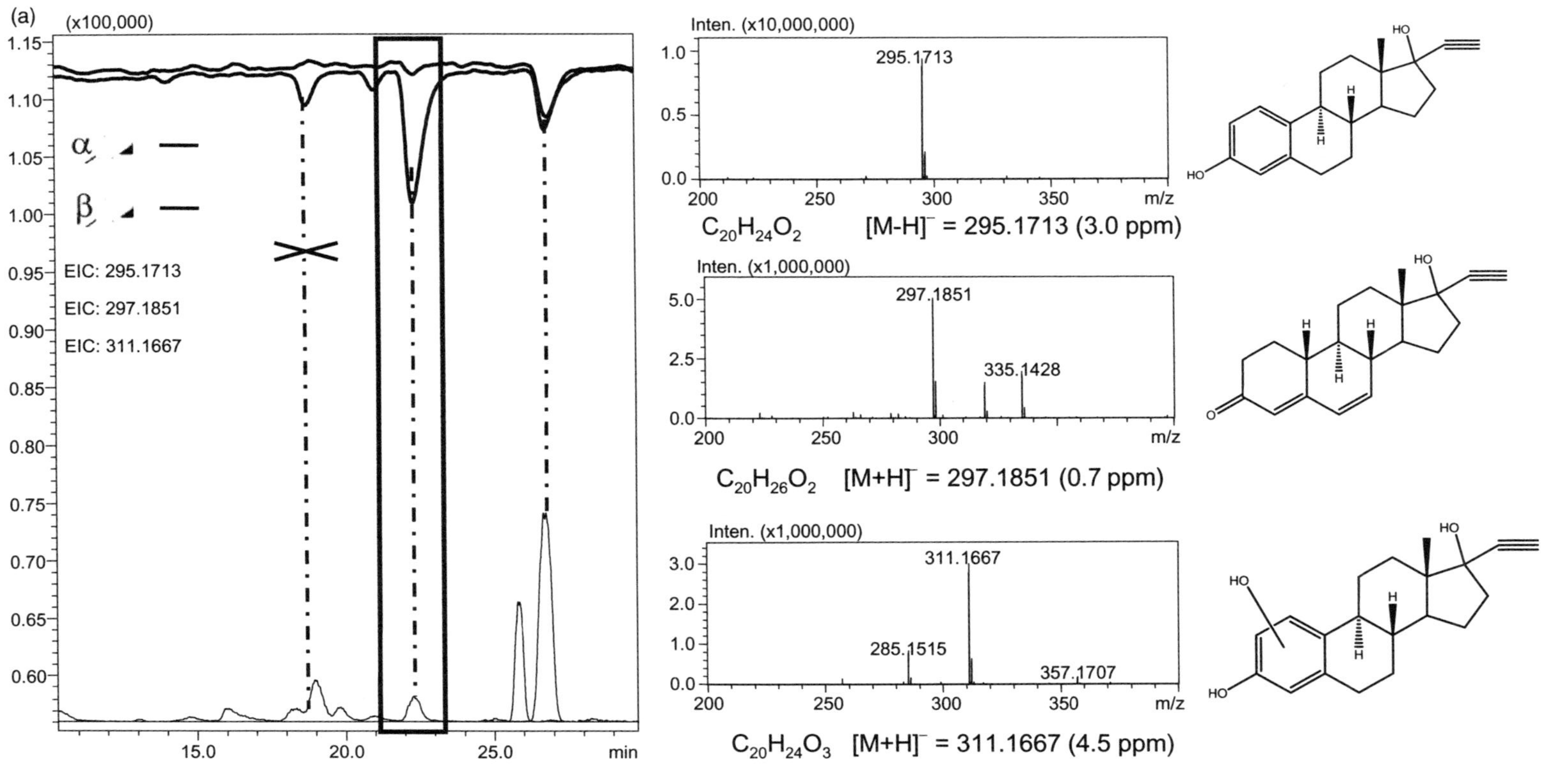

Figure 9.7 (A) LC-fluorescence and LC-MS chromatograms combined with negative ion MS spectra of high affinity compounds.

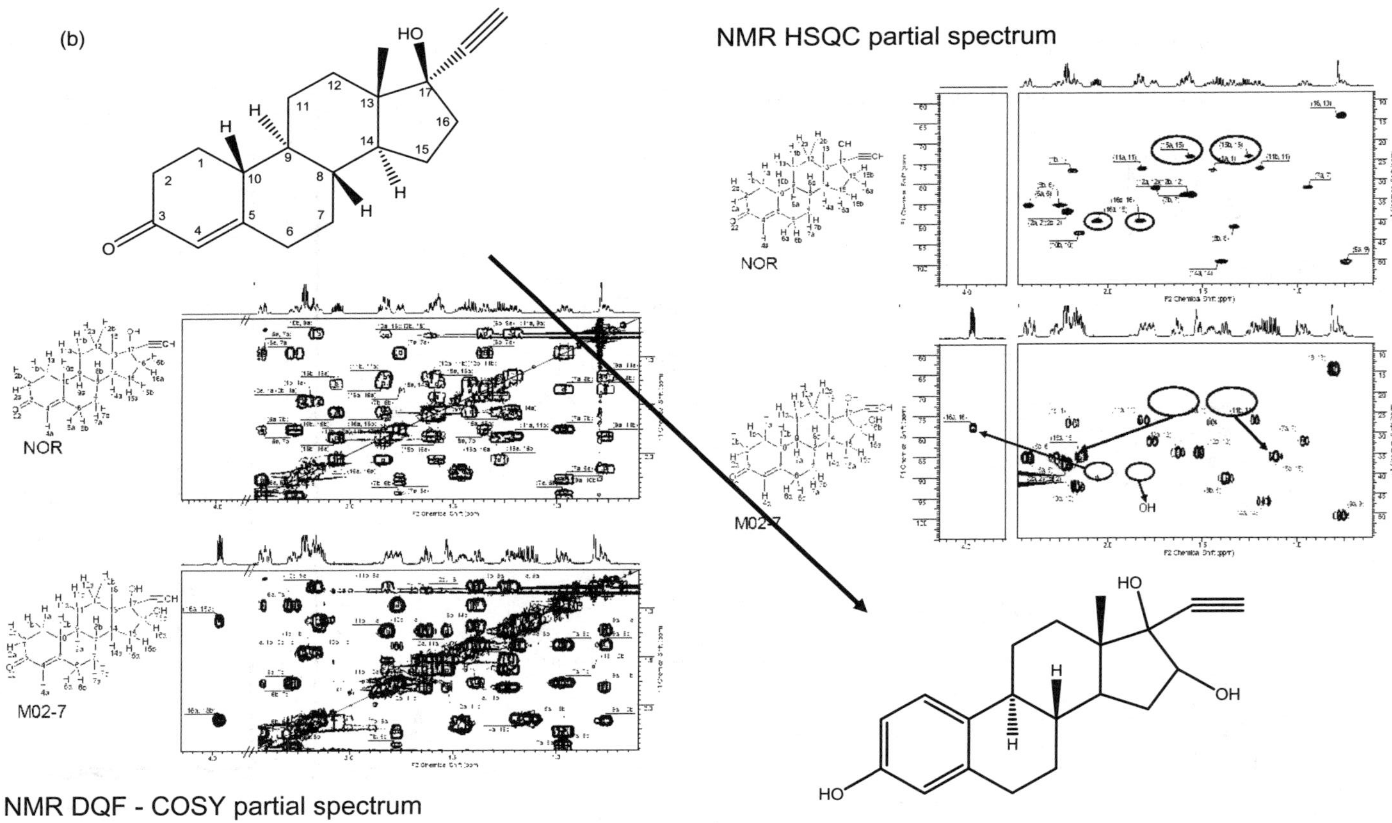

Figure 9.7 (Continued) (B) COSY and HSQC NMR spectra for the start compound and the high-affinity second-order biotransformation product.

REFERENCES

[1] Wanner KT, Hoefner G, editors. Mass Spectrometry in Medicinal Chemistry: Applications in Drug Discovery. Weinheim, Germany: Wiley-VHC Verlag GmbH & Co., 2007.

[2] Poulsen SA, Vu H. Dynamic combinatorial chemistry and mass spectrometry: a combined strategy for high performance lead discovery. In Miller BJ, editor. Dynamic Combinatorial Chemistry. West Sussex, England: John Wiley & Sons, 2010, pp. 201–27.

[3] Lipinski CA. Drug-like properties and the causes of poor solubility and poor permeability. J Pharmacol Toxicol Methods 2000;44:235–49.

[4] Korfmacher WA. Lead optimization strategies as part of a drug metabolism environment. Curr Opin Drug Discov Devel 2003;6:481–5.

[5] Andersson S, Armstrong A, Bjore A, Bowker S, Chapman S, Davies R, et al. Making medicinal chemistry more effective—application of Lean Sigma to improve processes, speed and quality. Drug Discov Today 2009;14:598–604.

[6] Ullman F, Boutellier R. A case study of lean drug discovery: from project driven research to innovation studios and process factories. Drug Discov Today 2008;13: 543–50.

[7] Reader JC. Automation in medicinal chemistry. Curr Top Med Chem 2004;4:671–86.

[8] Korfmacher W. Bioanalytical assays in a drug discovery environment. In Korfmacher W, editor. Using Mass Spectrometry for Drug Metabolism Studies. Boca Raton, FL: CRC Press, 2005, pp. 1–34.

[9] Korfmacher WA. Principles and applications of LC-MS in new drug discovery. Drug Discov Today 2005;10:1357–67.

[10] Korfmacher WA. Advances in the integration of drug metabolism into the lead optimization paradigm. Mini Rev Med Chem 2009;9:703–16.

[11] Korfmacher WA, Cox KA, Bryant MS, Veals J, Ng K, Lin CC. HPLC-API/MS/MS: a powerful tool for integrating drug metabolism into the drug discovery process. Drug Discov Today 1997;2:532–7.

[12] Lee MS, Kerns EH. LC/MS applications in drug development. Mass Spectrom Rev 1999;18:187–279.

[13] Ackermann B, Berna M, Eckstein JA, Ott LW, Chaudhary AG. Current applications of liquid chromatography/mass spectrometry in pharmaceutical discovery after a decade of innovation. Annu Rev Anal Chem 2008;1:357–96.

[14] Seraglia R, Traldi P. Mass spectrometry as test bench for medicinal chemistry studies. Anal Bioanal Chem 2011;399:2695–710.

[15] Atmanene C, Wagner-Rousset E, Malissard M, Chol B, Robert A, Corvaia N, et al. Extending mass spectrometry contribution to therapeutic monoclonal antibody lead optimization: characterization of immune complexes using noncovalent ESI-MS. Anal Chem 2009;81:6364–73.

[16] Pullen FS, Perkins GL, Burton KI, Ware RS, Teague MS, Kiplinger JP. Putting mass spectrometry in the hands of the end user. J Am Soc Mass Spectrom 1995;6:394–9.

[17] Mallis LM, Sarkahian AB, Kulishoff JM, Jr, Watts WL, Jr. Open-access liquid chromatography/mass spectrometry in a drug discovery environment. J Mass Spectrom 2002;37:889–96.

[18] Lee MS, editor. Integrated Strategies for Drug Discovery Using Mass Spectrometry. Hoboken, NJ: John Wiley & Sons, Inc, 2005.

[19] Lee H. Pharmaceutical applications of liquid chromatography coupled with mass spectrometry (LC/MS). J Liq Chromatogr Related Technol 2005;28:1161–202.

[20] Cai Y, Kingery D, McConnell O, Bach AC, 2nd. Advantages of atmospheric pressure photoionization mass spectrometry in support of drug discovery. Rapid Commun Mass Spectrom 2005;19:1717–24.

[21] Hsieh Y, Wang G. Integration of atmospheric pressure photoionization interfaces to HPLC-MS/MS for pharmaceutical analysis. Amer Pharm Rev 2004;7:88–93.

[22] Hsieh Y, Merkle K, Wang G, Brisson JM, Korfmacher WA. High-performance liquid chromatography-atmospheric pressure photoionization/tandem mass spectrometric analysis for small molecules in plasma. Anal Chem 2003;75:3122–7.

[23] Gallagher RT, Balogh MP, Davey P, Jackson MR, Sinclair I, Southern LJ. Combined electrospray ionization-atmospheric pressure chemical ionization source for use in high-throughput LC-MS applications. Anal Chem 2003;75:973–7.

[24] Keski-Hynnila H, Kurkela M, Elovaara E, Antonio L, Magdalou J, Luukkanen L, et al. Comparison of electrospray, atmospheric pressure chemical ionization, and atmospheric pressure photoionization in the identification of apomorphine, dobutamine, and entacapone phase II metabolites in biological samples. Anal Chem 2002;74: 3449–57.

[25] Yu K, Di L, Kerns E, Li SQ, Alden P, Plumb RS. Ultra-performance liquid chromatography/tandem mass spectrometric quantification of structurally diverse drug mixtures using an ESI-APCI multimode ionization source. Rapid Commun Mass Spectrom 2007;21:893–902.

[26] Coddington A, van Antwerp J, Ramjit H. Critical considerations for high-reliability open access LC/MS. J Liq Chromatogr Related Technol 2003;26:2839–59.

[27] Yu K, Little D, Plumb R, Smith B. High-throughput quantification for a drug mixture in rat plasma—a comparison of ultra performance liquid chromatography/tandem mass spectrometry with high-performance liquid chromatography/tandem mass spectrometry. Rapid Commun Mass Spectrom 2006;20:544–52.

[28] Sherma J. UPLC: ultra-performance liquid chromatography. J AOAC Int 2005;88: 63A–7A.

[29] Wainhaus S, Nardo C, Anstatt R, Wang S, Dunn-Meynell K, Korfmacher W. Ultra fast liquid chromatography-MS/MS for pharmacokinetic and metabolic profiling within drug discovery. Amer Drug Discov 2007;2:6–12.

[30] Hsieh Y. Potential of HILIC-MS in quantitative bioanalysis of drugs and drug metabolites. J Sep Sci 2008;31:1481–91.

[31] Thomas SR, Gerhard U. Open-access high-resolution mass spectrometry in early drug discovery. J Mass Spectrom 2004;39:942–8.

[32] Alexander AJ, Staab A. Use of achiral/chiral SFC/MS for the profiling of isomeric cinnamonitrile/hydrocinnamonitrile products in chiral drug synthesis. Anal Chem 2006;78:3835–8.

[33] Barnhart WW, Gahm KH, Thomas S, Notari S, Semin D, Cheetham J. Supercritical fluid chromatography tandem-column method development in pharmaceutical sciences for a mixture of four stereoisomers. J Sep Sci 2005;28:619–26.

[34] Hsieh Y, Favreau L, Schwerdt J, Cheng KC. Supercritical fluid chromatography/tandem mass spectrometric method for analysis of pharmaceutical compounds in metabolic stability samples. J Pharm Biomed Anal 2006;40:799–804.

[35] Laskar DB, Zeng L, Xu R, Kassel DB. Parallel SFC/MS-MUX screening to assess enantiomeric purity. Chirality 2008;20:885–95.

[36] Isbell J, Xu R, Cai Z, Kassel DB. Realities of high-throughput liquid chromatography/mass spectrometry purification of large combinatorial libraries: a report on overall sample throughput using parallel purification. J Comb Chem 2002;4:600–11.

[37] Isbell J. Changing requirements of purification as drug discovery programs evolve from hit discovery. J Comb Chem 2008;10:150–7.

[38] Guth O, Krewer D, Freudenberg B, Paulitz C, Hauser M, Ilg K. Automated modular preparative HPLC-MS purification laboratory with enhanced efficiency. J Comb Chem 2008;10:875–82.

[39] FitzGibbons J, Op S, Hobson A, Schaffter L. Novel approach to optimization of a high-throughput semipreparative LC/MS system. J Comb Chem 2009;11:592–7.

[40] Font LM, Fontana A, Galceran MT, Iturrino L, Perez V. Orthogonal analytical screening for liquid chromatography-mass spectrometry method development and preparative scale-up. J Chromatogr A 2011;1218:74–82.

[41] Weller HN, Ebinger K, Bullock W, Edinger KJ, Hermsmeier MA, Hoffman SL, et al. Orthogonality of SFC versus HPLC for small molecule library separation. J Comb Chem 2010;12:877–82.

[42] Bhatt HS, Patel GF, Vekariya NV, Jadav SK. Super critical fluid chromatography—an overview. J. Pharm Res 2009;2:1606–11.

[43] Sekhon B. Enantioseparation of chiral drugs—an overview. Int J Pharm Tech Res 2010;2:1584–94.

[44] Wenda C, Haghpanah R, Rajendran A, Amanullah M. Optimization of isocratic supercritical fluid chromatography for enantiomer separation. J Chromatogr A 2011;1218: 162–70.

[45] Pirzada Z, Personick M, Biba M, Gong X, Zhou L, Schafer W, et al. Systematic evaluation of new chiral stationary phases for supercritical fluid chromatography using a standard racemate library. J Chromatogr A 2010;1217:1134–8.

[46] Aurigemma C, Farrell W. FastTrack to supercritical fluid chromatographic purification: implementation of a walk-up analytical supercritical fluid chromatography/mass spectrometry screening system in the medicinal chemistry laboratory. J Chromatogr A 2010;1217:6110–4.

[47] Zeng L, Xu R, Laskar DB, Kassel DB. Parallel supercritical fluid chromatography/mass spectrometry system for high-throughput enantioselective optimization and separation. J Chromatogr A 2007;1169:193–204.

[48] Thite MA, Boughtflower R, Caldwell J, Hitzel L, Holyoak C, Lane SJ, et al. Ionisation in the absence of high voltage using supercritical fluid chromatography: a possible route to increased signal. Rapid Commun Mass Spectrom 2008;22:3673–82.

[49] Pinkston JD, Wen D, Morand KL, Tirey DA, Stanton DT. Comparison of LC/MS and SFC/MS for screening of a large and diverse library of pharmaceutically relevant compounds. Anal Chem 2006;78:7467–72.

[50] Pinkston JD. Advantages and drawbacks of popular supercritical fluid chromatography/mass interfacing approaches—a user's perspective. Eur J Mass Spectrom (Chichester, Eng) 2005;11:189–97.

[51] Smith RD, Udseth HR, Kalinoski HT. Capillary SFC/MS with electron impact ionization. Anal Chem 1984;56:2971–3.

[52] Huang E, Henion J, Covey T. Packed-column supercritical fluid chromatography-mass spectrometry and supercritical fluid chromatography tandem mass spectrometry with ionization at atmospheric pressure. J Chromatogr A 1990;511:257–70.

[53] Vasanti S, Sulabha S. Impurity profile—a review. Drug Inven Today 2009;1:81–8.

[54] Liu DQ, Sun M, Kord AS. Recent advances in trace analysis of pharmaceutical genotoxic impurities. J Pharm Biomed Anal 2010;51:999–1014.

[55] Pilaniya K, Chandrawanshi HK, Pilaniya U, Manchandani P, Jain P, Singh N. Recent trends in the impurity profile of pharmaceuticals. J Adv Pharma Technol Res 2011; 1:302–10.

[56] Gorog S, Lauko A, Herenyi B. Estimation of impurity profiles in drugs and related materials. J Pharm Biomed Anal 1988;6:697–705.

[57] Ermer J, Vogel M. Applications of hyphenated LC-MS techniques in pharmaceutical analysis. Biomed Chromatogr 2000;14:373–83.

[58] Wu Y. The use of liquid chromatography-mass spectrometry for the identification of drug degradation products in pharmaceutical formulations. Biomed Chromatogr 2000;14:384–96.

[59] Niessen WM. Progress in liquid chromatography-mass spectrometry instrumentation and its impact on high-throughput screening. J Chromatogr A 2003;1000:413–36.

[60] Molina-Martin M, Marin A, Rivera-Sagredo A, Espada A. Liquid chromatography-mass spectrometry and related techniques for purity assessment in early drug discovery. J Sep Sci 2005;28:1742–50.

[61] Cai SS, Hanold KA, Syage JA. Comparison of atmospheric pressure photoionization and atmospheric pressure chemical ionization for normal-phase LC/MS chiral analysis of pharmaceuticals. Anal Chem 2007;79:2491–8.

[62] Short LC, Hanold KA, Cai SS, Syage JA. Electrospray ionization/atmospheric pressure photoionization multimode source for low-flow liquid chromatography/mass spectrometric analysis. Rapid Commun Mass Spectrom 2007;21:1561–6.

[63] Arao T, Fuke C, Takaesu H, Nakamoto M, Morinaga Y, Miyazaki T. Simultaneous determination of cardenolides by sonic spray ionization liquid chromatography-ion trap mass spectrometry—a fatal case of oleander poisoning. J Anal Toxicol 2002;26: 222–7.

[64] Dams R, Benijts T, Gunther W, Lambert W, De Leenheer A. Sonic spray ionization technology: performance study and application to a LC/MS analysis on a monolithic silica column for heroin impurity profiling. Anal Chem 2002;74:3206–12.

[65] Dams R, Benijts T, Gunther W, Lambert W, De Leenheer A. Influence of the eluent composition on the ionization efficiency for morphine of pneumatically assisted electrospray, atmospheric-pressure chemical ionization and sonic spray. Rapid Commun Mass Spectrom 2002;16:1072–7.

[66] Niessen WM. Liquid Chromatography-Mass Spectrometry, 3rd Edition. Boca Raton, FL: CRC Press, 2006.

[67] Balimane PV, Pace E, Chong S, Zhu M, Jemal M, Pelt CK. A novel high-throughput automated chip-based nanoelectrospray tandem mass spectrometric method for PAMPA sample analysis. J Pharm Biomed Anal 2005;39:8–16.

[68] Van Pelt CK, Zhang S, Fung E, Chu I, Liu T, Li C, et al. A fully automated nanoelectrospray tandem mass spectrometric method for analysis of Caco-2 samples. Rapid Commun Mass Spectrom 2003;17:1573–8.

[69] Wickremsinhe ER, Singh G, Ackermann BL, Gillespie TA, Chaudhary AK. A review of nanoelectrospray ionization applications for drug metabolism and pharmacokinetics. Curr Drug Metab 2006;7:913–28.

[70] Zhang S, Van Pelt CK, Henion JD. Automated chip-based nanoelectrospray-mass spectrometry for rapid identification of proteins separated by two-dimensional gel electrophoresis. Electrophoresis 2003;24:3620–32.

[71] Niessen WM. Group-specific fragmentation of pesticides and related compounds in liquid chromatography-tandem mass spectrometry. J Chromatogr A 2010;1217: 4061–70.

[72] Honing M, Bockxmeer E, Beekman D. Adduct formation of steroids in APCI and its relation to structure identification. Analusis 2000;28:921–4.

[73] Hammond J, Jones I. The use of chromatography and online structure elucidation using spectroscopy. In Smith RJ, Webb ML, editors. Analysis of Drug Impurities. Oxford, UK: Blackwell Publishing, 2007, pp. 156–214.

[74] Wu Q. A "basket in a basket" approach for structure elucidation and its application to a compound from combinatorial synthesis. Anal Chem 1998;70:865–72.

[75] Seco JM, Quinoa E, Riguera R. The assignment of absolute configuration by NMR. Chem Rev 2004;104:17–117.

[76] Eldridge GR, Vervoort HC, Lee CM, Cremin PA, Williams CT, Hart SM, et al. High-throughput method for the production and analysis of large natural product libraries for drug discovery. Anal Chem 2002;74:3963–71.

[77] Qui F, Norwood DL. Identification of pharmaceutical impurities. J Liq Chromatogr Related Technol 2007;30:877–935.

[78] Cesar J. Structural analysis of new potential pharmacologically active compounds in combinatorial chemistry. In Injac R, editor. The Analysis of Pharmacologically Active Compounds and Biomolecules in Real Samples. Kerala: Transworld Research Network, 2009, pp. 239–61.

[79] Stephens PJ, Devlin FJ, Cheeseman JR, Frisch MJ, Rosini C. Determination of absolute configuration using optical rotation calculated using density functional theory. Org Lett 2002;4:4595–8.

[80] Freedman TB, Cao X, Dukor RK, Nafie LA. Absolute configuration determination of chiral molecules in the solution state using vibrational circular dichroism. Chirality 2003;15:743–58.

[81] Niessen WM. Fragmentation of toxicologically relevant drugs in positive-ion liquid chromatography-tandem mass spectrometry. Mass Spectrom Rev 2011;30:626–63.

[82] Holcapek M, Jirasko R, Lisa M. Basic rules for the interpretation of atmospheric pressure ionization mass spectra of small molecules. J Chromatogr A 2010;1217: 3908–21.

[83] Levsen K, Schiebel HM, Terlouw JK, Jobst KJ, Elend M, Preiss A, et al. Even-electron ions: a systematic study of the neutral species lost in the dissociation of quasi-molecular ions. J Mass Spectrom 2007;42:1024–44.

[84] Levsen K, Schiebel HM, Behnke B, Dotzer R, Dreher W, Elend M, et al. Structure elucidation of phase II metabolites by tandem mass spectrometry: an overview. J Chromatogr A 2005;1067:55–72.

[85] Plumb RS, Johnson KA, Rainville P, Smith BW, Wilson ID, Castro-Perez JM, et al. UPLC/MS(E); a new approach for generating molecular fragment information for biomarker structure elucidation. Rapid Commun Mass Spectrom 2006;20:1989–94.

[86] Campbell JL. Using a dual inlet atmospheric pressure ionization source as a dynamic reaction vessel. Rapid Commun Mass Spectrom 2010;24:3527–30.

[87] Nassar AE, Talaat R. Strategies for dealing with metabolite elucidation in drug discovery and development. Drug Discov Today 2004;9:317–27.

[88] Tao WA, Gozzo FC, Cooks RG. Mass spectrometric quantitation of chiral drugs by the kinetic method. Anal Chem 2001;73:1692–8.

[89] Djerassi C, Mutzenbecher G, Fajkos J, Williams DH, Budzikiewicz H. Mass spectrometry in structural and stereochemical problems. LXV. Synthesis and fragmentation behavior of 15-keto steroids. The importance of interatomic distance in the McLafferty rearrangement. J Am Chem Soc 1965;87:817–26.

[90] Cooks RG, Howe I, Williams DH. Structure and fragmentation mechanisms of organic ions in the mass spectrometer. Org Mass Spectrom 1969;2:137–56.

[91] Rourick RA, Volk KJ, Klohr SE, Spears T, Kerns EH, Lee MS. Predictive strategy for the rapid structure elucidation of drug degradants. J Pharm Biomed Anal 1996;14: 1743–52.

[92] Peng SX. Hyphenated HPLC-NMR and its applications in drug discovery. Biomed Chromatogr 2000;14:430–41.

[93] Wolfender JL, Ndjoko K, Hostettmann K. Liquid chromatography with ultraviolet absorbance-mass spectrometric detection and with nuclear magnetic resonance spectroscopy: a powerful combination for the on-line structural investigation of plant metabolites. J Chromatogr A 2003;1000:437–55.

[94] Lindon JC, Nicholson JK, Wilson ID. Directly coupled HPLC-NMR and HPLC-NMR-MS in pharmaceutical research and development. J Chromatogr B Biomed Sci Appl 2000;748:233–58.

[95] McConnell O, Bach A, 2nd, Balibar C, Byrne N, Cai Y, Carter G, et al. Enantiomeric separation and determination of absolute stereochemistry of asymmetric molecules in drug discovery: building chiral technology toolboxes. Chirality 2007;19:658–82.

[96] Barron D, Barbosa J, Pascual JA, Segura J. Direct determination of anabolic steroids in human urine by on-line solid-phase extraction/liquid chromatography/mass spectrometry. J Mass Spectrom 1996;31:309–19.

[97] Bloom J, Lehman P, Israel M, Rosario O, Korfmacher WA. Mass spectral characterization of three anthracycline antibiotics: a comparison of thermospray mass spectrometry to particle beam mass spectrometry. J Anal Toxicol 1992;16:223–7.

[98] Heimark L, Shipkova P, Greene J, Munayyer H, Yarosh-Tomaine T, DiDomenico B, et al. Mechanism of azole antifungal activity as determined by liquid chromatographic/mass spectrometric monitoring of ergosterol biosynthesis. J Mass Spectrom 2002;37: 265–9.

[99] Wolff JC, Hawtin PN, Monte S, Balogh M, Jones T. The use of particle beam mass spectrometry for the measurement of impurities in a nabumetone drug substance, not easily amenable to atmospheric pressure ionisation techniques. Rapid Commun Mass Spectrom 2001;15:265–72.

[100] Balbach S, Korn C. Pharmaceutical evaluation of early development candidates "the 100 mg-approach." Int J Pharm 2004;275:1–12.

[101] Kerns E, Di L. Drug-like Properties: Concepts, Structure Design and Methods from ADME to Toxicity Optimization. New York: Elsevier/Academic Press, 2008.

[102] Kerns EH, Di L. Pharmaceutical profiling in drug discovery. Drug Discov Today 2003;8:316–23.

[103] Lipinski CA, Lombardo F, Dominy BW, Feeney PJ. Experimental and computational approaches to estimate solubility and permeability in drug discovery and development settings. Adv Drug Deliv Rev 2001;46:3–26.

[104] Ku MS, Dulin W. A biopharmaceutical classification-based Right-First-Time formulation approach to reduce human pharmacokinetic variability and project cycle time from First-In-Human to clinical Proof-Of-Concept. Pharm Dev Technol 2012;17(3): 285–302.

[105] Wilson AGE, Nouraldeen A, Gopinathan S. A new paradigm for improving oral absorption of drugs in discovery: role of physical chemical properties, different excipients and the pharmaceutical scientist. Future Med Chem 2010;2:1–5.

[106] Waring MJ. Lipophilicity in drug discovery. Expert Opin Drug Discov 2010;5:235–48.

[107] Strickley RG. Formulation in Drug Discovery. Annual Reports in Medicinal Chemistry. New York, NY: Elsevier Inc, 2008.

[108] Nicoli R, Martel S, Rudaz S, Wolfender JL, Veuthey JL, Carrupt PA, et al. Advances in LC platforms for drug discovery. Expert Opin Drug Discov 2010;5:475–89.

[109] Kerns EH. Current drug metabolism. Curr Drug Metab 2008;9:845–6.

[110] Kerns EH, Di L, Petusky S, Kleintop T, Huryn D, McConnell O, et al. Pharmaceutical profiling method for lipophilicity and integrity using liquid chromatography-mass spectrometry. J Chromatogr B Analyt Technol Biomed Life Sci 2003;791:381–8.

[111] Kerns E, Di L. Utility of mass spectrometry for pharmaceutical profiling applications. Curr Drug Metab 2006;7:457–66.

[112] Volpe DA. Application of method suitability for drug permeability classification. AAPS J 2010;12:670–8.

[113] Polli JW, Wring SA, Humphreys JE, Huang L, Morgan JB, Webster LO, et al. Rational use of in vitro P-glycoprotein assays in drug discovery. J Pharmacol Exp Ther 2001; 299:620–8.

[114] Mensch J, Noppe M, Adriaensen J, Melis A, Mackie C, Augustijns P, et al. Novel generic UPLC/MS/MS method for high throughput analysis applied to permeability assessment in early drug discovery. J Chromatogr B Analyt Technol Biomed Life Sci 2007; 847:182–7.

[115] Fung EN, Chu I, Li C, Liu T, Soares A, Morrison R, et al. Higher-throughput screening for Caco-2 permeability utilizing a multiple sprayer liquid chromatography/tandem mass spectrometry system. Rapid Commun Mass Spectrom 2003;17:2147–52.

[116] Chu I, Liu F, Soares A, Kumari P, Nomeir AA. Generic fast gradient liquid chromatography/tandem mass spectrometry techniques for the assessment of the *in vitro* permeability across the blood-brain barrier in drug discovery. Rapid Commun Mass Spectrom 2002;16:1501–5.

[117] Masucci JJ, Caldwell GG, Jones WW, Juzwin SS, Sasso PP, Evangelisto M. The use of on-line and off-line chromatographic extraction techniques coupled with mass spectrometry for support of *in vivo* and *in vitro* assays in drug discovery. Curr Top Med Chem 2001;1:463–71.

[118] Liu H, Sabus C, Carter GT, Du C, Avdeef A, Tischler M. *In vitro* permeability of poorly aqueous soluble compounds using different solubilizers in the PAMPA assay with liquid chromatography/mass spectrometry detection. Pharm Res 2003;20:1820–6.

[119] Alelyunas YW, Liu R, Pelosi-Kilby L, Shen C. Application of a dried-DMSO rapid throughput 24-h equilibrium solubility in advancing discovery candidates. Eur J Pharm Sci 2009;37:172–82.

[120] Babu VR, Areefulla SH, Mallikarjun V. Solubility and dissolution enhancement: an overview. J Pharm Res 2010;3:141–5.

[121] Hill AP, Young RJ. Getting physical in drug discovery: a contemporary perspective on solubility and hydrophobicity. Drug Discov Today 2010;15:648–55.

[122] Kerns EH, Di L, Carter GT. *In vitro* solubility assays in drug discovery. Curr Drug Metab 2008;9:879–85.

[123] Kerns E, Di L. Chemical stability. In Triggle DJ, Taylor JB, editors. Comprehensive Medicinal Chemistry II. Oxford: Elsevier, 2007, pp. 489–507.

[124] Di L, Kerns EH. Stability challenges in drug discovery. Chem Biodivers 2009;6: 1875–86.

[125] Di L, Kerns EH. Solution stability—plasma, gastrointestinal, bioassay. Curr Drug Metab 2008;9:860–8.

[126] Raijada DK, Prasad B, Paudel A, Shah RP, Singh S. Characterization of degradation products of amorphous and polymorphic forms of clopidogrel bisulphate under solid state stress conditions. J Pharm Biomed Anal 2010;52:332–44.

[127] Shah RP, Singh S. Identification and characterization of a photolytic degradation product of telmisartan using LC-MS/TOF, LC-MSn, LC-NMR and on-line H/D exchange mass studies. J Pharm Biomed Anal 2010;53:755–61.

[128] Bedse G, Kumar V, Singh S. Study of forced decomposition behavior of lamivudine using LC, LC-MS/TOF and MS(n). J Pharm Biomed Anal 2009;49:55–63.

[129] Shah RP, Sahu A, Singh S. Identification and characterization of degradation products of irbesartan using LC-MS/TOF, MS(n), on-line H/D exchange and LC-NMR. J Pharm Biomed Anal 2010;51:1037–46.

[130] Mehta S, Shah RP, Singh S. Strategy for identification and characterization of small quantities of drug degradation products using LC and LC-MS: application to valsartan, a model drug. Drug Test Anal 2010;2:82–90.

[131] Tiller PR, Romanyshyn LA. Liquid chromatography/tandem mass spectrometric quantification with metabolite screening as a strategy to enhance the early drug discovery process. Rapid Commun Mass Spectrom 2002;16:1225–31.

[132] Clarke NJ, Rindgen D, Korfmacher WA, Cox KA. Systematic LC/MS metabolite identification in drug discovery. Anal Chem 2001;73:430A–9A.

[133] Yao M, Ma L, Humphreys WG, Zhu M. Rapid screening and characterization of drug metabolites using a multiple ion monitoring-dependent MS/MS acquisition method on a hybrid triple quadrupole-linear ion trap mass spectrometer. J Mass Spectrom 2008; 43:1364–75.

[134] Shou M. The impact of cytochrome P450 allosterism on pharmacokinetics and drug-drug interactions. Drug Discov Today 2004;9:636–7.

[135] Plumb RS, Mather J, Little D, Rainville PD, Twohig M, Harland G, et al. A novel LC-MS approach for the detection of metabolites in DMPK studies. Bioanalysis 2010;2: 1767–78.

[136] Gunaratna C, Zhang T. Application of liquid chromatography-electrospray ionization-ion trap mass spectrometry to investigate the metabolism of silibinin in human liver microsomes. J Chromatogr B Analyt Technol Biomed Life Sci 2003;794:303–10.

[137] Peterman SM, Duczak N, Jr, Kalgutkar AS, Lame ME, Soglia JR. Application of a linear ion trap/Orbitrap mass spectrometer in metabolite characterization studies: examination of the human liver microsomal metabolism of the non-tricyclic anti-depressant nefazodone using data-dependent accurate mass measurements. J Am Soc Mass Spectrom 2006;17:363–75.

[138] Zhang N, Fountain ST, Bi H, Rossi DT. Quantification and rapid metabolite identification in drug discovery using API time-of-flight LC/MS. Anal Chem 2000;72:800–6.

[139] Bateman KP, Castro-Perez J, Wrona M, Shockcor JP, Yu K, Oballa R, et al. MSE with mass defect filtering for *in vitro* and *in vivo* metabolite identification. Rapid Commun Mass Spectrom 2007;21:1485–96.

[140] Zhang H, Zhang D, Ray K, Zhu M. Mass defect filter technique and its applications to drug metabolite identification by high-resolution mass spectrometry. J Mass Spectrom 2009;44:999–1016.

[141] Mortishire-Smith RJ, Castro-Perez JM, Yu K, Shockcor JP, Goshawk J, Hartshorn MJ, et al. Generic dealkylation: a tool for increasing the hit-rate of metabolite rationalization, and automatic customization of mass defect filters. Rapid Commun Mass Spectrom 2009;23:939–48.

[142] Klopman G, Dimayuga M, Talafous J. META. 1. A program for the evaluation of metabolic transformation of chemicals. J Chem Inf Comput Sci 1994;34:1320–5.

[143] Trunzer M, Faller B, Zimmerlin A. Metabolic soft spot identification and compound optimization in early discovery phases using MetaSite and LC-MS/MS validation. J Med Chem 2009;52:329–35.

[144] Button WG, Judson PN, Long A, Vessey JD. Using absolute and relative reasoning in the prediction of the potential metabolism of xenobiotics. J Chem Inf Comput Sci 2003;43:1371–7.

[145] Ridder L, Wagener M. SyGMa: combining expert knowledge and empirical scoring in the prediction of metabolites. ChemMedChem 2008;3:821–32.

[146] Jacobs PL, Van Der Muelen E, Ridder L, Wagener M. Proceedings of the 55th Annual ASMS Conference on Mass Spectrometry and Allied Topics. Indianapolis, IN: ASMS, 2007.

[147] Smith DA, Obach RS. Metabolites: have we MIST out the importance of structure and physicochemistry? Bioanalysis 2010;2:1223–33.

[148] Evans DC, Watt AP, Nicoll-Griffith DA, Baillie TA. Drug-protein adducts: an industry perspective on minimizing the potential for drug bioactivation in drug discovery and development. Chem Res Toxicol 2004;17:3–16.

[149] Yan Z, Maher N, Torres R, Huebert N. Use of a trapping agent for simultaneous capturing and high-throughput screening of both "soft" and "hard" reactive metabolites. Anal Chem 2007;79:4206–14.

[150] Hopfgartner G. Reactive metabolite screening and covalent-binding assays. In Korfmacher W, editor. Using Mass Spectrometry for Drug Metabolism Studies, 2nd Edition. Boca Raton, FL: CRC Press, 2009, pp. 205–28.

[151] Wen B, Fitch WL. Analytical strategies for the screening and evaluation of chemically reactive drug metabolites. Expert Opin Drug Metab Toxicol 2009;5:39–55.

[152] Mitrea N, LeBlanc AS, Onge M, Sleno L. Assessing covalent binding of reactive drug metabolites by complete protein digestion and LC-MS analysis. Bioanalysis 2010;2: 1211–21.

[153] Geoghegan KF, Kelly MA. Biochemical applications of mass spectrometry in pharmaceutical drug discovery. Mass Spectrom Rev 2005;24:347–66.

[154] Greis KD. Mass spectrometry for enzyme assays and inhibitor screening: an emerging application in pharmaceutical research. Mass Spectrom Rev 2007;26:324–39.

[155] Deng G, Sanyal G. Applications of mass spectrometry in early stages of target based drug discovery. J Pharm Biomed Anal 2006;40:528–38.

[156] Hofstadler SA, Griffey RH. Mass spectrometry as a drug discovery platform against RNA targets. Curr Opin Drug Discov Devel 2000;3:423–31.

[157] Hofstadler SA, Sannes-Lowery KA. Applications of ESI-MS in drug discovery: interrogation of noncovalent complexes. Nat Rev Drug Discov 2006;5:585–95.

[158] Freire E. Do enthalpy and entropy distinguish first in class from best in class? Drug Discov Today 2008;13:869–74.

[159] Copeland RA, Pompliano DL, Meek TD. Drug-target residence time and its implications for lead optimization. Nat Rev Drug Discov 2006;5:730–9.

[160] Zhang R, Monsma F. The importance of drug-target residence time. Curr Opin Drug Discov Devel 2009;12:488–96.

[161] Wang W, Kitova EN, Klassen JS. Influence of solution and gas phase processes on protein-carbohydrate binding affinities determined by nanoelectrospray Fourier transform ion cyclotron resonance mass spectrometry. Anal Chem 2003;75:4945–55.

[162] Mathur S, Badertscher M, Scott M, Zenobi R. Critical evaluation of mass spectrometric measurement of dissociation constants: accuracy and cross-validation against surface plasmon resonance and circular dichroism for the calmodulin-melittin system. Phys Chem Chem Phys 2007;9:6187–98.

[163] Jecklin MC, Touboul D, Jain R, Toole EN, Tallarico J, Drueckes P, et al. Affinity classification of kinase inhibitors by mass spectrometric methods and validation using standard IC(50) measurements. Anal Chem 2009;81:408–19.

[164] Wortmann A, Jecklin MC, Touboul D, Badertscher M, Zenobi R. Binding constant determination of high-affinity protein-ligand complexes by electrospray ionization mass spectrometry and ligand competition. J Mass Spectrom 2008;43:600–8.

[165] Tjernberg A, Carno S, Oliv F, Benkestock K, Edlund PO, Griffiths WJ, et al. Determination of dissociation constants for protein-ligand complexes by electrospray ionization mass spectrometry. Anal Chem 2004;76:4325–31.

[166] Garcia RA, Pantazatos D, Villarreal FJ. Hydrogen/deuterium exchange mass spectrometry for investigating protein-ligand interactions. Assay Drug Dev Technol 2004;2: 81–91.

[167] Frego L, Gautschi E, Martin L, Davidson W. The determination of high-affinity protein/ inhibitor binding constants by electrospray ionization hydrogen/deuterium exchange mass spectrometry. Rapid Commun Mass Spectrom 2006;20:2478–82.

[168] Tang L, Hopper ED, Tong Y, Sadowsky JD, Peterson KJ, Gellman SH, et al. H/D exchange- and mass spectrometry-based strategy for the thermodynamic analysis of protein-ligand binding. Anal Chem 2007;79:5869–77.

[169] Powell KD, Ghaemmaghami S, Wang MZ, Ma L, Oas TG, Fitzgerald MC. A general mass spectrometry-based assay for the quantitation of protein-ligand binding interactions in solution. J Am Chem Soc 2002;124:10256–7.

[170] Chalmers MJ, Busby SA, Pascal BD, He Y, Hendrickson CL, Marshall AG, et al. Probing protein ligand interactions by automated hydrogen/deuterium exchange mass spectrometry. Anal Chem 2006;78:1005–14.

[171] Annis DA, Nazef N, Chuang CC, Scott MP, Nash HM. A general technique to rank protein-ligand binding affinities and determine allosteric versus direct binding site competition in compound mixtures. J Am Chem Soc 2004;126:15495–503.

[172] Annis DA, Cheng CC, Chuang CC, McCarter JD, Nash HM, Nazef N, et al. Inhibitors of the lipid phosphatase SHIP2 discovered by high-throughput affinity selection-mass spectrometry screening of combinatorial libraries. Comb Chem High Throughput Screen 2009;12:760–71.

[173] Annis DA, Nickbarg E, Yang X, Ziebell MR, Whitehurst CE. Affinity selection-mass spectrometry screening techniques for small molecule drug discovery. Curr Opin Chem Biol 2007;11:518–26.

[174] Annis DA, Shipps GW, Jr, Deng Y, Popovici-Muller J, Siddiqui MA, Curran PJ, et al. Method for quantitative protein-ligand affinity measurements in compound mixtures. Anal Chem 2007;79:4538–42.

[175] Jonker N, Kool J, Irth H, Niessen WM. Recent developments in protein-ligand affinity mass spectrometry. Anal Bioanal Chem 2010;399:2669–81.

[176] Besanger TR, Hodgson RJ, Green JR, Brennan JD. Immobilized enzyme reactor chromatography: optimization of protein retention and enzyme activity in monolithic silica stationary phases. Anal Chim Acta 2006;564:106–15.

[177] Calleri E, Temporini C, Caccialanza G, Massolini G. Target-based drug discovery: the emerging success of frontal affinity chromatography coupled to mass spectrometry. ChemMedChem 2009;4:905–16.

[178] Schriemer DC. Biosensor alternative: frontal affinity chromatography. Anal Chem 2004;76:440A–8A.

[179] Marchesini GR, Buijs J, Haasnoot W, Hooijerink D, Jansson O, Nielen MW. Nanoscale affinity chip interface for coupling inhibition SPR immunosensor screening with Nano-LC TOF MS. Anal Chem 2008;80:1159–68.

[180] Marchesini GR, Haasnoot W, Delahaut P, Gercek H, Nielen MW. Dual biosensor immunoassay-directed identification of fluoroquinolones in chicken muscle by liquid chromatography electrospray time-of-flight mass spectrometry. Anal Chim Acta 2007; 586:259–68.

[181] de Vlieger JS, Kolkman AJ, Ampt KA, Commandeur JN, Vermeulen NP, Kool J, et al. Determination and identification of estrogenic compounds generated with biosynthetic enzymes using hyphenated screening assays, high resolution mass spectrometry and off-line NMR. J Chromatogr B Analyt Technol Biomed Life Sci 2010;878:667–74.

[182] Falck D, de Vlieger JS, Niessen WM, Kool J, Honing M, Giera M, et al. Development of an online p38alpha mitogen-activated protein kinase binding assay and integration of LC-HR-MS. Anal Bioanal Chem 2010;398:1771–80.

[183] Kool J, Giera M, Irth H, Niessen WM. Advances in mass spectrometry-based post-column bioaffinity profiling of mixtures. Anal Bioanal Chem 2010;399:2655–68.

[184] van Elswijk DA, Diefenbach O, van der Berg S, Irth H, Tjaden UR, van der Greef J. Rapid detection and identification of angiotensin-converting enzyme inhibitors by on-line liquid chromatography-biochemical detection, coupled to electrospray mass spectrometry. J Chromatogr A 2003;1020:45–58.

[185] Schobel U, Frenay M, van Elswijk DA, McAndrews JM, Long KR, Olson LM, et al. High resolution screening of plant natural product extracts for estrogen receptor alpha and beta binding activity using an online HPLC-MS biochemical detection system. J Biomol Screen 2001;6:291–303.

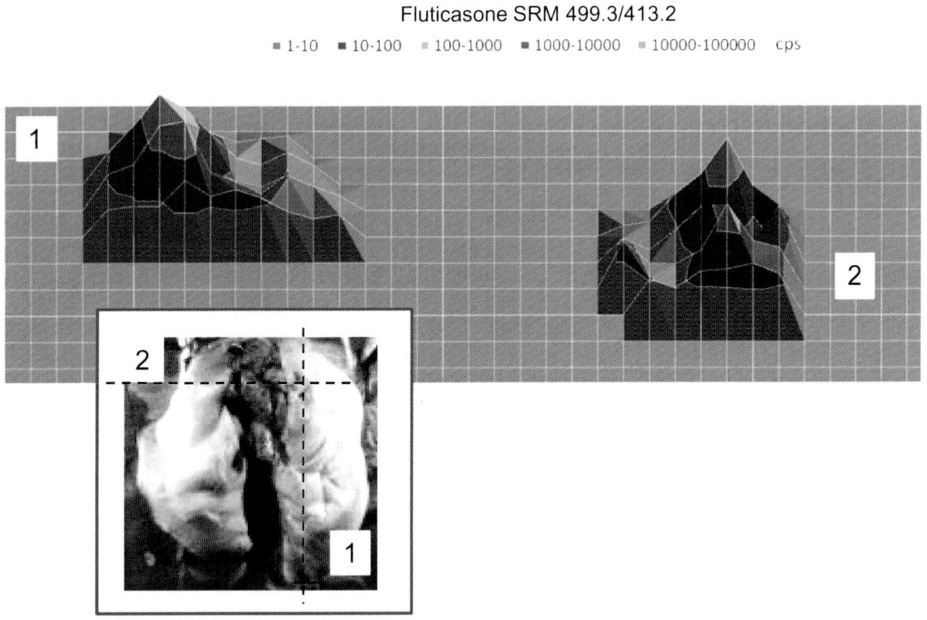

Figure 8.5 LESA-MS-SRM in negative ion mode nESI profiling of fluticasone (SRM m/z 499.3 > m/z 413.2) in lung tissue after aerosol dosage into a mechanically vented and perfusated guinea pig lung model. (See text for full caption.)

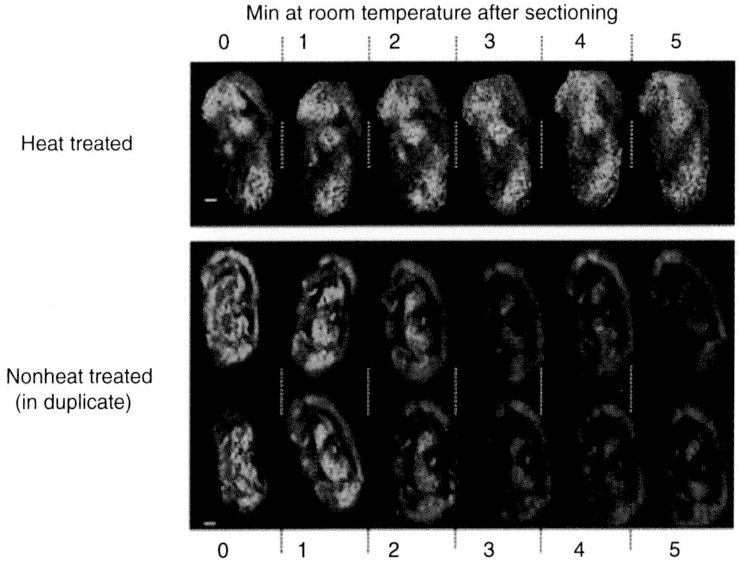

Figure 10.1 Effect of heat stabilization on the intensity of m/z 6723.5 from mouse brain tissue. The intensity of m/z 6723.5 remains constant at 0–5 min at room temperature after sectioning, while the same signal decreases rapidly in the snap frozen sections after only 2 min at room temperature. (See text for full caption.)

Mass Spectrometry for Drug Discovery and Drug Development, First Edition. Edited by Walter A. Korfmacher.

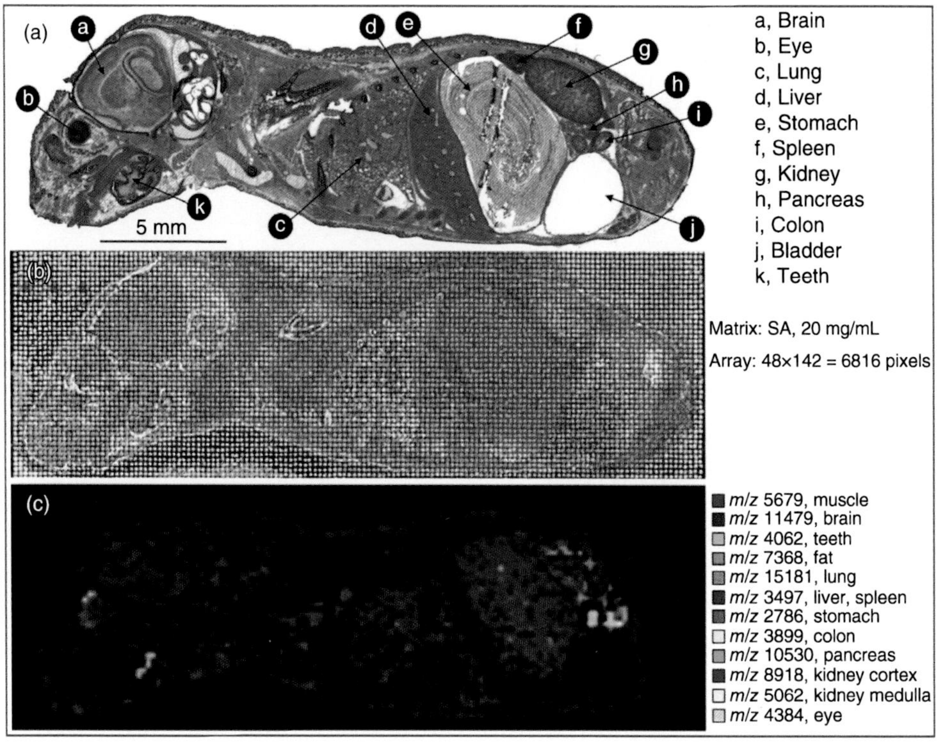

Figure 10.2 Imaging of proteins from a spotted matrix array on a whole-body mouse pup section. (A) H&E stained section showing many different organs throughout the section. (B) Optical picture of a serial section with the matrix array printed on top. A Labcyte Portrait 630 was used to deposit sinapinic acid in a 200 μm spot-to-spot array over the entire section. A total of 6816 spots were printed. (C) Overlay of 12 organ-specific protein images obtained from the section. (Adapted from Reference 31 with permission from Springer Science+ Business Media.)

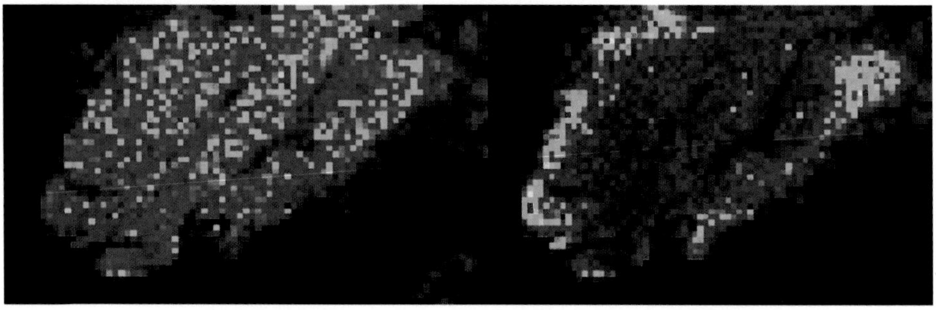

Figure 10.3 Ion mobility separation of isobaric tryptic peptides from formalin-fixed paraffin embedded rat brain sections. Ions with nominal mass *m/z* 1039 were fragmented after ion mobility on a Waters Synapt Q-TOF instrument. Tubulin-related peptides and fragments had a higher ion mobility and showed a different distribution from ubiquitin-related peptides and fragments. (Adapted from Reference 35 with permission from Elsevier.)

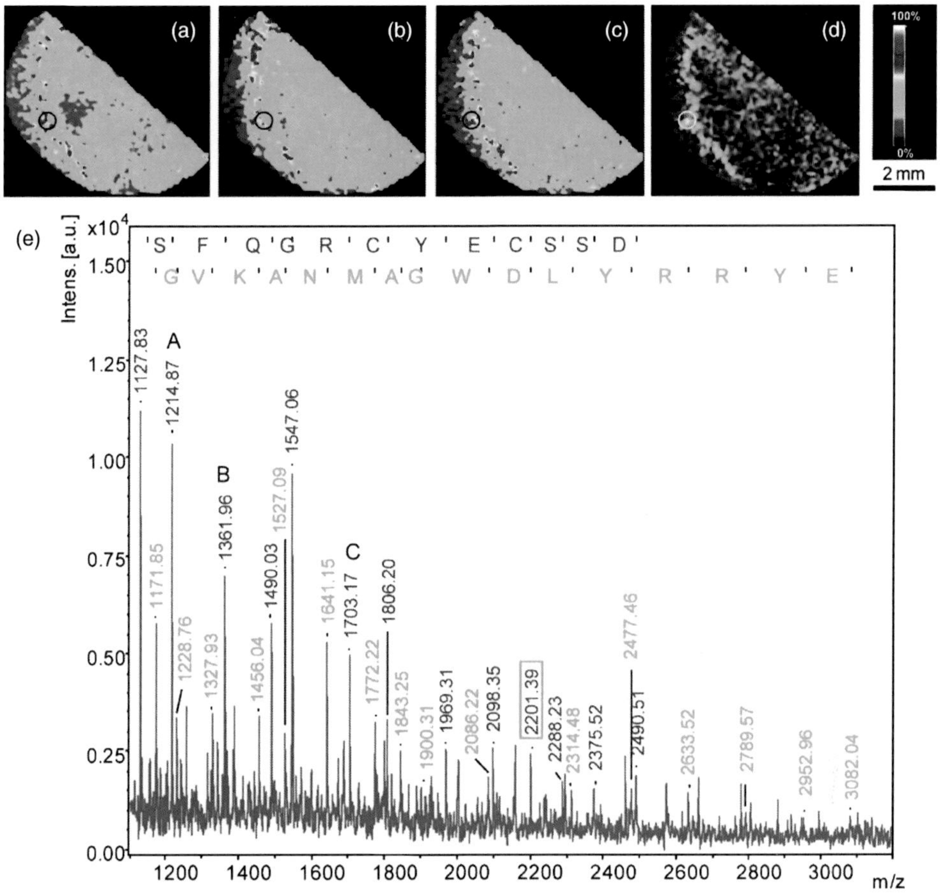

Figure 10.4 Imaging and identification of proteins from in-source decay (ISD) imaging analysis. (See text for full caption.)

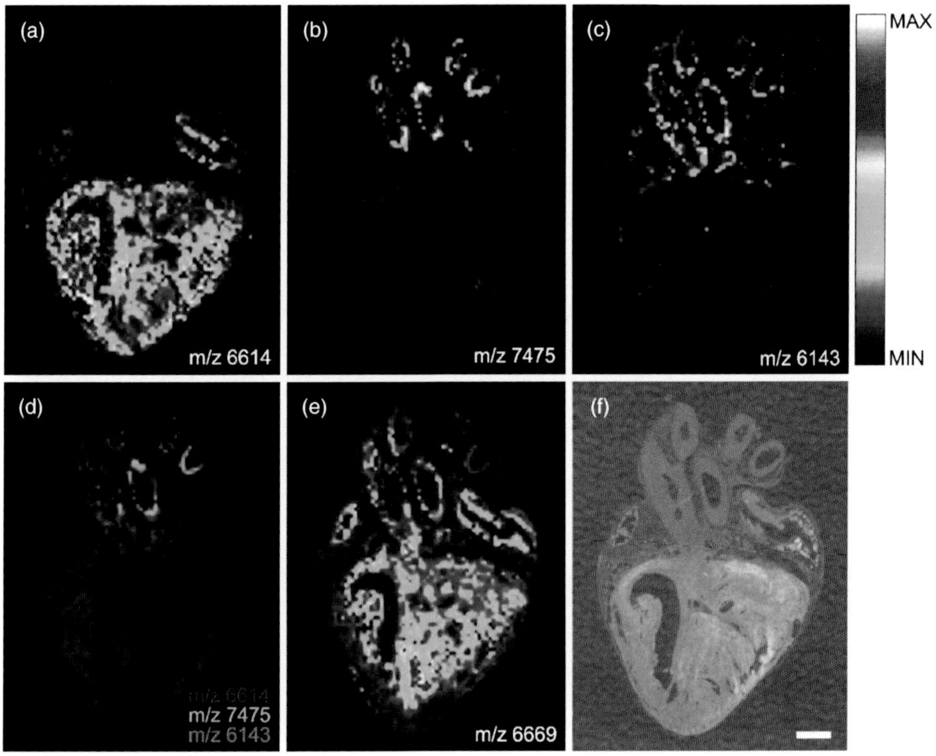

Figure 10.5 Imaging of proteins from a chick heart with DHB. (See text for full caption.)

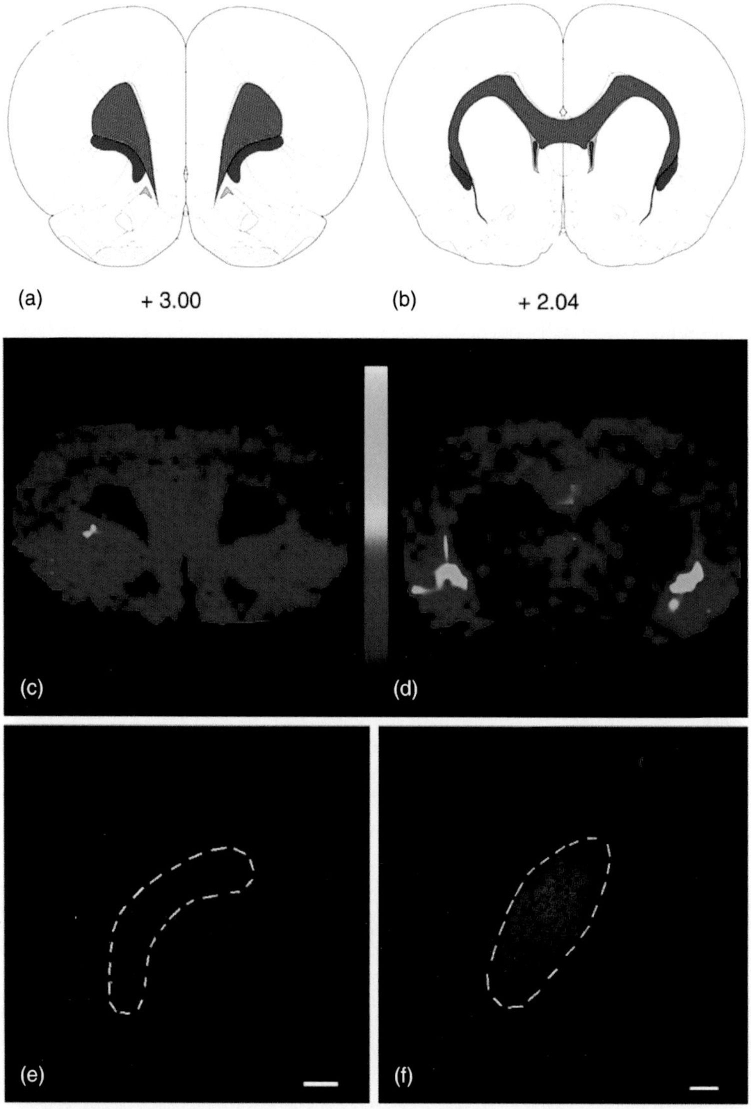

Figure 10.6 Imaging of proteins from a rat brain reveals a protein enriched in the claustrum that serves to redefine its boundaries. (A) and (B) show the claustrum as currently defined (in red) in two brain levels from a rat brain atlas (Paxinos and Watson). Ion images of *m/z* 7725 are shown at the frontal level (C) and at the striatal level (D) indicated in (A) and (B). (E) and (F) show IHC localization of Gng2 at the same two levels. (Reproduced from Reference 48 with permission from Oxford University Press.)

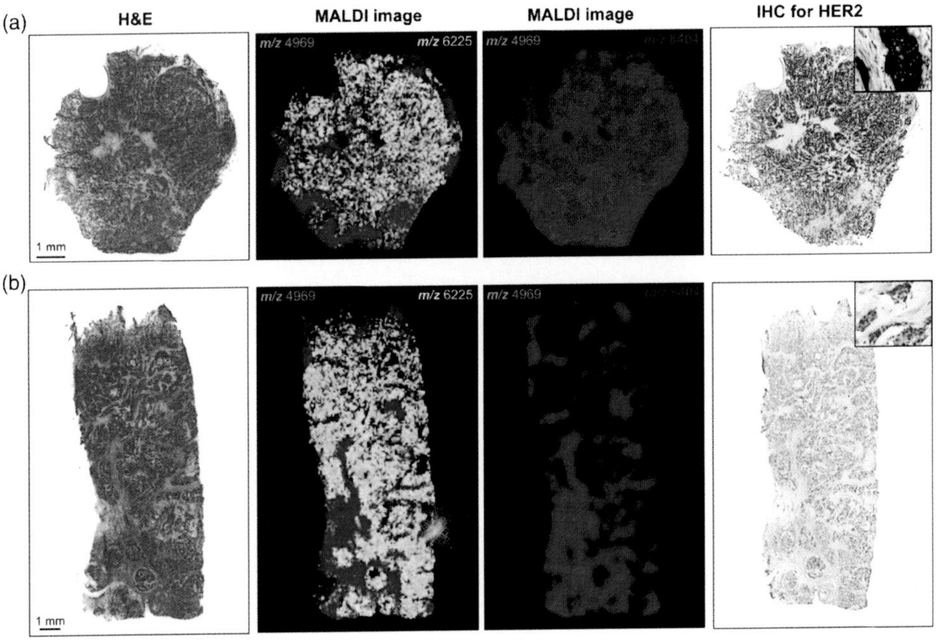

Figure 10.7 Imaging of breast cancer tissues reveals proteins that can distinguish HER2 receptor status. Images of (A) HER2-positive and (B) HER2-negative human breast cancer tissues. The signal at *m/z* 4969 (blue) is specific to tumor stroma; *m/z* 4969 for cancer cells; and *m/z* 8404 for HER2 positive cancer cells. (Reproduced with permission from Reference 29. Copyright 2010 American Chemical Society.)

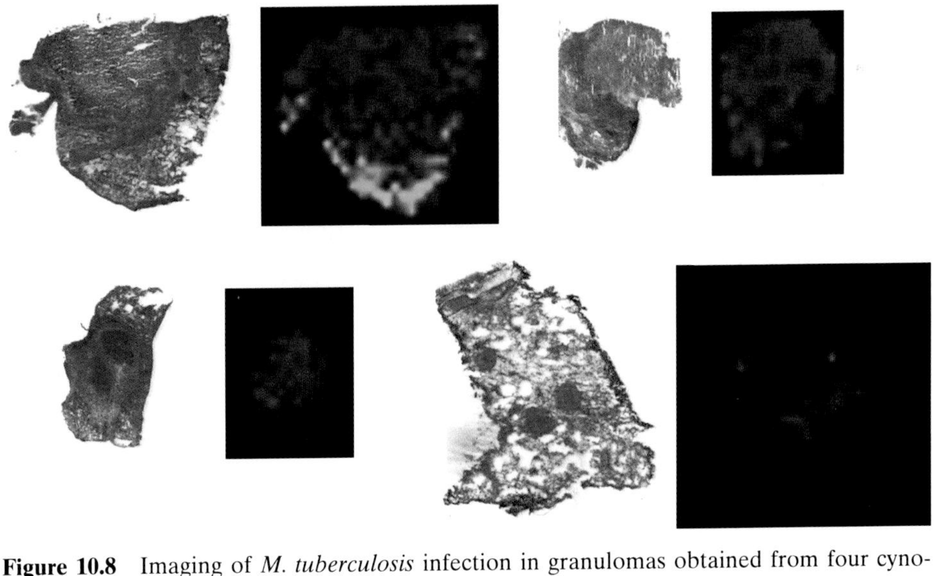

Figure 10.8 Imaging of *M. tuberculosis* infection in granulomas obtained from four cynomolgus monkeys. Granulomas show intense localization of *m/z* 3445 (red), surrounded by *m/z* 8451 (blue), and *m/z* 4046 (green) further out.

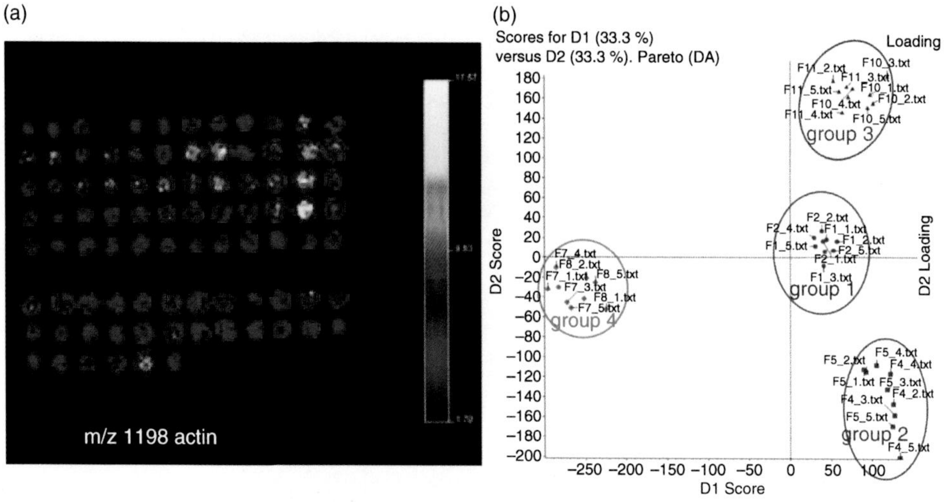

Figure 10.9 IMS analysis of a pancreatic cancer tissue microarray after *in situ* tryptic digestion. (A) The distribution of an actin peptide at *m/z* 1198 within the TMA samples. The TMA is assembled with duplicate cores from a single biopsy next to each other, followed by a single core from adjacent nontumor tissue from the same patient. (B) Principal component analysis–discriminant analysis (PCA–DA) separated a test set of TMA samples into four different tumor groups corresponding to stages IIB, IIA (separated into two groups), and III. (Adapted from Reference 30 with permission.)

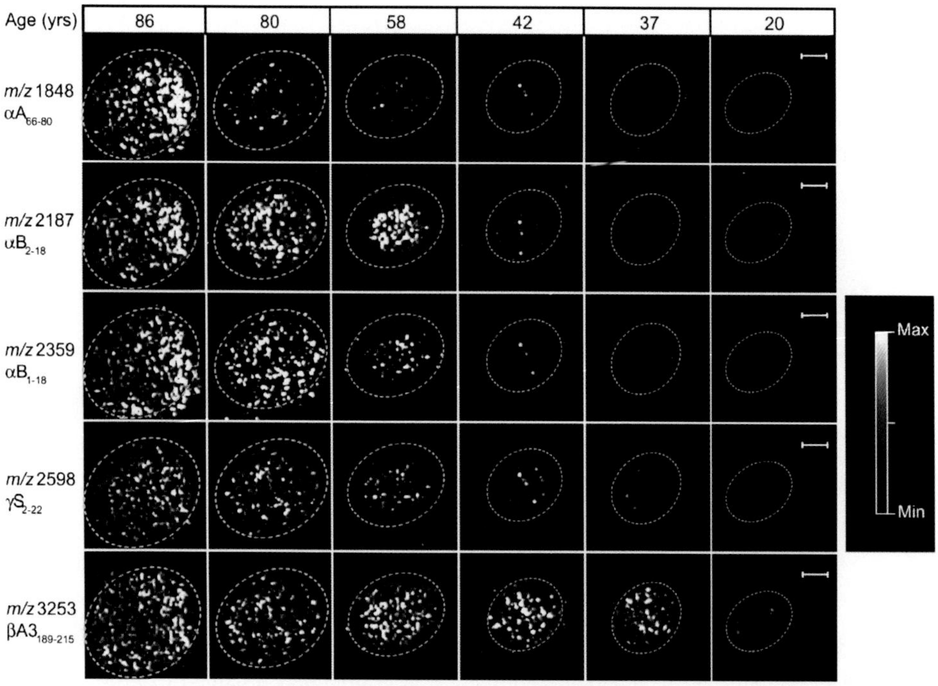

Figure 10.10 Imaging of crystallin peptides in human lens tissue. There is a clear increase in the presence of many crystallin fragments in the lens with age. Scale bars = 2 mm. (Reproduced from Reference 85 with permission from Elsevier.)

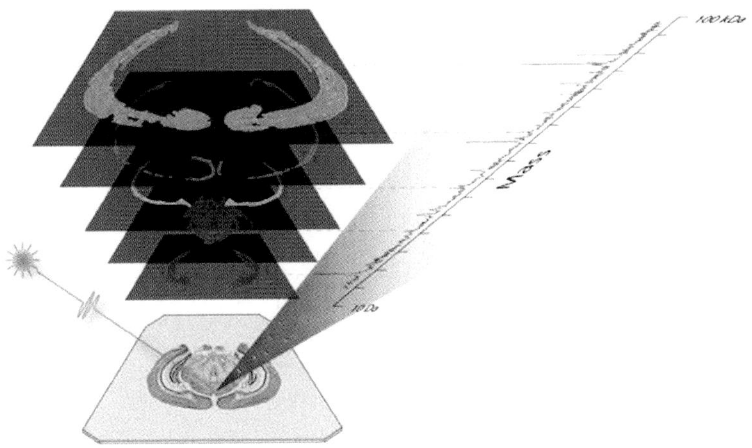

Figure 11.1 MALDI mass spectrometric imaging. A laser is rastered over a tissue sample while acquiring a complete mass spectrum from each position, resulting in molecular images for multiple analytes. Reprinted with permission from Rohner et al. [5]. Copyright 2005 Elsevier.

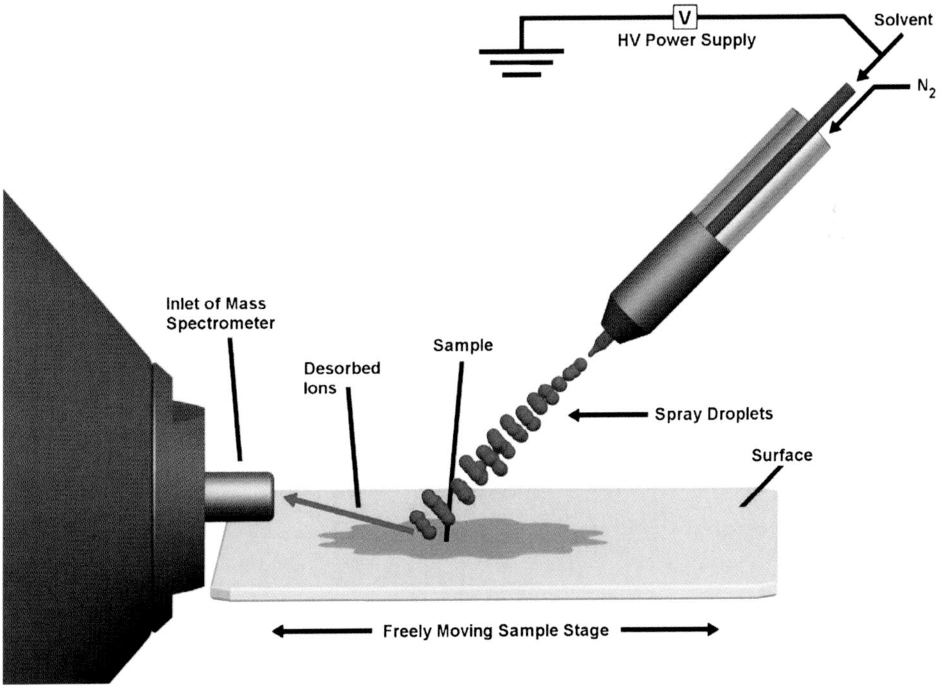

Figure 11.3 Schematic of direct tissue analysis by DESI. Contributed by Justin M. Wiseman, Prosolia Inc.

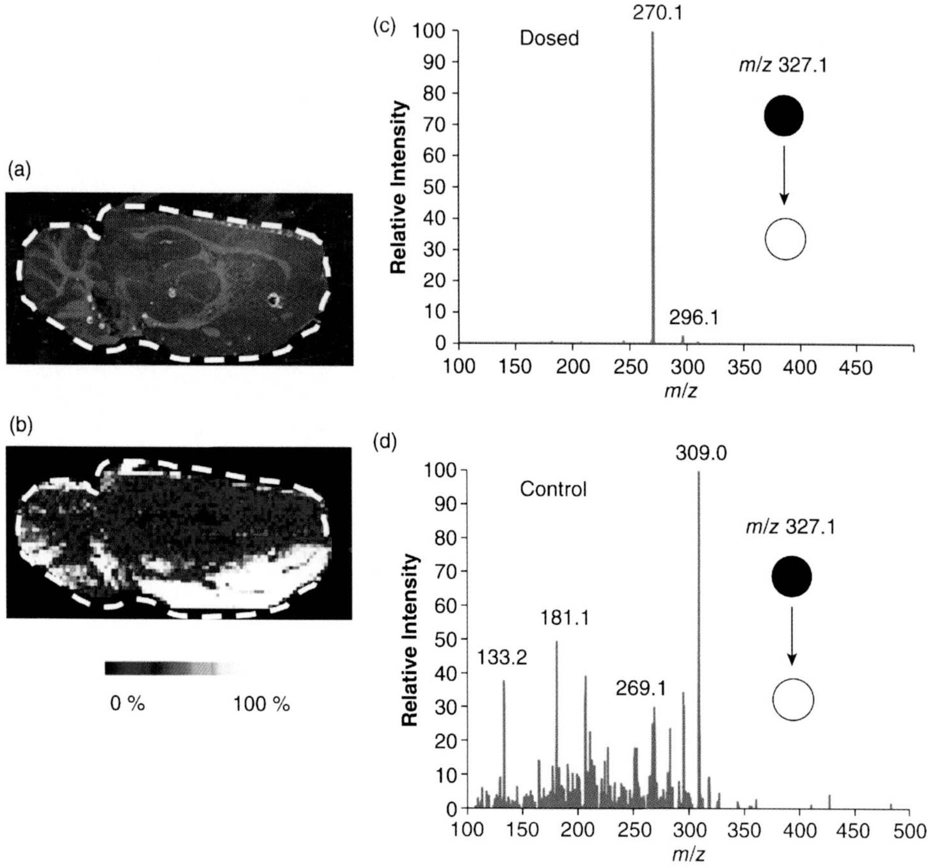

Figure 11.4 Optical image of a 22 × 11 mm^2 sagittal rat brain section and the corresponding selected ion image. (A) Optical image of a sagittal rat brain section taken from animal 992 (0.5 h after dose). CB, cerebellum; Cbc, cerebral cortex; Cpu, caudate-putamen; Hpc, hippocampus; SNr, substantia nigra. (B) DESI mass spectral image of clozapine in the brain section recorded in the MS/MS mode. The image of the fragment ion at *m/z* 270.1 is shown by using false colors in raw pixel format. (C) Average product ion mass spectrum of *m/z* 327.1 ± 2 in B. (D) Average product ion mass spectrum of *m/z* 327.1 ± 2 in control sample 984. Reprinted with permission from Wiseman et al. [56]. Copyright 2008 National Academy of Sciences, U.S.A.

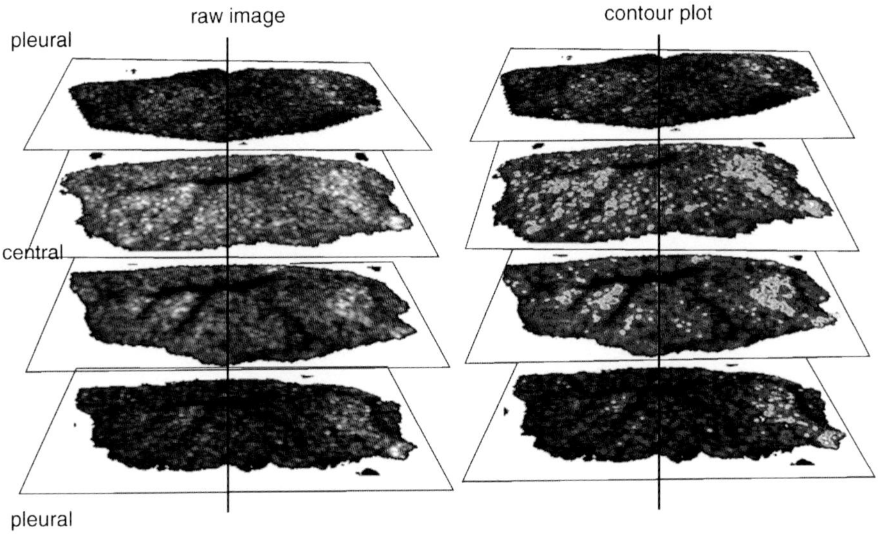

Figure 11.5 The spatial distribution of inhaled tiotropium (TTP) in serially sectioned rat lung tissue. Raw MS images (left column) showing MS localization of inhaled tiotropium (TTP) ion (*m/z* 392) by pixel location in serial sections of whole lung moving anatomically in lung volume from pleural to central to pleural. The arrow indicates approximate carinal entry point of the drug into the central conducting airways. Contour plots (right) of the relative concentration gradients of TPP found in the various lung segments showed a rapid and homogeneous transport of the drug from airways into the parenchyma within 15 min after exposure. The MALDI matrix CHCA was applied by an automatic sprayer device. Reprinted with permission from Nilsson et al. [96].

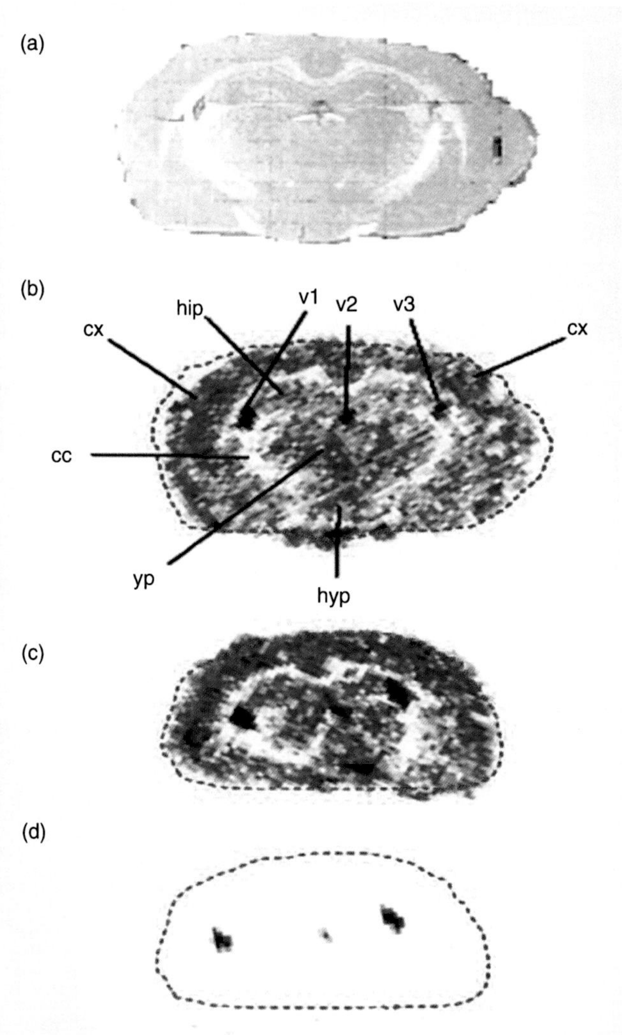

Figure 11.6 (A) The optical image of a rat brain from a coronal section. (B) Matrix-assisted laser desorption/ionization (MALDI)-mass spectrometry (MS)/MS images of astemizole in the rat brain slice without perfusion and (C) with perfusion. Cortex, hippocampus, corpus callosum, hypothalamic region, thalamus region, choroid plexus, dorsal third ventricle, and lateral ventricle are indicated by arrows. (D) MALDI-MS/MS images of M-14 metabolite of astemizole in the rat brain slice. Reprinted with permission from Li et al. [81]. Copyright 2009 Future Science Ltd.

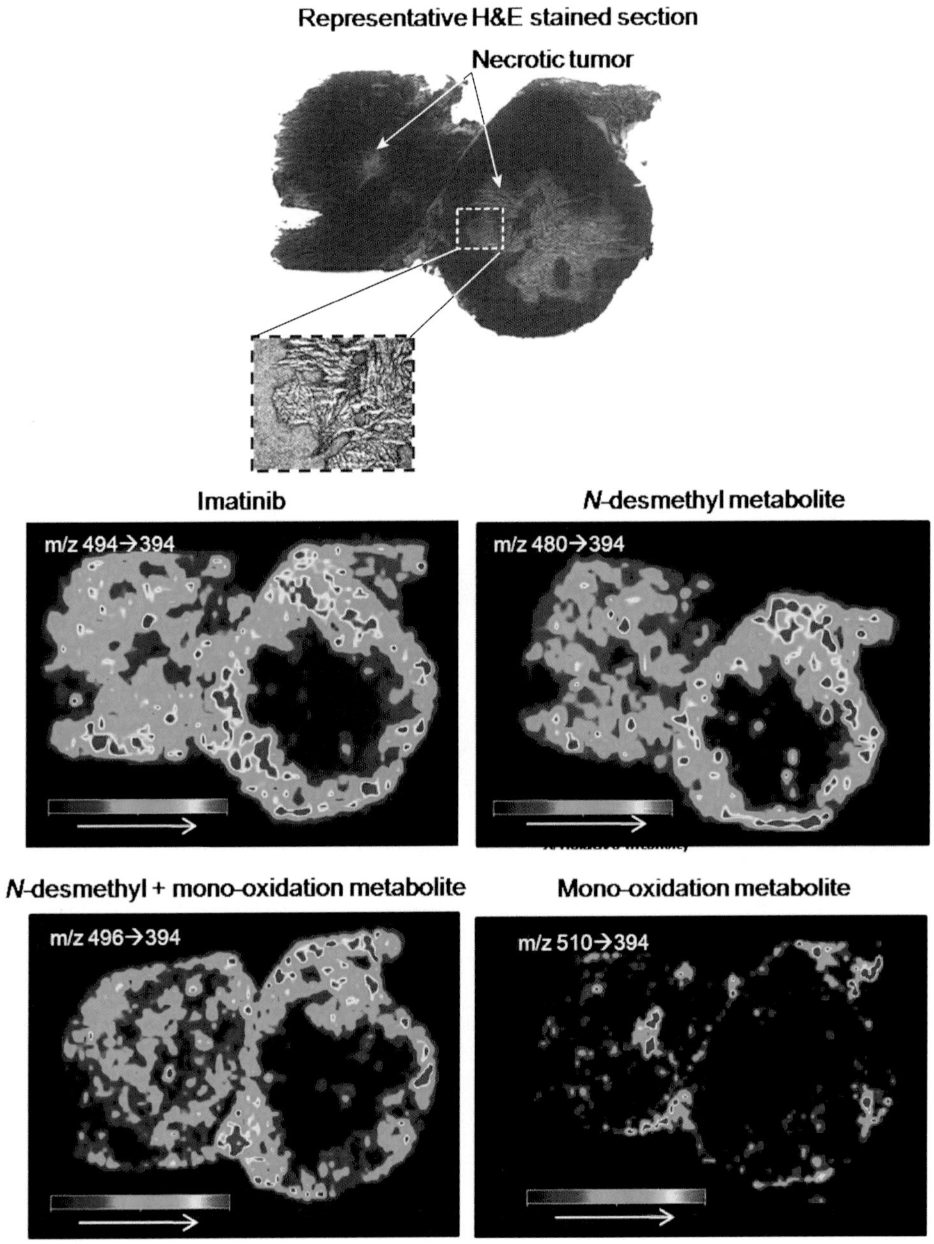

Figure 11.7 MALDI images of a parent drug and metabolite distribution in tumors. Comparison of images with the histological reference suggests that the parent drug and two of its metabolites are predominately localized to viable tumor and not as efficient at reaching necrotic regions of tumor tissue. The third metabolite has a unique distribution, primarily localized to the host tissue outside of the tumor. Reprinted with permission from Oppenheimer et al. [99].

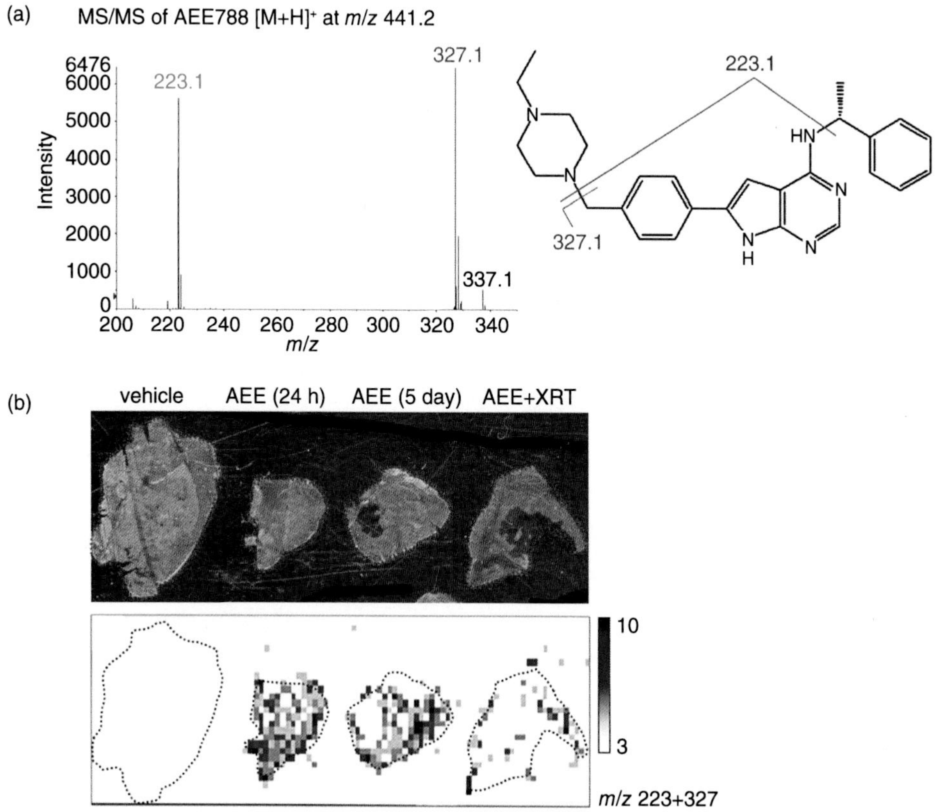

Figure 11.8 Imaging mass spectroscopy analysis of AEE788 biodistribution. (A) Mass spectrometry analysis for AEE788. Ionized peaks detected at 327 and 223 mass to charge (*m/z*) ratios represent AEE788. (B) Spatial biodistribution of AEE788 in prostate cancer xenograft sections through matrix-assisted laser desorption ionization (MALDI)-Q-TOF-MS. (Bottom panels). MALDI-MS images of frozen DU145 prostate tumor tissue sections harvested at multiple times after oral administration of AEE788 compound, as indicated: Lane 1, vehicle control treatment; Lane 2, 24 h after treatment with 25 mg/kg of AEE788; Lane 3, AEE788 (25 mg/kg) for 5 consecutive days; and Lane 4, AEE788 (25 mg/kg) + radiation therapy (XRT; 3 Gy) for 5 consecutive days. Adapted with permission from Huamani et al. [100]. Copyright 2008 Elsevier.

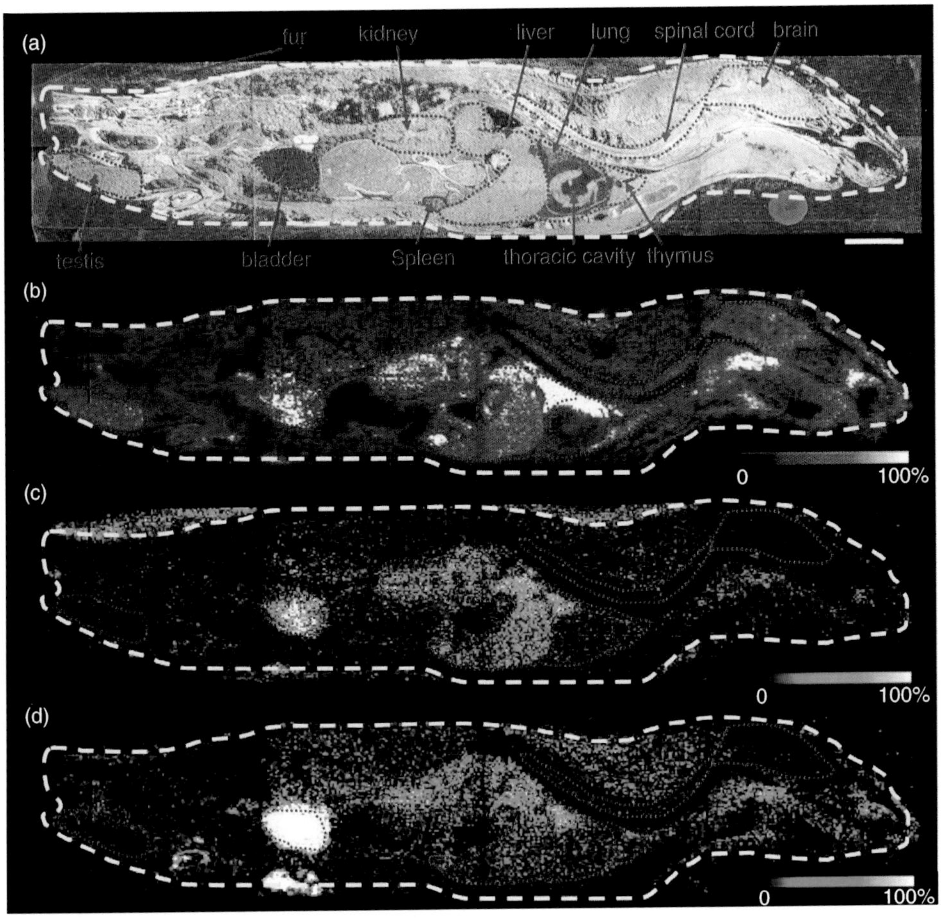

Figure 11.9 Detection of drug and metabolite distribution at 2 h postdose in a whole rat sagittal tissue section by a single IMS analysis. Optical image of a 2 h post OLZ dosed rat tissue section across four gold MALDI target plates (A). Organs outlined in red. Pink dot used as time-point label. MS/MS ion image of OLZ (*m/z* 256) (B). MS/MS ion image of N-desmethyl metabolite (*m/z* 256) (C). MS/MS ion image of 2-hydroxymethyl metabolite (*m/z* 272) (D). Bar, 1 cm. Adapted with permission from Khatib-Shahidi et al. [86]. Copyright 2006 American Chemical Society.

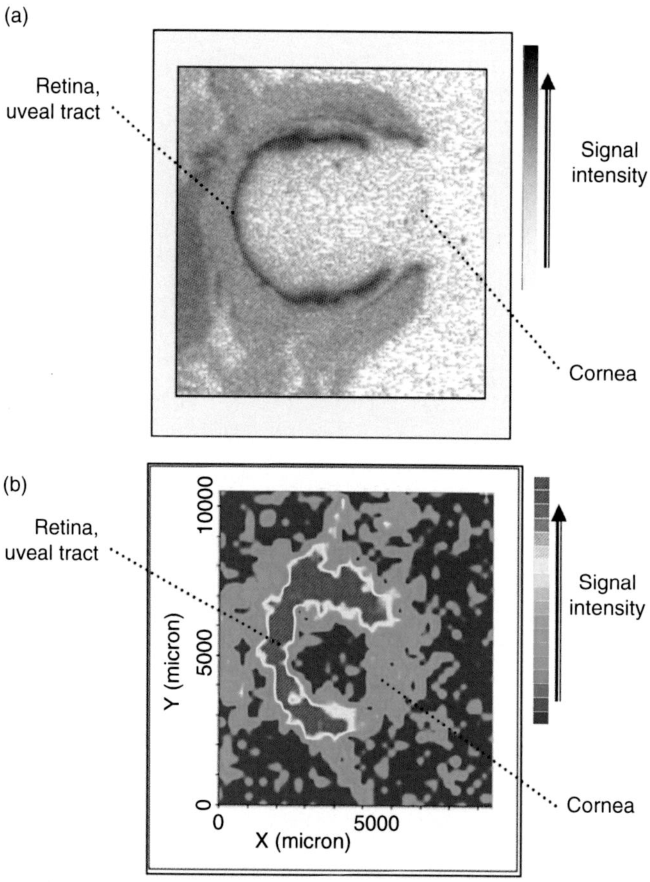

Figure 11.10 (A) Autoradiogram of an eye section indicating the preferential distribution of the radiolabeled analyte(s) to the back of the eye (retina and/or uveal tract). (B) Ion map of the drug with the highest levels present in the back of the eye (retina and/or uveal tract). Reprinted with permission from Drexler et al. [102]. Copyright 2010 Elsevier.

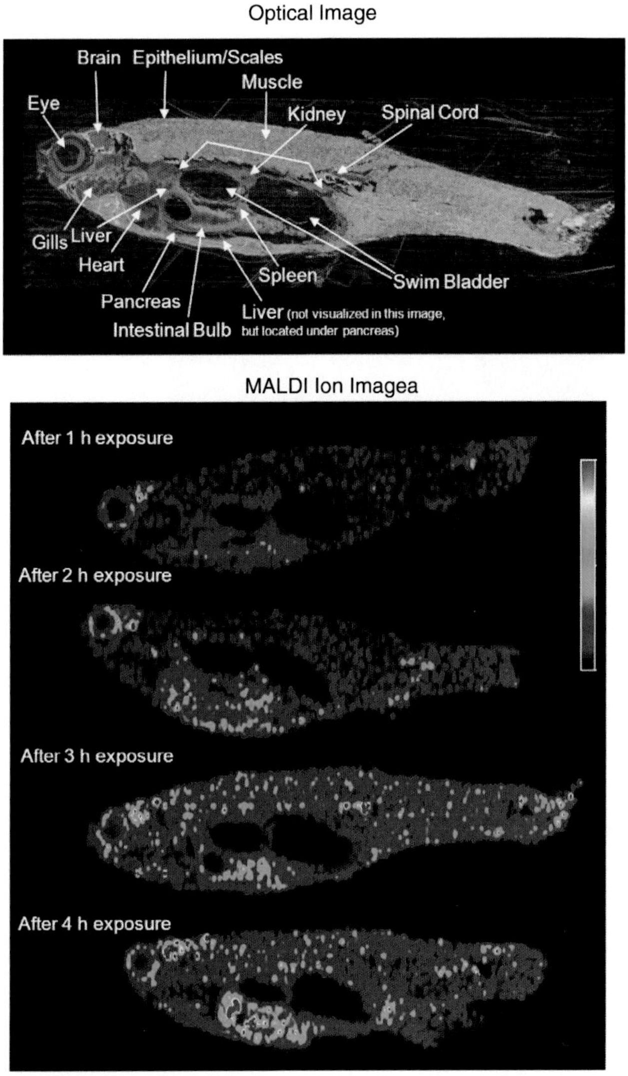

Figure 11.11 Exposure time-course study in zebrafish. Zebrafish were placed in separate holding containers for the duration of drug exposure. Fish were dosed with clozapine at 2 mg/L for 1, 2, 3, or 4 h while fasting. MALDI images were acquired with a QSTAR XL at $250 \times 250\,\mu m$ resolution. The m/z 327 → 270 transition was monitored for clozapine and the m/z 313 → 256 transition was monitored for the internal standard, olanzapine. The control zebrafish, not shown, confirmed that no endogenous signal was present at the transition monitored. Reprinted with permission from Oppenheimer et al. [103].

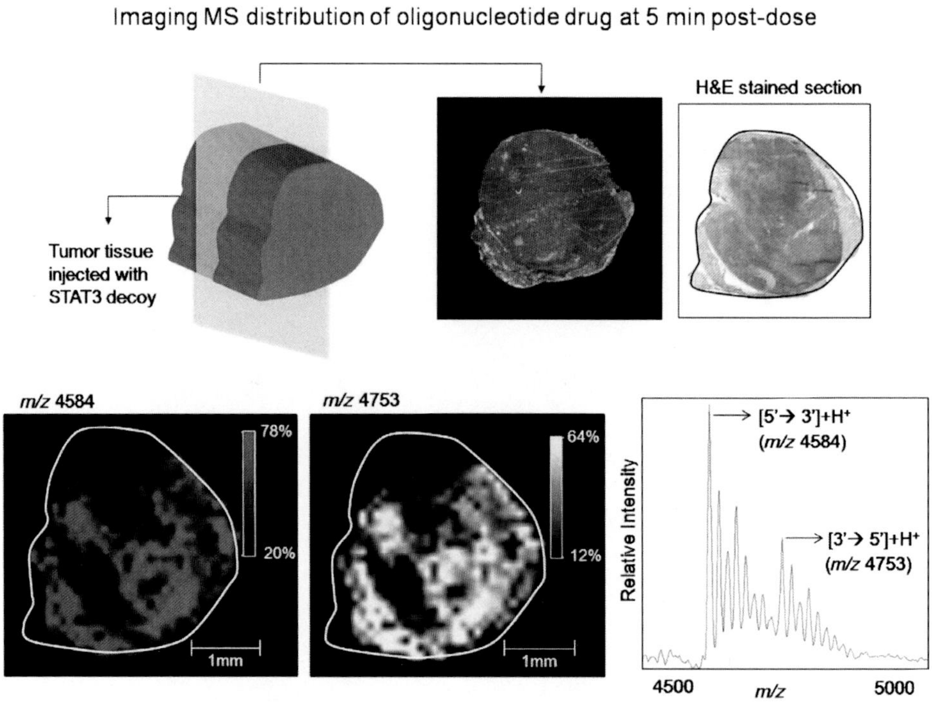

Figure 11.12 Imaging MS distribution of oligonucleotide drug at 5 min postdose. Images illustrate STAT3 decoy distribution in the tumor. The mass spectrum shows that peaks representing the single strands at *m/z* 4584 (5′-3′ sequence), and 4753 (3′-5′ sequence), for the STAT3 decoy dominate the spectrum with few ion adducts and high accuracy (mass error < 0.1 Da). Reprinted with permission from Casadonte et al. [115].

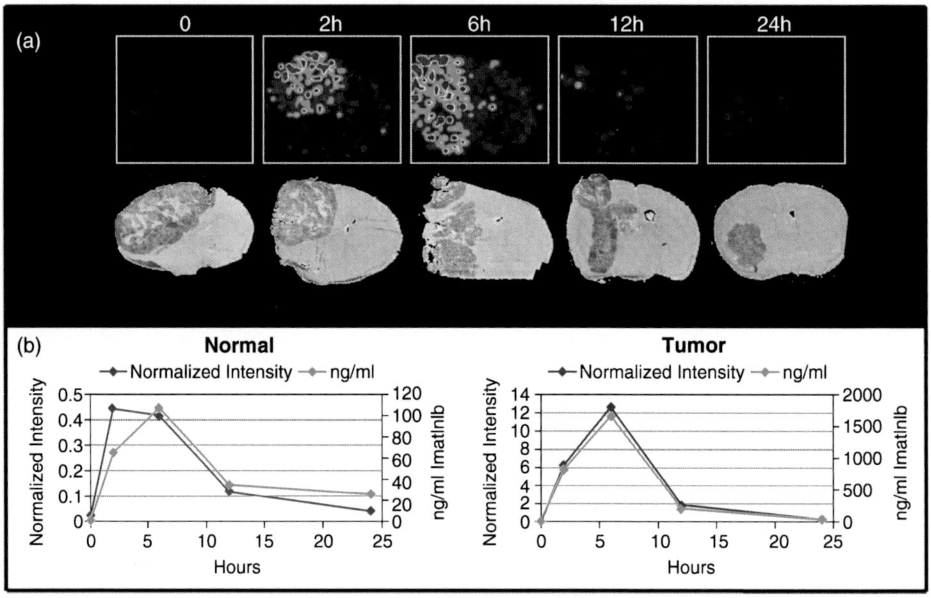

Figure 11.13 Imatinib distribution and quantitation results for imatinib-dosed, tumor bearing, mouse brains. (a) Imatinib distribution for a representative brain at each time-point post a single dose of drug (images acquired on the Thermo MALDI LTQ) with their corresponding H&E-stained section for the representative brains. (b) The average normalized drug intensity acquired by MALDI imaging at each time-point along with the amount of imatinib (ng/mL) found to be present in the tissue using traditional drug extraction techniques followed by LC-MS/MS. The MALDI and LC-MS/MS data show the same imatinib trends across the time-points for both the tumor bearing region and the normal tissue region. Reprinted with permission from Frappier et al. [120].

10

IMAGING MASS SPECTROMETRY OF PROTEINS AND PEPTIDES

MICHELLE L. REYZER AND RICHARD M. CAPRIOLI

10.1 INTRODUCTION

The introduction of electrospray ionization (ESI) [1] and matrix-assisted laser desorption/ionization (MALDI) [2–4] in the late 1980s revolutionized the study of proteins in the gas phase by mass spectrometry. This allowed accurate molecular weights to be determined up to 5 ppm with high-resolution instruments, molecular structures to be probed via hydrogen/deuterium exchange or ion mobility technology, and unknown proteins to be identified through MS/MS sequencing of peptides in enzymatic digests, all of which are now standard procedures in modern proteomics.

In the late 1990s, MALDI imaging mass spectrometry (IMS) was reported for proteins desorbed directly from thin tissue sections [5–7]. Prior to this, imaging using mass spectrometry was primarily performed with secondary ion mass spectrometry (SIMS), which could achieve high-resolution images (sub-μm), but only for relatively low molecular weight compounds (<2000 Da), typically of elemental ions and fragment ions [8]. MALDI allowed high molecular weight peptides and proteins to be visualized *in situ* for the first time by mass spectrometry.

In the intervening years, significant progress has been made in optimizing and advancing imaging technology. From instrumental improvements, protocol developments, integration of novel technologies, and an expanding diversity of applications, IMS is making a significant impact on the study of proteins and peptides. This chapter will provide a summary of the use of IMS for protein and peptide assays, with an emphasis on more recent applications and developments.

Mass Spectrometry for Drug Discovery and Drug Development, First Edition. Edited by Walter A. Korfmacher.

10.2 METHODOLOGY

IMS can be used effectively in two ways: full tissue image analysis and targeted profiling. Full tissue imaging can produce a spatial resolution of 10–200 μm for individual signals across the entire tissue sample. Individual mass spectra are acquired in an ordered array over the tissue surface and the intensity of each ion may be plotted as a function of its position. This enables a visual examination of any signal detected in the mass spectrum over the entire surface of the imaged area. Many hundreds of images can thus be recreated in a single experiment. Analyte localizations unique to sample histology (i.e., epithelium vs. stroma) or sample type (i.e., diseased vs. control) may quickly be assessed. In addition, analytes that specifically colocalize or show inverse localizations may also be visually determined. In the profiling mode, mass spectra are acquired from discrete user targeted areas on the tissue surface that are highly enriched in specific cell types (normal epithelium, tumor epithelium, stroma, etc.) [9]. These spectra are then processed and subjected to statistical analysis in order to determine differences between or among different groups.

The mode of analysis chosen will ultimately depend on the nature of the samples and the question that the experiment is designed to answer. Imaging experiments may be composed of tens of thousands of individual pixels (depending on sample size and spatial resolution), while profiling may only sample 10–50 spots per sample. The experimental question being asked will best dictate which mode is the most efficient and effective for a particular sample.

10.2.1 Sample Preparation

Thin sections of tissue (~3–20 μm) are cut on a cryostat and thaw-mounted onto a MALDI target plate or microscope slide, depending on instrument configuration and compatibility. The sample may then be subjected to washing, fixing, and/or staining protocols. Next, matrix is applied either as discrete matrix spots or as a homogeneous coating as described later. Mass spectra are acquired over the desired areas and spectral preprocessing and image generation are performed [10]. Specific considerations for each step are detailed in the following sections.

10.2.1.1 Sample Pretreatment Sample pretreatment, such as immersing the tissue sections in alcohol or organic solvents, serves not only to preserve tissues but also to improve the signal response for proteins and peptides [10–12]. Washing tissue sections with isopropanol after thaw-mounting has been shown to preserve MALDI signal quality for up to 7 days [11]. Organic solvents may be used to remove abundant lipids from tissues [12], while alcohol-based solvents remove both lipids and salts [11] that can interfere with matrix crystallization as well as induce ion suppression. Water washes have been used to improve the detectability of hydrophobic, membrane-bound proteins in lens tissues by removing the more abundant, water-soluble crystallin proteins [13, 14]. Improving the solubility of proteins, especially high mass proteins (~25–100 kDa), by addition of detergent or other solvents (such as hexafluoroisopropanol or hydrogen peroxide), has also been shown to improve their detection from tissue sections [15–17].

Sample pretreatment must be undertaken with care. While improvements in the number of signals detected and their intensities have been reported for proteins after washing, immersion of a tissue section in liquid may also induce significant delocalization or solubilization of analytes. The sample preparation for every tissue and analyte must therefore be uniquely optimized to ensure that spatial integrity is maintained while signal intensity is maximized.

10.2.1.2 Fresh Frozen or Minimally Fixed Tissue Typically fresh frozen samples give optimal results; the samples are excised and immediately immersed in liquid nitrogen. This significantly slows *in situ* enzymatic activity while preserving the native shape of the tissue. Tissues may then be stored at –80°C until analysis. However, flash freezing is not always sufficient to preserve the molecular integrity of all samples. For example, it has been shown that proteins and peptides in the brain can undergo significant degradation in the amount of time it takes to excise and freeze the specimens [18–20]. In these cases, extra care must be taken when first handling the sample, as enzymatic activity may significantly alter the protein content. Protein degradation results not only in the loss of intact proteins, but also in an increase of proteolytically produced peptides. Changes in authentic post-translational modifications may also result from *ex vivo* enzymatic activity.

Svennson et al. [19] reported on the results of a novel heat stabilizing instrument to retard proteolytic degradation in mouse brain tissues. Snap frozen tissue was compared with heat-stabilized tissue or *in vivo* fixed tissue (via microwave irradiation). The number of peptides detected (via nano-liquid chromatography (LC)-ESI MS) from extracted snap frozen tissue was ~30% higher than the number detected from heat-stabilized tissue, and ~45% higher than microwave-stabilized or *in vivo* fixed tissue, indicating considerable proteolytic degradation of the tissue prior to analysis [19]. In addition, a significant alteration in phosphorylated peptides was detected in snap frozen tissues compared with heat-stabilized or *in vivo* fixed tissue.

Recently, Goodwin et al. [20] expanded on these results with a full IMS analysis of mouse brain either with or without heat stabilization. Figure 10.1 shows the degradation of a signal at *m/z* 6723 after only 2 min at room temperature, compared with the heat-stabilized sections, which are shown to be constant up to 5 min at room temperature. Indeed, the heat-stabilized sections showed no loss in the intensity of *m/z* 6723 even after 20 min. A similar effect was reported for peptides, where an increase in peptides was observed in nonstabilized sections with time. These effects can be difficult to uncover, especially after experimentally inducing enzymatic digestion by addition of trypsin. Nonetheless, the approach of heat stabilization has been shown to significantly reduce proteomic degradation, especially for the study of neuroproteomics.

10.2.1.3 Fixed Tissue Proteolytic degradation may also be reduced by classical fixation procedures, whereby excised tissue is immersed in a solution of fixative. Formalin (formaldeyhde in phosphate buffered saline) is the most commonly used fixative used to preserve molecular integrity and maintain tissue architecture. However, because it acts by cross-linking proteins and induces other chemical modifications [21], it is not directly amenable to MALDI protein profiling.

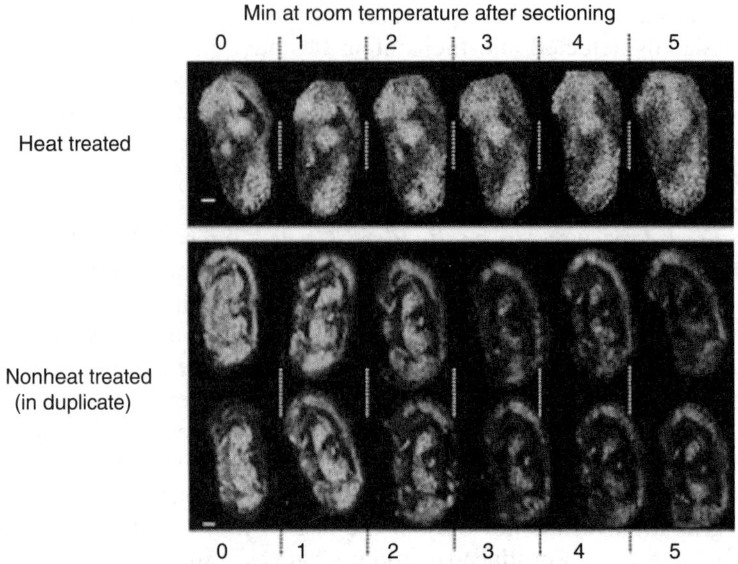

Figure 10.1 See color insert. Effect of heat stabilization on the intensity of *m/z* 6723.5 from mouse brain tissue. The intensity of *m/z* 6723.5 remains constant at 0–5 min at room temperature after sectioning, while the same signal decreases rapidly in the snap frozen sections after only 2 min at room temperature. (Adapted from Reference 20 with permission. Copyright Wiley-VCH Verlag GmbH & Co. KGaA.)

Nonetheless, formalin-fixed tissue sections have been analyzed by profiling after the addition of an enzyme (typically trypsin) to tissue sections [22–24], whereby peptides produced by enzymatic digestion of the cross-linked proteins are readily detected. This will be discussed in more detail in Section 10.4.

Additionally, alternative fixatives that do not cross-link proteins have been examined for their suitability for both tissue preservation and IMS. Ethanol alone [25] and two non-cross-linking alcohol-based fixatives (RCL2/CS100 [26] and PAXgene [27]) have been evaluated for their compatibility with MALDI imaging. Protein profiles obtained from tissues fixed with ethanol, RCL2/CS100, or PAXgene fixatives were shown to be quite similar to fresh frozen samples.

10.2.2 Matrix Application

Matrix added to tissue sections in preparation for MALDI-MS analysis serves several purposes, including extracting analytes, absorbing the energy of the laser, and aiding analyte desorption and ionization (as a source of protons). Traditional MALDI matrices, including 2,5-dihydroxybenzoic acid (DHB), a-cyano-4-hydroxycinnamic acid (CHCA), and sinapinic acid (SA), are commonly dissolved in an organic:water mixture and applied to the surface of the tissue. The several matrices commonly used show different affinities for different analytes, with CHCA usually used for peptides, DHB for lipids, and SA for proteins. Depending on the identity of the matrix and the composition of the solvent, different analytes may be preferentially extracted and co-crystallized. Thus, the choice of matrix parameters as well as how the matrix is applied has a substantial impact on the quality and integrity of the resulting image.

Matrix can be applied in basically two ways, depending on the experiment to be performed. For imaging analyses, a homogeneous coating of matrix is desired so that the diameter of the laser beam on target determines the spatial resolution. This can be done manually with an airbrush [28] or glass nebulizer [10], or automatically with one of several commercially available instruments (such as the ImagePrep by Bruker (Billerica, MA) [29], the TM Sprayer by HTX Technologies (Carrboro, NC), or the SunCollect by SunChrom (Friedrichsdorf, Germany) [30]). The basic approach is to apply the matrix as a fine mist/spray in several cycles, letting the matrix crystals build up over time, while minimizing how wet the tissue gets in order to reduce delocalization.

Matrix may also be deposited on tissues in discrete areas, which form well-defined spots when dry. This also may be done manually (e.g., with a pipette) or with a commercially available instrument (such as the Portrait 630 by Labcyte (Sunnyvale, CA) [31] or the ChIP by Shimadzu (Columbia, MD) [32]). Application of droplets typically makes the tissue wetter, leading to better extraction of analytes, but at the same time limits extraction to within the matrix spots, thereby limiting delocalization. Commercially available spotters allow spots to be precisely placed over defined, targeted areas for profiling studies. An array of spots may be printed over an entire tissue section and spectra acquired at each spot. Images may then be reconstructed in basically the same manner as for a homogeneous coating. The advantages of a microspotted array for imaging (compared with a homogeneous coating) are improved spectral quality and well-defined analyte localizations. The main disadvantage is that with current technologies the typical spot size is ~100–200 μm (depending on the final droplet diameter, tissue, number of passes, etc.). Spotters utilizing fine capillaries are susceptible to clogging, and in some cases, the small volume regime requires further refinement and optimization of matrix and solvent conditions.

One example of using a spotted array for imaging is shown in Figure 10.2. In this case, a Portrait 630 acoustic reagent microspotter was used to deposit sinapinic acid on a whole-body section of a mouse pup. Individual droplets of ~170 pL were deposited over the entire section in one cycle, with the pattern repeated 30 times. This resulted in ~5 nL matrix deposited per spot and final spot diameters of ~180 μm. The spot-to-spot spacing (and the resulting image resolution) was 200 μm. As shown, many protein signals were detected (up to *m/z* 15,181) in unique localizations over the section [31].

10.2.3 Instrumentation

MALDI typically forms intact, singly charged ions, and consequently, high molecular weight proteins form high *m/z* value ions. These ions are best analyzed by time-of-flight (TOF) mass spectrometers that in general have relatively high sensitivity. TOF/TOF analyzers are commonly used for MALDI analyses of peptides and proteins, as they may be used to generate sequence information via either high-energy collision-induced dissociation (CID) [33, 34] or laser-induced fragmentation [34]. Other mass analyzers have been successfully coupled to MALDI sources for imaging of lower molecular weight compounds, such as drugs, metabolites, and peptides, and include linear ion trap, Fourier transform-ion cyclotron resonance (FT-ICR), and hybrid-quadrupole-time-of-flight (QqTOF) instruments.

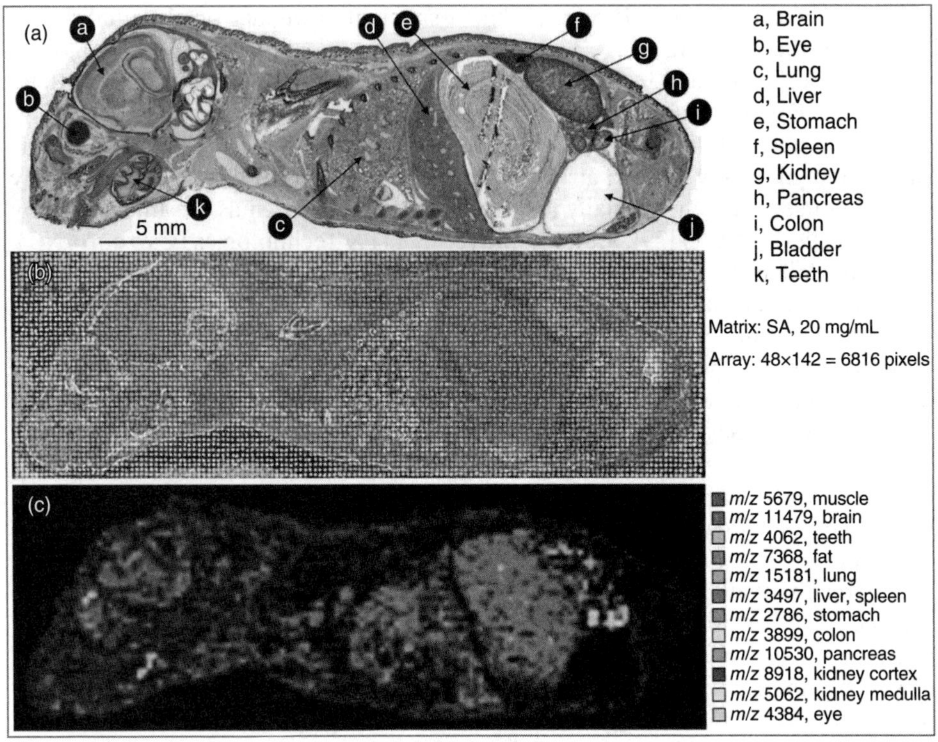

Figure 10.2 See color insert. Imaging of proteins from a spotted matrix array on a whole-body mouse pup section. (A) H&E stained section showing many different organs throughout the section. (B) Optical picture of a serial section with the matrix array printed on top. A Labcyte Portrait 630 was used to deposit sinapinic acid in a 200 μm spot-to-spot array over the entire section. A total of 6816 spots were printed. (C) Overlay of 12 organ-specific protein images obtained from the section. (Adapted from Reference 31 with permission from Springer Science+Business Media.)

Ion mobility instruments have also been utilized for IMS experiments. Ion mobility separates ions based on their collisional cross-section; thus, isobaric ions with distinctly different structures may be separated. One example of this has been shown for an on-tissue tryptic digest of a rat brain. Two peptides with the nominal mass of *m/z* 1039 show different distributions in the brain section after ion mobility separation and subsequent fragmentation. As shown in Figure 10.3, the peptides correspond to tubulin, which has a higher ion mobility, and ubiquitin, which has a lower ion mobility. The resulting images show distinctly different localizations, which would have been superimposed based solely on separation based on their nominal *m/z* values [35].

Traditionally, MALDI analyses have been performed with nitrogen lasers, which can be pulsed at up to ~60 Hz. Since one laser shot typically results in one spectrum, the speed of analysis is limited by the repetition rate of the laser. Recently, solid-state lasers (Nd:YAG, Nd:YLF) capable of being pulsed at up to 1 kHz have been introduced with commercial TOF instruments [36]. In addition, one report described the incorporation of an Nd:YVO_4 laser, capable of being pulsed at 20 kHz, to a

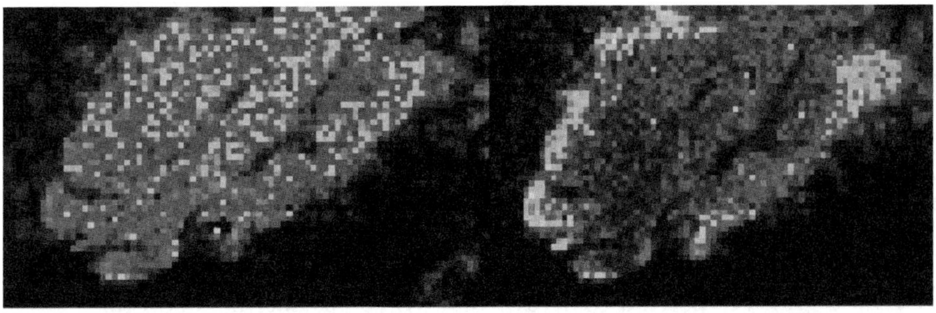

Figure 10.3 See color insert. Ion mobility separation of isobaric tryptic peptides from formalin-fixed paraffin embedded rat brain sections. Ions with nominal mass *m/z* 1039 were fragmented after ion mobility on a Waters Synapt Q-TOF instrument. Tubulin-related peptides and fragments had a higher ion mobility and showed a different distribution from ubiquitin-related peptides and fragments. (Adapted from Reference 35 with permission from Elsevier.)

hybrid QqTOF instrument [37]. Trim et al. used this instrument for the acquisition of peptide images from a section of pancreatic tumor tissue after *in situ* tryptic digestion. With the laser running at 5 kHz, images were acquired over the entire section in just 1 h 30 min, compared with the 8–10 h it would take with the 200-Hz laser ordinarily utilized with the instrument. Spraggins et al. [38] described a prototype MALDI-TOF instrument equipped with an Nd:YLF laser for high-speed imaging. Lipid images of sagittal rat brain sections at 100-µm spatial resolution were obtained in about 10 min using continuous laser raster sampling using a 3-kHz laser repetition rate. High-speed lasers thus have the ability to significantly reduce image acquisition time.

10.2.4 Protein Identification

A consequence of imaging intact proteins by IMS is that the direct identification of the detected signals cannot be done. It is difficult to match an empirically obtained molecular weight (functional protein) to a theoretical molecular weight (predicted protein) in a database without additional information. Proteins detected *in situ* may be processed, cleaved, or modified, or they may contain sequence errors. Because these modifications result in a change of molecular weight, additional information is required for unequivocal protein identification.

Conventional proteomic experiments may be utilized in order to identify proteins of interest, including tissue homogenization, chromatographic or electrophoretic separation, MS analysis of the fraction of interest, further purification of that fraction, enzymatic digestion, and peptide fingerprinting via MS (ESI or MALDI) and/or further MS/MS sequencing of unique peptides [39]. This approach is fairly time-consuming and may not always provide the identification of a targeted protein of interest. Further, even if an identification is made, matching that protein to a molecular signal obtained via MALDI of a tissue section is not straightforward. Additional validation procedures, such as immunohistochemistry (IHC), may still be warranted [40].

There are several approaches to the identification of protein signals directly from tissue sections. Direct MS/MS may be performed on some low molecular weight ions of interest (up to ~4–5 kDa) in order to obtain sequence information. In-source decay (ISD) uses the energy of the laser in conjunction with an "ISD favorable" matrix to produce fragments in the source region. Because the fragments are produced in the source, they appear at their correct *m/z* values in the TOF (as opposed to postsource decay, PSD). While this does not allow for precursor ion selection, it does promote the formation of c- and z-series ions from which a protein sequence can be obtained. This approach has recently been used in conjunction with IMS to identify proteins directly from tissue sections [41]. For example, Figure 10.4 shows several images of protein fragments obtained from a porcine eye lens section after ISD that were found to exhibit similar localizations. It was assumed that they came from the same protein. The mass spectrum shown in Figure 10.4 was obtained from a high-intensity pixel in the images and the high-intensity signals were sequenced and found to correspond to c- and z-ions of the protein γ-crystallin B.

Another approach involves application of an enzyme directly to the tissue section to perform *in situ* digestion [32, 42]. Trypsin is commonly used, and this has been shown to be successful, not only for the identification of proteins, but also for the detection of high molecular weight or more hydrophobic proteins that may not be directly amenable to MALDI analysis. This will be discussed further in Section 10.4. Recently, a derivatization agent was reportedly added to the matrix in order to direct the fragmentation to make the spectra easier to interpret [43]. These strategies have been successful; however, they do result in the digestion of many proteins and still can produce ambiguous results. This is an active area of research, and further improvements will substantially benefit the field.

10.3 INTACT PROTEIN ANALYSIS: APPLICATIONS

10.3.1 Anatomy and Development

The application of IMS to histology is a natural fit, in effect allowing a molecular atlas of a given tissue to be created. Images of protein distribution may be correlated with known morphology or function, and further studied as a function of time. For example, the expression of secreted proteins along the mouse epididymis has been studied in this way. Immature spermatozoa travel along the epididymis and interact with the protein-rich epididymal fluid in order to complete their maturation. IMS revealed the spatially localized distribution of a number of proteins along the length of the mouse epididymis, thus helping to elucidate this process [28, 44]. A similar report applied IMS to the study of rat spermatogenesis in the testes. Differential protein expression in the seminiferous tubules was found and correlated to different stages of germ cell development [45].

Protein changes during embryo implantation in mouse uterus has been studied using IMS [46]. The proteomic changes involved during embryo implantation were

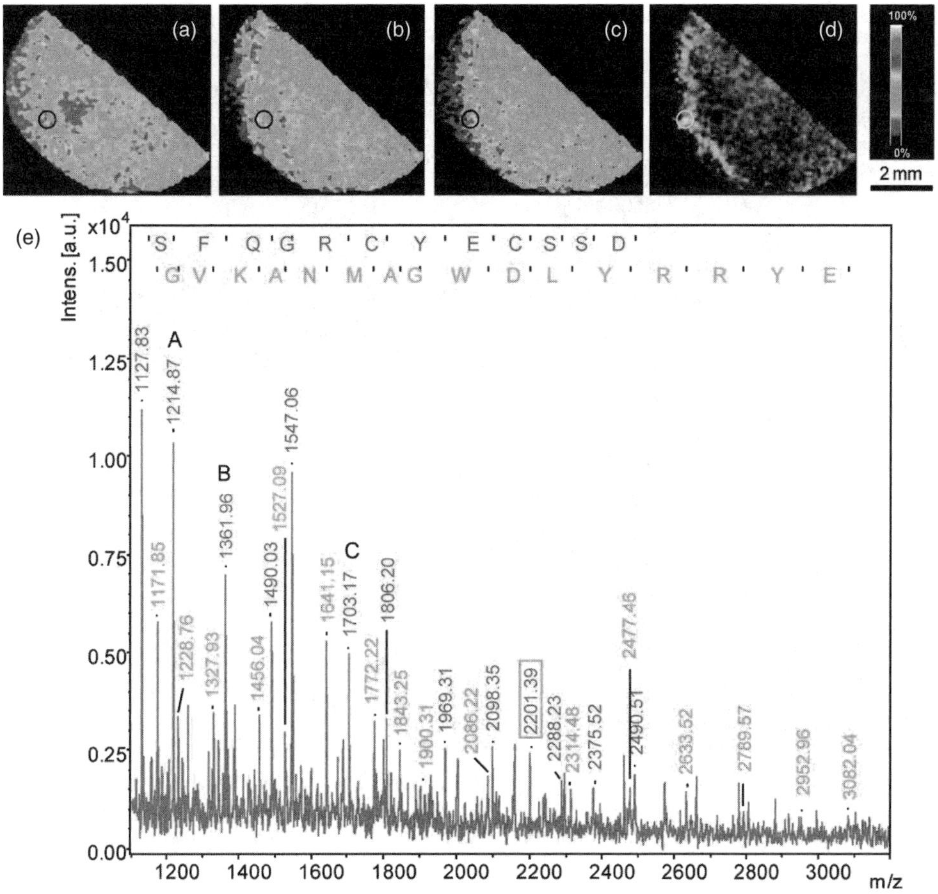

Figure 10.4 See color insert. Imaging and identification of proteins from in-source decay (ISD) imaging analysis. Images of (A) *m/z* 1214.87, (B) *m/z* 1361.96, (C) *m/z* 1703.17, and (D) an overlay obtained directly from a porcine eye lens section in ISD mode. (E) An ISD mass spectrum from a high-intensity pixel on the lens section highlighting the c-ion (red) and z-ion (green) series. The protein was identified as γ-crystallin B. (Adapted with permission from Reference 41. Copyright 2010 American Chemical Society.)

found to be dramatic over a 4-day period within the implantation site and also compared with the interimplantation site (site on the uterus with no embryo). The distributions obtained for ubiquitin, calcyclin, calgizzarin, and transthyretin were found to be similar to their respective mRNA expressions by *in situ* hybridization experiments, thus validating the approach [46].

Grey et al. [47] reported on the visualization of the molecular morphology of the chick heart using IMS. Tissue sections from term chicks were obtained and coated with either CHCA, DHB, or SA. Different protein signals were observed from each preparation, due to both the identity of the matrix and the different anatomical locations from which the sections were obtained. Figure 10.5 shows several images of proteins found to localize to different morphological regions. For example, *m/z* 6614 localizes to the myocardium, while *m/z* 7475 and *m/z* 6143 localize to different

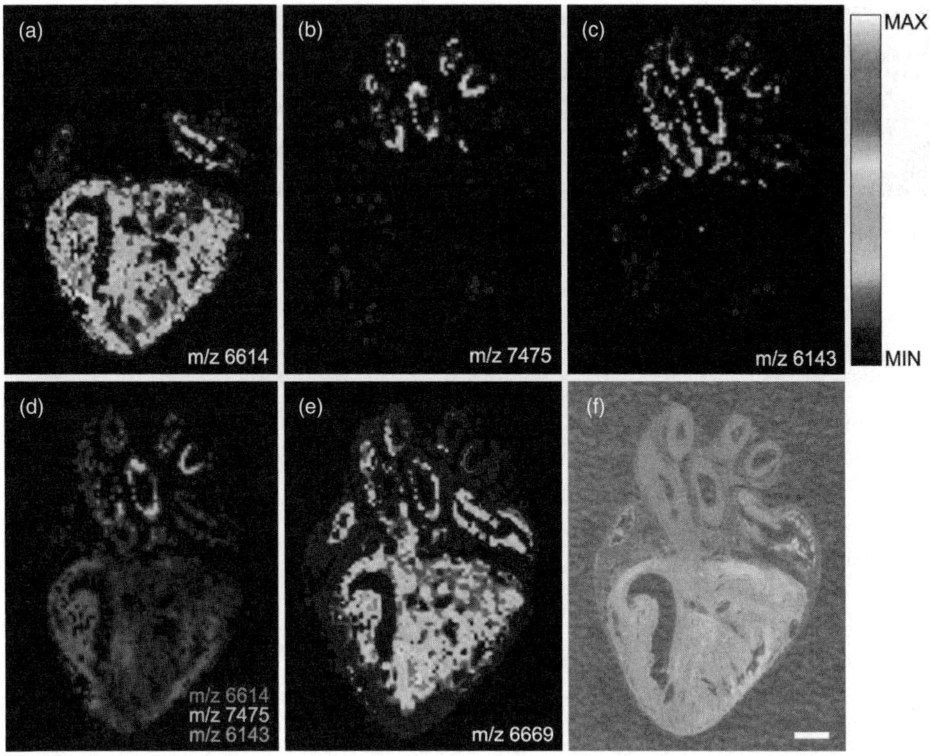

Figure 10.5 See color insert. Imaging of proteins from a chick heart with DHB. Images of (A) *m/z* 6614, (B) *m/z* 7475, (C) *m/z* 6143, (D) an overlay of *m/z* 6614, 7475, and 6143, and (E) *m/z* 6669 show differential localization within the heart tissue. (F) Optical image of heart section prior to matrix deposition. Scale bar = 1 mm. (Reproduced from Reference 47 with permission.)

regions of the vasculature. In contrast, the signal at *m/z* 6669 is present throughout the chick heart. These lower molecular weight signals are not often detectable by commercially available antibodies, which are typically targeted to much higher molecular weight proteins. MALDI imaging may provide a useful complement for the study of this subset of lower molecular weight proteins (<30,000 Da).

IMS was recently used to redefine a small region of the brain called the claustrum [48]. The claustrum is a prominent but ill-defined forebrain structure thought to be involved with integration of multisensory information; however, its vague borders have made experimental assessment of its functions difficult. IMS revealed a protein that appeared enriched in the claustum but not in adjacent structures. The results are shown in Figure 10.6. Figure 10.6A,B shows the shape of the claustrum as currently defined in red (in Paxino and Watson's atlas) at two different levels of the brain. Figure 10.6C,D shows the ion images for *m/z* 7725 showing its localization in the claustrum corresponding to bregma +2.04 but not at bregma +3.00. This protein was subsequently identified (via traditional tissue homogenization, fractionation, and tryptic digestion) as G-protein gamma2 subunit (Gng2). Figure 10.6E,F shows the IHC validation of the MS data, showing Gng is enriched in the claustrum only at striatal levels (F) and not at frontal levels (E) [48]. One advantage illustrated by

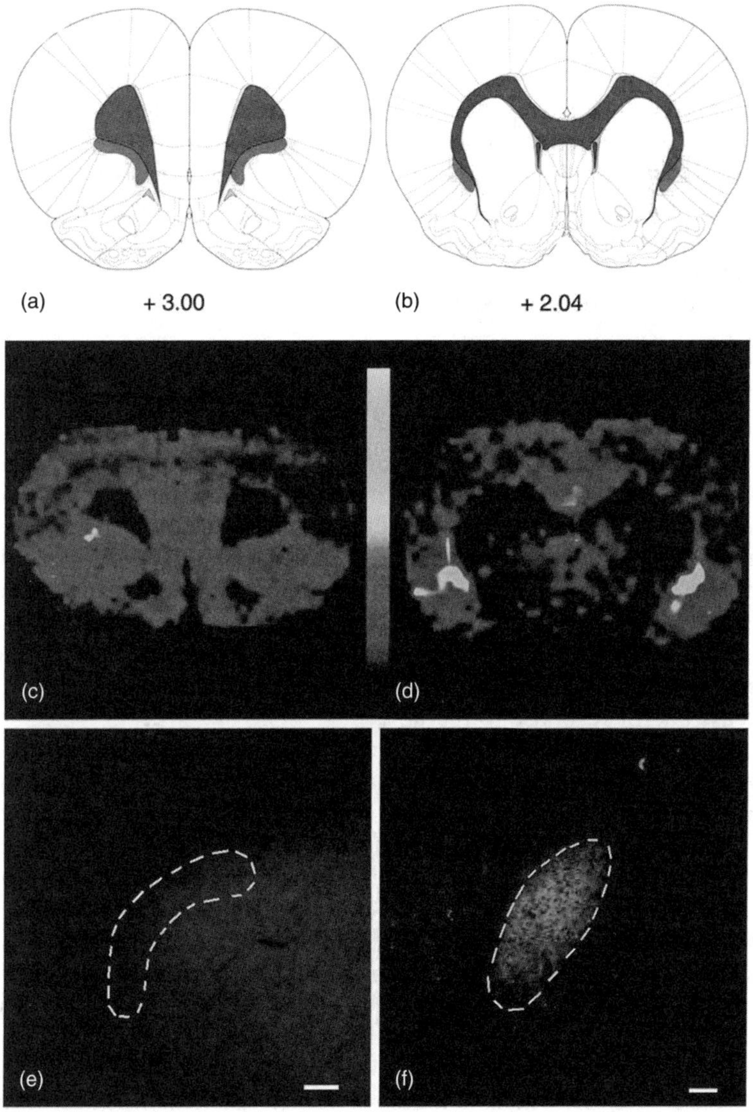

Figure 10.6 See color insert. Imaging of proteins from a rat brain reveals a protein enriched in the claustrum that serves to redefine its boundaries. (A) and (B) show the claustrum as currently defined (in red) in two brain levels from a rat brain atlas (Paxinos and Watson). Ion images of *m/z* 7725 are shown at the frontal level (C) and at the striatal level (D) indicated in (A) and (B). (E) and (F) show IHC localization of Gng2 at the same two levels. (Reproduced from Reference 48 with permission from Oxford University Press.)

this example is that IMS is an excellent discovery tool, where knowledge of discriminating proteins is not required in advance.

The distribution of lens proteins has also been investigated with IMS [14, 49–51]. Crystallin proteins (α, β, and γ) make up the bulk of proteins found in the lens. They are known to undergo many modifications with age (including truncation and phosphorylation) as there is no protein turnover in differentiated lens fibers [50]. The distribution of intact αA- and αB-crystallin have been determined in calf and adult bovine lens tissue, along with many age-related truncation and phosphorylation products [49]. Similar studies have been performed with rabbit lenses [51]. Age-related changes in the distribution of α-crystallin products in human lens were also studied [50]. A marked reduction in the presence of intact α-crystallin was noted in the nucleus of the lens in 51-and 75-year-old lenses, compared with 7- and 29-year old lenses. In addition, the localization of the singly and doubly phosphorylated forms of αB-crystallin moves from the nucleus to the periphery of lenses with age. This information will further our understanding of crystallin protein processing and how it may affect age-related vision changes.

10.3.2 Cancer

One of the first applications of IMS was to cancer biology [52]. Protein distributions in sections of a human glioblastoma tumor from a mouse xenograft were determined. Areas of tumor proliferation, ischemia, and necrosis were analyzed [52]. Further early studies utilized mouse models of tumors, including prostate cancer and colon cancer, to assess proteomic changes in cancer development [53, 54]. In all cases, proteomic differences were observed between normal and tumor tissue, which likely reflect the alterations that occur as cells undergo a neoplastic transformation. Proteins that are involved in this process may thus be further studied in order to better understand the biology, to find novel therapeutic targets, and to find biomarkers of disease and/or response to therapy.

These early investigations were followed up by those involving human cancer tissue. Two early studies involved lung cancer [55] and brain cancer (glioma) [40, 56]. In both cases, protein profiles were obtained from a series of tissues from patients undergoing resective surgery for lung or brain cancer. Statistical analysis of the data resulted in a series of signals that could distinguish nontumor from tumor tissue, as well as classify different subtypes or grades of tumors. In both cases, protein profiles were also correlated with patient survival, where a set of signals was able to distinguish a group of patients with poor prognosis from a group with good prognosis, independently of other factors (such as tumor grade and patient age). While most tumors are accurately classified by grade and stage based on histology, generally prognostic indications are not directly observable by a pathologist. Classifications based on molecular features, whether for diagnostic or prognostic purposes, such as those derived from IMS profiling, have tremendous clinical potential.

The work that resulted from the analysis of solid lung tumor tissue has been extended so that useful protein profiles can be obtained from fine needle aspirates from lung cancer patients [57]. In addition, it has been used to discover proteomic signatures that classify survival of lung cancer patients after targeted chemotherapy [58] and to determine its applicability for noninvasive screening of unfractionated serum [59].

The application of imaging mass spectrometry to cancer research is continuing to expand, as evidenced by several recent reviews that summarize the current state-of-the, art as well as highlight various applications and technological developments [60–62]. In addition to the early lung and glioma studies, IMS experiments have been reported on human breast [9, 63], ovarian [64–66], prostate [42, 67], renal [68], liver [69], gastric [39], sarcoma [70, 71], and oral cancers [72].

There is a great deal of interest in using the technology for the discovery of biomarkers which may allow clinicians to detect tumors at an earlier stage and better characterize individual tumors in terms of prognosis and likely response to therapy. Several biomarker candidates have been discovered through IMS experiments. The neutrophil defensins have been found to be upregulated in several different cancers [39], and other proteins including the S100 family of calcium-binding proteins and thymosin β 4 have also been found upregulated in cancer tissues. Defensins are known to be upregulated as a function of inflammation [73]. In another study, 48 human ovarian tissue samples (25 cancer, 23 benign) were found to have a signal at *m/z* 9744 that was present in ~80% of cancer samples compared with normal [66]. This signal was subsequently identified as the reg alpha fragment of the 11S proteasome activator complex, and it was further validated by IHC and Western blotting.

Another investigation involved 75 human prostate tissue samples (34 cancer, 41 uninvolved) where a marker at *m/z* 4355 was found that could not only distinguish normal from uninvolved tissue, but also appeared to be reduced in higher state or grade disease [42]. This signal was identified as a fragment of MEKK2, a MAP3K protein that regulates the MEK5/ERK5 pathway. This signal was found to be highly expressed in the cancer tissue compared with adjacent uninvolved tissue. As many clinical decisions are based on pathological grading or staging of tumors, the discovery of an objective molecular marker for establishing tumor grades could have significant clinical utility.

IMS was recently shown to be able to classify the HER2 status of human breast cancer tissues [29]. A total of 48 breast cancer tissues were analyzed, with 30 used as a discovery set and 18 as a training set. IHC and fluorescence *in situ* hybridization (FISH) were used to classify HER2 status prior to IMS. Spectra derived from specific HER2-positive or HER2-negative tumor cell populations revealed seven ions that could distinguish HER2 status. The distribution of one significant signal, *m/z* 8408, is shown in Figure 10.7. Figure 10.7A depicts serial sections of an HER2-positive sample, while sections of an HER2-negative sample are shown in Figure 10.7B. As shown, two ions at *m/z* 4969 (blue) and *m/z* 6225 (yellow) are present in both tissue sections and do not discriminate between HER2 status. In contrast, the signal at *m/z* 8408 (red) is widely present in tumor cells of the HER2-positive tissue, but absent from the HER2-negative case. A serial section of both tissues that was subjected to HER2 IHC staining is also shown. This signal was subsequently identified as cysteine-rich intestinal protein 1 (CRIP1) [29].

Several recent reports have described the use of IMS for the study of tumor margins [64, 68]. Local recurrence of a cancer after surgical removal of the main tumor is a significant clinical issue. This suggests that the histologically normal-appearing tissue remaining after resection has already been molecularly compromised. Analyses of renal cell carcinoma tissues found a number of signals that maintained reduced tumor levels well beyond the visible histological margin. These included several cytochrome c oxidase polypeptides, cytochrome c, and NADH-ubiquinone oxidoreductase MLRQ subunit [68]. Many of these proteins are involved in the mitochondrial electron transport system, the deficiency of which has

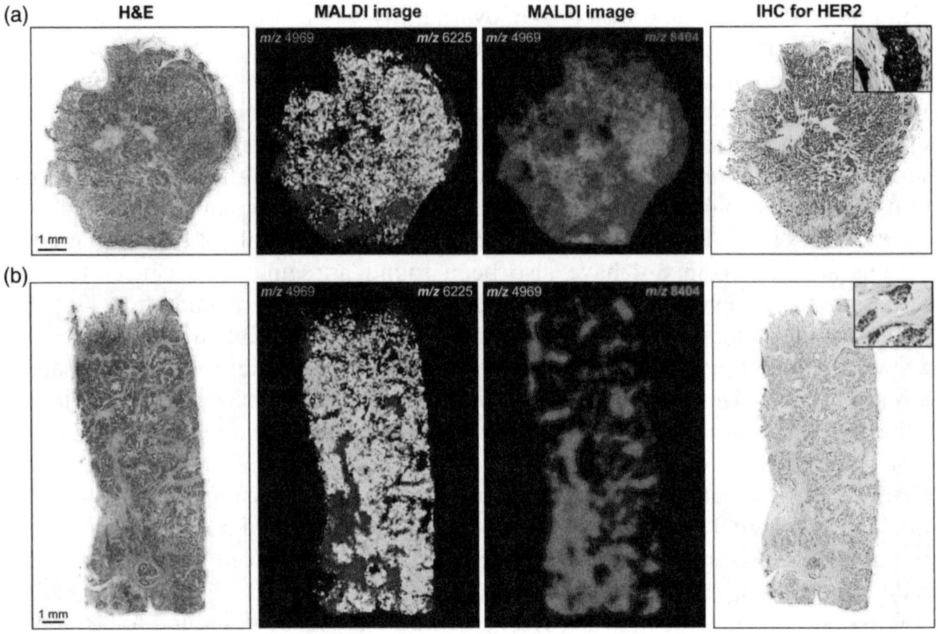

Figure 10.7 See color insert. Imaging of breast cancer tissues reveals proteins that can distinguish HER2 receptor status. Images of (A) HER2-positive and (B) HER2-negative human breast cancer tissues. The signal at *m/z* 4969 (blue) is specific to tumor stroma; *m/z* 4969 for cancer cells; and *m/z* 8404 for HER2 positive cancer cells. (Reproduced with permission from Reference 29. Copyright 2010 American Chemical Society.)

previously been implicated in cancer. However, the finding that the deficiency may persist into adjacent "normal" tissue could be important for understanding how tumors progress and recur. A similar study of ovarian tumors found an overexpression of two signals, subsequently identified as plastin 2 and peroxiredoxin 1, in the interface zone between normal and tumor tissue [64]. While these findings have tremendous potential to impact clinical cancer diagnosis and treatment, there is a significant need for any candidate biomarkers to be validated with much larger sample sets.

10.3.3 Bacterial Infection

Recently there have been several illustrations of the use of IMS as applied to the study of bacteria, with the aim of further understanding bacterial infection as well as aiding the discovery of new antibacterial compounds. With the increase in bacterial resistance to conventional antibiotics, there is a critical need for the development of new compounds with antibacterial activities. In addition, for some infections, such as with *Mycobacterium tuberculosis*, there is a poor understanding of the mode of action of current chemotherapeutics. IMS offers a new approach to this relatively old problem, and it has provided some new avenues for further research.

One approach to the discovery of novel antibacterial compounds is the study of bacterial colonies as they interact with themselves and each other. Bacteria are known to manufacture numerous secondary metabolites that they produce in response to their environment, and some of these compounds target cells from their own colony for destruction to provide nutrients for the remaining cells. These so-called cannibalistic factors may also be potent antibacterial compounds toward other bacterial strains as well. The cannibalistic factors of *Bacillus subtilis* only present themselves on solid media (as opposed to liquid), and there are few analytical tools capable of spatially characterizing the metabolic output on solid surfaces. Thus, IMS was utilized to visualize the production and spatial distribution of sporulation delaying protein (SDP) and sporulation killing factor (SKF) on *B. subtilis* colonies grown directly on a MALDI target plate [74]. Two ions were detected at *m/z* 4350 and *m/z* 2782 that were localized to one strain of *B. subtilis* and that seemed to correlate with a region of decreased growth on the other strain. These two ions were subsequently identified as SDP and SKF, respectively. They were further purified and tested against other pathogens, with SDP exhibiting potent antibacterial activity against several *Staphylococcus aureus* strains, including a methicillin-resistant strain [74].

The emergence of antibiotic-resistant strains of clinically relevant bacteria, such as *S. aureus*, is a growing clinical problem. Further understanding how cells and organs fight off an infection may eventually lead to the development of novel strategies for antibiotic therapy. For example, abscesses formed in the kidney during a mouse model of *S. aureus* infection were examined by IMS [75]. Abscesses are formed by the host organism's immune system as a way to confine the bacteria and prevent the spread of infection. One signal at *m/z* 10,165 was found localized to the abscesses of infected kidneys. This was subsequently identified as S100A8, a calcium-binding protein. S100A8, along with S100A9, forms the heterodimer calprotectin, which is known to exhibit antibacterial activity. Further examination of the mode of action of calprotectin showed that it inhibited bacterial growth by chelation of Mn^{2+} and Zn^{2+} metal ions. Thus, chelation of nutrient metals may serve as an alternative strategy for combating bacterial infections [75].

A similar study of *Mycobacterium tuberculosis* infection in animals has been undertaken in our laboratory. *M. tuberculosis* infection results in the production of a highly heterogeneous population of pulmonary granulomas. It typically takes 6–9 months of daily therapy to cure an active tuberculosis infection in humans. It may take up to 2 years if a resistant strain is involved. The factors involved in the slow rate of bacterial killing are poorly understood; however one factor may be poor penetration of drugs into the granulomas. We utilized IMS to examine the protein compositions of granulomas in a monkey model of tuberculosis infection. Figure 10.8 shows the results obtained from granulomas from four different animals. Hematoxylin and eosin (H&E)-stained serial sections are shown on the left, with granuloma regions highlighted by arrows. To the right of each H&E section is an ion image overlay of three signals found differentially distributed around granuloma areas. One ion at *m/z* 3445, shown in red, is localized to the center, necrotic area of the granulomas. The signal at *m/z* 8451 (in blue) is predominantly localized in the area directly surrounding the granuloma. An ion at *m/z* 4046 (in green) appears localized to the farther outer edges of the granuloma, but it is not observed in all the samples. This may be due to how the granulomas were excised and how much surrounding tissue was extracted, but it also may be due to inherent differences in granuloma subtypes, sizes, or unique microenvironment in the lung. In addition, the

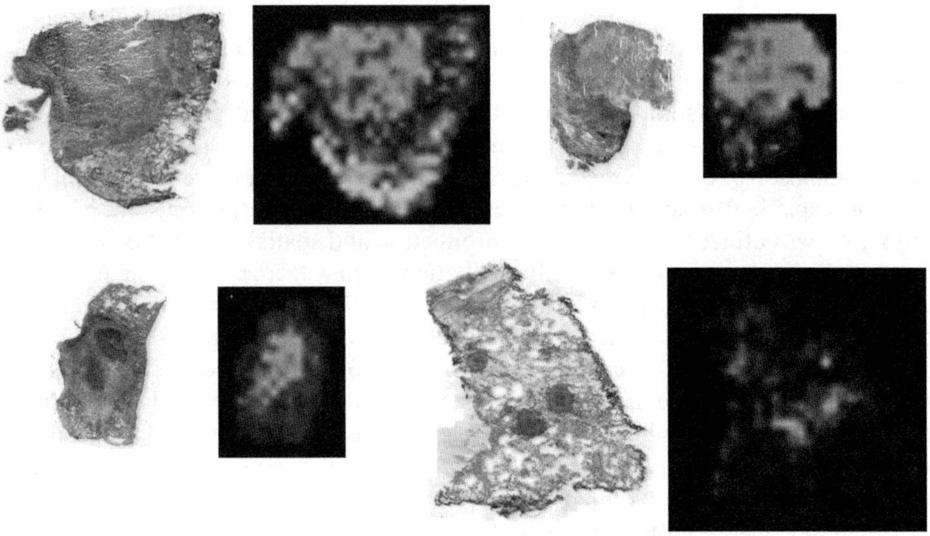

Figure 10.8 See color insert. Imaging of *M. tuberculosis* infection in granulomas obtained from four cynomolgus monkeys. Granulomas show intense localization of *m/z* 3445 (red), surrounded by *m/z* 8451 (blue), and *m/z* 4046 (green) further out.

distributions of small molecule species in and around the granulomas, including lipids and the chemotherapeutic drugs rifampicin and isoniazid, have also been obtained via IMS. It has yet to be determined what significance these patterns have, if any, on the penetration of individual drugs into the granulomas. Further study is ongoing regarding the identification of these proteins, how these distribution patterns correlate with drug and lipid distributions, and how all of the distributions correlate with the histology of the granulomas. Nonetheless, this information will certainly be helpful in furthering our understanding of the *in vivo* mechanism of action of the existing antituberculosis drugs and will facilitate the development of next-generation therapies.

10.4 PEPTIDE ANALYSIS: DIGESTED PEPTIDES

As briefly stated before, one strategy for the identification of proteins directly from tissue sections is the use of *in situ* digestion [32]. This approach generates many fragment peptides from which MS/MS sequence information may be obtained. While trypsin is most commonly employed, in principle other enzymes with different specific activities and cleavage sites may be used. The enzymes are commonly applied via a robotic spotting device, with some extra time allotted for the reaction to occur. Matrix is usually applied on top of the tryptic spots with the same spotting device, although it may be coated as well.

In addition to the direct identification of proteins from tissue sections, analysis of digested peptides also allows proteins not readily amenable to the MALDI process (e.g., large membrane-bound proteins or hydrophobic proteins) to be sampled via their smaller, more hydrophilic peptides. Also, with the use of antigen retrieval, formalin-fixed paraffin embedded (FFPE) samples may be analyzed via their peptides. This allows the vast stores of archival FFPE clinical samples to be interrogated via IMS. Not only will this dramatically increase the overall number of samples available for analysis, it will also allow rare conditions (for which it may take many years to accrue sufficient cases for statistical analysis) to be analyzed. Tissue microarrays (TMAs), which are commonly made by combining biopsies of many individual patient samples onto one paraffin block, allow these analyses to occur for multiple patients in one imaging run, thus making this a more high-throughput approach.

For example, TMAs have been analyzed after *in situ* digestion for lung cancer [22], follicular lymphoma [76], gastric cancer [77], and pancreatic cancer [30]. In the case of lung cancer, a single TMA containing tissue from 100 needle core biopsies from 50 lung tumor patients (along with 10 adjacent normal lung punches) was analyzed. Only squamous cell-carcinoma and adenocarcinoma samples were subjected to further analysis, as the numbers for other cancer types (such as bronchioloalveolar carcinoma and metastatic colon cancer) were too small. Trypsin was applied via a robotic spotter, followed by application of the matrix overlaid on the trypsin spots. A serial section of the TMA was obtained and subjected to H&E staining, and histological areas of interest on each sample were marked by a pathologist. The resulting peptide spectra were subjected to statistical analysis, and classification algorithms were constructed that were able to correctly classify all 18 adenocarcinoma biopsies and all 22 squamous cell carcinoma biopsies. Several individual spectra in the total set were misclassified; however, the algorithms were quite robust overall. In addition, the ions used as classifiers could be readily subjected to MS/MS sequencing. Over 50 proteins were identified directly from the TMA in this manner [22].

This approach was recently applied to a human pancreatic cancer TMA containing duplicate biopsies from 30 patients with pancreatic adenocarcinoma, along with adjacent nontumor tissue [30]. In this case, both trypsin and matrix were deposited via an automated spray-coating device as opposed to discrete droplets. Images were acquired from the TMA using an ion mobility mass spectrometer (Synapt HDMS, Waters, Milford, MA). The spectra were then subjected to principal component analysis (PCA) followed by discriminant analysis (DA) in order to classify the tumor tissues. The distribution of one peptide at *m/z* 1198, identified as an actin peptide, is shown across the entire TMA in Figure 10.9A. The TMA is organized as two duplicate tumor tissue cores followed by one adjacent nontumor core. The actin peptide appears more abundant in the tumor tissues than in the nontumor tissue, and it was found to be statistically significant upon PCA-DA in distinguishing three tumor classes. Figure 10.9B shows the PCA-DA results of a test set of spectra (obtained from the same TMA from different biopsies), which show a clear separation of the samples into four distinct groups. Interestingly, samples classified as stage IIA by pathology were consistently separated into two groups (groups 2 and 3) by the PCA-DA analysis [30]. The general goal of most of these experiments is similar to those performed on intact proteins, and that is to discover a molecular signature that can confirm current diagnoses or that can provide additional diagnostic or prognostic information to the clinician.

Other disease models with FFPE tissues have been studied in this way. For example, rat brains from an experimental model of Parkinson's disease were fixed

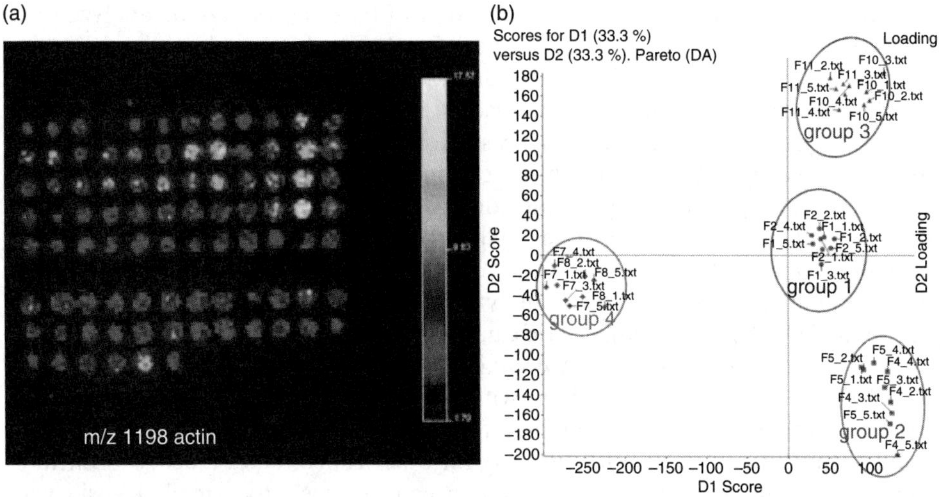

Figure 10.9 See color insert. IMS analysis of a pancreatic cancer tissue microarray after *in situ* tryptic digestion. (A) The distribution of an actin peptide at *m/z* 1198 within the TMA samples. The TMA is assembled with duplicate cores from a single biopsy next to each other, followed by a single core from adjacent nontumor tissue from the same patient. (B) Principal component analysis–discriminant analysis (PCA–DA) separated a test set of TMA samples into four different tumor groups corresponding to stages IIB, IIA (separated into two groups), and III. (Adapted from Reference 30 with permission.)

with paraformaldehyde, embedded with paraffin, and stored for 9 years. These tissues were then subjected to IMS after dewaxing and *in situ* tryptic digestion [23]. The spectra from five Parkinson brains and five control brains were compared to discover peptides that could distinguish the two conditions. Several peptides were found to be significantly different between the groups that corresponded to proteins known to be affected in Parkinson's disease, including the upregulation of peroxidoxin 6, F1 ATPase, and α-enolase and the downregulation of hexokinase and neurofilament M protein. In addition, two novel candidate biomarkers were discovered. A peptide at *m/z* 1025.5 was found downregulated in the Parkinson brains, which was subsequently identified as a tryptic peptide from trans-elongation factor 1. Another peptide at *m/z* 1083, identified as a fragment of the collapsin response mediator protein 2 (CRMP2), was found upregulated the Parkinson brains. Further imaging of CRMP2 peptides on Parkinson rat brains showed a distinct localization to the corpus callosum, which normally does not contain this protein. As the corpus callosum is known to be implicated in many neurodegenerative diseases, this is fitting with its involvement in Parkinson's disease [23].

10.5 PEPTIDE ANALYSIS: INTACT ENDOGENOUS PEPTIDES

The sample preparation for IMS analysis of peptides is somewhat different from that for protein imaging. Peptide analyses typically employ CHCA or DHB as a matrix, rather than SA, which is optimal for high molecular weight proteins. In addition, peptides are often detected on a TOF with a reflector for improved mass resolution. Direct sequencing via tandem MS of small peptides is often possible, leading to the unambiguous identification of the peptide and/or any post-translational modifications that may be present.

For example, neuropeptides are a class of endogenous polypeptides typically <3000 Da that are expressed by neurons. They are involved in the regulation of physiological processes throughout development and have been implicated in many psychiatric disorders [78]. There has been a great deal of direct analysis of neuropeptides from single neurons, often from simple model organisms, such as crustaceans, mollusks, and insects [79]. Single cells are often microdissected and then profiled, so a true distribution of neuropeptides is not obtained. However, the spatial distributions of neuropeptides from cultured *Aplysia californica* neurons were determined with a spatial resolution of 50 μm [80]. The small size regime of individual cells, from up to 1 mm for *Aplysia* neurons down to ~10 μm for human cells, requires particular attention to sample preparation and matrix application. This study required initial stabilization of the neurons in a glycerol/artificial sea water mixture, followed by matrix application and quick-drying on a hot plate or microwave oven [80]. Additionally, the distribution of neuropeptides from larger organs, including lobster brain [81] and crab brain [79], have been reported.

Mammalian systems have also been investigated for neuropeptide distribution [82–84]. The distribution of several neuropeptides, including substance P (*m/z* 1347.8) and two of its endogenous metabolites at *m/z* 900.4 and *m/z* 1104.6, were determined in serial sections of rat brain. These peptides were found localized to regions of the substantia nigra and interpeduncular nucleus [82]. This study also utilized serial sections to determine protein distributions in the same region, and co-registered both with a three-dimensional reconstruction of the rat brain region analyzed. Another study reported the distribution of many peptides in sections of rat spinal cord. The colocalization of several neuropeptides, including somatostatin-14 and substance P, in the first few laminae of the dorsal horn were shown [83]. In addition, without the use of trypsin, ions identified as fragments of larger proteins, including myelin basic protein and synapsin, were also detected. Taban et al. reported the distribution of peptides in a rat brain section using an FT-ICR mass spectrometer [84]. In this case, the ability of the FT-ICR to obtain accurate masses, along with its ability to do multistage MS analysis for fragmentation, proved beneficial for the identification of the peptides. The distribution of vasopressin at *m/z* 1084.4346 was shown to be localized around the third ventricle between the supraoptic nuclei, in good agreement with previous studies [84].

Natural degradation of proteins via enzymatic or chemical means occurs in tissues, as illustrated for the age-related degeneration of crystallin proteins in the lens. In addition to studies of the intact proteins as discussed previously, distributions of small peptide fragments have also been reported. Su et al. [85] identified 26 crystallin fragment peptides in human lens tissue, ranging in size from 1070 to 3389 Da. The presence of these fragment peptides increases with age, and the distribution shifts from the nucleus of the lens in middle age to the cortex in old age. The results for five of these peptides from human lenses from 20 years to 86 years old are shown in Figure 10.10. As shown, for all of the peptides there is a sharp increase in abundance with increasing age, along with a shift in distribution from the nucleus expanding into the cortex. The authors suggest that this points to the

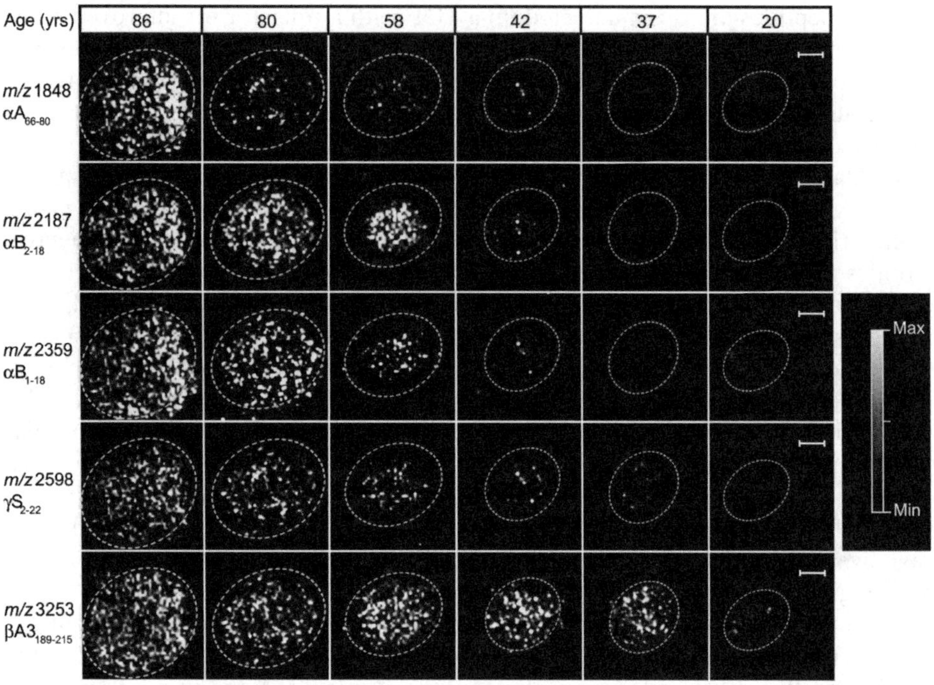

Figure 10.10 See color insert. Imaging of crystallin peptides in human lens tissue. There is a clear increase in the presence of many crystallin fragments in the lens with age. Scale bars = 2 mm. (Reproduced from Reference 85 with permission from Elsevier.)

time of middle age as a potentially important period in lens aging. They further suggest that these peptides arise from nonenzymatic processes that are facilitated by age-related changes to the structure of the lens, such as the formation of a barrier at the nucleo-cortical interface as well as a slightly more acidic environment in older lens nuclear fiber cells.

10.6 CONCLUSIONS

IMS technology is a powerful tool for *in situ* molecular assays because it allows molecular maps to be generated from analysis of the analytes themselves. Fresh frozen and archived FFPE samples may be analyzed with this technology, allowing stores of archived clinical tissue samples to be interrogated with regards to their molecular state. This technology has been applied to diverse fields such as cancer research, developmental biology, neurology, and aging. While validation and clinical implementation of some of the candidate biomarker proteins or peptides detected via IMS are necessary before the full potential of these results can be recognized, these experiments are nonetheless important steps toward understanding the molecular basis of disease. Further, the use of IMS in molecular pathology promises to bring a new perspective to this field to better aid improvement of human health.

ACKNOWLEDGMENTS

The authors thank Clif Barry III, JoAnne Flynn, and Laura Via for the collaboration on tuberculosis imaging. Funding from NIH (5R01 GM058008), DOD (W81XWH-05-1-0179), and the Bill and Melinda Gates Foundation is acknowledged.

REFERENCES

[1] Fenn JB, Mann M, Meng CK, Wong SF, Whitehouse CM. Electrospray ionization for mass spectrometry of large biomolecules. Science 1989;246:64–71.

[2] Karas M, Bachmann D, Bahr U, Hillenkamp F. Matrix-assisted ultraviolet laser desorption of non-volatile compounds. Int J Mass Spectrom Ion Process 1987;78:53–68.

[3] Karas M, Hillenkamp F. Laser desorption ionization of proteins with molecular masses exceeding 10,000 daltons. Anal Chem 1988;60:2299–301.

[4] Tanaka K, Waki H, Ido Y, Akita S, Yoshida Y, Yoshida T, et al. Protein and polymer analyses up to *m/z* 100,000 by laser ionization time-of-flight mass spectrometry. Rapid Commun Mass Spectrom 1988;2:151–3.

[5] Caprioli RM, Farmer TB, Gile J. Molecular imaging of biological samples: localization of peptides and proteins using MALDI-TOF MS. Anal Chem 1997;69:4751–60.

[6] Chaurand P, Stoeckli M, Caprioli RM. Direct profiling of proteins in biological tissue sections by MALDI mass spectrometry. Anal Chem 1999;71:5263–70.

[7] Stoeckli M, Farmer TB, Caprioli RM. Automated mass spectrometry imaging with a matrix-assisted laser desorption ionization time-of-flight instrument. J Am Soc Mass Spectrom 1999;10:67–71.

[8] Pacholski ML, Winograd N. Imaging with mass spectrometry. Chem Rev 1999;99: 2977–3006.

[9] Cornett DS, Mobley JA, Dias EC, Andersson M, Arteaga CL, Sanders ME, et al. A novel histology-directed strategy for MALDI-MS tissue profiling that improves throughput and cellular specificity in human breast cancer. Mol Cell Proteomics 2006;5:1975–83.

[10] Schwartz SA, Reyzer ML, Caprioli RM. Direct tissue analysis using matrix-assisted laser desorption/ionization mass spectrometry: practical aspects of sample preparation. J Mass Spectrom 2003;38:699–708.

[11] Seeley EH, Oppenheimer SR, Mi D, Chaurand P, Caprioli RM. Enhancement of protein sensitivity for MALDI imaging mass spectrometry after chemical treatment of tissue sections. J Am Soc Mass Spectrom 2008;19:1069–77.

[12] Lemaire R, Wisztorski M, Desmons A, Tabet JC, Day R, Salzet M, et al. MALDI-MS direct tissue analysis of proteins: improving signal sensitivity using organic treatments. Anal Chem 2006;78:7145–53.

[13] Thibault DB, Gillam CJ, Grey AC, Han J, Schey KL. MALDI tissue profiling of integral membrane proteins from ocular tissues. J Am Soc Mass Spectrom 2008;19:814–22.

[14] Grey AC, Chaurand P, Caprioli RM, Schey KL. MALDI imaging mass spectrometry of integral membrane proteins from ocular lens and retinal tissue. J Proteome Res 2009;8:3278–83.

[15] Leinweber BD, Tsaprailis G, Monks TJ, Lau SS. Improved MALDI-TOF imaging yields increased protein signals at high molecular mass. J Am Soc Mass Spectrom 2009;20: 89–95.

[16] Mainini V, Angel PM, Magni F, Caprioli RM. Detergent enhancement of on-tissue protein analysis by matrix-assisted laser desorption/ionization imaging mass spectrometry. Rapid Commun Mass Spectrom 2010;25:199–204.

[17] van Remoortere A, van Zeijl RJM, van den Oever N, Franck J, Longuespée R, Wisztorski M, et al. MALDI imaging and profiling MS of higher mass proteins from tissue. J Am Soc Mass Spectrom 2010;21:1922–9.

[18] Ferrer I, Santpere G, Arzberger T, Bell J, Blanco R, Boluda S, et al. Brain protein preservation largely depends on the postmortem storage temperature: implications for study of proteins in human neurologic diseases and management of brain banks: a BrainNet Europe study. J Neuropathol Exp Neurol 2007;66:35–46.

[19] Svensson M, Boren M, Skold K, Falth M, Sjogren B, Andersson M, et al. Heat stabilization of the tissue proteome: a new technology for improved proteomics. J Proteome Res 2009;8:974–81.

[20] Goodwin RJA, Lang AM, Allingham H, Borén M, Pitt AR. Stopping the clock on proteomic degradation by heat treatment at the point of tissue excision. Proteomics 2010;10:1751–61.

[21] Fowler CB, Chesnick IE, Moore CD, O'Leary TJ, Mason JT. Elevated pressure improves the extraction and identification of proteins recovered from formalin-fixed, paraffin-embedded tissue surrogates. PLoS ONE 2010;5:e14253.

[22] Groseclose MR, Massion PP, Chaurand P, Caprioli RM. High-throughput proteomic analysis of formalin-fixed paraffin-embedded tissue microarrays using MALDI imaging mass spectrometry. Proteomics 2008;8:3715–24.

[23] Stauber J, Lemaire R, Franck J, Bonnel D, Croix D, Day R, et al. MALDI imaging of formalin-fixed paraffin-embedded tissues: application to model animals of Parkinson disease for biomarker hunting. J Proteome Res 2008;7:969–78.

[24] Lemaire R, Desmons A, Tabet JC, Day R, Salzet M, Fournier I. Direct analysis and MALDI imaging of formalin-fixed, paraffin-embedded tissue sections. J Proteome Res 2007;6:1295–305.

[25] Chaurand P, Latham JC, Lane KB, Mobley JA, Polosukhin VV, Wirth PS, et al. Imaging mass spectrometry of intact proteins from alcohol-preserved tissue specimens: bypassing formalin fixation. J Proteome Res 2008;7:3543–55.

[26] Mange A, Chaurand P, Perrochia H, Roger P, Caprioli RM, Solassol J. Liquid chromatography-tandem and MALDI imaging mass spectrometry analyses of RCL2/CS100-fixed, paraffin-embedded tissues: proteomics evaluation of an alternate fixative for biomarker discovery. J Proteome Res 2009;8:5619–28.

[27] Ergin B, Meding S, Langer R, Kap M, Viertler C, Schott C, et al. Proteomic analysis of PAXgene-fixed tissues. J Proteome Res 2010;9:5188–96.

[28] Chaurand P, Fouchecourt S, DaGue BB, Xu BJ, Reyzer ML, Orgebin-Crist M-C, et al. Profiling and imaging proteins in the mouse epididymis by imaging mass spectrometry. Proteomics 2003;3:2221–39.

[29] Rauser S, Marquardt C, Balluff B, Deininger S-O, Albers C, Belau E, et al. Classification of HER2 receptor status in breast cancer tissues by MALDI imaging mass spectrometry. J Proteome Res 2010;9:1854–63.

[30] Djidja M-C, Claude E, Snel M, Francese S, Scriven P, Carolan V, et al. Novel molecular tumour classification using MALDI–mass spectrometry imaging of tissue micro-array. Anal Bioanal Chem 2010;397:587–601.

[31] Reyzer ML, Chaurand P, Angel PM, Caprioli RM. Direct molecular analysis of whole-body animal tissue sections by MALDI imaging mass spectrometry. In Rubakhin SS, Sweedler JV, editors. Mass Spectrometry Imaging: Principles and Protocols. New York: Humana Press, 2010, pp. 285–301.

[32] Groseclose MR, Andersson M, Hardesty WM, Caprioli RM. Identification of proteins directly from tissue: *in situ* tryptic digestions coupled with imaging mass spectrometry. J Mass Spectrom 2007;42:254–62.

[33] Rejtar T, Chen H-S, Andreev V, Moskovets E, Karger BL. Increased identification of peptides by enhanced data processing of high-resolution MALDI TOF/TOF mass spectra prior to database searching. Anal Chem 2004;76:6017–28.

[34] Suckau D, Resemann A, Schuerenberg M, Hufnagel P, Franzen J, Holle A. A novel MALDI LIFT-TOF/TOF mass spectrometer for proteomics. Anal Bioanal Chem 2003;376:952–65.

[35] Stauber J, MacAleese L, Franck J, Claude E, Snel M, Kaletas BK, et al. On-tissue protein identification and imaging by MALDI-ion mobility mass spectrometry. J Am Soc Mass Spectrom 2010;21:338–47.

[36] Holle A, Haase A, Kayser M, Höhndorf J. Optimizing UV laser focus profiles for improved MALDI performance. J Mass Spectrom 2006;41:705–16.

[37] Trim P, Djidja M-C, Atkinson S, Oakes K, Cole L, Anderson D, et al. Introduction of a 20 kHz Nd:YVO4 laser into a hybrid quadrupole time-of-flight mass spectrometer for MALDI-MS imaging. Anal Bioanal Chem 2010;397:3409–19.

[38] Spraggins JM, Caprioli RM. High-speed MALDI-TOF imaging mass spectrometry: rapid ion image acquisition and considerations for next generation instrumentation. J Am Soc Mass Spectrom 2011;22:1022–31.

[39] Kim HK, Reyzer ML, Choi IJ, Kim CG, Kim HS, Oshima A, et al. Gastric cancer-specific protein profile identified using endoscopic biopsy samples via MALDI mass spectrometry. J Proteome Res 2010;9:4123–30.

[40] Schwartz SA, Weil RJ, Thompson RC, Shyr Y, Moore JH, Toms SA, et al. Proteomic-based prognosis of brain tumor patients using direct-tissue matrix-assisted laser desorption ionization mass spectrometry. Cancer Res 2005;65:7674–81.

[41] Debois D, Bertrand V, Quinton L, De Pauw-Gillet M-C, De Pauw E. MALDI-in source decay applied to mass spectrometry imaging: a new tool for protein identification. Anal Chem 2010;82:4036–45.

[42] Cazares LH, Troyer D, Mendrinos S, Lance RA, Nyalwidhe JO, Beydoun HA, et al. Imaging mass spectrometry of a specific fragment of mitogen-activated protein kinase/extracellular signal-regulated kinase kinase kinase 2 discriminates cancer from uninvolved prostate tissue. Clin Cancer Res 2009;15:5541–51.

[43] Franck J, El Ayed M, Wisztorski M, Salzet M, Fournier I. On-tissue N-terminal peptide derivatizations for enhancing protein identification in MALDI mass spectrometric imaging strategies. Anal Chem 2009;81:8305–17.

[44] Cornett DS, Reyzer ML, Chaurand P, Caprioli RM. MALDI imaging mass spectrometry: molecular snapshots of biochemical systems. Nat Methods 2007;4:828–33.

[45] Lagarrigue M, Becker M, Lavigne R, Deininger SO, Walch A, Aubry F, et al. Revisiting rat spermatogenesis with MALDI imaging at 20 μm resolution. Mol Cell Proteomics 2010;10(3):M110.005991, 1–11.

[46] Burnum KE, Tranguch S, Mi D, Daikoku T, Dey SK, Caprioli RM. Imaging mass spectrometry reveals unique protein profiles during embryo implantation. Endocrinology 2008;149:3274–8.

[47] Grey AC, Gelasco AK, Section J, Moreno-Rodriguez RA, Krug EL, Schey KL. Molecular morphology of the chick heart visualized by MALDI imaging mass spectrometry. Anat Rec (Hoboken) 2010;293:821–8.

[48] Mathur BN, Caprioli RM, Deutch AY. Proteomic analysis illuminates a novel structural definition of the claustrum and insula. Cereb Cortex 2009;19:2372–9.

[49] Han J, Schey KL. MALDI tissue imaging of ocular lens alpha-crystallin. Invest Ophthalmol Vis Sci 2006;47:2990–6.

[50] Grey AC, Schey KL. Age-related changes in the spatial distribution of human lens alpha-crystallin products by MALDI imaging mass spectrometry. Invest Ophthalmol Vis Sci 2009;50:4319–29.

[51] Grey AC, Schey KL. Distribution of bovine and rabbit lens alpha-crystallin products by MALDI imaging mass spectrometry. Mol Vis 2008;14:171–9.

[52] Stoeckli M, Chaurand P, Hallahan DE, Caprioli RM. Imaging mass spectrometry: a new technology for the analysis of protein expression in mammalian tissues. Nat Med 2001;7:493–6.

[53] Masumori N, Thomas TZ, Chaurand P, Case T, Paul M, Kasper S, et al. A probasin-large T antigen transgenic mouse line develops prostate adenocarcinoma and neuroendocrine carcinoma with metastatic potential. Cancer Res 2001;61:2239–49.

[54] Chaurand P, DaGue BB, Pearsall RS, Threadgill DW, Caprioli RM. Profiling proteins from azoxymethane-induced colon tumors at the molecular level by matrix-assisted laser desorption/ionization mass spectrometry. Proteomics 2001;1: 1320–6.

[55] Yanagisawa K, Shyr Y, Xu BJ, Massion PP, Larsen PH, White BC, et al. Proteomic patterns of tumour subsets in non-small-cell lung cancer. Lancet 2003;362:433–9.

[56] Schwartz SA, Weil RJ, Johnson MD, Toms SA, Caprioli RM. Protein profiling in brain tumors using mass spectrometry: feasibility of a new technique for the analysis of protein expression. Clin Cancer Res 2004;10:981–7.

[57] Amann JM, Chaurand P, Gonzalez A, Mobley JA, Massion PP, Carbone DP, et al. Selective profiling of proteins in lung cancer cells from fine-needle aspirates by matrix-assisted laser desorption ionization time-of-flight mass spectrometry. Clin Cancer Res 2006;12:5142–50.

[58] Taguchi F, Solomon B, Gregorc V, Roder H, Gray R, Kasahara K, et al. Mass spectrometry to classify non-small-cell lung cancer patients for clinical outcome after treatment with epidermal growth factor receptor tyrosine kinase inhibitors: a multicohort cross-institutional study. J Natl Cancer Inst 2007;99:838–46.

[59] Yildiz PB, Shyr Y, Rahman JS, Wardwell NR, Zimmerman LJ, Shakhtour B, et al. Diagnostic accuracy of MALDI mass spectrometric analysis of unfractionated serum in lung cancer. J Thorac Oncol 2007;2:893–901.

[60] Schwamborn K, Caprioli RM. Molecular imaging by mass spectrometry—looking beyond classical histology. Nat Rev Cancer 2010;10:639–46.

[61] McDonnell LA, Corthals GL, Willems SM, van Remoortere A, van Zeijl RJM, Deelder AM. Peptide and protein imaging mass spectrometry in cancer research. J Proteomics 2010;73:1921–44.

[62] Wong SCC, Chan CML, Ma BBY, Lam MYY, Choi GCG, Au TCC, et al. Advanced proteomic technologies for cancer biomarker discovery.(Report). Expert Rev Proteomics 2009;6(2):123–34.

[63] Seeley EH, Caprioli RM. Molecular imaging of proteins in tissues by mass spectrometry. Proc Natl Acad Sci U S A 2008;105:18126–31.

[64] Kang S, Shim HS, Lee JS, Kim DS, Kim HY, Hong SH, et al. Molecular proteomics imaging of tumor interfaces by mass spectrometry. J Proteome Res 2010;9:1157–64.

[65] El Ayed M, Bonnel D, Longuespée R, Castellier C, Franck J, Vergara D, et al. MALDI imaging mass spectrometry in ovarian cancer for tracking, identifying, and validating biomarkers. Med Sci Monit 2010;16:BR233–45.

[66] Lemaire R, Menguellet SA, Stauber J, Marchaudon V, Lucot JP, Collinet P, et al. Specific MALDI imaging and profiling for biomarker hunting and validation: fragment of the 11S proteasome activator complex, Reg alpha fragment, is a new potential ovary cancer biomarker. J Proteome Res 2007;6:4127–34.

[67] Schwamborn K, Krieg RC, Reska M, Jakse G, Knuechel R, Wellmann A. Identifying prostate carcinoma by MALDI-Imaging. Int J Mol Med 2007;20:155–9.

[68] Oppenheimer SR, Mi D, Sanders ME, Caprioli RM. Molecular analysis of tumor margins by MALDI mass spectrometry in renal carcinoma. J Proteome Res 2010;9: 2182–90.

[69] Han EC, Lee Y-S, Liao W-S, Liu Y-C, Liao H-Y, Jeng L-B. Direct tissue analysis by MALDI-TOF mass spectrometry in human hepatocellular carcinoma. Clin Chim Acta 2011;412:230–9.

[70] Holt GE, Schwartz HS, Caldwell RL. Proteomic profiling in musculoskeletal oncology by MALDI mass spectrometry. Clin Orthop Relat Res 2006;450:105–10.

[71] Willems SM, van Remoortere A, van Zeijl R, Deelder AM, McDonnell LA, Hogendoorn PCW. Imaging mass spectrometry of myxoid sarcomas identifies proteins and lipids specific to tumour type and grade, and reveals biochemical intratumour heterogeneity. J Pathol 2010;222:400–9.

[72] Patel SA, Barnes A, Loftus N, Martin R, Sloan P, Thakker N, et al. Imaging mass spectrometry using chemical inkjet printing reveals differential protein expression in human oral squamous cell carcinoma. Analyst 2009;134:301–7.

[73] Droin N, Hendra JB, Ducoroy P, Solary E. Human defensins as cancer biomarkers and antitumour molecules. J Proteomics 2009;72:918–27.

[74] Liu WT, Yang YL, Xu Y, Lamsa A, Haste NM, Yang JY, et al. Imaging mass spectrometry of intraspecies metabolic exchange revealed the cannibalistic factors of *Bacillus subtilis*. Proc Natl Acad Sci U S A 2010;107:16286–90.

[75] Corbin BD, Seeley EH, Raab A, Feldmann J, Miller MR, Torres VJ, et al. Metal chelation and inhibition of bacterial growth in tissue abscesses. Science 2008;319: 962–5.

[76] Samsi SS, Krishnamurthy AK, Groseclose MR, Caprioli RM, Lozanski G, Gurcan MN. Imaging mass spectrometry analysis for follicular lymphoma grading. Engineering in Medicine and Biology Society, EMBC, Annual International Conference of the IEEE, 2009, pp. 6969–72.

[77] Morita Y, Ikegami K, Goto-Inoue N, Hayasaka T, Zaima N, Tanaka H, et al. Imaging mass spectrometry of gastric carcinoma in formalin-fixed paraffin-embedded tissue microarray. Cancer Sci 2010;101:267–73.

[78] Belzung C, Yalcin I, Griebel G, Surget A, Leman S. Neuropeptides in psychiatric diseases: an overview with a particular focus on depression and anxiety disorders. CNS Neurol Disord Drug Targets 2006;5:135–45.

[79] Chen R, Li L. Mass spectral imaging and profiling of neuropeptides at the organ and cellular domains. Anal Bioanal Chem 2010;397:3185–93.

[80] Rubakhin SS, Greenough WT, Sweedler JV. Spatial profiling with MALDI MS: distribution of neuropeptides within single neurons. Anal Chem 2003;75:5374–80.

[81] Chen R, Jiang X, Conaway MC, Mohtashemi I, Hui L, Viner R, et al. Mass spectral analysis of neuropeptide expression and distribution in the nervous system of the lobster *Homarus americanus*. J Proteome Res 2010;9:818–32.

[82] Andersson M, Groseclose MR, Deutch AY, Caprioli RM. Imaging mass spectrometry of proteins and peptides: 3D volume reconstruction. Nat Methods 2008;5:101–8.

[83] Monroe EB, Annangudi SP, Hatcher NG, Gutstein HB, Rubakhin SS, Sweedler JV. SIMS and MALDI MS imaging of the spinal cord. Proteomics 2008;8:3746–54.

[84] Taban IM, Altelaar AF, van der Burgt YE, McDonnell LA, Heeren RM, Fuchser J, et al. Imaging of peptides in the rat brain using MALDI-FTICR mass spectrometry. J Am Soc Mass Spectrom 2007;18:145–51.

[85] Su S-P, McArthur JD, Aquilina JA. Localization of low molecular weight crystallin peptides in the aging human lens using a MALDI mass spectrometry imaging approach. Exp Eye Res 2010;91:97–103.

11

IMAGING MASS SPECTROMETRY FOR DRUGS AND METABOLITES

STACEY R. OPPENHEIMER

11.1 INTRODUCTION

Mass spectrometry (MS) plays an intricate role in the pharmaceutical research and development (R&D) process of bringing new medicines to the clinic for improving quality of life. While liquid chromatography-tandem mass spectrometry (LC-MS/MS) platforms have become an invaluable tool at all levels in the R&D process, they tell us little about the spatial distribution of the drug compounds analyzed; therefore, one of the major challenges in the R&D process today is proving successful achievement of target exposure for potential drug candidates. While valuable imaging modalities such as magnetic resonance imaging (MRI), positron emission tomography (PET), and autoradiography enable *in vivo* and *ex vivo* monitoring of drug localization and distribution, these approaches require a drug with special isotopes, and they do not allow the differentiation between parent drugs and their metabolites. This type of information can mean the difference between continuing and abandoning a compound or chemical series. In contrast to label-dependent technologies, MS offers the advantage of molecular specificity, enabling the ability to differentiate parent drug from metabolites based on molecular weight. MS-based imaging methods have emerged as powerful tools enabling the spatial localization to be achieved along with molecular specificity directly from tissue surfaces [1]. Both exogenous and endogenous compounds, such as xenobiotics, peptides, and proteins, have been imaged directly from mammalian tissue sections with MS [2–5]. This chapter will describe the use of MS imaging (MSI) as a tool for the pharmaceutical analysis of drugs and metabolites in the drug discovery and development process. Since tissue imaging with matrix-assisted laser desorption/ionization (MALDI) is the primary MSI tool used today for the analysis of pharmaceutical compounds, this

Mass Spectrometry for Drug Discovery and Drug Development, First Edition. Edited by Walter A. Korfmacher.

will be the primary focus of the chapter. Other imaging-capable MS platforms such as secondary ion MS (SIMS) and desorption electrospray ionization (DESI) have shown potential in this field and will be introduced. In the pharmaceutical field, this semi-quantitative imaging technology provides an advantageous approach for drug distribution studies because it does not require a radiolabel, and it does not require the tissue homogenization and extraction steps needed for LC-MS/MS studies. The unique ability of MSI to differentiate between parent drug and metabolite without a label enables this technology to be used earlier in the drug discovery process.

Drug distribution studies in mammals play a crucial role in the drug discovery and development process, providing information about where the drug accumulates in the body and ensuring that the drug is metabolized and excreted as expected. These studies help researchers determine if a drug candidate has accumulated in the target organ or if it selectively accumulates in other tissues, which could lead to toxicity and other unwanted side effects. Accurate assessment of preclinical absorption/distribution/metabolism/excretion/toxicity (ADMET) plays a critical role in the assessment of new chemical entities and helps weed out compounds that should not be developed. In this chapter, the MALDI MS platform as applied to imaging drug and metabolite distribution in tissues will be discussed, and examples of its use in pharmaceutical R&D will be presented. Lastly, future directions to further develop the technology for its application to pharmaceutical analysis will be addressed.

11.2 CONVENTIONAL IMAGING TECHNOLOGIES

The imaging technologies most commonly used to determine drug distribution in the body are MRI, PET, and autoradiography. MRI [6–10] and PET [11–17] are *in vivo* imaging technologies that offer a real-time view of a drug moving throughout the body or region of interest. The noninvasive nature of these two platforms enables their use in preclinical animal studies and human clinical studies. In contrast, autoradiography imaging [18–21] is an *ex vivo* technology requiring excision of the tissue/region of interest. This is the most commonly used method in drug development studies to determine the drug distribution in the body or individual organ at a fixed point in time.

A common feature among these three technologies is the requirement that the drug contain special isotopes. Visualization of a substance with PET requires that the compound contain a positron-emitting radionuclide. Labeling with the radionuclides ^{11}C and ^{13}N can be performed by isotopic substitution and therefore have little to no impact on the physiochemical and biochemical properties of the drug. The insertion of the radionuclide ^{18}F is sometimes a nonisotopic substitution, depending on whether or not the drug contains a fluorine atom. If not, ^{18}F is often inserted in place of a hydrogen atom or hydroxyl group. Other positron-emitting nuclides include ^{15}O, ^{76}Br, ^{124}I, ^{68}Ga, and ^{64}U, each having a unique half-life. Out of all positron-emitters employed by PET, ^{18}F-labeled pharmaceuticals have the most desirable properties: low positron energy and an optimal physical half-life of 110 min, which allows for more complex radiosynthesis and longer *in vivo* experiments and monitoring [22]. While MRI does not require a radioisotope, it can only detect isotopes with nuclear spins such as ^{1}H, ^{13}C, and ^{19}F. Because the sensitivity of MRI depends on the magnetic properties of the monitored nucleus and its natural abundance, drugs must typically contain ^{19}F or be enriched in ^{13}C in order to be effectively monitored in the body. Paramagnetic atoms such as gadolinium, iron, and manganese are utilized as contrast agents to increase the contrast in MRI images. They are also frequently used as models for drugs that cannot be imaged on their own [9]. Autoradiography, which can be performed on whole body sections or excised individual tissues of interest, requires a beta-emitting radioactive isotope such as ^{14}C, ^{3}H, and ^{125}I [23]. In this case, an animal is dosed with a radiolabeled drug and after a specific time, the animal is sacrificed, flash-frozen, and sectioned. The radioactivity in the sections is then analyzed, resulting in an image indicating where the drug is distributed and to what extent it has accumulated. Individual tissues are analyzed in the same manner, but at higher resolutions to view substructures.

While these technologies provide invaluable information for drug discovery and development studies, there are some major limitations. The most significant of these limitations is the requirement for a label or tracer, which renders the technologies incapable of distinguishing between parent drug and its metabolites because only the label is monitored. This characteristic can give rise to misleading results if metabolism has occurred and the label has been retained on parent drug and/or metabolite species. In PET, for example, only coincident photons are detected, which provide no information concerning the chemical structure to which the radionuclide is attached. In order to validate the distribution observed in an image or validate a new radioligand for its intended target, additional technologies such as LC-MS are used to ensure that the radioactive emissions are derived primarily from the parent drug/radioligand [24]. The spatial resolution of the *in vivo* technologies is poor, which limits structural specificity within substructures of organs. The range of resolution for PET and MRI is on the 1- to 6-mm scale [9, 10, 25], although with major instrument improvements, the MRI resolution can be improved to ~700 μm [26]. Resolution in autoradiography is better, however, and varies depending on the imaging system used. The β-imager permits ~30–50 μm resolution, while the microchannel plates are ~70 μm, and the phosphor-imaging systems allow ~10–100 μm [27–29]. Although autoradiography is one of the most commonly used techniques and provides a significant advantage over PET and MRI with respect to resolution, it suffers from excessively long acquisition times. It takes weeks or months to acquire an image on film or 8–12 h to acquire an image on the β-imager [23, 27, 28]. Because of the invasiveness, this technology can only be used on lab rodents, not for real-time analysis in preclinical or clinical studies. The extra cost and time required for additional synthesis of isotope labels remain to be limitations.

11.3 MASS SPECTROMETRY IMAGING (MSI) TECHNOLOGIES FOR DRUG DISTRIBUTION

In contrast to conventional imaging technologies, MS provides molecular specificity, negating the need for a label. Surface sampling ionization MS platforms such as MALDI have enabled the *in situ* analysis of pharmaceuticals while maintaining the spatial integrity of samples. This provides complementary information to pharmacokinetic data acquired by LC-MS/MS platforms, which require tissue homogenization. In mass spectrometry-based imaging, the probe (e.g., laser) performs a raster

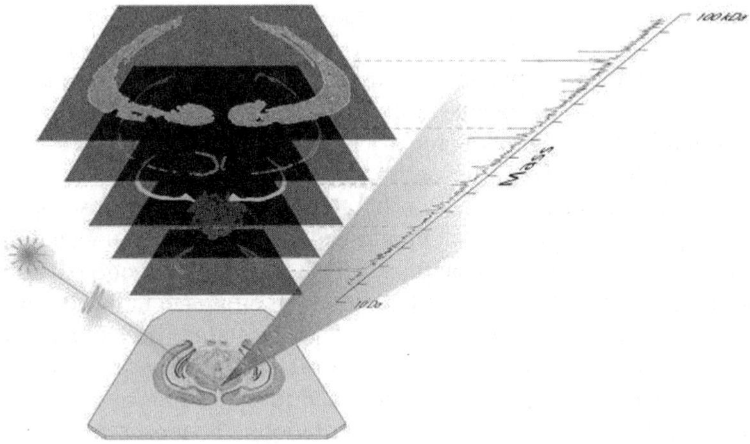

Figure 11.1 See color insert. MALDI mass spectrometric imaging. A laser is rastered over a tissue sample while acquiring a complete mass spectrum from each position, resulting in molecular images for multiple analytes. Reprinted with permission from Rohner et al. [5]. Copyright 2005 Elsevier.

over the tissue surface in a predefined two-dimensional array/grid, generating a mass spectrum at each grid coordinate. As illustrated in Figure 11.1, the coordinates of the irradiated spots are used to generate two-dimensional ion density maps, or images, that represent individual *m/z* values with their corresponding intensities and distribution within the tissue or area analyzed. This has led to a heightened interest in the utility of imaging mass spectrometry for drug distribution studies. While MALDI imaging is the most widely recognized platform for MSI, there are two other ionization methods that have illustrated utility for imaging drugs and metabolites in tissues. These technologies can be grouped according to the mechanism by which ions are generated from a surface: irradiation by a pulsed laser as in MALDI, irradiation by energetic ions as in SIMS, or desorption by projectile microdroplets as in DESI.

11.3.1 Secondary Ion MS

The concept of SIMS imaging has been around for nearly 50 years, but it was not until more recently that it was successfully applied to the analysis of biological tissue and cell surfaces. Historically, SIMS was primarily used for the surface analysis and imaging of films and polymers, as it requires very little sample preparation. SIMS utilizes a high energy beam of primary ions to bombard the sample surface, resulting in the desorption of ionized atoms, clusters, and molecules. These secondary ions are released from the surface as positive or negative ions that are conventionally analyzed by time-of-flight [30]. An advantage of this technology is the submicron resolution achievable with SIMS that is not yet available with laser desorption or DESI techniques. The disadvantage of SIMS with respect to MALDI is that it is a hard ionization technique that induces sample damage and extreme fragmentation of the desorbed ions, thus limiting the mass range. Early SIMS ion sources limited the mass range to around *m/z* 100, rendering applicability solely to elemental analysis. Since then, the introduction of cluster ion sources such as C_{60}^+ [31], Bi_3 [32], and Au_3 [33, 34] have successfully produced higher secondary ion yields, induced less sample damage, and increased the mass range to approximately *m/z* 1000, making it ideal for the analysis of small molecules, drugs, and lipids [35]. Although limited sample preparation (i.e., no matrix required) is an added advantage of SIMS over MALDI, studies have shown that the addition of a matrix for matrix-enhanced SIMS (ME-SIMS) or a metal compound for metal-enhanced SIMS (MetA-SIMS) can further enhance the secondary ion yield and help increase the mass range to larger lipids and small peptides. However, at a mass range above ~*m/z* 5000, MALDI, remains far superior [3].

While much development has occurred for the application of SIMS to imaging endogenous components, or mostly molecular fragment ions, of tissues, the use of SIMS for drug and metabolite imaging has been limited [36–39]. Most of the work done related to the pharmaceutical field has been accomplished by atomic imaging of drugs with SIMS ion microscopy [40–44]. This requires the drug of interest to be isotopically enriched with stable labels or labeled with an atom not natively found in cells or tissues, eliminating the ability to differentiate between a label on the parent or metabolite. In the early 1990s, Clerc et al. studied the intratumor biodistribution of the ^{127}I-labeled radiotherapeutic agent, metaiodobenzylguanidine (MIBG), in a neuroblastoma mouse model [45]. Imaging results illustrated a heterogeneous distribution of the drug in the tumor, enabling authors to suggest further optimization strategies for use of this therapeutic agent. Chandra et al. and Smith et al. have used SIMS to study boronphenylalanine-mediated boron neutron capture therapy in human glioblastoma cells injected with ^{10}B-labeled boronphenylalanine (^{10}BPA) and $^{13}C^{15}N$-labeled phenylalanine at 500-nm resolution [41, 42, 44, 46]. The SIMS image in Figure 11.2 illustrates the subcellular distribution of ^{10}B distribution in freeze-dried human glioblastoma cells after a 2-h treatment with ^{10}BPA [42]. Images of intracellular potassium and sodium were obtained to validate that the cells were cryogenically preserved properly prior to analysis. Images indicate that ^{10}B concentrations in the perinuclear mitochondria-rich cytoplasmic region have a lower concentration of ^{10}B from BPA than the cytoplasm or nucleus. Together with subcellular imaging data from $^{13}C^{15}N$-labeled phenylalanine dosed cells, the authors concluded that BPA may enter the cell in a pathway similar to that of phenylalanine. More in-depth reviews on the application of SIMS to tissue imaging of endogenous and exogenous molecules as well as molecular depth profiling can be found elsewhere [35, 36, 47, 48]. While SIMS has been relatively undeveloped for drug imaging applications, recent developments in new ion sources to achieve molecular, nonfragmented, ions are a good sign of improvements in the technology. Several groups are working to improve SIMS capabilities for molecular drug and metabolite imaging and will provide a significant path forward for the use of SIMS imaging in the pharmaceutical industry.

11.3.2 DESI

Desorption electrospray ionization is a surface sampling platform introduced more recently that has potential utility for imaging drugs and their metabolites in tissues. Rather than using an ultraviolet (UV) laser and a source under vacuum conditions, DESI is carried out under ambient conditions. High-velocity electrosprayed

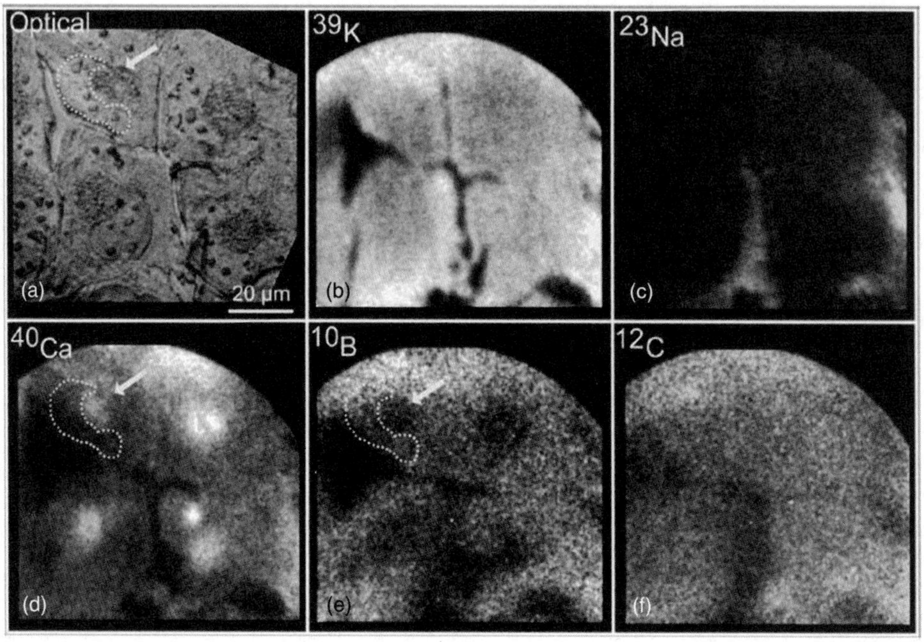

Figure 11.2 SIMS isotope images of T98G human glioblastoma cells treated with 110 µg/mL boron equivalent of ^{10}BPA for 2 h. In the reflected light image (a) several fractured freeze-dried cells are shown with discernible nuclei (the nucleus of one cell is shown with a dotted line) and a perinuclear mitochondria-rich cytoplasmic region (arrow). SIMS analysis of the same cells revealing subcellular isotopic distributions of ^{39}K (b), ^{23}Na (c), ^{40}Ca (d), ^{10}B (e), and ^{12}C (f) are shown. The areas within the dotted lines show the position of the nucleus and arrows indicate the mitochondria-rich perinuclear cytoplasmic region in SIMS images. Image integration times on the CCD camera for ^{39}K and ^{23}Na images were 0.4 s each and 120 s each for ^{40}Ca, ^{12}C, and ^{10}B images. Reprinted with permission from Chandra et al. [42]. Copyright 2007 Elsevier.

droplets (e.g., aqueous/organic, methanol/chloroform, or methylene chloride) are aimed at a surface of interest, and these projectile microdroplets desorb analytes from the surface due to pneumatic and some electrostatic forces. These gas-phase ions are then transferred to a mass spectrometer through an atmospheric transfer ion-transfer line [49].

With respect to MALDI, less development has occurred since this technology was introduced more recently; nonetheless, the DESI technology holds promise in many different areas of research, including drug distribution analyses [50–54]. Currently, the most widely recognized application of DESI by the pharmaceutical community is in the analysis of drugs and metabolites in dried blood spots [55]. Less known, however, is the ability to use this technology for the direct analysis of tissue surfaces [56–59]. In 2005, it was demonstrated that DESI could be used to desorb molecules directly from tissue surfaces, as illustrated in the Figure 11.3 schematic [59]. Shortly after this work, images were successfully acquired using a motorized two-dimensional (x,y) stage, paving the way for future drug distribution studies [58]. During a DESI tissue imaging experiment, a section of tissue is thaw-mounted onto

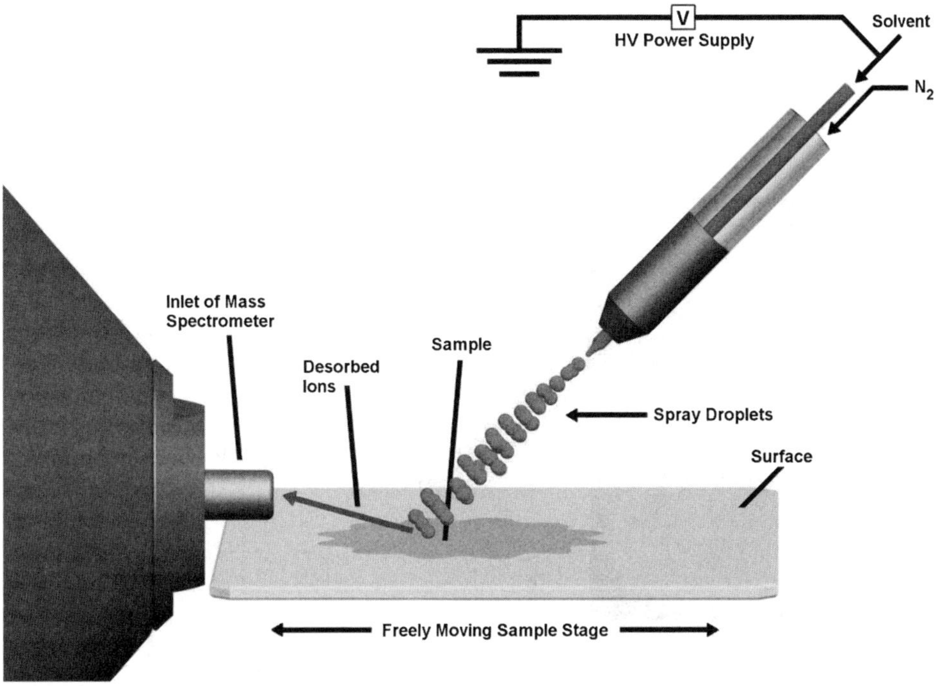

Figure 11.3 See color insert. Schematic of direct tissue analysis by DESI. Contributed by Justin M. Wiseman, Prosolia Inc.

a microscope slide (or it can be measured directly from an organ's surface without sectioning), and then the section or area of interest is moved under the fixed spray nozzle. By rastering the surface at a fixed velocity in two dimensions beneath the spray, ion images are constructed.

Although there have been only a few publications to date on the application of DESI to tissue distribution studies of pharmaceutical compounds, these examples should heighten interest in the utility of this technology for imaging drugs and metabolites. In the first example, Wiseman and colleagues examined the feasibility of using DESI for drug imaging studies by measuring clozapine and its desmethyl metabolite in various tissue types including brain, lung, kidney, and testis [56]. In this study, they used an automated DESI source connected to a linear ion trap mass spectrometer and a spray solvent of 70:30 methanol : water. Acquired image resolution varied, depending on the experimental objective, with the highest being 245 × 245 μm, although more recent experiments have demonstrated the capacity to achieve ~50 × 50 μm [60]. An example image of clozapine in brain is shown in Figure 11.4. In a parallel experiment with half of each organ, drug extracts from tissue homogenates were analyzed via LC-MS/MS to show that relative intensities from the DESI imaging results correlated well with the conventional drug quantification approach.

In another study, Kertesz et al. further expanded the applicability of DESI by illustrating that, like MALDI, whole body images of drug distribution could be obtained that were comparable to whole body autoradiography (WBA) results

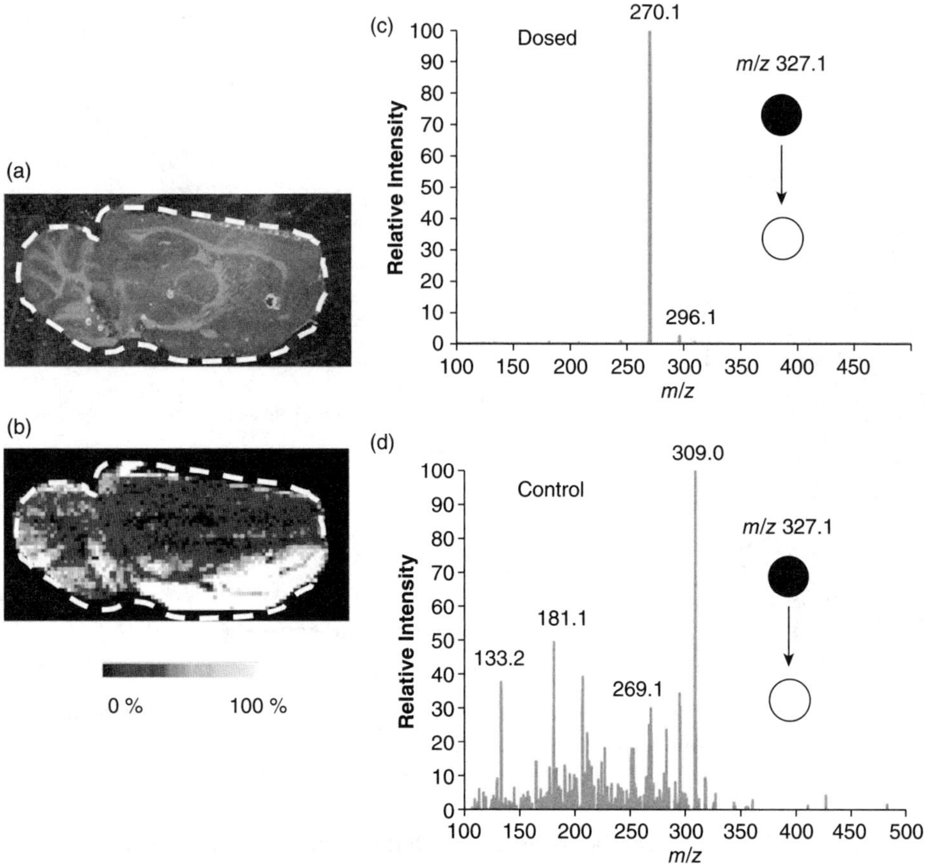

Figure 11.4 See color insert. Optical image of a 22 × 11 mm^2 sagittal rat brain section and the corresponding selected ion image. (A) Optical image of a sagittal rat brain section taken from animal 992 (0.5 h after dose). CB, cerebellum; Cbc, cerebral cortex; Cpu, caudate-putamen; Hpc, hippocampus; SNr, substantia nigra. (B) DESI mass spectral image of clozapine in the brain section recorded in the MS/MS mode. The image of the fragment ion at *m/z* 270.1 is shown by using false colors in raw pixel format. (C) Average product ion mass spectrum of *m/z* 327.1 ± 2 in B. (D) Average product ion mass spectrum of *m/z* 327.1 ± 2 in control sample 984. Reprinted with permission from Wiseman et al. [56]. Copyright 2008 National Academy of Sciences, U.S.A.

[61]. In this study, mice were dosed intravenously with propanolol at 7.5 mg/kg and sacrificed 20 and 60 min postdose. The sample preparation procedure was performed similarly to that used for whole body sections prior to MALDI imaging, but without the need to deposit MALDI matrix: frozen mice were embedded in a block of 2% carboxymethyl cellulose, sectioned sagittally at 40-μm thickness, and transferred to a glass slide using a tape-transfer system. All DESI experiments were carried out using a hybrid triple quadrupole linear ion trap mass spectrometer equipped with a particle discriminator interface [62–65]. Fragmentation of propanolol was performed from the tissue. Depending on the desired resolution,

the surface sampling scan rate ranged from 0.1, 0.5, 2, and 7 mm/s. While signal decreased with increasing scan rate, the overall analysis time was much less (i.e., 79 min) for a faster scan rate. In a similar experiment, an equivalent dose of [^{3}H] propranolol was administered to mice for WBA, which required exposing the sections to the phosphor imager plates for 7 days. Relative intensities of propanolol from kidney, lung, brain, and liver from DESI imaging were compared with the radiolabel signal attributed to propanolol by WBA imaging. Visual comparison of the two image techniques reveals drug (and/or metabolites in the case of WBA) in the brain, lung, kidney, and stomach regions. Results were in agreement between the two techniques, with the exception of kidney tissue, and potential reasons for this discrepancy were elucidated.

To date, there has not been a direct comparison of MALDI and DESI imaging, but there are few aspects of each technology that offer an obvious advantage of one over the other. Less sample preparation is required for DESI analysis since precoating of tissue with a matrix compound is not mandatory for the desorption/ionization process. The surface requirements are far more lenient with DESI because it is performed at atmospheric pressure and no lasers are used, whereas MALDI requires specially sized, conductive plates, depending on the instrument vendor. For most DESI imaging experiments, placing the sample on a standard microscope slide is sufficient. One drawback to analyzing tissues with DESI is the destructive nature of the surface probing process, but continued research with varying solvent compositions may help lessen this issue. MALDI, however, is much less destructive, and if the tissue slices are analyzed on conductive glass MALDI plates, additional experiments (e.g., histostaining, immunohistochemistry, SIMS analysis) can be performed on the same tissue section after washing off the remaining matrix material [2, 66]. Historically, one of the major criticisms of MALDI imaging has been resolution. Advances have occurred over the last few years that enable MALDI imaging at a raster width of approximately 20 μm, and less in some cases [67, 68]. Faster lasers have enabled high-resolution imaging without a significantly longer analysis time. Published images by DESI have been approximately 230 μm, but improvements are under way to enhance the image resolution for this platform, making it more competitive with the 20–100 μm achievable with MALDI [69]. Lastly, without a direct comparison to MALDI, it is difficult to ascertain the sensitivity variations between the two techniques. It is possible that they are similar, although some compounds may ionize from tissue surfaces more efficiently under ambient conditions than others. As pointed out by Kertesz et al., DESI was unexpectedly incapable of detecting major propanolol metabolites from tissues and the parent drug itself had very low signal intensities [61]. Wiseman et al., however, were able to detect metabolites from clozapine-dosed tissues. Lastly, as pointed out by Kertesz et al., there needs to be a standard procedure for determining the ionization efficiency of DESI from different tissues in order to normalize and make proper comparisons of drug/metabolite images from different tissue types, which is currently done in MALDI imaging experiments [4]. While DESI imaging is still in its infancy compared with the level of experimentation and improvements that have occurred for MALDI imaging, a direct comparison of the two techniques may shed light on some additional advantages or caveats of these two platforms for imaging of pharmaceutical compounds.

11.3.3 MALDI

Matrix-assisted laser desportion (MALDI) has been in existence since the 1980s, but the concept of imaging tissues with MS was demonstrated in 1997 by Caprioli and coworkers with the use of a MALDI-TOF instrument [1, 3, 70–72]. Although SIMS imaging is an older technology than MALDI imaging, the development of MALDI for tissue distribution studies of pharmaceuticals is far more advanced due to the softer ionization involved in the MALDI process and ability to ionize larger molecules without fragmentation. In contrast to SIMS, MALDI employs the use of a laser to initiate the ionization process. MALDI lasers range from UV (e.g., nitrogen or frequency tripled Nd:YAG) to IR, both able to desorb high molecular weight molecular ($[M+H]^+$) ions. While, like SIMS, some ions can be generated without the addition of matrix, it is most commonly added to enhance the ionization of molecules, as is also done in SIMS. Troendle et al. were the first to report the ability to detect drugs directly from tissue surfaces, and the first successful imaging of drugs in tissues came from Todd et al., who imaged clozapine distribution in a rat brain section after a tail vein injection of 3 mg/kg [36]. Soon after, in 2003, Reyzer et al. successfully imaged the tumor distribution of an anticancer drug in mouse tumor tissue [73]. Since then a lot of instrument and method development has occurred. Because of its molecular specificity and the ability to generate tissue distribution images of nonlabeled pharmaceutical material, this technology is now being used by the pharmaceutical industry to examine the distribution of unlabeled drugs and their metabolites, unambiguously, in individual tissues and whole body sections. The principles of pharmaceutical mapping in tissues by MALDI imaging and examples of its application to the pharmaceutical industry will be the subject of the remaining sections in this chapter.

11.4 FUNDAMENTALS OF IMAGING MS

There are various aspects to consider for imaging drugs and metabolites in tissues by MALDI MS. The sample preparation procedure is every bit as crucial as the mass spectrometer acquisition and data analysis for accurate determination of drug localization and the semi-quantitation of each drug entity. This section will cover the various aspects of these parameters for MALDI imaging.

11.4.1 Instrumentation

Due to the vast improvements in MS instrumentation since the introduction of tissue imaging and the compatibility of MALDI with various mass analyzer and hybrid instrument platforms, there are now many different instrument options available for the direct analysis of tissue sections. The choice of instrument depends on the analyte of interest; most commonly this choice is between small and large molecules. Most drug and metabolite imaging involves small molecule pharmaceuticals, which require an instrument designed for small molecule analysis. The first MALDI tissue images, both protein and drug, were acquired in the late 1990s and early 2000s from the MALDI-TOF platform [1, 3, 36, 72, 74]. Although MALDI-TOF instruments are capable of imaging small molecules, they suffer from lower sensitivity due to spectral noise in the low mass region that is generated from the matrix, matrix clusters, and some fragment ions [75]. Second, matrix compounds are extremely effective at self-protonation and are present in a much greater excess than the analyte of interest. Third, endogenous molecules may produce interfering signals, which cannot be differentiated from the analyte of interest under MALDI-TOF conditions. Lastly, there is a high concentration of salt in tissues that gets incorporated into the matrix crystals, resulting in the formation of many matrix cluster ions that produce signals at many *m/z* values across the low mass region of the mass spectrum, often generating ions that are isobaric with the analyte. Many tissue washing procedures have been examined to eliminate the salt, but these washing procedures are likely to delocalize the drug of interest and are therefore not used for drug and metabolite imaging [2, 76–78].

To circumvent the issues regarding low mass matrix interferences, tandem mass spectrometry (MS/MS) is most commonly used to generate fragment ions of the drug and/or its metabolite of interest. Reyzer et al. demonstrated the advantage of MS/MS analysis for small molecule drugs over the MALDI-TOF platform [73]. When doing MS/MS experiments, the use of a drug standard to optimize and obtain fragmentation spectra from enables researchers to create spectral patterns that can easily be monitored during experiments from the tissue. Since this work in 2003, there has been a rapid advancement in the mass analyzer options for small molecule imaging. Additionally, most commercial instruments have foregone the nitrogen laser for a longer lasting and higher repetition rate Nd:YAG system. There are now numerous instruments available for MALDI imaging, some equipped with both ESI and MALDI sources. Most of these instruments are amenable to many other types of research, which should be taken into consideration when choosing a platform. Examples of these platforms include, but are not limited to, MS-TOF (with or without ion mobility), hybrid triple quadrupole linear ion trap, linear ion trap, and Fourier transform-ion cyclotron resonance (FT-ICR) [79–85].

11.4.2 Sample Preparation for MS Imaging

The sample preparation for drug imaging is relatively simple, yet special care is required during the procurement process. After removal from the animal, tissues are flash frozen in a manner that does not create tissue fractures. Special procedures are required for sensitive tissue such as brain. The manner in which freezing and storage is performed for tissues to be analyzed via LC-MS/MS platforms does not require as much care, since tissues will be homogenized. For MSI, however, the samples are wrapped after the freezing process in a way that eliminates smashing of the tissue, which destroys its three-dimensional integrity. In the case of whole-body imaging, animals are frozen slowly in a mixture of dry ice and hexane, followed by freezing in a block of water or 3% carboxymethyl cellulose [82, 86].

The choice of cryo-sectioning depends on whether the tissue(s) to be analyzed are individual organs or a whole-body section. Single organs are sectioned in a cryo-microtome at a thickness of between 10 and 20 μm, depending on the application or the ease of handling of a particular tissue. Optimal cutting temperature polymer (OCT) is the medium used for cryo-sectioning frozen samples. Since it is a polymer, contamination of the sample with OCT will suppress ion formation in the MALDI source. It is therefore recommended that the sample be mounted onto the sample holder so that the OCT touches only the bottom portion of the sample, avoiding any contact with the surface being sectioned. Frozen sections are then placed onto the MALDI plate and thaw-mounted, followed by desiccation at least 0.5–1 h prior to matrix application [76]. Whole-body sections are cut in a cryomacrotome at approximately 20 μm in thickness. Tape is used to collect the sections and then mounted onto the MALDI plate with the help of double-sided adhesive tape or a tape transfer system. Prior to mounting onto the MALDI plate, sections are desiccated at least 1–2 h [4, 82, 86].

During method development it is essential to find the matrix and solvent system that is appropriate for the analyte and maximizes its signal from tissue. There are many options for the matrix/solvent combination; the goal of the matrix solution is to solubilize the analyte of interest. Solvent conditions will then vary depending on the analyte of interest or the tissue type being analyzed. The matrices most commonly used for drug analysis are DHB (2,5-dihydroxybenzoic acid), CHCA (α-cyano-4-hydroxycinnamic acid), and sometimes SA (sinapinic acid). The solvent used must be able to solubilize both the matrix compound and the drug. Solvents frequently used are ethanol, methanol, and acetonitrile, typically 50–60% in water. The addition of trifluoroacetic acid (TFA) to the matrix solution (0.1–1.0%) often aids in the ionization process for biomolecules. While it is not often required for ionizing low molecular weight compounds, it can sometimes have a positive effect on the drug signal and should be considered [73, 76].

Prior to analysis by MS, matrix must be applied onto the tissue surface in a reproducible manner to achieve homogeneous crystallization. For direct tissue analysis by MALDI MS, there are two conventional strategies used to apply matrix to the tissue surface: depositing small droplets of matrix in an array/grid pattern, and coating the entire tissue surface with matrix solution. While manual pipetting of matrix solution onto the tissue is possible, it is only used for method optimization and method testing procedures as the spot size is ~1 mm. Robotic devices capable of depositing very small droplets of matrix in a defined array are one approach to coating the tissue for imaging. Such devices provide noncontact deposition so that there is no physical contact with the sample, avoiding cross-contamination and destruction of tissue integrity. Examples of these mechanisms include piezoelectric, ink jet, and nozzle-less acoustic-wave devices. Such devices generally produce matrix spots in the range of 150–250 μm in diameter [87–90]. Through sophisticated software parameters, the laser rasters only at the matrix spot location, which results in an image resolution defined by the spacing between the matrix spots in the array. Spray coating is the second option for covering the tissue with matrix. The advantage of this technique is that there is more freedom in choosing the image resolution, unless the laser beam is smaller than the individual crystals. This could become an issue as lasers are continuously improved to achieve cellular resolution. Spray-coating can be done manually or robotically. The cheapest option is manual application using a glass spray nebulizer (TLC reagent sprayer) [76]. Robotic devices include vibration vaporization and heated capillary nebulization [91, 92]. In each case, several cycles of coating are performed, allowing the matrix to coat the tissue surface each time before allowing it to dry 30 s–2 min per cycle. The objective is to get the tissue wet enough to solubilize the analyte(s), but not wet enough to delocalize molecules or too dry to hinder analyte(s) solubilization. After sufficient and homogeneous crystal coverage has been achieved, the sample is ready for MS analysis. A detailed review on the sample preparation criteria for both MALDI and matrix-enhanced SIMS can be found elsewhere [93].

11.4.3 Data Acquisition and Analysis

Once in the mass spectrometer, software is used to define the area of interest to be analyzed. The sample stage moves the sample under the laser (fixed) at discrete steps as defined by the image resolution, and then a mass spectrum is acquired at each location. An internal standard, ideally a stable-isotope-labeled standard, or a compound with a similar structure or ionization efficiency, is spiked into the matrix for normalization of the data [94, 95]. This helps remove signal differences due to minor matrix variabilities across the sample, but does not correct for extraction efficiency differences between tissue types. Stoeckli and coworkers provide an explanation to approaching normalization between tissue types [4]. The MS/MS fragmentation pattern, or intact mass when MS/MS is not employed, is acquired for both the analyte and the internal standard in each spot measured. After the image is complete, software tools (e.g., provided by instrument vendor, free-ware, or custom programs) are used to recreate the image by plotting the intensities of the ion of interest at each spot as a function of the location on the tissue surface, thus creating a two-dimensional ion density map or image.

11.5 APPLICATIONS OF IMAGING MS TO DRUG AND METABOLITE IMAGING

MALDI imaging mass spectrometry is an invaluable tool for visualizing the distribution of drugs and their metabolites in tissues. Not only does it provide the unambiguous identification of the drug entity as opposed to conventional imaging technologies requiring a label, but it also provides a more high-throughput and more cost-effective option than designing and disposing of a radiolabeled molecule. MALDI imaging has been used in pharmaceutical R&D to image drugs and their metabolites from individual tissues as well as whole-body sections. Examples applications of this technology will be discussed, and in some cases, how it has been used in combination with other imaging platforms.

11.5.1 Imaging Drugs and Metabolites in Individual Organs/Tissues

There are many examples of research employing MALDI imaging to determine tissue-specific distribution patterns of drugs and their metabolites in individual tissues. Nilsson et al. followed the *in vivo* transport of the inhaled compound, tiotropium, a bronchodilator for the treatment of asthma and chronic obstructive pulmonary disease (COPD), within the lungs of dosed rats (1.1 mg/kg) [96]. By collecting 12-μm-thick sections of lung tissue at 200 mm intervals, they compared drug distribution patterns through the lung. Results showed that within 15 min following exposure, the tiotropium parent MS ion (*m/z* 392.1) and fragmented daughter MS/MS ions (*m/z* 170.1 and 152.1) were dispersed in a concentration gradient away from the central airways into the lung parenchyma and pleura as

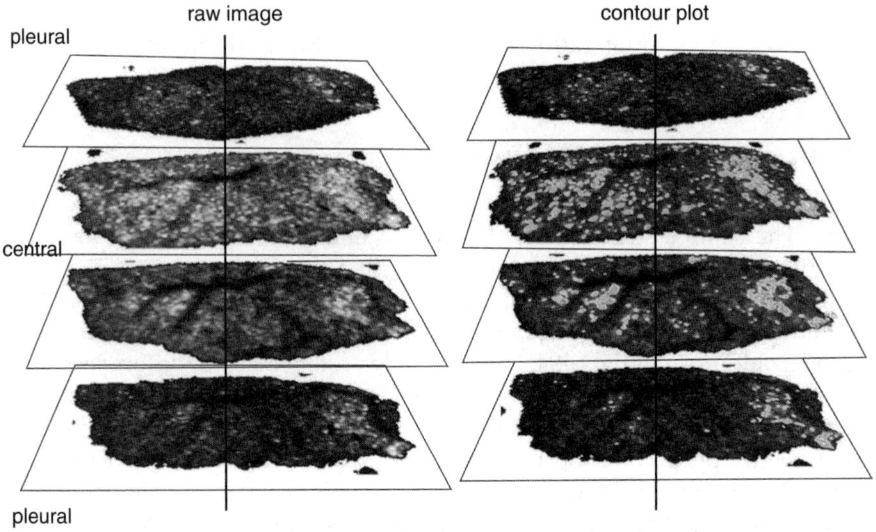

Figure 11.5 See color insert. The spatial distribution of inhaled tiotropium (TTP) in serially sectioned rat lung tissue. Raw MS images (left column) showing MS localization of inhaled tiotropium (TTP) ion (*m/z* 392) by pixel location in serial sections of whole lung moving anatomically in lung volume from pleural to central to pleural. The arrow indicates approximate carinal entry point of the drug into the central conducting airways. Contour plots (right) of the relative concentration gradients of TPP found in the various lung segments showed a rapid and homogeneous transport of the drug from airways into the parenchyma within 15 min after exposure. The MALDI matrix CHCA was applied by an automatic sprayer device. Reprinted with permission from Nilsson et al. [96].

seen in Figure 11.5. These levels agreed well with amounts detected in lung compartments by chemical extraction and conventional LC-MS/MS. In another paper, Li and colleagues determined the distribution of astemizole and its metabolites by MS^n in perfused and nonperfused brains to elucidate the possible causes of its central nervous system side effects [81]. Rats were dosed at 100 mg/kg and sacrificed 2 h postdose. Figure 11.6 shows the MALDI imaging results for astemizole and a metabolite in perfused and nonperfused brain sections. As seen in the images, no differences in drug distribution or intensities were observed from the perfused and nonperfused brain, suggesting that the drug level observed was likely due to tissue-binding activities and not likely due to residual biological fluids from the systemic circulation. The M-14 metabolite image showed localization isolated to the three ventricle sites, suggesting that this metabolite would likely stay in the cerebrospinal fluid rather than to cross the blood–brain barrier. This enabled researchers to conclude that the drug, astemizole, did reach the target site of action and was likely to be the major cause of efficacy and central nervous system side effects following drug administration in rats. In a third example, Marshall et al. demonstrated the use of MALDI imaging to determine the degree of skin penetration for the three novel nonsteroid glucocorticoid receptor (GR) agonists [97]. Glucocorticoids are known to cause skin blanching when applied topically to human skin, so the researchers aimed to determine the distribution of these agonists in porcine

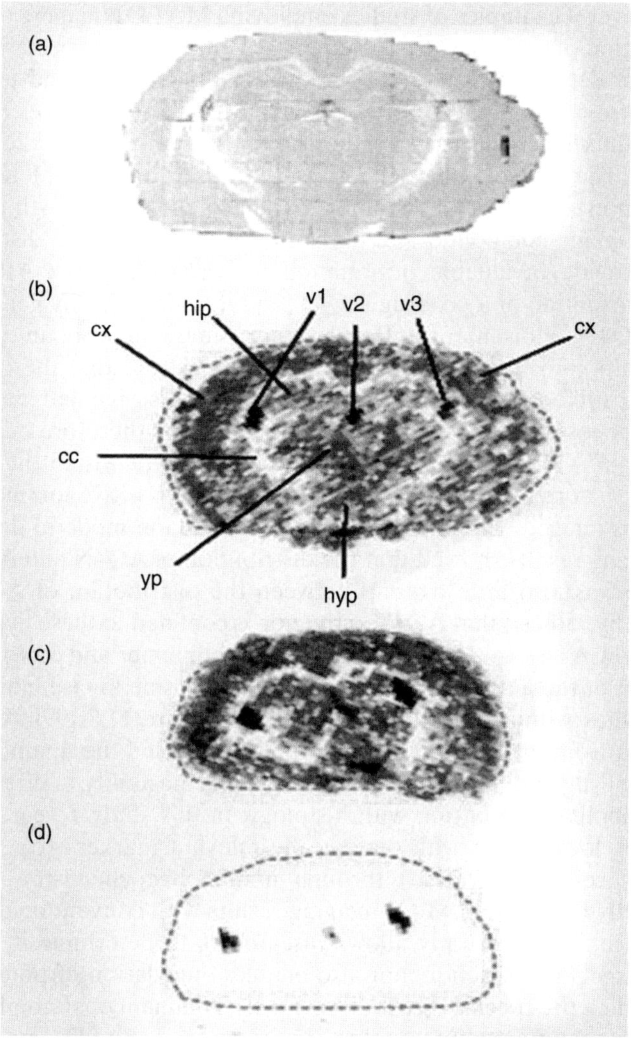

Figure 11.6 See color insert. (A) The optical image of a rat brain from a coronal section. (B) Matrix-assisted laser desorption/ionization (MALDI)-mass spectrometry (MS)/MS images of astemizole in the rat brain slice without perfusion and (C) with perfusion. Cortex, hippocampus, corpus callosum, hypothalamic region, thalamus region, choroid plexus, dorsal third ventricle, and lateral ventricle are indicated by arrows. (D) MALDI-MS/MS images of M-14 metabolite of astemizole in the rat brain slice. Reprinted with permission from Li et al. [81]. Copyright 2009 Future Science Ltd.

skin tissue in order to understand the different responses obtained from skin blanching and *in vitro* potency assays. The results showed that two of the three compounds penetrated through the epidermis layer into the dermis, while the third compound demonstrated limited penetration past the epidermis. The results agreed with liquid extraction surface analysis (LESA) analyses. MALDI imaging and LESA data correlated well with the extent of skin blanching obtained for the same three compounds in human volunteers.

There are several examples of studies employing MALDI imaging to understand tumor penetration of oncology drugs. Tumors are inhomogeneous masses that grow rapidly, without defined architecture. The lack of organization and understanding of their structure can complicate drug design. Many tumor features (e.g., vasculature, acidity, efflux mechanisms) impede drug diffusion, leading to reduced therapeutic efficacy. The lack of drug delivery to the tumor or incomplete penetration of the drug can eventually lead to tumor resistance and/or regrowth of the tumor. The ability to image compound and metabolite penetration and distribution into tumors can provide invaluable information for drug discovery. Atkinson et al. imaged the distribution of a prodrug, AQ4N, and its metabolically generated active compound, AQ4 (topoisomerase II inhibitor), simultaneously in H460 human tumor xenografts [98]. The authors aimed to demonstrate that the conversion of AQ4N can be linked to regions of hypoxia, which is signaled in part by the depletion of adenosine triphosphate (ATP). The authors therefore expected to see reduced levels of ATP in regions where AQ4 was present within the tumor. First, AQ4N and AQ4 were measured simultaneously in the positive ion mode, then the same tissue was imaged a second time in the negative ion mode to detect ATP. As expected, imaging results showed that the distribution of AQ4N and AQ4 had little overlap. There was also little overlap between the distribution of AQ4 and ATP, supporting the hypothesis that AQ4N activation is confined to these hypoxic regions of tissue (i.e., low ATP). In a second example, Oppenheimer and coworkers imaged the distribution of the anticancer drug, imatinib, a tyrosine kinase inhibitor, and its major metabolites within solid tumors as seen in Figure 11.7 [99]. Acquisition of MS/MS spectra from imatinib and its metabolites enabled the unambiguous identification of the drug entities. An advantage of MSI is the ability to directly correlate drug and metabolite distribution with histology. In this study, researchers directly correlated drug localization with biologically relevant markers (e.g., markers for vasculature or cell proliferation) through immunohistochemistry staining. The ability to directly compare MALDI imaging results with conventional immunohistochemistry techniques not only allows researchers to determine if the drug has reached the target site of action, but also enables elucidation of potential mechanisms surrounding the specific distribution. Lastly, Huamani et al. combined the use of Doppler ultrasonography, immunohistochemistry, and MALDI imaging to determine the efficacy of combining radiation (XRT) with a dual epidermal growth factor receptor (EGFR)/vascular endothelial growth factor receptor inhibitor, AEE788, in prostate cancer models with different levels of EGFR expression [100]. Four treatment groups were examined: control, AEE788-dosed tumors, XRT-treated tumors, and tumors treated with a combination of AEE788 and XRT. Two prostate cancer models were used in the study: one derived from DU145 cells and the other from PC-3 cells. Of the two models studied, the DU145-based model exhibited significant tumor growth delay when treated with the combined-modality therapy. In addition, Doppler sonography demonstrated a significant decrease in tumor blood flow and immunohistochemistry showed tumor blood vessel destruction. The authors sought to use MALDI imaging to confirm AEE788 bioavailability in the tumors and whether this correlated with a decrease in the tumor blood flow. Figure 11.8 shows representative results from the MALDI imaging data. As expected, no drug signal was observed in the vehicle control. The next two images, one of a 24-h postdose sample and one after 5 days of consecutive therapy, show

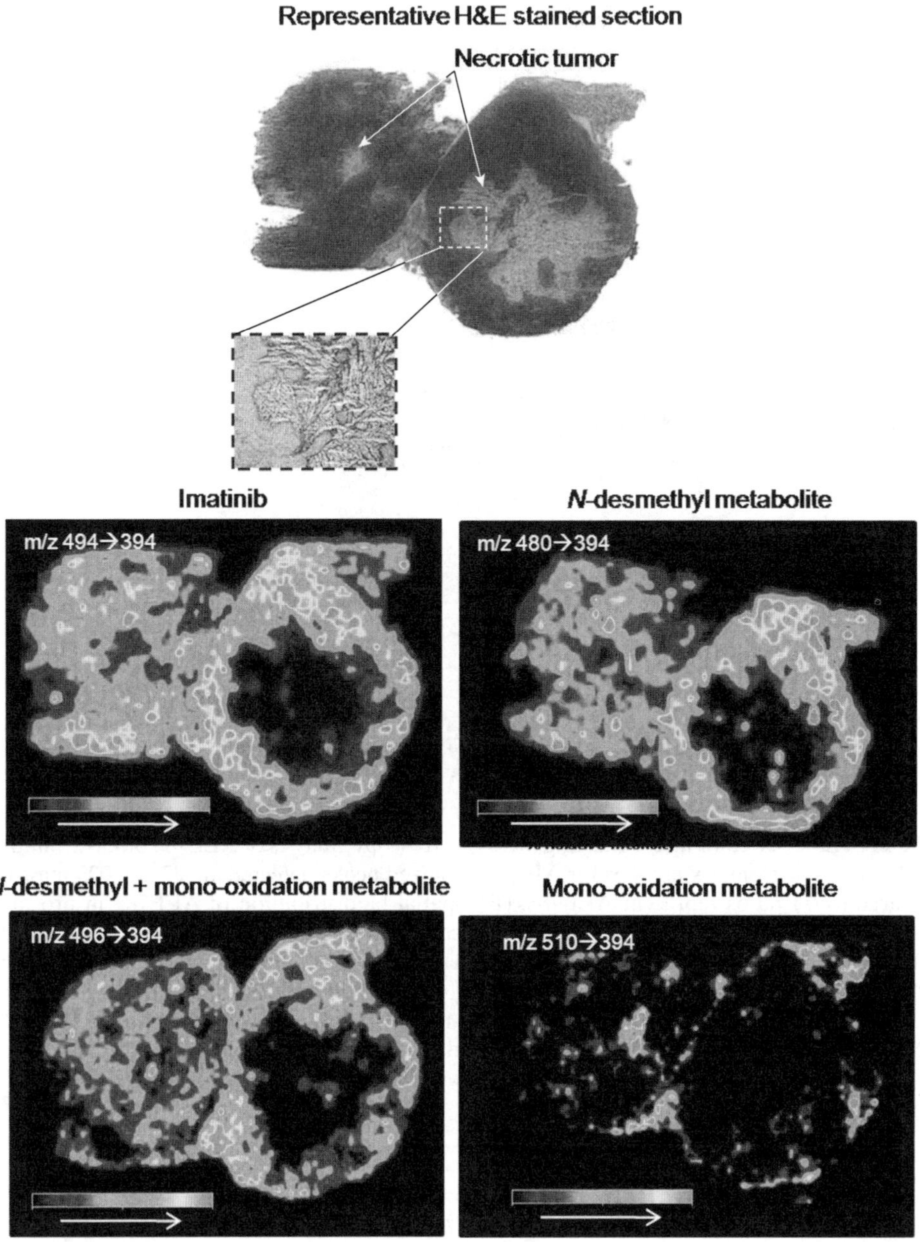

Figure 11.7 See color insert. MALDI images of a parent drug and metabolite distribution in tumors. Comparison of images with the histological reference suggests that the parent drug and two of its metabolites are predominately localized to viable tumor and not as efficient at reaching necrotic regions of tumor tissue. The third metabolite has a unique distribution, primarily localized to the host tissue outside of the tumor. Reprinted with permission from Oppenheimer et al. [99].

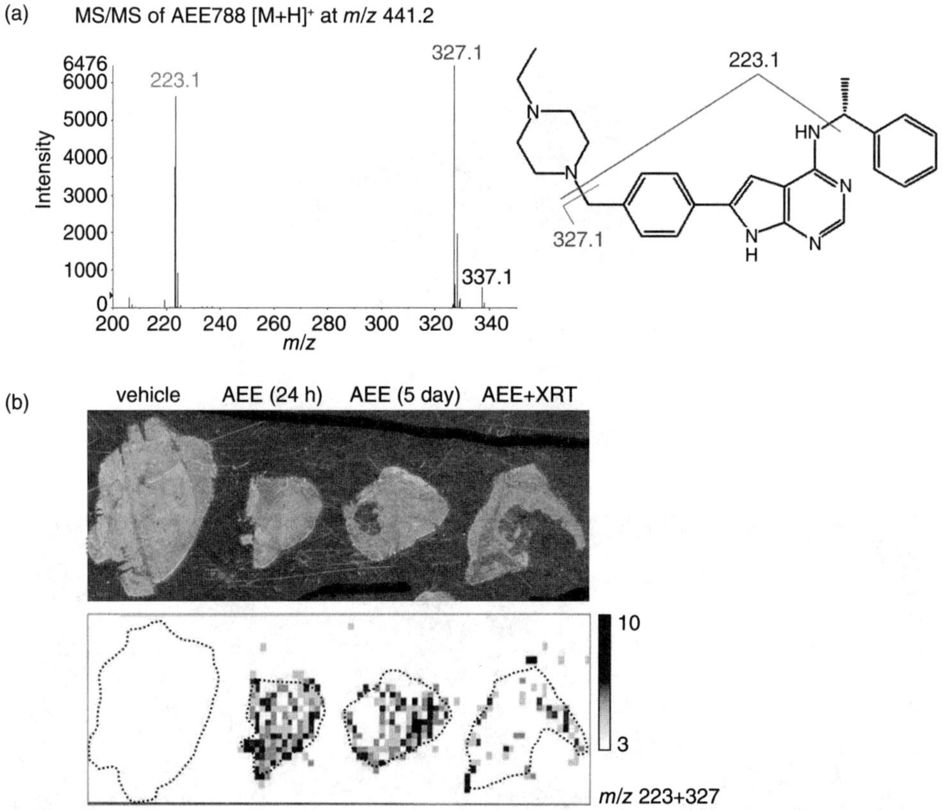

Figure 11.8 See color insert. Imaging mass spectroscopy analysis of AEE788 biodistribution. (A) Mass spectrometry analysis for AEE788. Ionized peaks detected at 327 and 223 mass to charge (*m/z*) ratios represent AEE788. (B) Spatial biodistribution of AEE788 in prostate cancer xenograft sections through matrix-assisted laser desorption ionization (MALDI)-Q-TOF-MS. (Bottom panels). MALDI-MS images of frozen DU145 prostate tumor tissue sections harvested at multiple times after oral administration of AEE788 compound, as indicated: Lane 1, vehicle control treatment; Lane 2, 24 h after treatment with 25 mg/kg of AEE788; Lane 3, AEE788 (25 mg/kg) for 5 consecutive days; and Lane 4, AEE788 (25 mg/kg) + radiation therapy (XRT; 3 Gy) for 5 consecutive days. Adapted with permission from Huamani et al. [100]. Copyright 2008 Elsevier.

heterogeneous distribution of the AEE788 compound. The sustained exposure after 24 h of dosing demonstrated favorable pharmacokinetics for use in combination with XRT; thus, the last sample imaged was treated with combined AEE788 and XRT. As illustrated by immunohistochemistry, this regimen led to tumor blood vessel destruction. The supplemental MALDI data supported these findings by illustrating decreased biodistribution of AEE788 in the combination-treated prostate tumors. Together, their data supported the conclusion that AEE788 was efficacious in DU145 prostate tumors and that low doses of AEE788 combined with XRT can lead to significant tumor growth delay in highly EGFR-expressing DU145 prostate tumors.

11.5.2 Imaging Drugs and Metabolites in Whole-Body Sections

About three years after MALDI imaging of pharmaceuticals was introduced, researchers began extending the application of this technology from individual organs/tissues to whole-body sections [5, 86]. This approach is complementary to quantitative whole-body autoradiography imaging (QWBA) because it enables the simultaneous detection of a drug and its metabolites in a cross-section of the body. In the first detailed paper of whole-body imaging, Shahidi et al. administered the antipsychotic drug olanzapine at 8 mg/kg to rats and sacrificed 2 and 6 h postdose. Whole-body sections were prepared as described earlier in this chapter. MALDI MS/MS analysis was performed on the sagittal whole-body sections to simultaneously detect olanzapine and its *N*-desmethyl and 2-hydroxymethyl metabolites. As shown in Figure 11.9, the three compounds can be detected throughout the body at 2 h postdose, including the target organs of brain and spinal cord. The paper also shows that olanzapine and its metabolites are present 6 h postdose, but decreased by 66% in the brain and spinal cord. No metabolites were detected in the target organs, as expected, but were visible in the liver and bladder. Although parallel QWBA experiments were not performed in this study, researchers pointed to earlier QWBA results, and the MALDI imaging data were in agreement with that data and the known metabolic pathways of this compound in rats. Stoeckli et al. performed a parallel MALDI imaging and WBA experiments to directly compare the agreement between the two techniques for whole-body imaging [4]. In this study, an equivalent concentration of unlabeled and radiolabeled compound was administered for MALDI imaging and WBA experiments, respectively. Comparisons of WBA and MALDI imaging data from whole-body sections at 5 min and 1 h postdose revealed remarkable similarities between the two techniques, showing high levels of drug in the trachea, lung, and stomach, with lower drug levels in the blood. Relative quantification levels from each of the methods were compared using the stomach, lung, and trachea signals. These data also showed good correlation between the two technologies, with a relative standard deviation better than 10%. The value measured by WBA in lung was 78% of the trachea concentration, with corresponding MALDI data at 69%. The relative concentrations of the stomach compared with the trachea were 90% for WBA and 73% for MALDI imaging. A third paper, from Trim et al., provides an additional whole-body imaging study combined with WBA. The drug of interest, vinblastine, was administered at 6 mg/kg for MALDI imaging, and 6 mg/kg of radiolabeled drug was administered for WBA. Since the *m/z* of vinblastine falls in a mass range rich with phospholipid signals, the authors used ion mobility separation combined with MS/MS. The resulting MALDI images of vinblastine were in good correlation with WBA results for the radiolabeled drug. A fourth paper, by Chen et al., used whole-body imaging to study the first pass metabolism in whole-body sections dosed with 50 mg/kg of tertfenadine, an antihistamine [101]. Simultaneous detection of the drug and its active metabolite, fexofenadine, was performed by MS/MS analysis. At 1 h postdose, the drug was mostly in the stomach and intestine, whereas its active metabolite was homogeneously detected in the liver, intestine, and stomach. The absence of parent drug in the systemic circulation indicated a high extraction ratio, suggesting that most of the dosed compound entering the liver was metabolized before reaching general blood circulation. At 4 h postdose, fexofenadine was detected in the small and large intestines. Authors

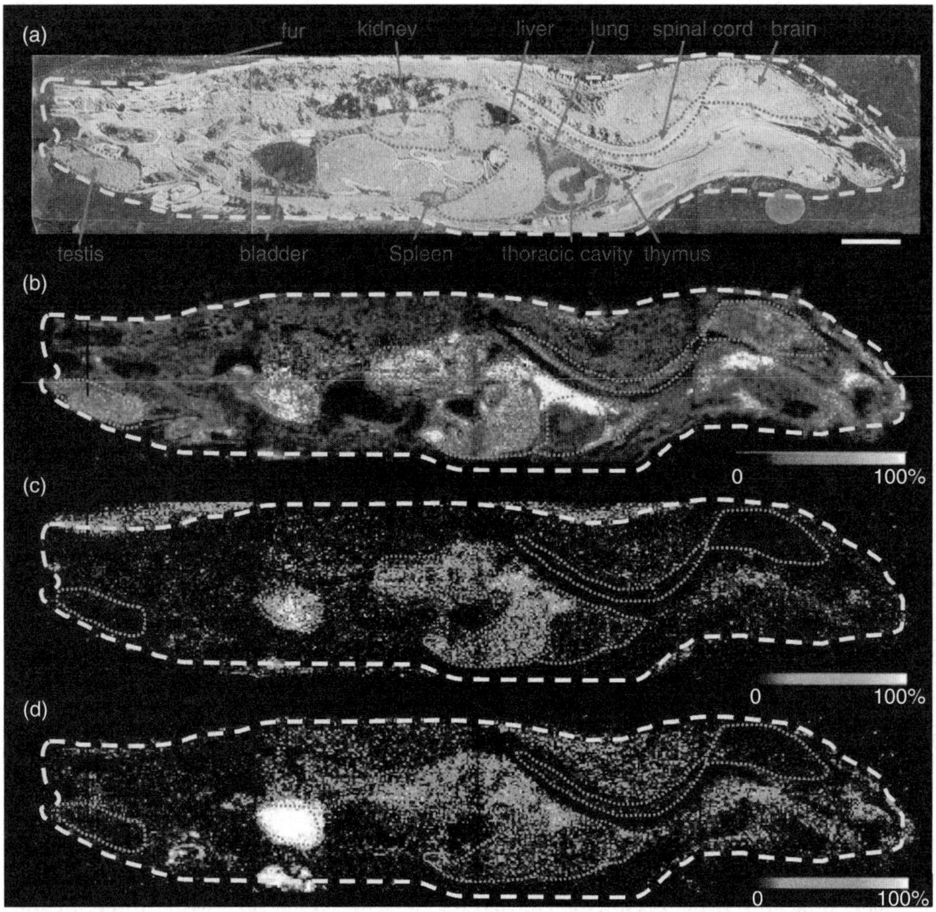

Figure 11.9 See color insert. Detection of drug and metabolite distribution at 2 h postdose in a whole rat sagittal tissue section by a single IMS analysis. Optical image of a 2 h post OLZ dosed rat tissue section across four gold MALDI target plates (A). Organs outlined in red. Pink dot used as time-point label. MS/MS ion image of OLZ (*m/z* 256) (B). MS/MS ion image of N-desmethyl metabolite (*m/z* 256) (C). MS/MS ion image of 2-hydroxymethyl metabolite (*m/z* 272) (D). Bar, 1 cm. Adapted with permission from Khatib-Shahidi et al. [86]. Copyright 2006 American Chemical Society.

concluded that this was potentially originating from metabolism within the gastrointestinal (GI) tract or direct excretion to the GI tract from the systemic circulation since fexofenadine is a P-glycoprotein (Pgp) transporter substrate. The whole-body imaging examples discussed here illustrate that with proper sample preparation and normalization procedures in place, MALDI imaging data from multiple tissue types correspond well to conventional autoradiography, indicating that MALDI imaging is a valid method to measure the whole-body distribution of pharmaceutical compounds. A second advantage illustrated is the ability to study the first-pass metabolism of drug compounds, since MALDI can differentiate between parent and metabolite distribution and WBA cannot provide this information.

11.5.3 MALDI Imaging as a Complement to Autoradiography

Combining complementary techniques like MALDI imaging with autoradiography studies has the potential to be of extremely high value in the pharmaceutical industry. While the autoradiography imaging data can provide absolute quantitation, and higher sensitivity, the MALDI imaging data can determine which autoradiography signals are from the drug and which are solely from the metabolite. This type of information is invaluable when drug retention has occurred; MALDI imaging can decipher whether this accumulation is from the parent or drug metabolite, which can mean the difference between continuing and halting a program. One example using MALDI imaging to clarify autoradiography data is provided by Hsieh et al. [80]. In this study, distribution of clozapine in sagittal rat brain sections was determined by MS/MS fragmentation. In a parallel experiment, the equivalent dose of radiolabeled clozapine was administered for autoradiographic imaging. Results from the autoradiography study suggested that clozapine is distributed throughout the brain, with the highest concentration found in the lateral ventricle. This same differential localization was also illustrated by the MALDI imaging results, where the parent drug was readily detected in the entire brain, with the most intense signal also observed in the ventricle area. To further validate both imaging results, two regions from the rat brain tissue sections (low and high signal) were isolated for simultaneous determination of clozapine by a more classic LC-MS/MS approach. These data correlated with the two imaging approaches, and also confirmed that no measurable clozapine metabolites were present in the brain sections. Another example illustrating the use of MALDI imaging to validate autoradiography results is illustrated by Drexler et al. in an ocular study [102]. A radiolabeled test compound was administered for QWBA analysis and an unlabeled version of the same compound was administered for MALDI imaging analysis, both at an equivalent concentration of 12 mg/kg. As illustrated in Figure 11.10A, autoradiograms of the eye indicated that the radiolabeled entity preferentially distributed to the uveal tract and not the cornea, but these results were unable to elucidate whether the entity detected in the uveal tract was parent drug or metabolite. The MALDI imaging experiments, performed via MS/MS fragmentation, provided unambiguous identification of the analyte as the dosed parent drug and its semi-quantitative localization within the uveal tract as shown in Figure 11.10B. Combining the information obtained by both QWBA and MALDI imaging experiments, the authors confirmed that neither the parent drug nor any metabolites accumulated in the cornea, which reduced concerns regarding ocular phototoxicity for this compound.

11.5.4 Unconventional Applications for Imaging

The utility of MALDI imaging in pharmaceutical R&D is not limited to studying drugs and their metabolites in whole-body sections, target tissues, or sites of toxicity. Its applicability spans much wider, with many of its potential uses yet to be realized. In this section three examples are provided to illustrate other ways in which this technology can be utilized: environmental safety, imaging tablets for impurity testing, and validating potential imaging agents for conventional imaging modalities. In the first example, Oppenheimer et al. used MALDI imaging to determine the distribution of pharmaceuticals in zebrafish [103]. Interest in zebrafish research has rapidly

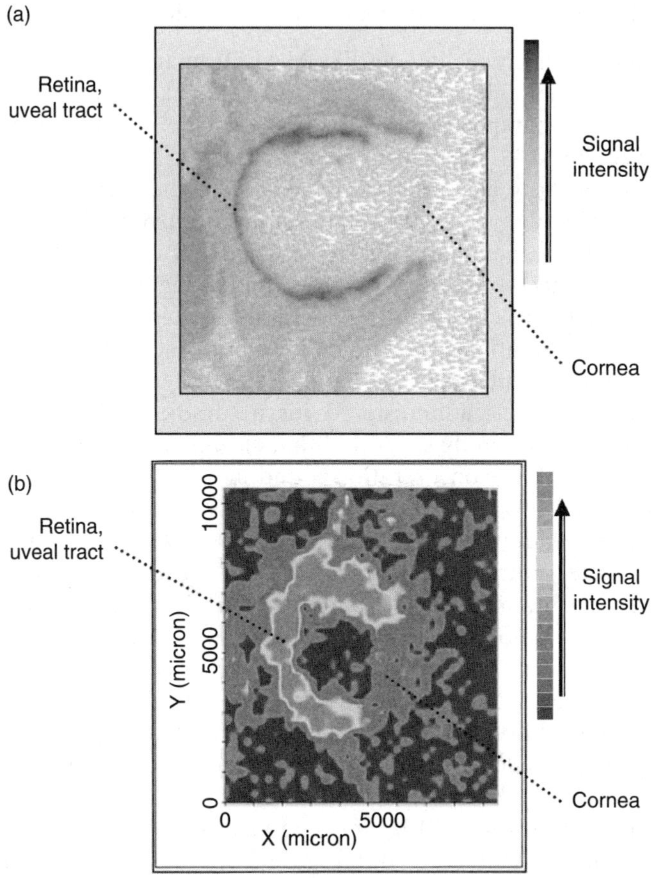

Figure 11.10 See color insert. (A) Autoradiogram of an eye section indicating the preferential distribution of the radiolabeled analyte(s) to the back of the eye (retina and/or uveal tract). (B) Ion map of the drug with the highest levels present in the back of the eye (retina and/or uveal tract). Reprinted with permission from Drexler et al. [102]. Copyright 2010 Elsevier.

increased over the years for their potential benefits as screening models for toxicity and human disease [104–106]. In this study, researchers examined the uptake and distribution of clozapine in zebrafish exposed to the drug in their water environment. The distribution of clozapine was determined by MS/MS fragmentation. Figure 11.11 shows the uptake and distribution of clozapine each hour over a 4-h time-course, illustrating increased uptake over time. This approach has been used by Lisa Constantine's Environmental Safety group at Pfizer to examine the uptake, distribution, and clearance of drug candidates. This method enables rapid screening of potential new drugs for their bioconcentration potential in aquatic species and subsequent biomagnification up the food chain. A second unique application is illustrated by Earnshaw et al., who used MALDI imaging to assess the homogeneity of the active drug compound throughout the excipients contained within the tablets [107]. Successful images were obtained from four of the tablets studied, but two of them were unable to be analyzed because of the crumbly nature of the tablets. The

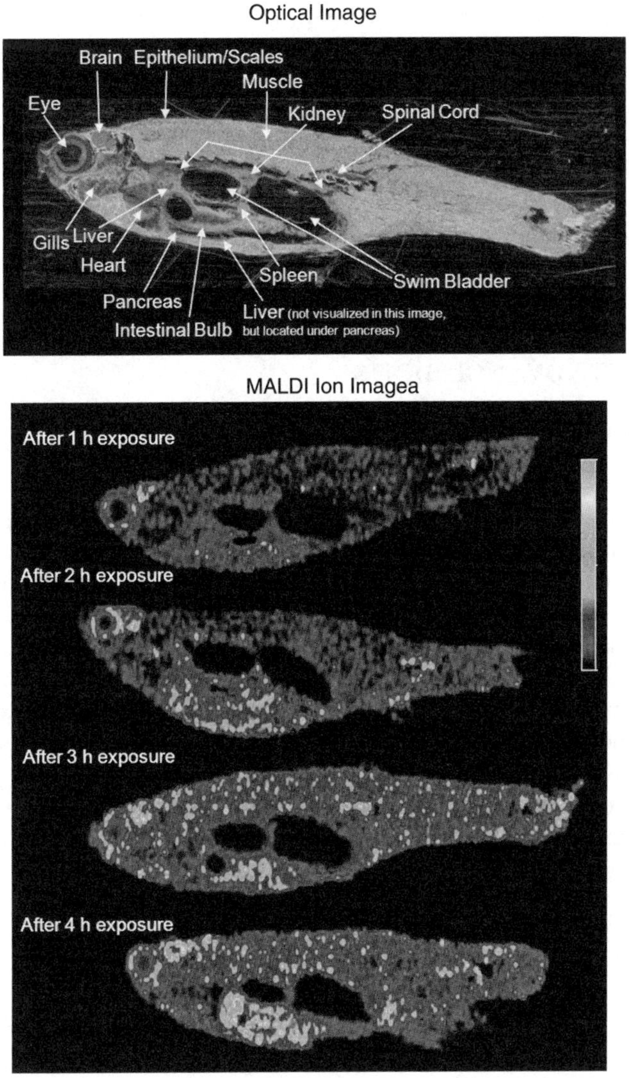

Figure 11.11 See color insert. Exposure time-course study in zebrafish. Zebrafish were placed in separate holding containers for the duration of drug exposure. Fish were dosed with clozapine at 2 mg/L for 1, 2, 3, or 4 h while fasting. MALDI images were acquired with a QSTAR XL at 250 × 250 μm resolution. The *m/z* 327 → 270 transition was monitored for clozapine and the *m/z* 313 → 256 transition was monitored for the internal standard, olanzapine. The control zebrafish, not shown, confirmed that no endogenous signal was present at the transition monitored. Reprinted with permission from Oppenheimer et al. [103].

ability to image the distribution of the active pharmaceutical ingredient in tablets could be used to improve the manufacturing process, enable improvements to product quality for safer and more efficacious tablet formulations, and test for impurities [108]. In the third example, Acquadro et al. used MALDI imaging to study the distribution of an MRI contrast agent in mouse liver and validated the results with both *ex vivo* inductively coupled plasma-atomic emission spectroscopy

(ICP-AES) and *in vivo* MRI [109]. Image results from the gadolinium complex B22956/1 showed homogeneous distribution through the liver section. Over the five time-points studied, the relative signal-to-noise ratio (S/N) from the MALDI imaging experiments was determined and compared with signals from the MRI and ICP-AES, although both MRI and ICP-AES only measured gadolinium. Each technique revealed the same trend: the MRI contrast agent reaches its maximum around 30–90 min postdose followed by decreasing concentration before it is excreted approximately 8 h postdose. These data illustrate the ability to screen potential MRI contrast agents and possibly PET tracers with MALDI imaging.

11.6 FUTURE ADVANCEMENTS

11.6.1 Analysis of Larger Molecule Therapeutics

An increasing proportion of pharmaceutical R&D focus is being spent on biologics, or biotherapeutic drugs. Biologics include proteins intended for therapeutic use, monoclonal antibodies, and oligonucleotides [110]. The development of MALDI imaging for drug distribution studies has primarily focused on small molecules, but as increasing research and development endeavors are spent in the realm of biologics, it is becoming increasingly important to understand the feasibility of this technique for imaging biologics in tissues. While extremely large biomolecules such as monoclonal antibodies may be out of the scope of practicality at this time without some prior modification to the molecule, the ability to image smaller therapeutics such as proteins and oligonucleotides are worth investigating. To date, little work has been carried out in this area, but Caprioli and coworkers at Vanderbilt University are the first researchers to show potential in this application of MALDI imaging [111].

"Decoy" oligodeoxynucleotides are a new class of therapeutic agents designed to regulate transcription factor activity in a selective and sequence-specific manner. The decoy used in the Caprioli study is a synthetic 15-mer double-stranded oligodeoxynucleotide that is used to target the action of signal transducer and activator of transcription 3 (STAT3) in cancer [112–115]. Tumors, derived from subcutaneous injection of a tumor cell line, were directly injected with 100 μg/50 μL of STAT3 decoy, and mice were sacrificed 5 min post-injection. The optimal MALDI matrix for this STAT3 compound was determined to be 3-hydroxypicolininc acid spiked with ammonium citrate to reduce sodium adduct formation. MALDI imaging results, acquired at a spatial resolution of 250 μm, show drug distribution within the tumor as illustrated in Figure 11.12 [115]. The mass spectrum shown in the figure shows that peaks representing the single strands at *m/z* 4584 (5′-3′ sequence) and *m/z* 4753 (3′-5′ sequence) for the STAT3 decoy dominate the spectrum with few ion adducts and high mass accuracy (mass error <0.1 Da). Results from this preliminary work demonstrate the future potential of imaging oligonucleotide therapeutics in tissues.

11.6.2 Derivatization

Efficient ionization of neutral molecules is difficult to achieve by MS, yet conventional LC-MS/MS methods for drug quantitation have found ways around this issue by derivatizing the pharmaceutical of interest during the extraction process in order

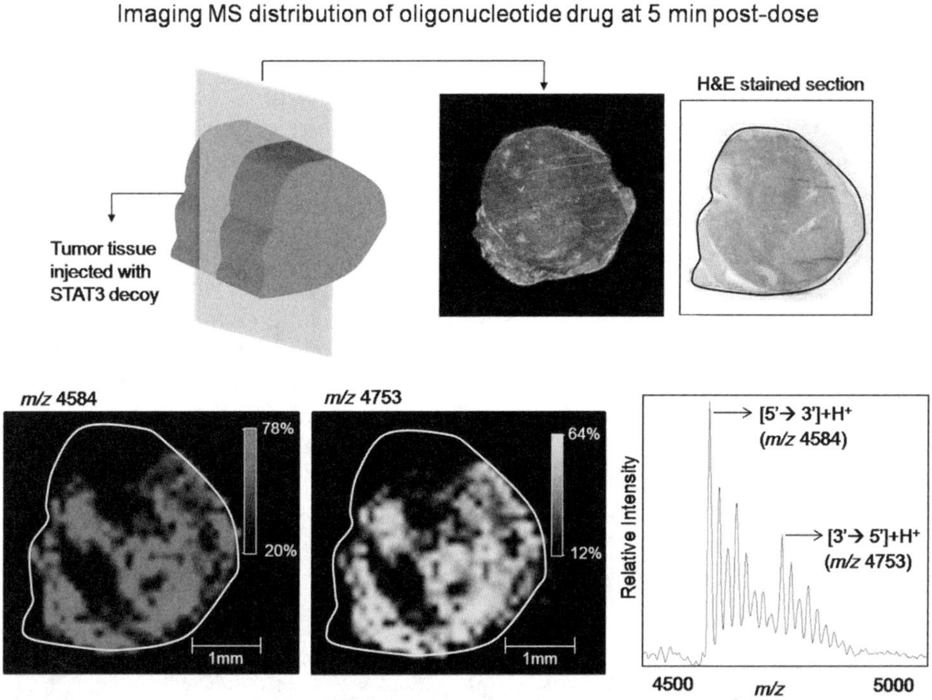

Figure 11.12 See color insert. Imaging MS distribution of oligonucleotide drug at 5 min postdose. Images illustrate STAT3 decoy distribution in the tumor. The mass spectrum shows that peaks representing the single strands at *m/z* 4584 (5′-3′ sequence), and 4753 (3′-5′ sequence), for the STAT3 decoy dominate the spectrum with few ion adducts and high accuracy (mass error <0.1 Da). Reprinted with permission from Casadonte et al. [115].

to gain more efficient ionization of the molecule. MALDI ionization is no exception to this rule, and while some research has been carried out in this area, little has been done to determine sample preparation methods in which derivatization can be performed for MALDI imaging, rendering it incapable of imaging entire drug classes. Two research labs, however, have begun investigating derivatization methods that will enable ionization of some of these difficult compounds. In a study from Richard Caprioli's lab (Vanderbilt University), Chacon et al. first showed that it is possible to derivatize a drug analyte directly from the tissue surface [116]. In a second preliminary study, Flynders and colleagues used a tailor made hydrazine-based derivative, 4-dimethylamino-6-(4-methoxy-1-naphthyl)-1,3,5-triazine-2-hydrazine (DMNTH), to react with steroids to form hydrazones directly on the tissue surface, which improved their mass spectral ionization efficiency and detection [117]. Since DMNTH served as the MALDI matrix, no special steps were required to derivatize the compounds in tissue. Although these studies are preliminary, both demonstrated the feasibility and potential ease of using derivatizing agents to perform reactions directly from the tissue surface to enhance the ionization efficiency of analytes prior to imaging their distribution. These early results are an important first step to enable the MALDI imaging of some compound classes that are currently not amenable to the technology.

11.6.3 Quantification of Imaging Data for Pharmacokinetics

Several researchers have been working on a quantitative approach for MALDI imaging, but a well-tested and validated method has yet to be delivered; however, it should be highlighted that the semi-quantitative nature of MALDI imaging data can serve as an invaluable complement to pharmacokinetic data [118, 119]. Richard Caprioli's lab (Vanderbilt University) has illustrated that imaging data correlate very well with traditional pharmacokinetic data approaches [120]. In this study, a glioma mouse model was administered imatinib (100 mg/kg) at 2, 6, 12, or 24 h prior to sacrifice. A reagent multispotter was used to coat the tissues with matrix prior to MS/MS analysis. Using conventional drug extraction techniques followed by LC-MS/MS, the amount of imatinib in both the tumor bearing regions and the normal regions of the brain was determined (ng of imatinib per mL of tissue homogenate). Figure 11.13 shows the resulting imatinib distribution and quantitation results in the tumor bearing mouse brains. Matching the MALDI images to the hematoxylin and eosin (H&E)-stained sections shows that the drug is most intense in the tumor. Below the images, normalized MALDI imaging intensities are graphed with the LC-MS/MS quantitation data. The graph illustrates how well the imaging trends correlate with the quantitation data across the time-points for both the tumor

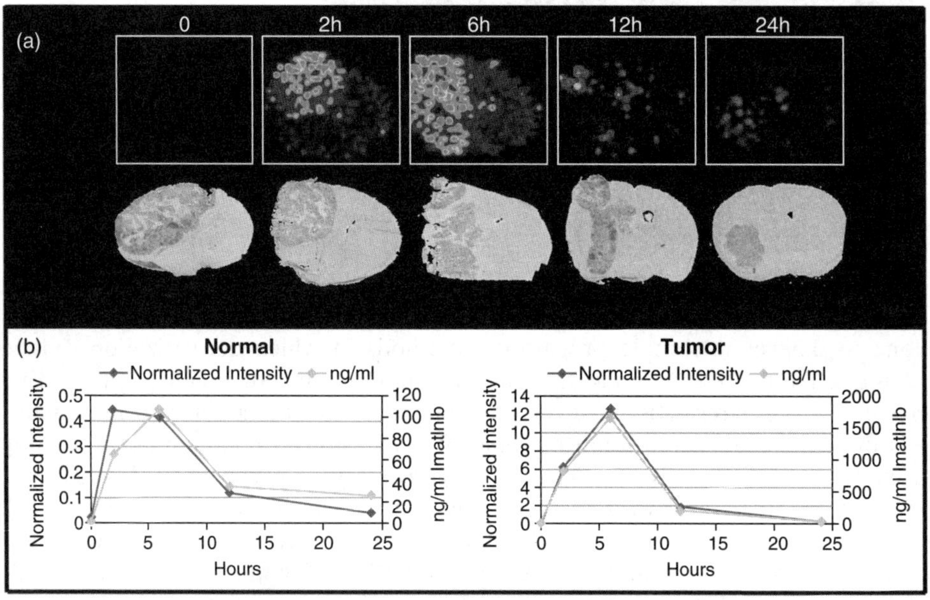

Figure 11.13 See color insert. Imatinib distribution and quantitation results for imatinib-dosed, tumor bearing, mouse brains. (a) Imatinib distribution for a representative brain at each time-point post a single dose of drug (images acquired on the Thermo MALDI LTQ) with their corresponding H&E-stained section for the representative brains. (b) The average normalized drug intensity acquired by MALDI imaging at each time-point along with the amount of imatinib (ng/mL) found to be present in the tissue using traditional drug extraction techniques followed by LC-MS/MS. The MALDI and LC-MS/MS data show the same imatinib trends across the time-points for both the tumor bearing region and the normal tissue region. Reprinted with permission from Frappier et al. [120].

bearing region and normal tissue regions. This type of data could also be used to explain discrepancies in LC-MS/MS quantitation data that may arise due to the loss of spatial information when the tissues are homogenized. In a second study, Koeniger et al. illustrate that several tissue sections collected adjacent to sections acquired for the MALDI imaging analysis can be pooled together and analyzed by conventional LC-MS/MS to quantitate the total level of the drug present [118]. Based on these results, more quantitative information can be derived for the imaging data by correlating the two sets of results. Labs such as Richard Yost's (University of Florida) are continuously working on a solution for the direct quantitation of drugs from imaging data. Early results illustrate that with special sample preparation steps and incorporation of deuterated internal standards, the level of inherent MALDI variability can be controlled [119, 121]. These procedures significantly reduce the % relative standard deviation (RSD) and increase the confidence of quantitation data. These studies show promise for the future potential of quantitative MALDI imaging, but many more experiments and validation steps are required to ensure these methods will work for the variety of small molecule pharmaceutical compounds.

11.7 CONCLUSIONS

Over the past several years, significant progress has been made to enable the application of MS to imaging drug and metabolite distributions in tissues, specifically with the MALDI platform. While the long imaging acquisition times were once a limitation of this technique in terms of high-throughput capacity, the development of instruments with faster lasers have reduced the turnaround time from days to hours. MALDI imaging is now a simple, quick, and molecularly specific platform that enables differentiation between parent drug and metabolites in tissue and whole-body sections, and can serve as a rapid screening tool for compound selection. While this technique has seen rapid improvement since the first drug imaging experiments nearly a decade ago, there is still a need for further improvement to instrumentation and the sample preparation methods. While SIMS imaging is capable of submicron resolution, MALDI lasers need to be improved in order to achieve this submicron resolution as well. This will enable more specific localization within tissues and cells to answer questions currently not achievable with the current resolution. New matrices and derivatization methods will enhance the range of compounds that are amenable to ionization platforms such as MALDI and DESI. As discussed herein, there are many advantages that MSI provides for pharmaceutical research. Continued advancements to MALDI, SIMS, and DESI technologies will only enhance these opportunities. Additionally, as MSI scientists continue to educate others on the benefits of this technology, there will be better collaborations between analytical and biological scientists, resulting in enhanced experimental design and result interpretation to exploit the advantages of these technologies for pharmaceutical R&D.

REFERENCES

[1] Caprioli RM, Farmer TB, Gile J. Molecular imaging of biological samples: localization of peptides and proteins using MALDI-TOF MS. Anal Chem 1997;69:4751–60.

[2] Chaurand P, Schwartz SA, Billheimer D, Xu BJ, Crecelius A, Caprioli RM. Integrating histology and imaging mass spectrometry. Anal Chem 2004;76:1145–55.

[3] Chaurand P, Stoeckli M, Caprioli RM. Direct profiling of proteins in biological tissue sections by MALDI mass spectrometry. Anal Chem 1999;71:5263–70.

[4] Stoeckli M, Staab D, Schweitzer A. Compound and metabolite distribution measured by MALDI mass spectrometric imaging in whole-body tissue sections. Int J Mass Spec 2007;260:195–202.

[5] Rohner TC, Staab D, Stoeckli M. MALDI mass spectrometric imaging of biological tissue sections. Mech Ageing Dev 2005;126:177–85.

[6] Caravan P. Protein-targeted gadolinium-based magnetic resonance imaging (MRI) contrast agents: design and mechanism of action. Acc Chem Res 2009;42:851–62.

[7] Hodgson RJ, Connolly S, Barnes T, Eyes B, Campbell RS, Moots R. Pharmacokinetic modeling of dynamic contrast-enhanced MRI of the hand and wrist in rheumatoid arthritis and the response to anti-tumor necrosis factor-alpha therapy. Magn Reson Med 2007;58:482–9.

[8] Jia G, Heverhagen JT, Henry H, Polzer H, Baudendistel KT, von Tengg-Kobligk H, et al. Pharmacokinetic parameters as a potential predictor of response to pharmacotherapy in benign prostatic hyperplasia: a preclinical trial using dynamic contrast-enhanced MRI. Magn Reson Imaging 2006;24:721–5.

[9] Singh M, Waluch V. Physics and instrumentation for imaging *in-vivo* drug distribution. Adv Drug Deliv Rev 2000;41:7–20.

[10] Port RE, Wolf W. Noninvasive methods to study drug distribution. Invest New Drugs 2003;21:157–68.

[11] Antoch G, Saoudi N, Kuehl H, Dahmen G, Mueller SP, Beyer T, et al. Accuracy of whole-body dual-modality fluorine-18-2-fluoro-2-deoxy-D-glucose positron emission tomography and computed tomography (FDG-PET/CT) for tumor staging in solid tumors: comparison with CT and PET. J Clin Oncol 2004;22:4357–68.

[12] Bockisch A, Beyer T, Antoch G, Freudenberg LS, Kuhl H, Debatin JF, et al. Positron emission tomography/computed tomography—imaging protocols, artifacts, and pitfalls. Mol Imaging Biol 2004;6:188–99.

[13] Propper DJ, de Bono J, Saleem A, Ellard S, Flanagan E, Paul J, et al. Use of positron emission tomography in pharmacokinetic studies to investigate therapeutic advantage in a phase I study of 120-hour intravenous infusion XR5000. J Clin Oncol 2003;21: 203–10.

[14] Kissel J, Brix G, Bellemann ME, Strauss LG, Dimitrakopoulou-Strauss A, Port R, et al. Pharmacokinetic analysis of 5-[18F]fluorouracil tissue concentrations measured with positron emission tomography in patients with liver metastases from colorectal adenocarcinoma. Cancer Res 1997;57:3415–23.

[15] Di Mascio M, Srinivasula S, Bhattacharjee A, Cheng L, Martiniova L, Herscovitch P, et al. Antiretroviral tissue kinetics: *in vivo* imaging using positron emission tomography. Antimicrob Agents Chemother 2009;53:4086–95.

[16] Besret L, Dolle F, Herard AS, Guillermier M, Demphel S, Hinnen F, et al. Dopamine D1 receptor imaging in the rodent and primate brain using the isoquinoline +-[11C] A-69024 and positron emission tomography. J Pharm Sci 2008;97:2811–9.

[17] Syvanen S, Blomquist G, Appel L, Hammarlund-Udenaes M, Langstrom B, Bergstrom M. Predicting brain concentrations of drug using positron emission tomography and venous input: modeling of arterial-venous concentration differences. Eur J Clin Pharmacol 2006;62:839–48.

[18] Varrone A, Steiger C, Schou M, Takano A, Finnema SJ, Guilloteau D, et al. *In vitro* autoradiography and *in vivo* evaluation in cynomolgus monkey of [18F]FE-PE2I, a new dopamine transporter PET radioligand. Synapse 2009;63:871–80.

[19] Ma KH, Huang WS, Kuo YY, Peng CJ, Liou NH, Liu RS, et al. Validation of 4-[18F]-ADAM as a SERT imaging agent using micro-PET and autoradiography. Neuroimage 2009;45:687–93.

[20] Bauer A, Holschbach MH, Cremer M, Weber S, Boy C, Shah NJ, et al. Evaluation of 18F-CPFPX, a novel adenosine A1 receptor ligand: *in vitro* autoradiography and high-resolution small animal PET. J Nucl Med 2003;44:1682–9.

[21] Ogasawara K, Ito H, Sasoh M, Okuguchi T, Kobayashi M, Yukawa H, et al. Quantitative measurement of regional cerebrovascular reactivity to acetazolamide using 123I-N-isopropyl-p-iodoamphetamine autoradiography with SPECT: validation study using H2 15O with PET. J Nucl Med 2003;44:520–5.

[22] Ametamey SM, Honer M, Schubiger PA. Molecular imaging with PET. Chem Rev 2008;108:1501–16.

[23] Whitby B. Quantitative whole-body autoradiography (QWBA). In Temple S, Lappin G, editors. Radiotracers in Drug Development. Boca Raton, FL: CRC Press, 2006, pp. 129–53.

[24] Ma Y, Kiesewetter DO, Lang L, Gu D, Chen X. Applications of LC-MS in PET radioligand development and metabolic elucidation. Curr Drug Metab 2010;11:483–93.

[25] Ekesbo A, Torstenson R, Hartvig P, Carlsson A, Sonesson C, Waters N, et al. Effects of the substituted (S)-3-phenylpiperidine (-)-OSU6162 on PET measurements of [11C] SCH23390 and [11C]raclopride binding in primate brains. Neuropharmacology 1999;38:331–8.

[26] Noworolski SM, Nelson SJ, Henry RG, Day MR, Wald LL, Star-Lack J, et al. High spatial resolution 1H-MRSI and segmented MRI of cortical gray matter and subcortical white matter in three regions of the human brain. Magn Reson Med 1999;41:21–9.

[27] Coe RA. Quantitative whole-body autoradiography. Regul Toxicol Pharmacol 2000;31:S1–S3.

[28] Langlois X, Te Riele P, Wintmolders C, Leysen JE, Jurzak M. Use of the beta-imager for rapid ex vivo autoradiography exemplified with central nervous system penetrating neurokinin 3 antagonists. J Pharmacol Exp Ther 2001;299:712–7.

[29] Tribollet E, Dreifuss JJ, Charpak G, Dominik W, Zaganidis N. Localization and quantitation of tritiated compounds in tissue sections with a gaseous detector of beta particles: comparison with film autoradiography. Proc Natl Acad Sci U S A 1991;88: 1466–8.

[30] Garrison BJ, Winograd N. Ion beam spectroscopy of solids and surfaces. Science 1982;216:805–12.

[31] Weibel D, Wong S, Lockyer N, Blenkinsopp P, Hill R, Vickerman JC. A C60 primary ion beam system for time of flight secondary ion mass spectrometry: its development and secondary ion yield characteristics. Anal Chem 2003;75:1754–64.

[32] Touboul D, Kollmer F, Niehuis E, Brunelle A, Laprevote O. Improvement of biological time-of-flight-secondary ion mass spectrometry imaging with a bismuth cluster ion source. J Am Soc Mass Spectrom 2005;16:1608–18.

[33] Davies N, Wiebel D, Blenkinsopp P, Lockyer N, Hill R, Vickerman JC. Development and experimentation application of a gold liquid metal ion source. Appl Surf Sci 2003;203:223–7.

[34] Walker A, Winograd N. Prospects for imaging with TOF-SIMS using gold liquid metal ion sources. Appl Surf Sci 2003;203:198–200.

[35] Winograd N, Garrison BJ. Biological cluster mass spectrometry. Annu Rev Phys Chem 2010;61:305–22.

[36] Todd PJ, Schaaff TG, Chaurand P, Caprioli RM. Organic ion imaging of biological tissue with secondary ion mass spectrometry and matrix-assisted laser desorption/ionization. J Mass Spectrom 2001;36:355–69.

[37] Todd PJ, McMahon JM, McCandlish CA, Jr. Secondary ion images of the developing rat brain. J Am Soc Mass Spectrom 2004;15:1116–22.

[38] McCandlish CA, McMahon JM, Todd PJ. Secondary ion images of the rodent brain. J Am Soc Mass Spectrom 2000;11:191–9.

[39] McMahon JM, Short RT, McCandlish CA, Brenna JT, Todd PJ. Identification and mapping of phosphocholine in animal tissue by static secondary ion mass spectrometry and tandem mass spectrometry. Rapid Commun Mass Spectrom 1996;10:335–40.

[40] Fragu P, Kahn E. Secondary ion mass spectrometry (SIMS) microscopy: a new tool for pharmacological studies in humans. Microsc Res Tech 1997;36:296–300.

[41] Smith DR, Chandra S, Barth RF, Yang W, Joel DD, Coderre JA. Quantitative imaging and microlocalization of boron-10 in brain tumors and infiltrating tumor cells by SIMS ion microscopy: relevance to neutron capture therapy. Cancer Res 2001;61: 8179–87.

[42] Chandra S, Lorey II DR. SIMS ion microscopy imaging of boronophenylalanine (BPA) and $^{13}C^{15}N$-labeled phenylalanine in human glioblastoma cells: relevance of subcellular scale observations ot BPA-mediated boron neutron capture therapy of cancer. Int J Mass Spec 2007;260:90–101.

[43] Guerquin-Kern JL, Coppey M, Carrez D, Brunet AC, Nguyen CH, Rivalle C, et al. Complementary advantages of fluorescence and SIMS microscopies in the study of cellular localization of two new antitumor drugs. Microsc Res Tech 1997;36:287–95.

[44] Smith DR, Chandra S, Coderre JA, Morrison GH. Ion microscopy imaging of 10B from p-boronophenylalanine in a brain tumor model for boron neutron capture therapy. Cancer Res 1996;56:4302–6.

[45] Clerc J, Halpern S, Fourre C, Omri F, Briancon C, Jeusset J, et al. SIMS microscopy imaging of the intratumor biodistribution of metaiodobenzylguanidine in the human SK-N-SH neuroblastoma cell line xenografted into nude mice. J Nucl Med 1993;34: 1565–70.

[46] Chandra S, Smith DR, Morrison GH. Subcellular imaging by dynamic SIMS ion microscopy. Anal Chem 2000;72:104A–14A.

[47] Solon EG, Schweitzer A, Stoeckli M, Prideaux B. Autoradiography, MALDI-MS, and SIMS-MS imaging in pharmaceutical discovery and development. AAPS J 2009;12: 11–26.

[48] Fletcher JS. Cellular imaging with secondary ion mass spectrometry. Analyst 2009;134: 2204–15.

[49] Takats Z, Wiseman JM, Gologan B, Cooks RG. Mass spectrometry sampling under ambient conditions with desorption electrospray ionization. Science 2004;306: 471–3.

[50] Esquenazi E, Dorrestein PC, Gerwick WH. Probing marine natural product defenses with DESI-imaging mass spectrometry. Proc Natl Acad Sci U S A 2009;106:7269–70.

[51] Miao Z, Chen H. Direct analysis of liquid samples by desorption electrospray ionization-mass spectrometry (DESI-MS). J Am Soc Mass Spectrom 2009;20:10–9.

[52] Nyadong L, Hohenstein EG, Galhena A, Lane AL, Kubanek J, Sherrill CD, et al. Reactive desorption electrospray ionization mass spectrometry (DESI-MS) of natural products of a marine alga. Anal Bioanal Chem 2009;394:245–54.

[53] Wells JM, Roth MJ, Keil AD, Grossenbacher JW, Justes DR, Patterson GE, et al. Implementation of DART and DESI ionization on a fieldable mass spectrometer. J Am Soc Mass Spectrom 2008;19:1419–24.

[54] Zhao M, Zhang S, Yang C, Xu Y, Wen Y, Sun L, et al. Desorption electrospray tandem MS (DESI-MSMS) analysis of methyl centralite and ethyl centralite as gunshot residues on skin and other surfaces. J Forensic Sci 2008;53:807–11.

[55] Wiseman JM, Evans CA, Bowen CL, Kennedy JH. Direct analysis of dried blood spots utilizing desorption electrospray ionization (DESI) mass spectrometry. Analyst 2010;135:720–5.

[56] Wiseman JM, Ifa DR, Zhu Y, Kissinger CB, Manicke NE, Kissinger PT, et al. Desorption electrospray ionization mass spectrometry: imaging drugs and metabolites in tissues. Proc Natl Acad Sci U S A 2008;105:18120–5.

[57] Wiseman JM, Ifa DR, Venter A, Cooks RG. Ambient molecular imaging by desorption electrospray ionization mass spectrometry. Nat Protoc 2008;3:517–24.

[58] Wiseman JM, Ifa DR, Song Q, Cooks RG. Tissue imaging at atmospheric pressure using desorption electrospray ionization (DESI) mass spectrometry. Angew Chem Int Ed Engl 2006;45:7188–92.

[59] Wiseman JM, Puolitaival SM, Takats Z, Cooks RG, Caprioli RM. Mass spectrometric profiling of intact biological tissue by using desorption electrospray ionization. Angew Chem Int Ed Engl 2005;44:7094–7.

[60] Ovchinnikova OS, Kertesz V, Van Berkel GJ. Molecular surface sampling and chemical imaging using proximal probe thermal desorption/secondary ionization mass spectrometry. Anal Chem 2011;83:598–603.

[61] Kertesz V, Van Berkel GJ, Vavrek M, Koeplinger KA, Schneider BB, Covey TR. Comparison of drug distribution images from whole-body thin tissue sections obtained using desorption electrospray ionization tandem mass spectrometry and autoradiography. Anal Chem 2008;80:5168–77.

[62] Leuthold LA, Mandscheff JF, Fathi M, Giroud C, Augsburger M, Varesio E, et al. Desorption electrospray ionization mass spectrometry: direct toxicological screening and analysis of illicit Ecstasy tablets. Rapid Commun Mass Spectrom 2006;20:103–10.

[63] Schneider BB, Baranov VI, Javaheri H, Covey TR. Particle discriminator interface for nanoflow ESI-MS. J Am Soc Mass Spectrom 2003;14:1236–46.

[64] Corkery LJ, Pang H, Schneider BB, Covey TR, Siu KW. Automated nanospray using chip-based emitters for the quantitative analysis of pharmaceutical compounds. J Am Soc Mass Spectrom 2005;16:363–9.

[65] Schneider BB, Lock C, Covey TR. AP and vacuum MALDI on a QqLIT instrument. J Am Soc Mass Spectrom 2005;16:176–82.

[66] Galicia MC, Vertes A, Callahan JH. Atmospheric pressure matrix-assisted laser desorption/ionization in transmission geometry. Anal Chem 2002;74:1891–5.

[67] Chaurand P, Schriver KE, Caprioli RM. Instrument design and characterization for high resolution MALDI-MS imaging of tissue sections. J Mass Spectrom 2007;42:476–89.

[68] Taira S, Sugiura Y, Moritake S, Shimma S, Ichiyanagi Y, Setou M. Nanoparticle-assisted laser desorption/ionization based mass imaging with cellular resolution. Anal Chem 2008;80:4761–6.

[69] Kertesz V, Van Berkel GJ. Improved imaging resolution in desorption electrospray ionization mass spectrometry. Rapid Commun Mass Spectrom 2008;22:2639–44.

[70] Karas M, Bachmann D, Bahr U, Hillenkamp F. Matrix-assisted ultraviolet laser desorption of non-volatile compounds. Int J Mass Spectrom Ion Process 1987;78:53–68.

[71] Karas M, Hillenkamp F. Laser desorption ionization of proteins with molecular masses exceeding 10,000 daltons. Anal Chem 1988;60:2299–301.

[72] Stoeckli M, Farmer TB, Caprioli RM. Automated mass spectrometry imaging with a matrix-assisted laser desorption ionization time-of-flight instrument. J Am Soc Mass Spectrom 1999;10:67–71.

[73] Reyzer ML, Hsieh Y, Ng K, Korfmacher WA, Caprioli RM. Direct analysis of drug candidates in tissue by matrix-assisted laser desorption/ionization mass spectrometry. J Mass Spectrom 2003;38:1081–92.

[74] Stoeckli M, Chaurand P, Hallahan DE, Caprioli RM. Imaging mass spectrometry: a new technology for the analysis of protein expression in mammalian tissues. Nat Med 2001;7:493–6.

[75] Krutchinsky AN, Chait BT. On the nature of the chemical noise in MALDI mass spectra. J Am Soc Mass Spectrom 2002;13:129–34.

[76] Schwartz SA, Reyzer ML, Caprioli RM. Direct tissue analysis using matrix-assisted laser desorption/ionization mass spectrometry: practical aspects of sample preparation. J Mass Spectrom 2003;38:699–708.

[77] Seeley EH, Oppenheimer SR, Mi D, Chaurand P, Caprioli RM. Enhancement of protein sensitivity for MALDI imaging mass spectrometry after chemical treatment of tissue sections. J Am Soc Mass Spectrom 2008;19:1069–77.

[78] Lemaire R, Wisztorski M, Desmons A, Tabet JC, Day R, Salzet M, et al. MALDI-MS direct tissue analysis of proteins: improving signal sensitivity using organic treatments. Anal Chem 2006;78:7145–53.

[79] Cornett DS, Frappier SL, Caprioli RM. MALDI-FTICR imaging mass spectrometry of drugs and metabolites in tissue. Anal Chem 2008;80:5648–53.

[80] Hsieh Y, Casale R, Fukuda E, Chen J, Knemeyer I, Wingate J, et al. Matrix-assisted laser desorption/ionization imaging mass spectrometry for direct measurement of clozapine in rat brain tissue. Rapid Commun Mass Spectrom 2006;20:965–72.

[81] Li F, Hsieh Y, Kang L, Sondey C, Lachowicz J, Korfmacher WA. MALDI-tandem mass spectrometry imaging of astemizole and its primary metabolite in rat brain sections. Bioanalysis 2009;1:299–307.

[82] Hopfgartner G, Varesio E, Stoeckli M. Matrix-assisted laser desorption/ionization mass spectrometric imaging of complete rat sections using a triple quadrupole linear ion trap. Rapid Commun Mass Spectrom 2009;23:733–6.

[83] Trim PJ, Henson CM, Avery JL, McEwen A, Snel MF, Claude E, et al. Matrix-assisted laser desorption/ionization-ion mobility separation-mass spectrometry imaging of vinblastine in whole body tissue sections. Anal Chem 2008;80:8628–34.

[84] Conaway M, Cao S, Durrani F, Rustum Y, Wang P, Marlar K, et al. The role of MALDI-enabled linear ion trap mass spectrometry as a sensitive tool in tissue imaging. Spectroscopy Europe, 2008, pp. 12–14.

[85] McLean JA, Ridenour WB, Caprioli RM. Profiling and imaging of tissues by imaging ion mobility-mass spectrometry. J Mass Spectrom 2007;42:1099–105.

[86] Khatib-Shahidi S, Andersson M, Herman JL, Gillespie TA, Caprioli RM. Direct molecular analysis of whole-body animal tissue sections by imaging MALDI mass spectrometry. Anal Chem 2006;78:6448–56.

[87] Aerni HR, Cornett DS, Caprioli RM. Automated acoustic matrix deposition for MALDI sample preparation. Anal Chem 2006;78:827–34.

[88] Sloane AJ, Duff JL, Wilson NL, Gandhi PS, Hill CJ, Hopwood FG, et al. High throughput peptide mass fingerprinting and protein macroarray analysis using chemical printing strategies. Mol Cell Proteomics 2002;1:490–9.

[89] Nakanishi T, Ohtsu I, Furuta M, Ando E, Nishimura O. Direct MS/MS analysis of proteins blotted on membranes by a matrix-assisted laser desorption/ionization-quadrupole ion trap-time-of-flight tandem mass spectrometer. J Proteome Res 2005; 4:743–7.

[90] Baluya DL, Garrett TJ, Yost RA. Automated MALDI matrix deposition method with inkjet printing for imaging mass spectrometry. Anal Chem 2007;79:6862–7.

[91] Van Dyck S, Flammang P, Meriaux C, Bonnel D, Salzet M, Fournier I, et al. Localization of secondary metabolites in marine invertebrates: contribution of MALDI MSI for the study of saponins in Cuvierian tubules of *H. forskali*. PLoS ONE 2010;5:e13923.

[92] Benabdellah F, Touboul D, Brunelle A, Laprevote O. In situ primary metabolites localization on a rat brain section by chemical mass spectrometry imaging. Anal Chem 2009;81:5557–60.

[93] Heeren RMA, Kukrer-Kaletas B, Taban IM, MacAleese L, McDonnell LA. Quality of surface: the influence of sample preparation on MS-based biomolecular tissue imaging with MALDI-MS and (ME-)SIMS. Appl Surf Sci 2008;255:1289–97.

[94] Sleno L, Volmer DA. Assessing the properties of internal standards for quantitative matrix-assisted laser desorption/ionization mass spectrometry of small molecules. Rapid Commun Mass Spectrom 2006;20:1517–24.

[95] Persike M, Karas M. Rapid simultaneous quantitative determination of different small pharmaceutical drugs using a conventional matrix-assisted laser desorption/ionization time-of-flight mass spectrometry system. Rapid Commun Mass Spectrom 2009;23: 3555–62.

[96] Nilsson A, Fehniger TE, Gustavsson L, Andersson M, Kenne K, Marko-Varga G, et al. Fine mapping the spatial distribution and concentration of unlabeled drugs within tissue micro-compartments using imaging mass spectrometry. PLoS ONE 2010; 5:e11411.

[97] Marshall P, Toteu-Djomte V, Bareille P, Perry H, Brown G, Baumert M, et al. Correlation of skin blanching and percutaneous absorption for glucocorticoid receptor agonists by matrix-assisted laser desorption ionization mass spectrometry imaging and liquid extraction surface analysis with nanoelectrospray ionization mass spectrometry. Anal Chem 2010;82:7787–94.

[98] Atkinson SJ, Loadman PM, Sutton C, Patterson LH, Clench MR. Examination of the distribution of the bioreductive drug AQ4N and its active metabolite AQ4 in solid tumours by imaging matrix-assisted laser desorption/ionisation mass spectrometry. Rapid Commun Mass Spectrom 2007;21:1271–6.

[99] Oppenheimer SR, Gale DC, Wilkie D, Obert LA. Utilizing complimentary *in-situ* imaging platforms to understand tumor penetration for oncology drug discovery. 58th Annual Conference on Mass Spectrometry and Allied Topics. Salt Lake City, UT, 2010.

[100] Huamani J, Willey C, Thotala D, Niermann KJ, Reyzer M, Leavitt L, et al. Differential efficacy of combined therapy with radiation and AEE788 in high and low EGFR-expressing androgen-independent prostate tumor models. Int J Radiat Oncol Biol Phys 2008;71:237–46.

[101] Chen J, Hsieh Y, Knemeyer I, Crossman L, Korfmacher WA. Visualization of first-pass drug metabolism of terfenadine by MALDI-imaging mass spectrometry. Drug Metab Lett 2008;2:1–4.

[102] Drexler DM, Tannehill-Gregg SH, Wang L, Brock BJ. Utility of quantitative whole-body autoradiography (QWBA) and imaging mass spectrometry (IMS) by matrix-assisted laser desorption/ionization (MALDI) in the assessment of ocular distribution of drugs. J Pharmacol Toxicol Methods 2010;2:205–8.

[103] Oppenheimer SR, Li J, Sciarra J. MALDI imaging of pharmaceuticals in zebrafish for discovery and drug safety screening. 57th Annual Conference on Mass Spectrometry and Allied Topics. Philadelphia, PA, 2009.

[104] Amatruda JF, Shepard JL, Stern HM, Zon LI. Zebrafish as a cancer model system. Cancer Cell 2002;1:229–31.

[105] Berry JP, Gantar M, Gibbs PD, Schmale MC. The zebrafish (Danio rerio) embryo as a model system for identification and characterization of developmental toxins from marine and freshwater microalgae. Comp Biochem Physiol C Toxicol Pharmacol 2007; 145:61–72.

[106] Hernandez PP, Allende ML. Zebrafish (Danio rerio) as a model for studying the genetic basis of copper toxicity, deficiency, and metabolism. Am J Clin Nutr 2008;88: 835S–9S.

[107] Earnshaw CJ, Carolan VA, Richards DS, Clench MR. Direct analysis of pharmaceutical tablet formulations using Matrix-Assisted Laser Desorption/Ionisation Mass Spectrometry Imaging. Rapid Commun Mass Spectrom 2010;24:1665–72.

[108] Miller T, Havrilla G. Elemental imaging for pharmaceutical tablet formulation analysis by mircro X-ray fluorescence. Adv X-Ray Anal 2005;48:274–83.

[109] Acquadro E, Cabella C, Ghiani S, Miragoli L, Bucci EM, Corpillo D. Matrix-assisted laser desorption ionization imaging mass spectrometry detection of a magnetic resonance imaging contrast agent in mouse liver. Anal Chem 2009;81:2779–84.

[110] Meibohm B. The role of pharmacokinetics and pharmacodynamics in the development of biotech drugs. In Meibohm B, editor. Pharmacokinetics and Pharmacodynamics of Biotech Drugs: Principles and Case Studies in Drug Development. Weinheim: Wiley-VCH Verlag GmbH & Co. KGaA, 2006, pp. 1–13.

[111] Casadonte R, Manier L, Carbone D, Grandis J, Caprioli RM. Decoy oligonucleotide: analysis and use in carcinogenesis. 56th Annual Conference on Mass Spectrometry and Allied Topics. Denver, CO, 2008.

[112] Xi S, Gooding WE, Grandis JR. *In vivo* antitumor efficacy of STAT3 blockade using a transcription factor decoy approach: implications for cancer therapy. Oncogene 2005;24:970–9.

[113] Leong PL, Andrews GA, Johnson DE, Dyer KF, Xi S, Mai JC, et al. Targeted inhibition of Stat3 with a decoy oligonucleotide abrogates head and neck cancer cell growth. Proc Natl Acad Sci U S A 2003;100:4138–43.

[114] Sen M, Tosca PJ, Zwayer C, Ryan MJ, Johnson JD, Knostman KA, et al. Lack of toxicity of a STAT3 decoy oligonucleotide. Cancer Chemother Pharmacol 2009;63: 983–95.

[115] Casadonte R, Amann JM, Carbone D, Grandis J, Caprioli RM. 3D spatial distribution of synthetic oligodeoxynucleotide in tumor-bearing mice: an imaging mass spectrometry study. 58th Annual Conference on Mass Spectrometry and Allied Topics. Salt Lake City, UT, 2010.

[116] Chacon A, Zagol-Ikapitte I, Amarnath V, Reyzer ML, Oates JA, Caprioli RM, et al. On-tissue chemical derivatization of 3-methoxysalicylamine for MALDI-imaging mass spectrometry. J Mass Spectrom 2011;46:840–6.

[117] Flinders B, Marshall PS, Morrell J, Ranshaw L, Khan M, Clench MR. The use of hydrazine base derivatisation reagents for improved sensitivity and detection of carbonyl containing compounds "on tissue" using MALDI-MS. 58th Annual Conference on Mass Spectrometry and Allied Topics. Salt Lake City, UT, 2010.

[118] Koeniger SL, Talaty N, Luo Y, Ready D, Voorbach M, Seifert T, et al. A quantitation method for mass spectrometry imaging. Rapid Commun Mass Spectrom 2011;25: 503–10.

[119] Reich RF, Cudzilo K, Levisky JA, Yost RA. Quantitative MALDI-MS(n) analysis of cocaine in the autopsied brain of a human cocaine user employing a wide isolation window and internal standards. J Am Soc Mass Spectrom 2010;21:564–71.

[120] Frappier SL, Caprioli R. Correlation of imatinib with proteome response in glioma mouse xenografts by MALDI imaging mass spectrometry. 58th Annual Conference on Mass Spectrometry and Allied Topics. Salt Lake City, UT, 2010.

[121] Pirman DA, Yost RA. Quantitative tandem mass spectrometric imaging of endogenous acetyl-L-carnitine from piglet brain tissue using an internal standard. Anal Chem 2011;83:8575–81.

12

SCREENING REACTIVE METABOLITES: ROLE OF LIQUID CHROMATOGRAPHY–HIGH-RESOLUTION MASS SPECTROMETRY IN COMBINATION WITH "INTELLIGENT" DATA MINING TOOLS

SHUGUANG MA AND SWAPAN K. CHOWDHURY

12.1 INTRODUCTION

Xenobiotics are metabolized by various oxidation and conjugation enzymes to more hydrophilic metabolites to facilitate elimination from the body. Thus, metabolism is generally considered a detoxification process. Although in many cases metabolites are less toxic than the parent drug, it is not uncommon that drugs undergo bioactivation to form reactive species that have intrinsic chemical reactivity toward cellular macromolecules (DNA and proteins), thus altering their biological functions, and resulting in serious adverse drug reactions (ADRs). In addition, reactive metabolites have been implicated in a number of off-target ADRs in humans [1]. Although there is no definitive proof of a causal relationship between metabolic activation and ADRs, a substantial amount of evidence implies that chemically reactive metabolites may play an important role as toxicity mediators [2–4]. A recent study reported that among 21 drugs that were either withdrawn from the US market due to hepatotoxicity or have a black box warning for hepatotoxicity, there was evidence for the formation of reactive metabolites found for 5 out of 6 drugs that were withdrawn, and 8 out of 15 drugs that have black box warnings [5]. Nakayama et al. [6] retrospectively investigated a potential relationship between ADRs and covalent binding (*in vitro* in human liver microsomes [HLMs] and hepatocytes, and *in vivo* in rat

---

Mass Spectrometry for Drug Discovery and Drug Development, First Edition. Edited by Walter A. Korfmacher.

liver). Using a zone classification system these authors reported a correlation of the extent of covalent binding in hepatocytes and daily administered dose with reported clinical safety profile. Due to the possible link between reactive metabolites and ADRs, it has been a widely adapted approach in the pharmaceutical industry to screen out drug candidates that have propensity to undergo metabolic activation as an important part of lead optimization process in drug discovery [7–9].

12.2 *IN VITRO* TRAPPING OF REACTIVE METABOLITES

Reactive metabolites may be broadly classified into electrophiles and free radicals. Most reactive metabolites are electrophiles that can react with nucleophiles. Electrophiles can be classified as "hard" or "soft"; a localized positive charge would make the electrophile "hard," while a delocalized charge would make it "soft." Similarly, nucleophiles can also be classified as "hard" or "soft." For example, a sulfur-containing nucleophile is considered softer than a nitrogen-containing nucleophile, because a sulfur atom is larger and the lone pair electrons are further away from the nucleus and therefore are more diffused. In general, "hard" electrophiles tend to react with "hard" nucleophiles, while "soft" electrophiles tend to react with "soft" nucleophiles.

Reactive metabolites are often short-lived and are not usually detectable in circulation; therefore, *in vitro* trapping approaches are generally employed to examine the bioactivation potential of drug candidates [10]. These experiments are often conducted in liver microsomes with reduced nicotinamide adenine dinucleotide phosphate (NADPH) and appropriate nucleophilic trapping agents, such as thiols (glutathione [GSH], its ethyl ester derivative, or N-acetylcysteine), amines (semicarbazide or methoxylamine), or cyanide anion. GSH contains a free sulfhydryl group, a "soft" nucleophile capable of reacting with a broad range of reactive electrophiles, including quinoneimines, nitrenium ions, arene oxides, quinones, imine methides, epoxides, alkyl halides, and Michael acceptors [10]. GSH is present virtually in all mammalian tissues and therefore serves as a natural scavenger for chemically reactive metabolites. Semicarbazide and methoxylamine are "hard" nucleophiles, which will preferentially react with "hard" electrophiles such as aldehydes. The cyanide anion is a "hard" nucleophile that can be used to effectively trap iminium species. Examples of different trapping reactions that are commonly used *in vitro* to capture reactive intermediates for structural characterization are displayed in Figure 12.1.

12.3 TRADITIONAL LIQUID CHROMATOGRAPHY-TANDEM MASS SPECTROMETRY (LC/MS/MS) APPROACHES FOR REACTIVE METABOLITE SCREENING

Since GSH is capable of reacting with a broad range of electrophiles, investigation of the formation of GSH conjugates could potentially identify if the drug has propensity for the formation of reactive metabolites [11, 12]. GSH conjugates when subjected to collision-induced dissociation (CID) in the positive ion mode produce a characteristic fragment ion corresponding to the loss of pyroglutamic acid (a loss of 129 Da, fragment *e*, as shown in Fig. 12.2a). Thus, the majority of GSH conjugates

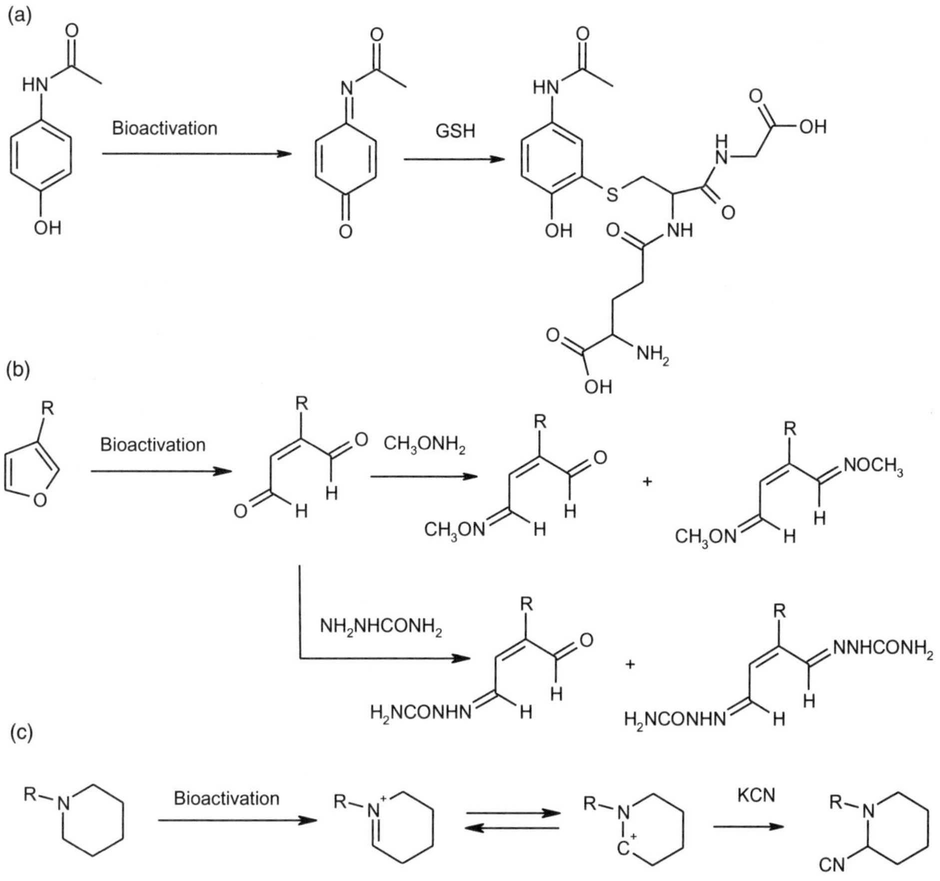

Figure 12.1 Reaction scheme for trapping quinoneimine-reactive metabolite by glutathione (a), aldehyde by methoxylamine or semicarbazide (b), and iminium ion intermediate by cyanide anion (c). Adapted from Reference 8 with permission.

can be detected by scanning for a neutral loss of 129 Da. However, the main disadvantage of the constant neutral loss scan (CNLS) is its poor selectivity, as many endogenous compounds present in biological matrices under MS/MS conditions give rise to a loss of 129 Da, which are not related to drug–GSH adducts. Therefore, false positives are routinely detected using CNLS for screening GSH conjugates. In addition, the detection by CNLS is not very sensitive. To transcend these deficiencies of CNLS, several techniques have been developed for the rapid, sensitive, and selective detection and characterization of reactive metabolites. Yan et al. trapped reactive metabolites using an equal molar ratio of GSH and $^{13}C_2$-^{15}N-labeled GSH in microsomal incubations. The resulting GSH conjugates were detected by the presence of a unique doublet isotopic peak with *m/z* differences of 3 Da in the mass spectra [13, 14]. Leblanc et al. improved selectivity by using a brominated analog of GSH, *N*-(2-bromocarbobenzyloxy)-GSH, for *in vitro* screening of reactive metabolites [15]. The incorporation of bromine in the trapping agent provided a distinct isotope pattern (^{79}Br: ^{81}Br = 1:1). Zheng and coworkers developed a selective and

(a)

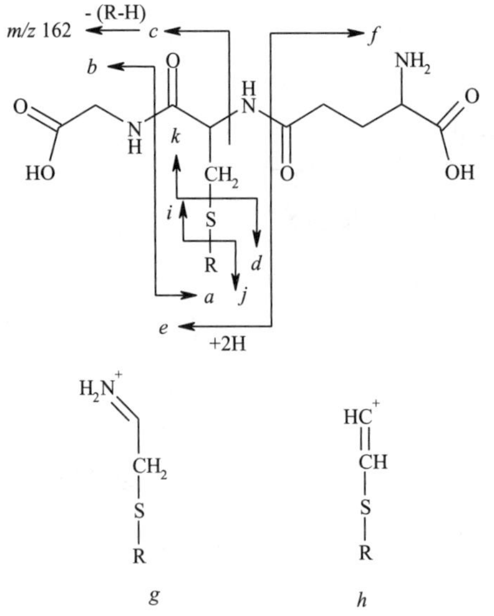

Positive Ion Mode	
Fragment Ion	*m/z*
a	MH^+ - 75
b	76
c	MH^+ - 146
c - (R-H)	162
c - H2O	MH^+ - 164
d	MH^+ - 273/275
e	MH^+ - 129
e - H2O	MH^+ - 147
f	130
g	MH^+ - 232
g - H2O	MH^+ - 250
h	MH^+ - 249
i	308
j	MH^+ - 305/307
k	274

(b)

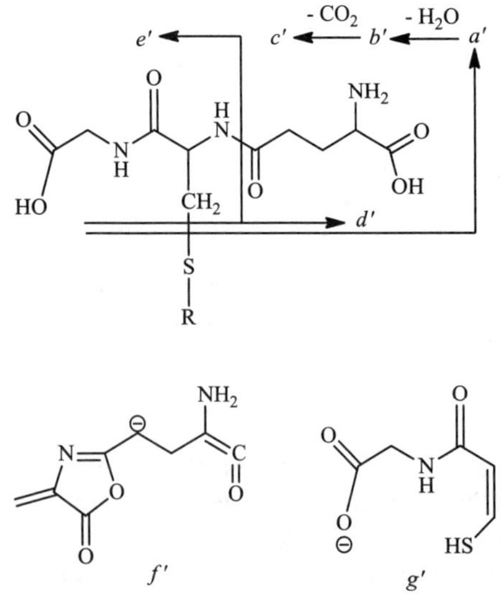

Negative Ion Mode	
Fragment Ion	*m/z*
a'	272
b'	254
c'	210
d'	128
e'	143
f'	179
g'	160

Figure 12.2 Characteristic fragment ions of glutathione conjugates under collision-induced dissociation in the (a) positive and (b) negative ion mode. Adapted from Reference 12 with permission.

sensitive approach using selected reaction monitoring (SRM) as the survey scan to trigger the acquisition of enhanced product ion spectra on a Q-trap mass spectrometer. SRM transitions were constructed from the protonated molecules of potential GSH adducts to their product ions derived from neutral losses of 129 and 307 Da [16]. This SRM-triggered detection approach was shown to be more selective and sensitive than the CNLS method.

Different classes of GSH conjugates appear to behave differently upon CID; not all yield a neutral loss of 129 Da in the positive ion mode as the primary fragmentation pathway [11]. In addition, many GSH adducts form doubly charged ions, which typically do not fragment by a neutral loss of 129 Da under CID but rather produce singly charged species (e.g., *m/z* 76, 84, 130, 162, 179) resulting from the GSH moiety [17]. Therefore, there is a need for a broader MS/MS survey scan for the detection of GSH adducts from different structural classes. The precursor ion scan of *m/z* 272 (deprotonated γ-glutamyl-dehydroalanyl-glycine, fragment *a'*, as shown in Fig. 12.2b) in the negative ion mode was demonstrated to provide a generally applicable screening way for the detection of benzylic-, aromatic-, thioester-, and aliphatic-GSH conjugates [17]. To further improve the selectivity, Mahajan and Evans extended this methodology by incorporating dual precursor ions at *m/z* 272 and 254 (another major fragment ion of GSH from the dehydration of *m/z* 272) [18]. These ions are characteristic fragment ions of GSH adducts in the negative ion mode and, when detected in parallel, achieved a further increase in selectivity.

All the methods described above rely on either the prior knowledge of the predicted GSH conjugates or the fragmentation patterns of the GSH conjugates. These methods would not work well for novel GSH conjugate structures via ring opening, cleavage, or rearrangement, neither for GSH downstream products (e.g., cys-conjugates, cys-gly-conjugates, or mercapturic acids). In addition, not all reactive metabolites can be trapped with GSH. Reactive iminium species are best trapped by cyanide, and aldehydes will preferentially react with semicarbazide or methoxylamine. Therefore, new approaches capable of detecting different classes of reactive metabolites with various trapping agents without prior knowledge of their molecular weights or fragmentation patterns are highly desired. The recent advances in high-resolution mass spectrometry (HRMS) technologies and the development of new computational data mining tools that perform objective searching/filtering of accurate mass-based LC-MS data have greatly improved the analytical capabilities for the detection of reactive metabolites with no prior knowledge of the structure or molecular weight of the trapped metabolites.

12.4 HRMS

12.4.1 Time-of-Flight (TOF)

Analysis in a TOF mass spectrometer is based on the principle that ions of different *m/z* values, when accelerated by the same kinetic energy (2–30 kV), possess different velocities after acceleration out of the ion source and into a field-free drift tube [19]. As a result, the time (t) required for each ion to traverse the flight tube is different and high-mass ions will take longer to reach the detector than low-mass ions. The equation relating the flight time of an ion with its *m/z* value is shown below:

$$t = \frac{L}{v} = \frac{L}{\sqrt{\frac{2zV}{m}}} = L\sqrt{\frac{m}{2zV}}$$

where L, v, and V are the ion drift length, the ion velocity, and the accelerating potential, respectively.

Due to the short flight time (50–100 μs) a high-quality spectrum can be generated within 100 ms over a large mass range. In theory, a TOF mass analyzer has the advantage of being able to perform complete spectral acquisition of all ions (no low mass cutoff) with "in principle" no upper *m/z* limit. The mass resolving power of TOF instruments is strongly dependent on the ability to produce a highly focused ion beam and to avoid kinetic energy dispersion in the ionization and acceleration region. Utilizing reflectron technology in combination with delayed extraction, TOF instruments have traditionally provided respectable mass resolving power (~8000–30,000) and mass accuracy (~2–10 ppm) with high spectral acquisition rates and expanded dynamic ranges (>3 orders of magnitude) [20]. To couple electrospray ionization (ESI) where ion beam is continuously generated, orthogonal acceleration was developed to improve the performance of TOF instruments. The replacement of time-to-digital converter (TDC) with analog-to-digital converter (ADC) in the newer TOF mass spectrometers eliminated the ion saturation problem in the TDC systems due to inherent dead time and thus improved linearity and mass accuracy. Better front-end ion sampling, high linear accelerated quadrupole frequencies, and accelerated transmission of the ions through the TOF region also contributed to high sensitivity across the mass range.

The hybrid quadrupole-time-of-flight instrument (Q-TOF) represents a powerful combination of mass range, resolution, sensitivity, and accurate mass measurements on both MS and MS/MS modes. Because the TOF mass analyzer has a low duty cycle, the placement of a quadrupole ion trap (QIT) or linear ion trap (LIT) in front of the TOF mass analyzer can improve its sensitivity. In addition, IT provides MS^n capability, and thus accurate mass measurements on multistage fragments can be achieved by a TOF mass analyzer in an IT-TOF hybrid mass spectrometer.

12.4.2 Fourier Transform-Ion Cyclotron Resonance (FT-ICR)

The main components of an FT-ICR mass analyzer include a superconductive magnet (4–12 Tesla) and a cube consisting of three pairs of parallel plates for trapping, excitation, and detection of ions. Ions are stored in the cube according to their cyclotron motion, which arises from the interaction of an ion with the magnetic field [21]. The cyclotron frequency (f) of an ion is inversely proportional to its *m/z* as shown below:

$$f = \mathrm{k}B\frac{z}{m}$$

where k is the constant and B is the magnetic field.

Trapped ions are detected by applying a frequency-sweep signal. When the applied frequency becomes equal to the cyclotron frequency of ions at a given *m/z*, the ions absorb energy and orbit at a larger radius. These translationally excited ions move coherently between the receiver plates. When the ion packet approaches the detection plates, an image current is generated. The time-dependent image current is a composite of different frequencies and amplitudes, which is subjected to Fourier transformation to resolve the components of ion currents and convert them into frequency domain to generate the mass spectrum [22]. Frequency is a physical parameter that is most easily and accurately measurable; therefore, FT-ICR provides extraordinarily good mass accuracy (~1 ppm) and high mass resolving power (>100,000). Mass resolution increases with magnetic field strength and decreases when mass increases. Resolution is also strongly dependent on the acquisition time.

FT-ICR requires ultra-high vacuum because the transient signal decreases with collision of ions with neutral gas molecules. The dynamic range of FT-ICR is relatively poor since the number of ions in the cell is required to be in a specific range to avoid the space-charge problem and to ensure the mass accuracy. The ICR cell is utilized for all ion manipulations including isolation, fragmentation, and mass measurement, which results in relatively slow scan speeds, especially when performing MS/MS experiments. To overcome these drawbacks, hybrid FT-ICR instruments, such as LIT-FT-ICR, were recently introduced, in which MS/MS can be performed outside the ICR cell and FT-ICR can be used only for mass measurements, resulting in fast scan time. In addition, the front LIT with automatic gain control can be used to regulate the packet of ions sent to the ICR cell.

12.4.3 Orbitrap

The Orbitrap™ mass spectrometer (Thermo Scientific, San Jose, CA) has rapidly gained popularity since it was first introduced in 2000 [23]. An Orbitrap mass analyzer consists of two electrodes: an outer barrel-shaped surface and an inner spindle-shaped electrode. A static electric field is imposed between these two electrodes. Ions are radially trapped around the central spindle electrode, rotate about the inner electrode, and oscillate harmonically along the central electrode with a frequency (ω_z) characteristic of their *m/z* values:

$$\omega_z = \sqrt{\frac{k}{m/z}}$$

where k is field curvature. The axial motion of the ions around the inner electrode produces an image current on split outer electrodes [24]. A broadband detection of this signal is followed by a fast Fourier transform (FT) to convert the time-domain signal into a frequency, and then into an *m/z* spectrum. Similar to FT-ICR, Orbitrap mass spectrometry provides very high resolving power (~30,000–100,000 in most Orbitraps and 240,000 in the Orbitrap Elite™), mass accuracy (1–5 ppm), and dynamic range (~5000) [25].

An Orbitrap mass analyzer typically precedes with an external injection device, called C-trap, for trapping ions in radiofrequency (RF)-only gas-filled curved quadrupole. The C-trap allows storage of a significant ion population before they are injected into the Orbitrap mass analyzer so that each *m/z* ion population forms a sub-microsecond pulse. The combination of the Orbitrap mass spectrometer with an external ion accumulation device such as an LIT allows accurate mass measurements on multiple stages of fragmentation for structural elucidation.

12.5 ACCURATE MASS-BASED DATA MINING TOOLS FOR SCREENING REACTIVE METABOLITES

As a result of the recent advances in HRMS technologies discussed above, several new approaches based on the accurate mass of the characteristic neutral loss of

GSH conjugates in the positive ion mode or the characteristic fragment ion of GSH-conjugates in the negative ion mode were developed to improve the selectivity in reactive metabolite screening. In addition, the development of new computational data mining tools allowed "intelligent" ways to selectively detect reactive metabolites.

12.5.1 "Pseudo" Neutral Loss Approach and Precursor Ion Approach

Castro-Perez et al. developed an exact mass "pseudo" neutral loss method as a way to eliminate false positives in screening for GSH conjugates [26]. In the analysis, exact neutral loss of 129.0426 Da (corresponding to the exact mass of pyroglutamic acid) was monitored from the full scan mass spectra obtained by alternating the collision energy between 5 and 20 eV in the positive ion mode. Whenever this exact neutral loss was detected in the high energy mass spectrum, the instrument automatically switched to MS/MS mode to acquire the product ion spectra. The specificity of this strategy was demonstrated from the analysis of a mixture of three incubation samples of acetaminophen, raloxifene, and troglitazone with liver microsomes supplemented with GSH. As shown in Figure 12.3, three distinct GSH adduct peaks were detected in the LC-MS/MS ion chromatogram while these peaks were masked among the matrix ions in the total LC-MS ion chromatogram. This exact mass neutral loss acquisition enabled extremely selective detection and identification of GSH conjugates.

Zhu and coworkers developed a selective and sensitive approach for screening GSH conjugates by monitoring the presence of the product ion at *m/z* 272.0888

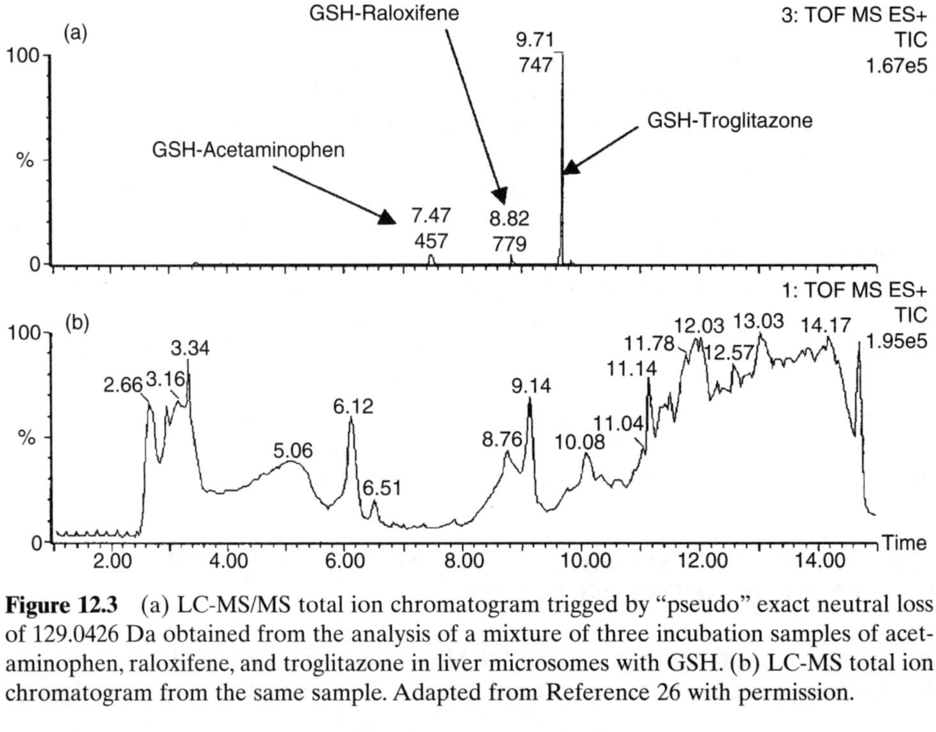

Figure 12.3 (a) LC-MS/MS total ion chromatogram trigged by "pseudo" exact neutral loss of 129.0426 Da obtained from the analysis of a mixture of three incubation samples of acetaminophen, raloxifene, and troglitazone in liver microsomes with GSH. (b) LC-MS total ion chromatogram from the same sample. Adapted from Reference 26 with permission.

(corresponding to the accurate mass of deprotanated γ-glutamyl-dehydroalanyl-glycine) under nonselective in-source CID in the negative ion mode on a LTQ-Orbitrap mass spectrometer [27]. The molecular weights of the GSH conjugates are then confirmed from the full scan MS and their structures are elucidated from the corresponding MS^2 data set in either positive or negative ion mode. The effectiveness of this approach was demonstrated with four model compounds (amodiaquine, clozapine, diclofenac, and fipexide).

12.5.2 Mass Defect Filter (MDF)

The mass of an element is based on a convention defining the mass of carbon $^{12}C = 12.0000$ Da. All other elements are either slightly above or below their integral value (e.g., $^{1}H = 1.007825$ Da and $^{16}O = 15.994910$ Da). The difference between the exact mass and the nominal mass is called mass defect. A mass defect filtering software [28, 29] was developed to facilitate the detection of drug-related ions by removing the majority of the background ions from biological matrices whose mass defects reside outside of the MDF tolerance window using high-resolution LC-MS data. MDF approach is also of great utility in screening and identification of reactive metabolites because the differences in mass defects between GSH adducts and GSH adduct filter template (MH^+ of the drug + GSH − 2H) are no greater than 0.04 Da even though GSH adducts represent a variety of changes in the structures of the drug moieties [30]. Therefore, MDF of ±0.04 Da will selectively remove nondrug-related ions whose mass defects fall outside of the GSH adduct MDF template windows. The sensitivity and selectivity of MDF approach was evaluated by analyzing reactive metabolites of seven model compounds (acetaminophen, diclofenac, carbamazepine, clozapine, *p*-cresol, 4-ethylphenol, and 3-methylindole) in human liver microsome incubations in the presence of GSH [30]. The processed full scan LC-MS chromatograms of these model compounds all displayed GSH adducts as major components with no or few interference peaks [30]. MDF also facilitated the identification of the molecular ions of GSH adducts in the mass spectra by removing interference ions.

Although the MDF approach can remove undesired matrix-related ions in most cases, it suffers from lack of specificity, resulting in appearance of false positive signals in the LC-MS chromatograms, especially in complex biological matrices. Therefore, multiple MDF templates are necessary when searching for uncommon GSH adducts that resulted from cleavage or rearrangement and for GSH adducts that form multiple charges in the mass spectra [31]. As shown in Figure 12.4a, the total ion chromatogram of the full scan LC-MS data from the incubation of ticlopidine with rat liver microsomes supplemented with GSH revealed three major GSH adducts (M2, M7, and M9) while most low abundant GSH adducts were invisible. After applying MDF with drug GSH adduct template, many matrix peaks were removed and, as a result, seven additional distinct GSH adduct peaks (M4, M11, M12, and M14-M17) were observed (Fig. 12.4b). When processing with doubly charged GSH filter template, additional four GSH adducts (M1, M5, M6, and M10) were detected (Fig. 12.4c). These adducts were predominantly ionized as doubly charged molecular ions in electrospray ionization conditions. Therefore, multiple MDF templates are necessary to provide more selective and comprehensive detection of GSH adducts.

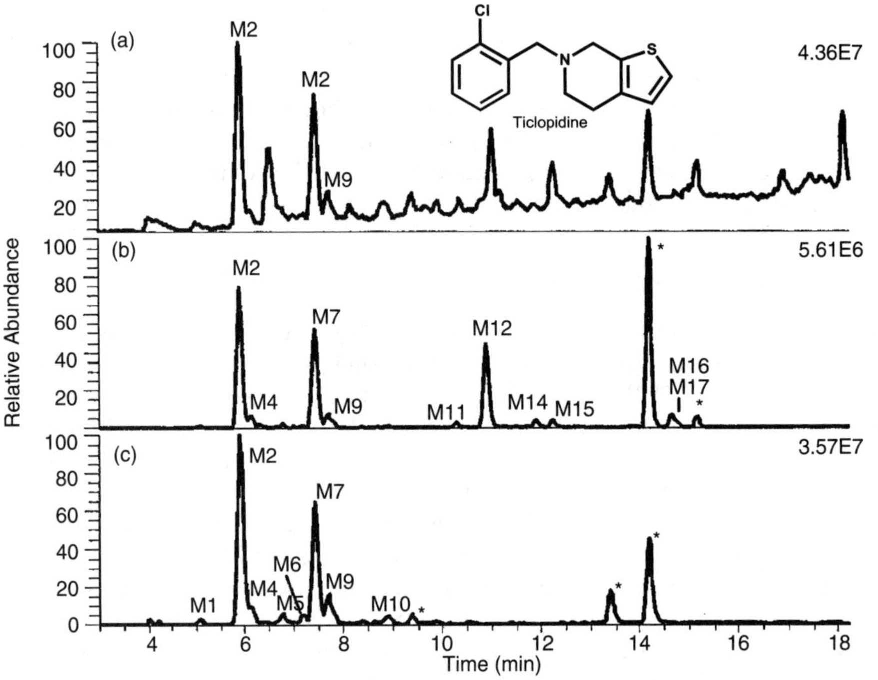

M1: P+GSH+O-4H	**M2: P+GSH+O**	**M3: P+GSH-C_6H_4ClCH+O**
M4: P+GSH-Cl+O-4H	**M5, M6, M7, M9: P+GSH+O**	**M8: P+GSH-C_7H_9NS+2O**
M10: P+GSH+2O	**M11, M17: P+GSH+3O-2H**	**M12, M14, M15, M16: P+GSH+2O-2H**
M13: P+GSH-C_6H_4ClCH-6H		

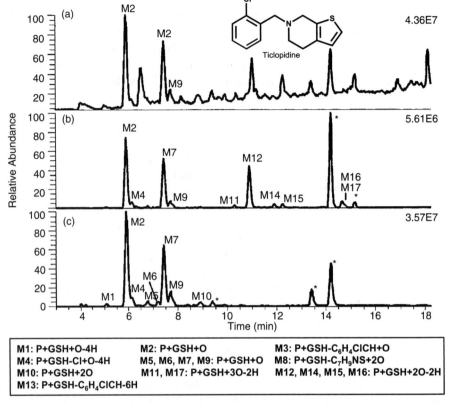

Figure 12.4 Ion chromatograms from incubations of ticlopidine with rat liver microsomes in the presence of 1 mM GSH. (a) Unprocessed full scan total ion chromatogram. (b) Total on chromatogram after MDF using singly charged drug–GSH filter template. (c) Total on chromatogram after MDF using doubly charged drug–GSH filter template. The asterisk indicates the false positive peaks. Adapted from Reference 31 with permission.

The use of MDF in combination with MS^E experiments (where E represents collision energy ramping without preselection of the parent ions) was also applied for GSH adduct screening. In addition to mass defect filtering of accurate full scan MS data sets to search for GSH adducts, the pseudo-MS/MS data sets can be further processed to generate product ion, precursor ion, and neutral loss chromatograms for searching GSH adducts and reconstituting product ion spectra of GSH adducts for structural characterization. MDF with heteroatom dealkylation algorithm in combination with MS^E and time alignment enabled detection of 53 GSH-trapped reactive metabolites formed from five drugs. This integrated approach was shown to minimize false positive results [32].

12.5.3 Background Subtraction with Noise Reduction Algorithm (BgS-NoRA)

An accurate mass-based and retention-time-shift-tolerant background subtraction (BgS) software was recently developed for extraction of drug metabolites in biological matrices [33, 34]. Zhu and coworkers improved the software by adding noise

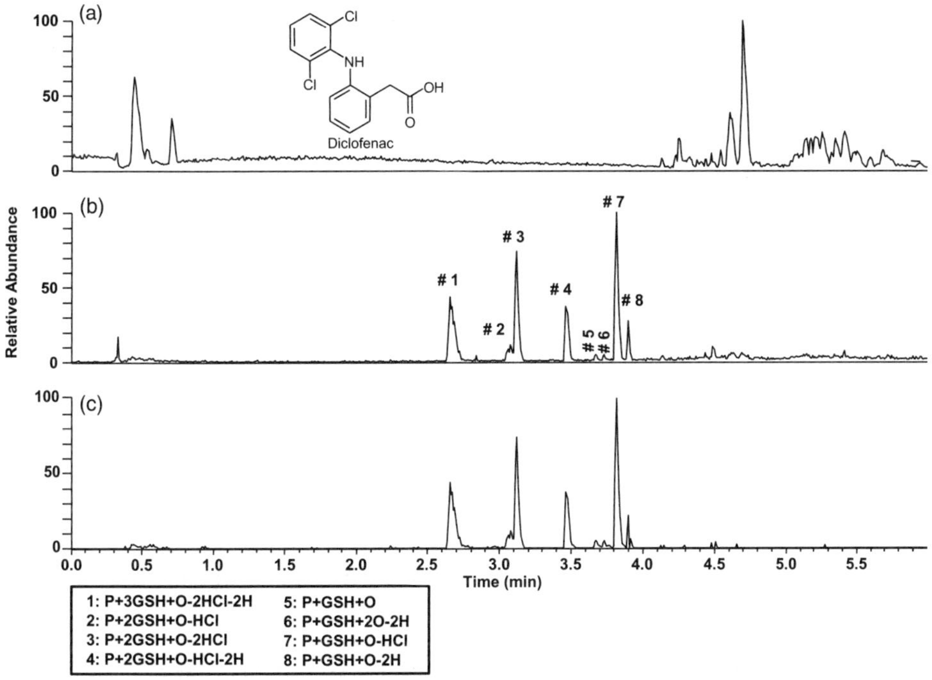

Figure 12.5 Total ion chromatograms (TIC) from the incubation of diclophenac with human liver microsomes and NADPH in the presence of 5 mM glutathione. (a) Unprocessed TIC. (b) TIC after background subtraction. (c) TIC after background subtraction and noise reduction. Adapted from Reference 35 with permission.

reduction algorithm (NoRA) to help further clean up the residual ion noises after background subtraction by removing ion signals that are not consistent across adjacent scans [34]. BgS-NoRA was successfully applied for rapid detection of GSH trapped adducts of diclofenac in the incubation with HLM [35]. The total ion chromatogram from the incubation of diclofenac with HLM and 5 mM GSH after thorough background subtraction revealed eight distinct GSH adduct peaks (Fig. 12.5b) at *m/z* 580.15338 (+2, P+3GSH+O-2HCl-2H), 445.60773 (+2, P+2GSH+O-HCl), 427.61969 (+2, P+2GSH+O-2HCl), 444.60025 (+2, P+2GSH+O-HCl-2H), 619.10211 (+1, P+GSH+O), 633.08081 (+1, P+GSH+2O-2H), 583.12531 (+1, P+GSH+O-HCl), and 617.08636 (+1, P+GSH+O-2H), respectively, where P represents the parent drug and +2, +1 represent the charge state of the GSH adducts. In contrast, these peaks were not readily discernible in the unprocessed chromatogram (Fig. 12.5a). Noise reduction further eliminated the matrix ions, leaving for the most part only GSH adduct peaks (Fig. 12.5c). The processed mass spectral data also facilitated the identification of the molecular ions of GSH adducts. The molecular ion of the trace level GSH conjugate (peak #2) at *m/z* 445.60773 was embedded among many other predominant endogenous ions and was barely detected in the unprocessed mass spectrum (Fig. 12.6a). In contrast, this ion was the only peak with no interference after applying background subtraction algorithm (Fig. 12.6b). Results from these experiments clearly demonstrated that BgS-NoRA is very effective for rapid detection of GSH-trapped reactive metabolites. This algorithm requires no prior

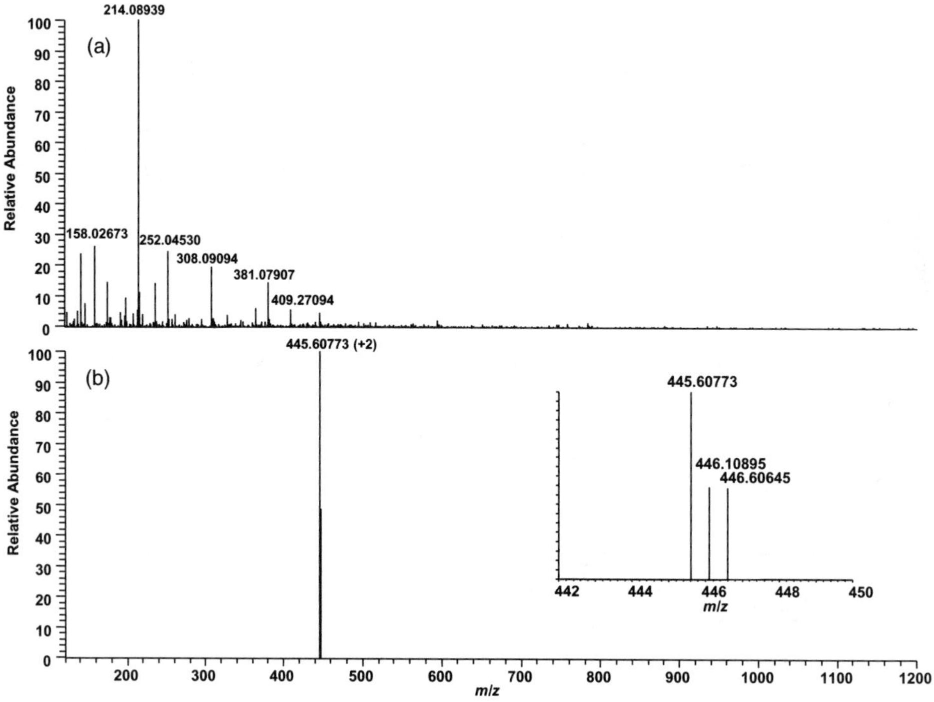

Figure 12.6 Unprocessed (a) and background subtraction and noise reduction processed (b) mass spectra of peak #2 (retention time 3.08 min). Adapted from Reference 35 with permission.

knowledge on the nature of reactive metabolites and no presumptions on their CID fragmentation patterns or mass defects. In addition, this approach does not require a radiolabeled drug or stable-isotope-labeled trapping agents and is capable of detecting reactive metabolites trapped by cyanide or methoxylamine [35].

12.5.4 Isotope Pattern Filter

Accurate mass-based isotope pattern filtering (IPF) algorithm was recently introduced to facilitate the detection of drug-derived material from complex biological matrices for analytes that possess diagnostic isotopic patterns (such as chlorine- or bromine-containing compounds) [36]. IPF is also applicable to compounds that contain synthetically incorporated isotopes (e.g., ^{13}C-, ^{15}N) generating a distinct isotope pattern. Therefore, IPF can potentially be very useful for detecting GSH adducts when a mixture of GSH and $^{13}C_2$-^{15}N- labeled GSH is used at a fixed ratio to trap reactive metabolites generated in microsomal incubations [14, 37]. Similarly, it is capable of searching GSH adducts when a brominated analog of glutathione, N-(2-bromocarbobenzyloxy)-GSH, is employed as a trapping agent in *in vitro* incubations [15]. IPF can be applied to selectively detect cyanide-trapped reactive iminium ions when a mixture of stable-isotope-labeled $K^{13}C^{15}N$ and natural KCN is used as trapping agent [38], and methoxylamine-trapped reactive aldehyde when a mixture of methoxylamine and methoxyl-d_3-amine is used in the incubations with liver microsomes.

Recently, MsMetrix (Maarssen, The Netherlands) developed a similar accurate mass-based isotope pattern filter, and it was applied for screening GSH conjugates when a mixture (1:1) of natural and stable-labeled GSH ($^{13}C_2$-^{15}N-GSH) was used for trapping reactive metabolites [39]. The software searches for GSH conjugates with a distinct doublet peak with an *m/z* difference of 3.0037. The selectivity is further improved by applying peak picking algorithm, in which the extracted ion chromatograms of light and heavy isotopic molecular ions must overlap with each other and at the same time satisfy a predefined intensity ratio. This approach was successfully applied in detecting GSH adducts of clozapine, diclofenac, imipramine, and ticlopidine in HLM incubations [39].

12.6 QUANTITATION OF REACTIVE METABOLITE FORMATION

Advances in HRMS technology and the development of accurate mass-based data mining tools have made it relatively easy to detect and identify reactive metabolites; however, it remains considerably challenging to obtain quantitative assessment of reactive metabolite levels without radiolabeled drugs. Radiolabeled drugs are rarely available at the early stage of drug discovery, and in many cases, a quantitative analysis is necessary to differentiate the effects of structure alteration on the degree of reactive metabolite formation; therefore, alternative ways of obtaining an accurate measurement of reactive metabolite formation are highly desirable.

Radiolabeled trapping agents, such as ^{35}S-GSH [40–42], ^{3}H-GSH [43], or ^{14}C-cyanide [44], and GSH derivatives [45, 46] have been utilized for quantitation of reactive metabolite levels *in vitro*. Gan and coworkers developed a quantitative method that used GSH labeled with a fluorescent tag for trapping and quantitation of the reactive metabolites. Dansyl-GSH displayed a maximum excitation at 340 nm and a maximum emission at 525 nm. These wavelengths were used for the detection and quantitation of GSH adducts formed in the incubations. Soglia et al. reported a semi-quantitative method for determining reactive metabolite levels using LC-MS/MS and a novel quaternary ammonium glutathione analog (QA-GSH) [46]. A fixed positive charge at QA-GSH significantly improves the ionization efficiency and therefore increases the limit of detection. It also equalizes the mass spectrometric response from different GSH conjugates. The mass spectrometric responses of three QA-GSH conjugate standards were within threefold even though the parent moiety responses differed by as much as 19-fold, suggesting that the MS response was based predominantly on the fixed charge of the QA-GSH moiety and that conjugation to other structurally diverse compounds resulted in a similar MS response. Therefore, the LC-MS/MS response factor for any QA-GSH conjugate standard could be used as an internal standard for semi-quantitation of reactive metabolite levels to assess the bioactivation potential of drug candidates *in vitro*. Further development of new ionization methods that minimize the differences in mass spectral responses from diverse classes of trapped reactive metabolites and other quantitative detection technologies is highly coveted.

12.7 CHALLENGES AND FUTURE PERSPECTIVES

With the advancement of HRMS technology and the development of "intelligent" data processing tools, the detection and identification of reactive metabolites from *in vitro* trapping experiments and *in vivo* biological samples are becoming a routine practice. The introduction of mass defect filtering technology, BgS-NoRA, and IPF allows detection and identification of reactive metabolites with ease and enables high-throughput screening for drug candidates' propensity to form reactive metabolites. However, it should be recognized that not all reactive metabolites can be trapped by small molecule nucleophiles. Bioactivation may lead to the formation of highly reactive species that covalently bind to the active site of the enzyme before having the opportunity to diffuse into the incubation medium and react with the chemical trapping agent [47].

Technologies that facilitate screening process will continue to be highly sought after. While trying to put chemically reactive metabolite data into perspective, it is important to recognize that the connection of reactive metabolite formation and toxicity endpoint is very loose [48]. Not all drugs that undergo bioactivation are associated with ADRs in the clinic. Recent studies showed that total adduct burden (total adduct burden = dose $\times F_a \times F_m \times F_{adduct}$, where F_a is the fraction absorbed, F_m is the fraction metabolized, and F_{adduct} is the ratio of covalent adduct/total metabolite) rather than the levels of adducts formed in the *in vitro* incubations should be taken into consideration for evaluation of the potential for drug-induced liver toxicity [6, 49]. Formation of reactive metabolites cannot always be completely eliminated, and therefore, in drug discovery and development caution and judgment (e.g., risk vs. benefit assessment, anticipated clinical dose and dose duration) are needed when assessing the potential impact of reactive metabolite formation in clinical safety of a therapeutic candidate.

REFERENCES

[1] Park BK, Boobis A, Clarke S, Goldring CE, Jones D, Kenna JG, et al. Managing the challenge of chemically reactive metabolites in drug development. Nat Rev Drug Discov 2011;10:292–306.

[2] Erve JC. Chemical toxicology: reactive intermediates and their role in pharmacology and toxicology. Expert Opin Drug Metab Toxicol 2006;2:923–46.

[3] Gan J, Ruan Q, He B, Zhu M, Shyu WC, Humphreys WG. In vitro screening of 50 highly prescribed drugs for thiol adduct formation—comparison of potential for drug-induced toxicity and extent of adduct formation. Chem Res Toxicol 2009;22:690–8.

[4] Uetrecht J. Idiosyncratic drug reactions: past, present, and future. Chem Res Toxicol 2008;21:84–92.

[5] Walgren JL, Mitchell MD, Thompson DC. Role of metabolism in drug-induced idiosyncratic hepatotoxicity. Crit Rev Toxicol 2005;35:325–61.

[6] Nakayama S, Atsumi R, Takakusa H, Kobayashi Y, Kurihara A, Nagai Y, et al. A zone classification system for risk assessment of idiosyncratic drug toxicity using daily dose and covalent binding. Drug Metab Dispos 2009;37:1970–7.

[7] Baillie TA. Metabolism and toxicity of drugs. Two decades of progress in industrial drug metabolism. Chem Res Toxicol 2008;21:129–37.

[8] Evans DC, Watt AP, Nicoll-Griffith DA, Baillie TA. Drug-protein adducts: an industry perspective on minimizing the potential for drug bioactivation in drug discovery and development. Chem Res Toxicol 2004;17:3–16.

[9] Wen B, Fitch WL. Analytical strategies for the screening and evaluation of chemically reactive drug metabolites. Expert Opin Drug Metab Toxicol 2009;5:39–55.

[10] Kalgutkar AS, Soglia JR. Minimising the potential for metabolic activation in drug discovery. Expert Opin Drug Metab Toxicol 2005;1:91–142.

[11] Ma S, Zhu M. Recent advances in applications of liquid chromatography-tandem mass spectrometry to the analysis of reactive drug metabolites. Chem Biol Interact 2009; 179:25–37.

[12] Ma S, Subramanian R. Detecting and characterizing reactive metabolites by liquid chromatography/tandem mass spectrometry. J Mass Spectrom 2006;41:1121–39.

[13] Yan Z, Caldwell GW, Maher N. Unbiased high-throughput screening of reactive metabolites on the linear ion trap mass spectrometer using polarity switch and mass tag triggered data-dependent acquisition. Anal Chem 2008;80:6410–22.

[14] Yan Z, Caldwell GW. Stable-isotope trapping and high-throughput screenings of reactive metabolites using the isotope MS signature. Anal Chem 2004;76:6835–47.

[15] Leblanc A, Shiao TC, Roy R, Sleno L. Improved detection of reactive metabolites with a bromine-containing glutathione analog using mass defect and isotope pattern matching. Rapid Commun Mass Spectrom 2010;24:1241–50.

[16] Zheng J, Ma L, Xin B, Olah T, Humphreys WG, Zhu M. Screening and identification of GSH-trapped reactive metabolites using hybrid triple quadrupole linear ion trap mass spectrometry. Chem Res Toxicol 2007;20:757–66.

[17] Dieckhaus CM, Fernandez-Metzler CL, King R, Krolikowski PH, Baillie TA. Negative ion tandem mass spectrometry for the detection of glutathione conjugates. Chem Res Toxicol 2005;18:630–8.

[18] Mahajan MK, Evans CA. Dual negative precursor ion scan approach for rapid detection of glutathione conjugates using liquid chromatography/tandem mass spectrometry. Rapid Commun Mass Spectrom 2008;22:1032–40.

[19] Lacorte S, Fernandez-Alba AR. Time of flight mass spectrometry applied to the liquid chromatographic analysis of pesticides in water and food. Mass Spectrom Rev 2006; 25:866–80.

[20] Andrews GL, Simons BL, Young JB, Hawkridge AM, Muddiman DC. Performance characteristics of a new hybrid quadrupole time-of-flight tandem mass spectrometer (TripleTOF 5600). Anal Chem 2011;83:5442–6.

[21] Scigelova M, Hornshaw M, Giannakopulos A, Makarov A. Fourier transform mass spectrometry. Mol Cell Proteomics 2011;10:M111–009431.

[22] Junot C, Madalinski G, Tabet JC, Ezan E. Fourier transform mass spectrometry for metabolome analysis. Analyst 2010;135:2203–19.

[23] Makarov A. Electrostatic axially harmonic orbital trapping: a high-performance technique of mass analysis. Anal Chem 2000;72:1156–62.

[24] Makarov A, Scigelova M. Coupling liquid chromatography to Orbitrap mass spectrometry. J Chromatogr A 2010;1217:3938–45.

[25] Scigelova M, Makarov A. Advances in bioanalytical LC-MS using the Orbitrap mass analyzer. Bioanalysis 2009;1:741–54.

[26] Castro-Perez J, Plumb R, Liang L, Yang E. A high-throughput liquid chromatography/tandem mass spectrometry method for screening glutathione conjugates using exact mass neutral loss acquisition. Rapid Commun Mass Spectrom 2005;19:798–804.

[27] Zhu X, Kalyanaraman N, Subramanian R. Enhanced screening of glutathione-trapped reactive metabolites by in-source collision-induced dissociation and extraction of product ion using UHPLC-high resolution mass spectrometry. Anal Chem 2011;83: 9516–23.

[28] Zhang H, Zhang D, Ray K, Zhu M. Mass defect filter technique and its applications to drug metabolite identification by high-resolution mass spectrometry. J Mass Spectrom 2009;44:999–1016.

[29] Zhang H, Zhang D, Ray K. A software filter to remove interference ions from drug metabolites in accurate mass liquid chromatography/mass spectrometric analyses. J Mass Spectrom 2003;38:1110–2.

[30] Zhu M, Ma L, Zhang H, Humphreys WG. Detection and structural characterization of glutathione-trapped reactive metabolites using liquid chromatography-high-resolution mass spectrometry and mass defect filtering. Anal Chem 2007;79:8333–41.

[31] Ruan Q, Zhu M. Investigation of bioactivation of ticlopidine using linear ion trap/orbitrap mass spectrometry and an improved mass defect filtering technique. Chem Res Toxicol 2010;23:909–17.

[32] Barbara JE, Castro-Perez JM. High-resolution chromatography/time-of-flight MS^E with *in silico* data mining is an information-rich approach to reactive metabolite screening. Rapid Commun Mass Spectrom 2011;25:3029–40.

[33] Zhang H, Yang Y. An algorithm for thorough background subtraction from high-resolution LC/MS data: application for detection of glutathione-trapped reactive metabolites. J Mass Spectrom 2008;43:1181–90.

[34] Zhu P, Ding W, Tong W, Ghosal A, Alton K, Chowdhury S. A retention-time-shift-tolerant background subtraction and noise reduction algorithm (BgS-NoRA) for extraction of drug metabolites in liquid chromatography/mass spectrometry data from biological matrices. Rapid Commun Mass Spectrom 2009;23:1563–72.

[35] Ma S, Chowdhury SK. Application of LC-high-resolution MS with "intelligent" data mining tools for screening reactive drug metabolites. Bioanalysis 2012;4:501–10.

[36] Zhu P, Tong W, Alton K, Chowdhury S. An accurate-mass-based spectral-averaging isotope-pattern-filtering algorithm for extraction of drug metabolites possessing a distinct isotope pattern from LC-MS data. Anal Chem 2009;81:5910–7.

[37] Ma L, Wen B, Ruan Q, Zhu M. Rapid screening of glutathione-trapped reactive metabolites by linear ion trap mass spectrometry with isotope pattern-dependent scanning and postacquisition data mining. Chem Res Toxicol 2008;21:1477–83.

[38] Rousu T, Pelkonen O, Tolonen A. Rapid detection and characterization of reactive drug metabolites in vitro using several isotope-labeled trapping agents and ultra-performance liquid chromatography/time-of-flight mass spectrometry. Rapid Commun Mass Spectrom 2009;23:843–55.

[39] Ruijken MMA. MsXelerator RM: a software platform for reactive metabolite detection using low and high resolution mass spectrometry data. 58th ASMS Conference on Mass Spectrometry and Allied Topics. Salt Lake City, UT, USA; 2010.

[40] Miyaji Y, Makino C, Kurihara A, Suzuki W, Okazaki O. *In vitro* evaluation of the potential for drug-induced toxicity based on ^{35}S-labeled glutathione adduct formation and daily dose. Bioanalysis 2012;4:263–9.

[41] Takakusa H, Masumoto H, Makino C, Okazaki O, Sudo K. Quantitative assessment of reactive metabolite formation using ^{35}S-labeled glutathione. Drug Metab Pharmacokinet 2009;24:100–7.

[42] Meneses-Lorente G, Sakatis MZ, Schulz-Utermoehl T, De Nardi C, Watt AP. A quantitative high-throughput trapping assay as a measurement of potential for bioactivation. Anal Biochem 2006;351:266–72.

[43] Thompson DC, Perera K, London R. Quinone methide formation from para isomers of methylphenol (cresol), ethylphenol, and isopropylphenol: relationship to toxicity. Chem Res Toxicol 1995;8:55–60.

[44] Gorrod JW, Whittlesea CM, Lam SP. Trapping of reactive intermediates by incorporation of 14C-sodium cyanide during microsomal oxidation. Adv Exp Med Biol 1991;283: 657–64.

[45] Gan J, Harper TW, Hsueh MM, Qu Q, Humphreys WG. Dansyl glutathione as a trapping agent for the quantitative estimation and identification of reactive metabolites. Chem Res Toxicol 2005;18:896–903.

[46] Soglia JR, Contillo LG, Kalgutkar AS, Zhao S, Hop CE, Boyd JG, et al. A semiquantitative method for the determination of reactive metabolite conjugate levels in vitro utilizing liquid chromatography-tandem mass spectrometry and novel quaternary ammonium glutathione analogues. Chem Res Toxicol 2006;19:480–90.

[47] Hollenberg PF, Kent UM, Bumpus NN. Mechanism-based inactivation of human cytochromes P450s: experimental characterization, reactive intermediates, and clinical implications. Chem Res Toxicol 2008;21:189–205.

[48] Humphreys WG. Overview of strategies for addressing BRIs in drug discovery: impact on optimization and design. Chem Biol Interact 2011;192:56–9.

[49] Usui T, Mise M, Hashizume T, Yabuki M, Komuro S. Evaluation of the potential for drug-induced liver injury based on in vitro covalent binding to human liver proteins. Drug Metab Disopos 2009;37:2383–92.

13

MASS SPECTROMETRY OF SIRNA

MARK T. CANCILLA AND W. MICHAEL FLANAGAN

13.1 INTRODUCTION

RNA interference, RNAi, is a conserved mechanism of post-transcriptional gene regulation. The first report of an RNAi-like mechanism in plants occurred by Napoli et al. in which they inserted a plasmid containing chalcone synthetase that is necessary for the biosynthesis of anthocyanin pigments that are responsible for the blue, purple, and red coloration of many flowers, into petunias to create petunia's with deep blue flowers [1]. The experiments resulted in 42% of all flowers having variegated blue and white or all white flowers due to the inadvertent suppression of the endogenous chalcone synthetase gene. These early investigations were followed by studies in *Neurospora crassa* [2]. At the time of these initial reports, the understanding of the RNAi mechanism of action and the ability to manipulate and control biological systems was not well understood until Fire et al. reported the use of double-stranded RNA to post-transcriptionally control gene expression [3]. A few years later, researchers demonstrated RNAi in mammalian cells, and that initiated a new technology platform for altering mammalian gene expression [4, 5]. Today, RNAi is a routine technique used to interrogate biological systems in laboratories across the world. RNAi also holds tremendous potential for specifically and safely treating human disease and has spawned a multibillion dollar biotechnology and pharmaceutical industry in seeing the research promise become a therapeutic reality.

Small interfering RNAs (siRNAs) are generally 19–23 nucleotides (nt) in length double-stranded RNA with a 2 nt overhangs on the 3′ ends. siRNAs exist endogenously in a cell or synthetic siRNAs may be delivered as a therapeutic by a variety of different methods (see recent reviews on siRNA delivery [6–10]). Once inside the cell, the siRNAs enter the RNA-induced silencing complex (RISC) as illustrated in Figure 13.1. The RISC complex is composed of numerous proteins that have been

Mass Spectrometry for Drug Discovery and Drug Development, First Edition. Edited by Walter A. Korfmacher.

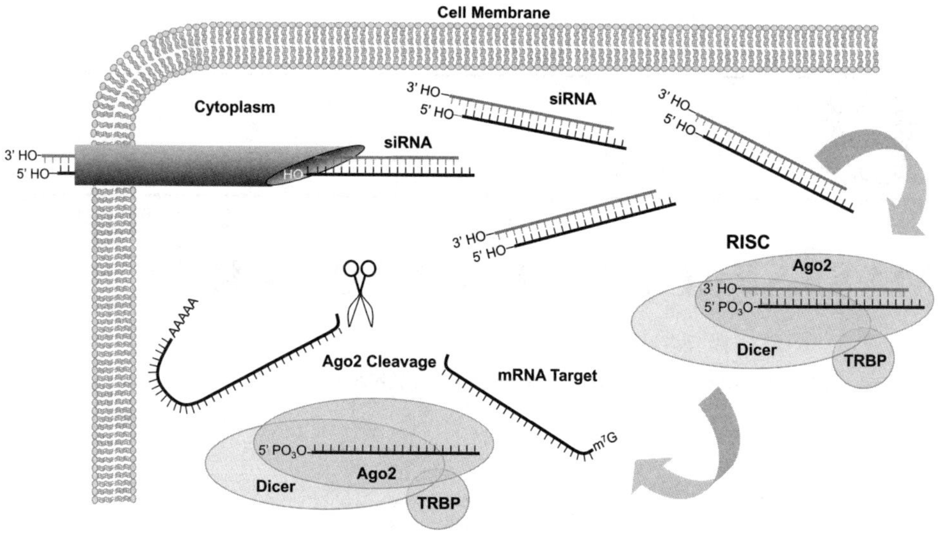

Figure 13.1 RNA interference mechanism. siRNA delivery to the cytoplasm of the cell is depicted by the needle-like tube. The gray and black strands of the siRNA duplex represent the passenger (sense) and guide (antisense) strand, respectively. The phosphorylated duplex enters the RISC complex composed of the HIV transactivating response (TAR) RNA-binding protein (TRBP), Dicer, which cleaves longer dsRNA to 21 base pairs, and Argonaute 2, which slices the target mRNA through its endonuclease activity. The RISC is catalytic and will continue to cleave target mRNA until the guide (antisense) strand dissociates from the complex.

identified by a variety of biochemical methods including immunoprecipitation of the RISC complex followed by mass spectrometry-based proteomics [11, 12]. The RISC core proteins are Dicer, TRBP, and Ago2. The duplex siRNA enters the complex and the guide strand (also referred to as the antisense strand), which is complementary to the mRNA target binds to Ago2. The passenger strand (also referred to as the sense strand) either is unwound by a yet-to-be-determined helicase or is cleaved by the RNAse activity of Ago2, rendering the guide strand available and competent to bind its complementary mRNA target [13–15]. The guide strand-directed RISC is catalytic and will continue to seek and destroy complementary RNA targets, thus downregulating a gene's protein levels until the guide strand dissociates from RISC.

13.2 COMMON CHEMICAL MODIFICATIONS OF SIRNA

To create safe and effective siRNA therapeutics, the siRNA must be chemically modified to enhance potency, extend duration of action, improve specificity, limit nuclease metabolism, and reduce immune stimulation [16, 17]. There are three potential sites where modification within an siRNA may exist: the ribose sugar, base, and backbone. Examples of common chemical modifications are shown in Figure 13.2. A wide variety of chemical modifications have been applied to siRNAs, and these frequently include 2′-deoxy (DNA), 2′-hydroxy (RNA), 2′-O-methyl (2′-OMe

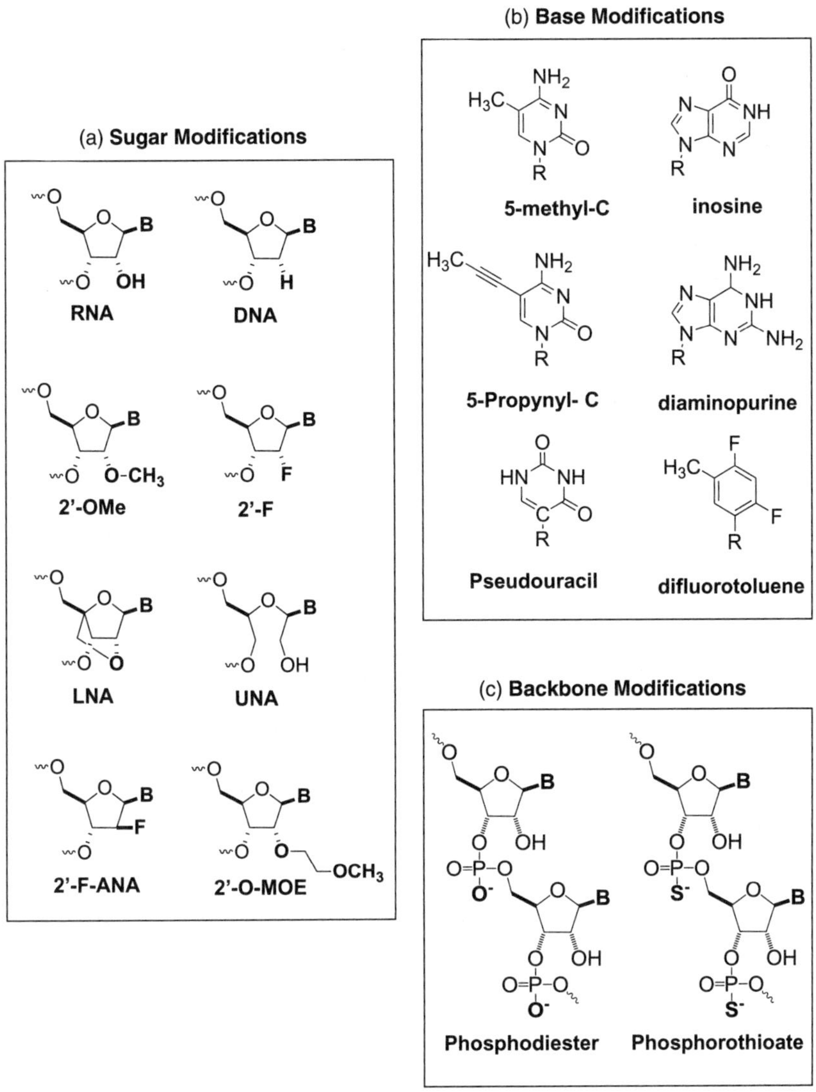

Figure 13.2 Structures of common chemical modifications applied to siRNAs to improve potency, stability, and specificity, and to avoid immune stimulation. (a) Sugar modifications (identified in black) that have been used in siRNAs including the naturally occurring RNA, DNA, 2′-O-Methyl (2′-OMe) and synthetic 2′-fluoro (2′-F), locked nucleic acids (LNA), unlocked nucleic acids (UNA), 2′-fluro-arabino nucleic acid (2′-F-ANA), and 2′-O-methoxyethyl (2′-O-MOE). (b) Nucleobase modifications incorporated into siRNAs. The ribose phosphate backbone of the siRNA is indicated by R. Most base modifications have been used to alter recognition of the mRNA target. Difluorotoluene is unusual since it cannot form hydrogen bonds with the opposing base on the other strand. (c) Common nucleotide linkages used in siRNAs. The naturally occurring phosphodiester linkage (oxygen shown in black) is the most commonly used backbone in siRNAs followed by the synthetic phosphorothioate linkage (sulfur shown in black) that improves nuclease stability. The nucleobase is represented by (b).

or 2′-methoxy), synthetic 2′-fluoro (2′-F), locked nucleic acids (LNA) that contain a bridging methylene between the 2′ and 4′ position, unlocked nucleic acids (UNA) that are acyclic between the 2′ and 3′ positions, 2′-fluoroarabinose (2′-FANA or 2′-fluoro-arabino nucleic acid), and 2′-O-methoxyethyl (2′-O-MOE). In general, 2′-ribose-modified siRNAs show improved stability against nucleases, enhanced specificity, retained activity compared with RNA alone, and, in some cases, increased potency.

In contrast to sugar modifications, the incorporation of nucleobase modifications into siRNAs has been more limited partially due to complex and costly syntheses, despite the potential ability, in some cases, to stabilize the RNA structure and improve binding affinity between the siRNA and mRNA target. The most common nucleobase modifications include the naturally occurring 5-methyl cytosine, inosine, pseudouridine, synthetically derived diaminopurine, 5-propynyl cytosine, and difluorotoluene. The most widely used modification to the phosphate backbone is the substitution of one of the nonbridging oxygen atoms with sulfur. Phosphorothioate-modified siRNAs retain silencing activity and also demonstrate enhanced nuclease stability. As noted, the majority of siRNA chemical modifications improve nuclease stability in biological matrices. Increase siRNA stability allows for less frequent and smaller doses that will lead to broader therapeutic utility. Despite the wide range of chemical modifications, the first generation of siRNAs entering the clinic usually contain phosphorothioates near the 3′ end of passenger and guide, and the incorporation of 2′-methoxy to improve siRNA stability from intracellular nucleases, and block immune recognition by the innate immune system. Second-generation siRNAs would be expected to contain a much wider range of novel modifications that not only stabilize the siRNA, avoid immune surveillance, and improve specificity, but may also improve Ago2 target cleavage and binding.

13.3 CURRENT ANALYTICAL METHODOLOGIES

As siRNAs progress into and through the clinic to marketable human therapeutics, there is a great need to evaluate siRNAs in plasma and monitor intracellular levels of siRNAs to establish pharmacokinetic (PK) and pharmacodynamic (PD) relationships in support of clinical trials. Furthermore, siRNA quantitation of full-length duplex parent and metabolites will be essential to understand siRNA exposure and its potential relationship to toxicity. siRNA quantitation methods must be robust, specific, sensitive, and accurate from biological matrices. Multiple methods have been develop to track siRNAs in biological fluids including enzyme-linked immunosorbent assays (ELISA), reverse transcriptase quantitative polymerase chain reaction (RT-qPCR), solution-based probe hybridization, capillary gel electrophoresis, and mass spectrometry. Each detection technique has its strengths and weaknesses.

Both sandwich-based and competitive ELISA demonstrated high sensitivity with minimal sample preparation; however, these assays suffer from cross-hybridization to metabolized siRNAs: 5′ and 3′ N-2 metabolites showed 48% cross-hybridization [18]. Improvements to ELISA methods have increased discrimination between parent and 3′ metabolites, although 5′ metabolites were still able to cross-react with the probe [19].

A multitude of RT-qPCR-based assays have been developed to quantitate siRNAs and microRNAs in biological matrices, including primer-extension qPCR [20], stem-loop PCR [21, 22], universal probe library RT-PCR [23], and others [24–26]. Quantitative PCR is a sensitive method with the ability to measure pg/mL concentrations of siRNAs with relative ease using commercially available reagents. Despite the wide dynamic range, qPCR techniques also fail to discriminate among distinct metabolites since the primers can hybridize and amplify partially 5′ and 3′ degraded siRNAs. Depending on the primer set and sequence of the siRNA, siRNA metabolites that are three nucleotides shorter on the 5′ and/or 3′ end may be detected as full-length products. In addition, qPCR polymerases may have difficulties progressing through second-generation siRNAs that are highly chemically modified. qPCR-based detection of siRNAs containing locked nucleic acids (LNAs) and unlocked nucleic acids (UNAs) or other potential novel modifications may fail completely.

Several reports have outlined the virtues of solution phase hybridization to quantify siRNA parents and metabolites [27]. Similar to ELISA and qPCR methods, the identification of metabolites can be confounded by siRNA chemistries that increase or decrease the ability to hybridize with a probe. For instance, 2′-OMe and 2′-F modifications can increase the melting temperature (Tm) between an RNA and DNA probe by 0.5–1.0°C per substitution compared with 2′-OH RNA. LNAs can increase the Tm by 2–8°C [28, 29], whereas UNAs can decrease Tm by 4–8°C per incorporation [30]. The overall arrangement of chemical modifications on the siRNA may have dramatic effects on how well potential metabolites of the siRNA hybridize to a probe.

To overcome the deficiencies in the current analytically methods and techniques, the field is turning to mass spectrometry in order to detect and quantify siRNA and metabolites in biological matrices. For decades, mass spectrometry has facilitated the analysis of oligonucleotide therapeutics, such as antisense, due to the advent of the soft ionization techniques electrospray ionization (ESI) and matrix-assisted laser desorption/ionization (MALDI), which allow the production of ions from high molecular weight, thermally labile, nonvolatile biomolecules [31, 32]. The same promise holds for siRNA analysis, where mass spectrometry-based techniques would become the standard analytical methodology due to their general applicability (e.g., independent of strand sequence), selectivity, sensitivity, and accuracy.

13.4 OBSTACLES ASSOCIATED WITH THE ANALYSIS OF SIRNA BY MASS SPECTROMETRY

13.4.1 Double-Stranded Characteristic of siRNA

Although similar in physical and chemical properties to antisense oligonucleotide therapeutics, the inherent duplex nature of siRNA creates a unique set of analytical challenges (where each strand weighs approximately 6000–7000 Da with around 20 negative charges per strand). In order to properly characterize the state of a siRNA, it is important to maintain the duplex during sample extraction and have a clear methodology to discern minor changes in each strand due to exterior factors, for example, synthesis impurities, storage degradation, or metabolite products. Chromatographic separation of duplex from such truncated duplex species has been shown to be possible, yet difficult [33]. Unfortunately, liquid chromatography-mass spectrometry (LC-MS) analysis of the intact duplex has demonstrated a decrease in sensitivity compared with the denatured individual strands due to increased cation adduction [34]. MALDI-time-of flight (TOF) MS has demonstrated the ability to detect intact siRNA duplexes using 6-aza2-thiothymine (ATT) containing diammonium hydrogen citrate (DAHC) as the matrix for quality control purposes [35], yet to date there have been no reports of MALDI being applied to analysis of intact siRNA duplexes isolated from biological matrices. Due to the challenges associated with analyzing the intact duplex by mass spectrometry, most analysts have decided to pursue denaturing the duplex using high-performance liquid chromatography (HPLC) column heaters [36] or MALDI matrices [35] to focus on characterizing potential changes in each single strand.

13.4.2 Lack of Mass Spectrometry Sensitivity

A reduction in ionization efficiency is primarily due to the production of cation (e.g., Na^+, K^+, Mg^{++}) adduct species that form with the phosphodiester backbone of oligonucleotides during ionization. This heterogeneous mixture of ionized adducts further separates the ion current of the siRNA for each charge state, broadens peaks, and may complicate molecular weight determinations for instruments with low to moderate resolving power. In some instances, high-resolution mass spectrometers are needed for accurate mass measurements to differentiate clearly the 1-Da difference between cytidine and uridine [36].

When analyzed by ESI, siRNAs produce an envelope of multiply charged ions in the gas phase that allow for the analysis of high molecular weight compounds such as the intact duplex or each individual strand based on chosen ionization conditions. The production of the charge state envelope reduces detection limits by diluting the concentration of the ion current to multiple *m/z* values. The formation of multiply charged ions also increases the complexity of ESI spectra and mass deconvolution software is needed to determine intact molecular weights (Fig. 13.3).

Current practice for the LC-MS of siRNA is to use ion-pairing (IP) reagents to increase reversed-phase retention and chromatographic resolution [37]. Although ion-pairing reagents are all but essential for optimal chromatography and many have been found to be compatible with mass spectrometry, they have been shown to be nonoptimal for ESI and decrease ESI sensitivity. Current siRNA limit of quantitation (LOQ) using IP-HPLC-MS/MS techniques are approximately 5–10 ng/mL [38], which is not adequate for thorough PK/PD studies, which require LOQs of less than the 1 ng/mL level. To reach the necessary sensitivity limits for *in vivo* studies, hybridization-ELISA and RT-qPCR techniques are presently utilized.

13.4.3 Oligonucleotide Sequence and Chemical Modifications Influence LC-MS Response

Ribonucleotides have different hydrophobicities based on the nucleobase (C vs. G vs. A vs. U) and chemical modifications (e.g., 2′-F vs. 2′-hydroxy). The overall oligonucleotide hydrophobicity influences the properties of both reversed-phase and

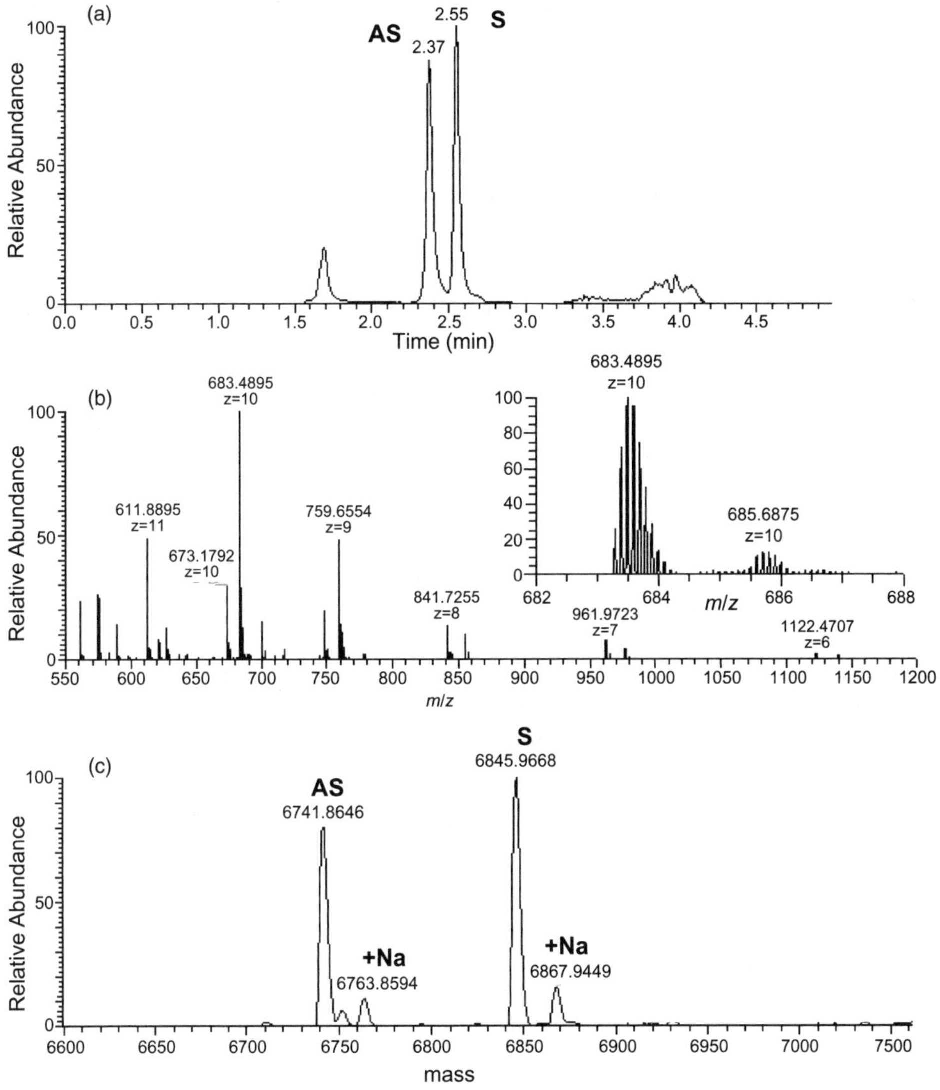

Figure 13.3 (a) Total ion chromatogram for the ion-pair LC-MS of an siRNA duplex separated into its component sense and antisense strands using 1.7 mM TEA/100 mM HFIP with a methanol elution gradient on a C_{18} column. Data generated in the negative mode using a hybrid LTQ-Orbitrap mass spectrometer. Peaks are labeled AS for antisense, S for sense. (b) Averaged mass spectrum from 2.3 min to 2.6 min representing the charge state envelopes for both the sense and antisense strands. Inset is a close-up around the –10 charge state of the sense strand. (c) Deconvoluted zero charge state mass spectrum of the strand's charge state envelopes to determine their molecular weights. Minor sodium adducts are also observed.

anion exchange chromatography, and resolving each species chromatographically requires optimal method development and in many instances extended HPLC run times [39–43]. Furthermore, ESI efficiency is also influenced by the siRNA hydrophobicity. As hydrophobicity increases, the ESI signal increases [44, 45]. This potential range of ionization characteristics based on nucleobase and chemical modification composition may lead to ion suppression of certain co-eluting species and further creates quantitative challenges when analyzing samples containing siRNA failure sequences, degradants, or metabolites.

13.4.4 Difficult Isolation from Biomatrices

As siRNA delivery technology continues to advance, the local delivery to target tissues and systemic delivery are becoming safer and more efficient [7]. The understanding of the siRNA absorption, distribution, metabolism, and excretion (ADME) properties along with the PK parameters *in vivo* is a necessary step toward developing an RNAi therapeutic. The primary depots for siRNAs delivered *in vivo* with a delivery vehicle, such as a clinically acceptable lipid nanoparticle (LNP), are to organs such as the liver and spleen [46], thus requiring tissue extraction protocols in order to identify and quantitate the minor amounts of siRNA in these matrices. Isolating siRNAs from plasma and tissue for MS analysis have proven to be difficult and laborious, usually requiring a two-step liquid–liquid, solid-phase extraction (SPE) [34, 47]. The high protein binding of oligonucleotides requires an extraction with phenol : chloroform to disrupt protein–RNA interactions, which impedes using automation and high-throughput sample preparation.

13.4.5 Inefficient Tandem Mass Spectrometry Fragmentation

The complete sequencing of siRNAs using a tandem mass spectrometry approach in an LC-MS experimental time frame is a challenging analytical goal. Currently, there are multiple hurdles to overcome before this achievement is met. Collision-induced dissociation (CID) of siRNA duplexes has not provided complete strand sequence information because the primary dissociation pathway of the duplex is separation into the individual sense and antisense strands [48]. Even if CID is performed on single-stranded siRNAs, the necessary chemical modification of the ribose hydroxyl groups significantly reduces the extent of CID fragmentation and does not always provide adequate product ions to locate modifications within a sequence [49]. In general, tandem mass spectrometry techniques applied to oligonucleotides produce complex spectra of multiple fragmentation pathways requiring advanced software and experienced analytical personnel for adequate interpretation.

Indeed, analyzing siRNAs by mass spectrometry poses many challenges, yet the reward is worth the effort. Unlike hybridization ELISAs or RT-qPCR, mass spectrometry offers the ability for the precise identification and selective monitoring of each intact siRNA strand or its metabolites or degradants from complex matrices. The following sections will focus on overcoming the current analytical challenges when analyzing siRNA using mass spectrometry. Fortunately, there are decades of literature on the mass spectrometry analysis of oligonucleotides that can be directly applied to the analytical challenges of siRNA.

13.5 SAMPLE PREPARATION

13.5.1 Desalting

Desalting (the removal of nonvolatile cations from oligonucleotides) is essential for increasing mass spectrometry sensitivity and producing simplified mass spectra. Initially, desalting of oligonucleotides for ESI is discussed and the sample preparation for MALDI will be covered in Section 13.7. Typically, the depletion of nonvolatile cations such as Na^+ is achieved by exchange with volatile ammonium ions. A proven desalting technique is ethanol precipitation from ammonium acetate [50–53] where repeated precipitations further reduce the amount nonvolatile cations. Sample recovery for the procedure is approximately 60–80% so consideration should be taken for samples of limited amount.

The addition of chelating agents such as ethylenediaminetetraacetic acid (EDTA) and *trans*-1,2-diaminocyclohexane-*N,N,N′,N′*-tetraacetic acid (CDTA) are commonly used to displace cations such as Na^+ and Mg^{++} from oligonucleotides [54], but high levels of chelators can be deleterious to ESI. Organic bases such as triethylamine (TEA), piperidine, and imidazole are also used in ESI buffers to displace nonvolatile cations from oligonucleotides [55, 56]. For example, the direct infusion of oligonucleotides in a buffer composed of (7 mM TEA and 3 mM ammonium formate)/methanol 50:50 (v/v) has demonstrated increased ESI intensities due to diminishing cation adducts [47].

Cation exchange has been employed to desalt samples prior to ESI, both on- and offline. For example, online microcolumns packed with cation-exchange resins situated before the ESI source have been successful for the removal of cations from oligonucleotides [57, 58]. As an offline technique, the robust desalting of PCR products employing a commercial pipette tip packed with anion-exchange resin has demonstrated recoveries of ~80% in an automated fashion for ESI [59], and the universal technique is transferable to desalting siRNA-type oligonucleotides. In any case, due to the inherent formation of oligonucleotide cation adducts in the gas phase, no one technique is efficient enough to remove all nonvolatile cations and, in most instances, a combination of techniques needs to be employed. A recommended review by Castleberry concisely reports oligonucleotide desalting methodologies including protocols [53].

Most recently, a novel oligonucleotide desalting technique has been reported. Vapors of weak acids such as formic and acetic acid were mixed with the ESI curtain gas before the instrument orifice to successfully reduce nonvolatile cation adducts from various oligonucleotides including siRNA [60]. This application could be a general desalting technique across multiple ESI-MS platforms and holds great promise.

13.5.2 Extracting siRNA from Biological Matrices

The efficient extraction of oligonucleotides from biomatrices is necessary because current desalting techniques alone or even in combination are not enough to effectively isolate oligonucleotides bound to proteins or remove matrix effects due to protein, salts, lipids, and genomic material. The current benchmark for extracting, purifying, and desalting siRNA from serum or plasma for LC-MS analysis is a combination extraction method of a liquid–liquid extraction (LLE) followed by SPE [36, 47]. The ammonia-based phenol/chloroform extraction is able to sufficiently release oligonucleotides from matrix proteins and retain them in the aqueous phase. Alone, LLE is not robust enough to entirely remove other matrix interferences for ESI. To address this, the SPE step is utilized to further wash and preconcentrate the samples before LC-MS analysis. Although the LLE-SPE methodology has provided high recoveries of oligonucleotides from plasma (>70%), the procedure is laborious, and it is not easily or cost-effectively amendable to high-throughput automated protocols.

To address the throughput limitations of the LLE-SPE procedure, an innovative one-step mixed-mode SPE has recently enabled the efficient extraction of oligonucleotides, including siRNA, from serum and plasma prior to LC-MS analysis [61]. First, an acidic (pH 5.5) chaotrope containing lysis-loading buffer is added to the sample to disrupt the protein–oligonucleotide interactions within a matrix such as serum. Next, samples are loaded onto a mixed-mode solid-phase sorbent that primarily retains oligonucleotides by anion exchange through the low pH of the conditioning, lysis-loading, and wash buffers. The procedure sufficiently removes salts, sugars, proteins, and lipids throughout the sequential buffer washing steps. Following sample cleanup, a basic (pH 8.5) ESI-compatible elution buffer efficiently releases the oligonucleotides from the solid phase. Unlike the benchmark LLE-SPE procedure, the single-step SPE protocol is applicable for high-throughput bioanalytical purposes and may be used for processing numerous clinical samples.

13.6 LC-MS TECHNIQUES

When using optimized conditions, high-performance liquid chromatography (HPLC) ESI-MS provides concurrent online desalting, separation, and characterization of siRNA duplexes, individual strands, impurities, degradants, and/or metabolites [37, 53, 62]. Oligonucleotide analysis by ion-pair reversed-phase HPLC (IP-RP-HPLC) is advantageous over conventional HPLC methods (acetonitrile or methanol organic solvents in ammonium acetate or ammonium formate buffers) due to the enhanced retention and resolution achieved of exceptionally polar compounds [63]. Originally, triethylammonium acetate (TEAA)/acetonitrile gradients at pH 7.0 were used for IP-RP-HPLC in conjunction with ESI, yet the TEAA concentrations needed to allow chromatographic separation (50–100 mM) diminished MS sensitivity by suppressing the electrospray signal [64]. The use of triethylammonium bicarbonate (TEAB) has also been extensively explored as an alternative to TEAA, and demonstrated increased mass spectrometry sensitivity when used in conjunction with the postcolumn addition of an organic sheath liquid [65–67].

Ion-pairing reagents with increasing hydrophobicity such as dimethylbutlyamine (DMBA) [68] and hexylamine (HA) [33] have also been successfully reported in the LC-MS analysis of oligonucleotides and siRNA. They increase HPLC performance by greater oligonucleotide retention and separation at lower ion-pairing concentrations but may be detrimental to ESI signal intensities at concentrations >25 mM. To provide a superlative balance between optimal HPLC separations, ion production for ESI, and minimizing adduct formation, Appfel introduced the addition of 1, 1, 1, 3, 3, 3-hexafluoro-2-propanol (HFIP) to triethylamine (TEA) as an IP-RP-HPLC mobile phase at neutral pH [64]. In this buffer system, TEA performs as the ion-pairing reagent while HFIP aids in the electrospray desolvation process due to its high volatility (bp = 57°C) and pK_a (~9), allowing it to remain charge neutral at neutral pH and therefore is more easily evaporated than TEA (bp = 89°C, pK_a = 11.01) or acetic acid (bp = 118°C, pK_a = 4.75) used in TEAA buffers. The HFIP/TEA ion-paring system is widely accepted as the method of choice for the LC-MS analysis of siRNA in the literature using either 16.3 mM TEA/400 mM HFIP (pH 7.0) or 1.7 mM TEA/100 mM HFIP (pH 7.5) with a methanol gradient elution [36, 45]. As a precautionary note, when using the TEA/HFIP buffer system, an increase in Na^+ and K^+ adduction has been observed at lower TEA concentrations (1–2 mM) as well as an increase in ESI back ground signal at the high 400 mM HFIP concentration, thus negatively raising detection limits and lowering signal-to-noise (S/N) levels compared with 50 mM TEAB [65].

As mentioned, a current limitation toward the improved characterization of ADME and the PK properties for siRNA *in vivo* by mass spectrometry is the lack of required sensitivity. For this reason, RT-qPCR and ELISA hybridization techniques are used even though they are not specific techniques and may mistakenly detect certain metabolites as parent species [69, 70]. For mass spectrometry to reach its full potential as a detection methodology for siRNA from *in vivo* matrices such as tissue, limits of detection and quantification for siRNA would at least need to be enhanced to the level of small molecules. To that end, the use of capillary chromatography with ESI in conjunction with optimized sample preparation and chromatography has detected oligonucleotides down to the attomole level [65]. Huber and coworkers have optimized the utility of monolithic capillary columns packed with octadecyl poly(styrene/divinylbenzene) particles as the stationary phase and gradients of acetonitrile in 25 mM TEAB as the mobile phase to analyze a diverse assortment of synthetic and biological oligonucleotides [67, 71]. Recently Taoka et al. reported the use of nanoelectrospray LC-MS for the analysis of synthetic siRNA [72]. They used a 50 mm × 150 μM i.d. reversed-phase packed tips at a flow rate of 100 nL/min using a methanol gradient and 20 mM TEA or 10 mM DMBA ion-paring reagents with or without 400 mM HFIP, which separated siRNA sense and antisense strands at a sub-femtomole level. Samples were sprayed online into an LTQ-Orbitrap hybrid mass spectrometer and the overall detection limits for this system were linear in the attomole to femtomole range with smaller synthetic oligonucleotides.

Although anion exchange is widely used for separating oligonucleotides, it is historically not used in conjunction with mass spectrometry because the salt gradients used to elute oligonucleotides from the solid phase support causes it to be incompatible for the direct coupling to ESI-MS [73]. Recently, an automated anion-exchange purification and desalting HPLC system has demonstrated the ability to properly desalt oligonucleotide containing anion-exchange fractions to produce suitable LC-MS data. A fraction-collecting autosampler was configured with a dual column selection valve using anion-exchange oligonucleotide purification as the first step followed by ion-pair reversed-phase desalting. The collected purified and desalted fractions were then injected into the mass spectrometer with a flow injection system to produce multiply charged species with minimal cation adducts [74]. Although purification and desalting examples were demonstrated with siRNA products from an enzymatic digestion reaction, the method should be applicable to analyzing siRNA from complex biological matrices.

13.7 MALDI-MS OF SIRNA

While much has been written and reviewed on the matrix-assisted laser desorption/ionization (MALDI) of oligonucleotides [37, 75], this discussion will primarily focus on MALDI of siRNA. As with ESI, the occurrence of cation adducts of oligonucleotides regularly deteriorates MALDI mass spectra. Thus, desalting is also required for oligonucleotide samples prior to MALDI-MS analysis in order to maintain unambiguous mass spectra [76]. The exchange of alkali ions such as Na^+ and K^+ with ammonium ions that readily vaporize as ammonia during the MALDI process is performed by either the addition of diammonium hydrogen citrate (DAHC) to the MALDI matrix [77] or by mixing the sample with ammonium form ion exchange beads prior to analysis [78, 79]. Common matrices used in the MALDI-MS analysis of siRNA are 2, 4, 6,-trihydroxyacetophenone (THAP), 3-hydroxypicolinic acid (3-HPA), and 6-aza-2-thiothymine (ATT), all of which are co-mixed with DAHC. Acidic matrices, such as 3-HPA, have been identified to denature double-stranded oligonucleotides and largely provide molecular weight information on each single strand. Neutral matrices, such as THAP and ATT, are appropriate for the analysis of noncovalent complexes when dissolved in water and will maintain the siRNA duplex for MALDI-MS [35]. Duplex intensities measured by MALDI-MS have been shown to maintain and reflect solution conditions. A beneficial use for MALDI-MS of double-stranded oligonucleotides over LC-MS is in the rapid quality control of synthetic siRNAs. The methodology has been shown to quickly gather information about sample purity and the relative quantification of the ratios between single and double strands when an internal standard single-stranded oligonucleotide was used.

13.8 SEQUENCING SIRNA BY MASS SPECTROMETRY

The demands for confirmation of the primary sequence of synthetic siRNAs increase as RNAi therapeutics continue to develop toward clinical applications. Oligonucleotide sequencing with mass spectrometry is performed by either analyzing the products of enzymatic or chemical reactions (termed "bottom-up") or using various tandem mass spectrometry techniques (termed "top-down") in the gas phase. To obtain complete sequence coverage by nuclease digestion, usually 3′- and 5′-exonucleases are used in combination to create a $3' \rightarrow 5'$ and $5' \rightarrow 3'$ "mass ladder" where nucleotide identification correlates with the mass difference between adjacent peaks in the mass spectrum [77, 80]. Gao et al. applied the combination exonuclease digestion strategy to chemically modified single-strand siRNA followed my MALDI-TOF MS analysis of the generated sequence mass ladders [81]. After optimizing the digestion conditions, it was concluded that while the sense and antisense strands were generally sequenced, there were gaps in the generated sequence ladders due to resistance of some modified nucleotides toward exonuclease digestion, such as a reverse abasic modification. The enzymatic resistance of chemically modified oligonucleotides has been previously reported [82–84] and in order to develop broadly applicable sequencing strategies of modified siRNA, the generation of sequence mass ladders by chemical degradation are alternatively employed. Farand et al. demonstrated the complete sequencing of both sense and antisense strands of highly modified siRNA using a series of chemical reactions [58]. Although no one chemical degradation reaction generated enough fragments to entirely sequence each strand alone, when the generated sequence information from multiple reactions was combined, the confirmation of the strands was effectively obtained.

Acid hydrolysis and MALDI-MS have also been used for sequencing siRNA [85]. The digestion of chemically modified RNA oligonucleotides with 3.75% trifluoroacetic acid generated almost complete 5′→3′ and 3′→5″ mass ladders. Potentially intrusive products from base losses or internal fragments that are created with conventional chemical degradation techniques were not observed. Hydrolysis products were analyzed by MALDI combined with a high-resolution Orbitrap mass spectrometer that has a resolving power of up to 100,000 with <3 ppm mass accuracy. Any remaining unhydrolized di- and trimer products were further characterized by tandem mass spectrometry. Combining the MS/MS data with the acid hydrolysis mass ladders permitted the unambiguous identification of all nucleotides in the tested sequences. The use of high-resolution instrumentation, such as Fourier transform (FT)-MS, Orbitrap, reflector TOFs, or orthogonal TOFs, for oligonucleotide sequencing is able to overcome the limited mass resolution and accuracies of linear MALDI-TOF instruments, which have been historically problematic in distinguishing the nucleobases uracil and cytosine.

The determination of an unknown chemically modified siRNA by *de novo* sequencing has also been demonstrated [86]. A combination of chemical degradation and tandem mass spectrometry was first used to sequence the oligonucleotide followed by a nucleoside composition analysis by enzymatic digestion to cross-reference the proposed *de novo* sequence.

Sequencing oligonucleotides by tandem mass spectrometry (MS/MS) as a top-down approach has the potential to be a more universal, rapid, and complete identification technique than bottom-up solution phase digestion strategies previously discussed. MS/MS is capable of providing primary sequence information and the identification of chemically modified nucleosides within siRNAs. In a typical MS/MS experiment, an ion of interest, in this case an oligonucleotide ion, is isolated in the gas phase, where it is dissociated to produce a series of fragment ions that are resolved and their *m/z* ratios are determined by the mass spectrometer. The mass of the intact parent ion along with its distinctive fragment ions provide an adequate amount of structural information to relate back to the oligonucleotide sequence. Wu et al. have thoroughly reviewed the fragmentation of oligonucleotide ions [87] and McLuckey et al. have provided a comprehensive description of their fragmentation behavior [88].

The collision-induced dissociation (CID) behavior of various charge states for both duplex and single-stranded siRNA have been investigated using a prototype nano-ESI-QqTOF [48]. Activation of the siRNA duplex lead primarily to the separation of the two individual sense and antisense strands and did not provide sequence information. Minimal strand fragmentation was able to be generated at higher collision energies, but sequence information was not readily obtained due to the inability to discriminate whether the origin of the fragment ions was from the sense or antisense strands. Low energy dissociation of the individual single-strand anions on a lower (5–) charge state precursor ion leads to predominately c/y backbone fragmentation, and full sequence coverage was nearly achieved. When higher excitation energies were used with a beam-type experiment on the QqTOF, increasingly complex spectra were produced due to the more prominent a–B/W dissociation channels, and made analysis of peaks difficult. In order to attain the complete sequence of each strand and simplify the product ion spectra, proton-transfer ion/ion reactions of the siRNA anions with benzoquinoline cations were carried out. After dissociation of the parent anion at a relatively low excitation energy, the charge states of the fragment ions were reduced to –1 and –2 by ion/ion proton-transfer reactions. A complete c/y ion series of each sense and antisense strand was generated to unambiguously obtain their sequence (Fig. 13.4). Taucher et al. have also discussed strategies to obtain complete sequence coverage of RNA oligonucleotides by minimizing the internal energy of primary fragment ions by collisional cooling and selecting relatively low charge states for activation in order to curtail unwanted base loss and internal fragmentation [89].

Infrared multiphoton dissociation (IRMPD) has also been demonstrated to be a feasible dissociation method for sequencing siRNA [90], where trapped ions are activated by IR irradiation from an external infrared laser. IRMPD of unmodified siRNA single-strand anions and cations typically yielded c/y fragment ions as well as neutral base loss, and was able to provide full sequence coverage. IRMPD of duplex siRNA using short irradiation times dissociated the sense–antisense noncovalent interactions and primarily caused strand separation, similar to the dissociation trends of CID. However, when longer irradiation times were used, both duplex strands underwent fragmentation to yield limited sequence information pertaining to the individual strands.

Fragmentation of chemically modified oligonucleotides by tandem mass spectrometry techniques, such as CID, is generally more demanding than unmodified oligonucleotides because substitution of the ribose hydroxyl group with such modifications as a 2′-O-methyl or 2′-fluoro considerably reduces the level of fragmentation at the modified nucleoside positions [49, 91]. This effect hampers the elucidation of modified oligonucleotides because MS/MS may not be able to provide sufficient information on the modification location within a sequence. As a strategy for elucidating chemically modified oligonucleotides, Smith and Brodbelt explored the use of combining a dissociation technique known for its ability to elucidate post-transcriptional modifications within peptides, electron-transfer dissociation (ETD) with CID, IRMPD, or ultraviolet photodissociation (UVPD) [92]. These hybrid ion activation techniques successfully provided extensive backbone cleavage of modified single-strand oligonucleotide cations containing both phosphorothioate linkages and 2′-O-methyl nucleosides to determine their location within the sequence.

Another analytical challenge is the ability to obtain complete sequence coverage of RNAi therapeutics using LC-MS/MS. Online gas phase MS/MS techniques need to operate on a time scale that is compatible with current IP-LC-MS chromatographic separations and provide sufficient sequence fragmentation information from limited sample amounts. To this end, Ivleva and coworkers investigated the use of ultra-performance liquid chromatography (UPLC)-MS/MS of unmodified

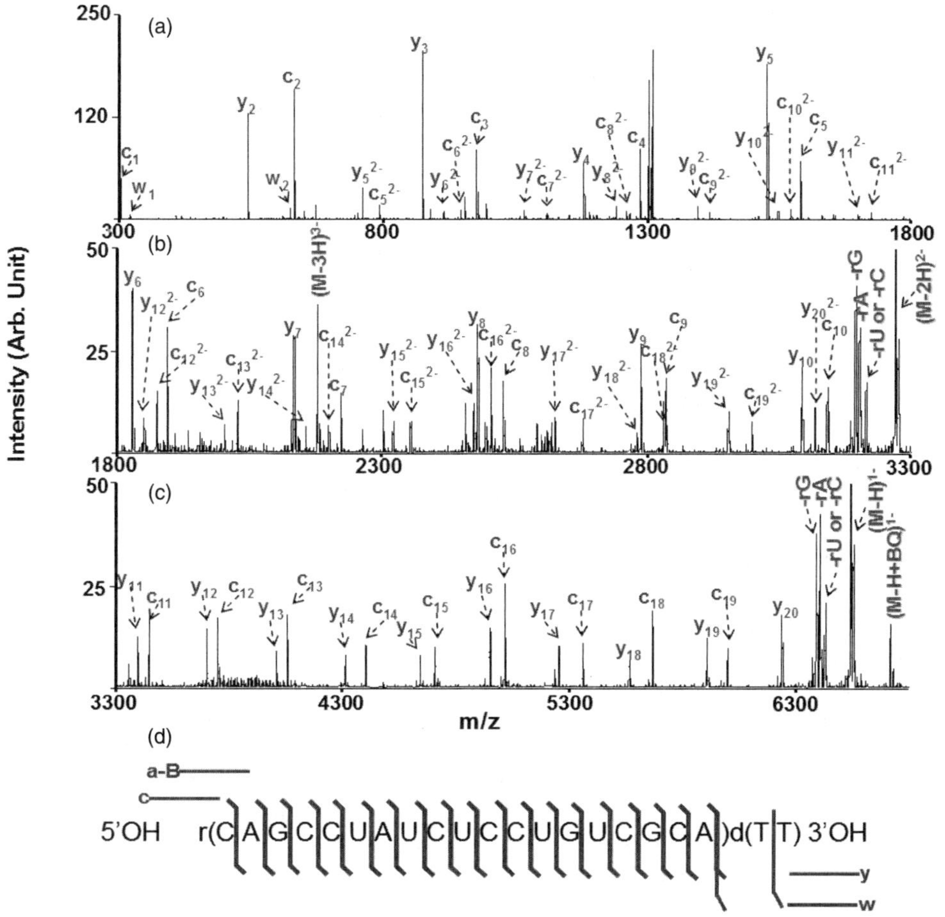

Figure 13.4 Post-ion/ion ion trap CID spectrum of a –5 charge state of an siRNA antisense strand excited at low excitation amplitude. Panels (a) to (c) correspond to mass range *m/z* 300–1800, 1800–3300, and 3300–6800, respectively. (d) Summary of backbone cleavages resulting from post-ion/ion ion trap CID demonstrating complete sequence coverage from the c/y ion series. Reproduced from Huang et al. [48] with permission from American Chemical Society (Copyright 2008).

and modified siRNA as a quality control technique of target sequence-related impurities [49]. Failed sequence and exonuclease digest products were unambiguously identified and sequenced by a combination of UPLC separation, exact mass determination, and MS/MS methods using a Q-TOF instrument. Data-dependent MS/MS were sufficient to validate the sequence of unmodified RNA strands up to 21 nt when performed on a lower (–4) charge state. An alternative MS/MS methodology, MS^E, was also employed as a more general and efficient way to produce oligonucleotide fragmentation. MS^E works in a data-independent fashion where it simultaneously fragments all ions during the course of the LC-MS run without isolating a precursor ion by ramping between low and high collision energies. An advantage is that it simultaneously fragments all the charge states of the oligonucleotide, thus improving the S/N of the product ion spectra. A disadvantage with the technique is that any species under investigation in a complex mixture needs to be chromatographically resolved in order to definitively interpret the MS^E data. The MS^E methodology was investigated with various chemically modified oligoribonucleotides of shorter length (8 nt) to determine its dissociation characteristics. Reduction in fragmentation due to the presence of various 2′-OH replacements and phosphorothioate linkage substitution was observed, yet these low-intensity fragments were clearly detected due to the enhanced S/N levels of the MS^E scanning function, thus allowing sequence confirmation.

13.9 QUANTITATIVE AND QUALITATIVE ANALYSIS OF siRNA FROM BIOLOGICAL MATRICES

Due to its robustness, selectivity and high-throughput capability, LC-MS/MS has become a principal and reliant analytical tool for the quantitative and qualitative analysis of small molecules from biological matrices. The development of validated LC-MS/MS methods has been more complicated to achieve for oligonucleotides due to the analytical challenges described previously such as lower sensitivity, low analyte recovery from biomatrices, and complex MS/MS fragmentation. Due to these deficiencies, ligation-based hybridization assays and quantitative PCR are used as sensitive oligonucleotide quantitation assays, although they have characteristic key limitations such as low selectivity and narrow dynamic range [69]. As therapeutic siRNAs mature through the drug discovery pipeline, their complete quantitative and qualitative characterization from biological matrices will be principal for determining their ADME and PK parameters. Toward this end, a number of key LC-MS/MS methods have been developed to support antisense oligonucleotide therapeutic discovery efforts [47, 93–95] and have been recently reviewed [63]. Although most of the literature on the quantitative analysis of therapeutic oligonucleotides is antisense focused, the protocols are directly applicable to the bioanalytical LC-MS/MS method development of both the sense and antisense strands of siRNA. To date, no work has been published on the bioanalytical quantitative LC-MS/MS of siRNA, yet methods are starting to appear at national conferences. Wheller demonstrated the quantitative analysis of siRNA from rat plasma using IP-LC-MS/MS [38]. Samples were extracted using the Clarity OTX SPE (Phenomenex, Torrance, CA) system with recoveries in excess of 85% from plasma. Chromatographic separation was performed with an Acquity OST (Waters Corp., Milford, MA) 50 × 2.1 mm, 1.7 μM particle size column held at 60°C in order to denature the duplex, with 15 mM TEA/400 mM HFIP as the aqueous phase and 50:50 15 mM TEA/400 mM HFIP : acetonitrile as the organic elution at 200 μL/min. Both sense and antisense strands were resolved with a 7-min runtime. A PE Sciex API 5000 (AB Sciex, Framingham, MA) triple quadrupole mass spectrometer was used in the selected reaction monitoring (SRM) mode where *m/z* –79 was chosen as the product ion to provide maximum sensitivity, although the choice of a nonspecific product ion may lower method selectivity. The linear range was from 5 to 5000 ng/mL for both the sense and antisense strands. Overall, the methodology supports the assumption that specific and high-throughput LC-MS/MS methods can be developed to quantify both siRNA strands from plasma with an LOQ approaching the levels needed for discovery ADME and PK studies.

Quantification of modified and unmodified siRNA in plasma samples has also been reported utilizing the accurate mass capabilities of an Orbitrap mass spectrometer [96]. siRNAs were extracted from plasma samples using an miRNA purification kit after ultrafiltration of the plasma with a 30-kD filter. Recovery values for the unmodified strands were in the range of ~54%–67% where the 2′-O-Me modified strands only had recoveries from ~23% to 28%. The isolated siRNA was analyzed by negative mode high-resolution mass spectrometry without chromatographic separation by flow injection in 80:20 5 mM ammonium acetate (pH 3.5) : acetonitrile at 20 μL/min with a 10-μL injection volume. The sample preparation and ionization conditions denatured the duplex so that the individual strand masses were measured. The developed method had limits of detection of 0.25–1 nmol/mL and precision from 11% to 21% utilizing the high-resolution and mass accuracy characteristics of the mass spectrometer as the foundation for the analysis.

As mentioned previously, following systemic administration, accumulation of current siRNA therapeutic formulations are found primarily in organs such as the liver, and currently, the ADME properties of the active pharmaceutical ingredient are measured using hybridization-based assays, RT-quantitative PCR and also quantitative whole-body autoradiography [97]. These techniques have limited selectivity due to the possibility that the metabolites cannot be differentiated from the full-length siRNA strands. The nondirect analytical methods provide sound primary ADME-type information but will always be of limited use in metabolism studies due to their lack of selectivity. For these reasons, little is currently known about the metabolic fate of siRNA *in vivo*. There has been extensive characterization of therapeutic single-strand antisense oligonucleotide metabolism both *in vitro* and *in vivo* using LC-MS and MS/MS [47, 93–95, 98–100]. Generally, the antisense oligonucleotides are extracted from plasma or tissue homogenates using a phenol/chloroform LLE followed by secondary purification step such as ethanol precipitation, anion exchange, or SPE before analysis by IP-LC-MS or MS/MS. The principal metabolic pathway of first- and second-generation (phosphorothioate containing) antisense oligonucleotides was via 3′ exonuclease cleavage followed by 5′ exonuclease degradation. Endonuclease activity has also been observed but to a lesser degree [63, 101].

Beverly et al. used IP-LC-MS to characterize metabolites of highly modified siRNA duplexes from rabbit ocular vitreous humor and retinal/choroid tissue after intravitreal injection [34, 45]. The vitreous samples were prepared using phenol/chloroform LLE followed by evaporation of the aqueous layer. The retina/choroid samples were further purified according to a modified Mirvana (Life Technologies, Carlsbad, CA) RNA extraction procedure, which included tissue homogenization and phenol/chloroform LLE. All samples were resuspended in mobile phase before injection onto a triple quadrupole LC-MS system. Separation of metabolites from intact single strands and duplex were carried out by IP-RP-HPLC using an Xterra MSC18 (Waters Corp.), 2.1 × 150 mm, 3.5 μm column with either 200 mM HFIP/8.15 mM TEA (pH 7.9) or 400 mM HFIP/16.5 mM TEA (pH 7.9) as mobile phase A and 200 mM HFIP/8.15 mM TEA in 50% methanol/water or 100% methanol as mobile phase B at 120 μL/min. The column was held at 25°C to preserve the intact duplex during the chromatography, and proper ESI tuning voltages allowed the analysis of the intact duplex in the gas phase. Currently, these studies provide the only critical insight into the metabolism of siRNA duplexes *in vivo*. Endonuclease cleavages of the duplex were detected from the vitreous extract after four days

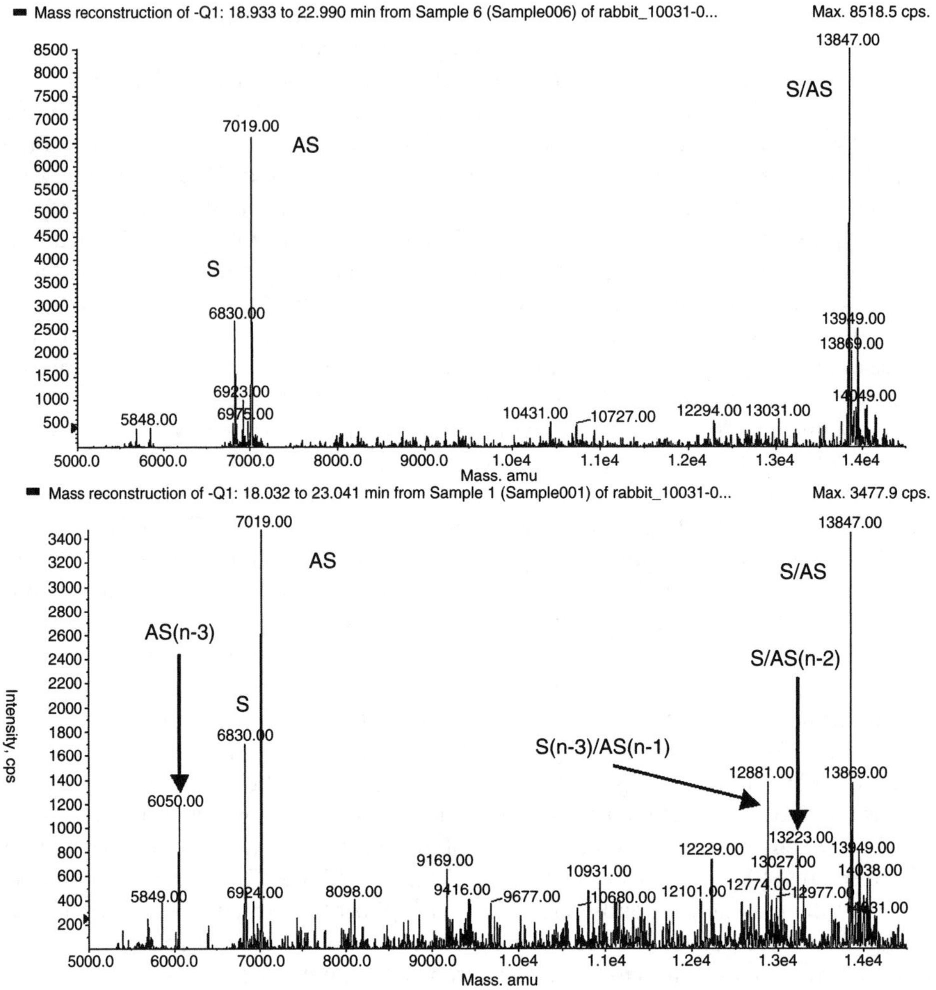

Figure 13.5 Deconvoluted ESI mass spectra of vitreous extract analyzed by LC-MS. Top spectrum showing vitreous sample collected at day 1. Bottom spectrum showing day 7 sample. Peaks are labeled AS for antisense, S for sense, with the duplex being designated as S/AS. Degradation species are shown with the number of nucleotides lost form each respective strand. $AS(N\text{-}3)^{3'}/S$ designates a duplex formed from the antisense strand, minus three nucleotides from the 3′ end, paired with the full-length sense strand. Reproduced from Beverly et al. [45] with permission from John Wiley & Sons, Inc. (Copyright 2005).

and the sites of metabolism were directed toward both the 5′ end of the sense strand and 3′ end of the antisense strand (Fig. 13.5). The sites of metabolism indicate that the duplex itself appears to provide increased nuclease protection as compared to single-strand metabolic profiles. It appears the ability of the duplex to partially unwind or "breath" toward the 5′ and 3′ ends plays an important role in duplex metabolism. As the duplex becomes more single stranded in nature, it also becomes more susceptible to both exo- and endonulcease degradation. This phenomenon has also been witnessed by the MALDI-TOF analysis of siRNA degradation in serum

[102]. It was hypothesized that the endo RNAse A-like activity found in serum would be unable to cleave duplex oligoribonucleotides but is able to do so during the momentary breaking of hydrogen bonds between the ends of the two strands.

The detection of the intact duplex does have its disadvantages. The LC-MS sensitivity of the duplex was 5–10 times less that that of the single strands due to extensive cation adduction. Fortuitously, the single-strand components of the duplex were also simultaneously detected due to slight denaturing caused by the ESI source. Appearance of the single-strand species was used to confirm the identity of the duplex strands and its metabolites.

The *in vitro* metabolism profiles of a highly modified siRNA duplex HBV263 targeting hepatitis B virus mRNA incubated in rat and human serum and liver microsomes have recently been compared [36]. The siRNA and its metabolites were extracted from all matrices using a combination phenol/chloroform LLE followed by SPE on Oasis HLB cartridges. SPE eluants were lyophilized and resuspended before IP-RP-HPLC separation using an XBridge OST C18 2.1 × 50 mm, 2.5-μm particle size column (heated to 85°C to ensure denaturing of the duplex) with 100 mM HFIP/1.7 mM TEA (pH 7.5) as mobile phase A and 100% methanol as mobile phase B at a flow rate of 150 μL/min. High-resolution mass spectrometry using an LTQ-Orbitrap mass spectrometer was utilized for the unambiguous metabolite identification by clearly differentiating species that differed by less than 1 Da. The application of accurate mass measurements was used as an alternative to MS/MS techniques for determining the composition of siRNA metabolites because, as discussed previously, the data interpretation of chemically modified siRNA MS/MS fragmentation is difficult and time-consuming. To demonstrate the ability of accurate mass measurements to differentiate between metabolites that would otherwise be isobaric using a typical resolving power mass spectrometer, an siRNA was incubated in rat liver microsomes that produced a metabolite that could have two possible assignments, sense (N-10)3′ or antisense (N-9)5′ + phosphate, whose theoretical masses differ by 0.9827 Da. To correctly assign the observed metabolite, the theoretical isotope envelopes of the −6 charge state for the two possible metabolites were aligned with the experimental isotope envelope (Fig. 13.6). The predicted isotope envelope coincided with the experimental envelope of the sense (N-10)3′ metabolite and not the antisense (N-9)5′ + phosphate, thus providing an accurate metabolite assignment without the need for additional verification from tandem mass spectrometry fragmentation.

The metabolites resulting from the *in vitro* serum and liver microsome incubations of HBV263 generated different sites of metabolism between the two matrices. In rat and human serum, the antisense strand was sequentially cleaved from the 3′ end due to 3′ exonuclease degradation. The sense strand exhibited a resistance to degradation in serum presumably due to the 5′ and 3′ inverted abasic end caps that successfully block exonuclease activity [103]. In contrast to the serum incubations, when the HBV263 duplex was incubated in rat and human liver microsomes, the sense strand showed increased endonuclease activity while the antisense strand demonstrated higher resistance to nuclease degradation (Fig. 13.7). Nuclease cleavage toward the 3′ end of the sense strand such as sense (N-3)3′ were observed without any obvious N-1 and N-2 cleavages from the 3′ end. This strongly suggests that the initial cleavage of the strand was by endonuclease activity, which may be then followed by sequential exonuclease degradation. Overall, siRNA sites of

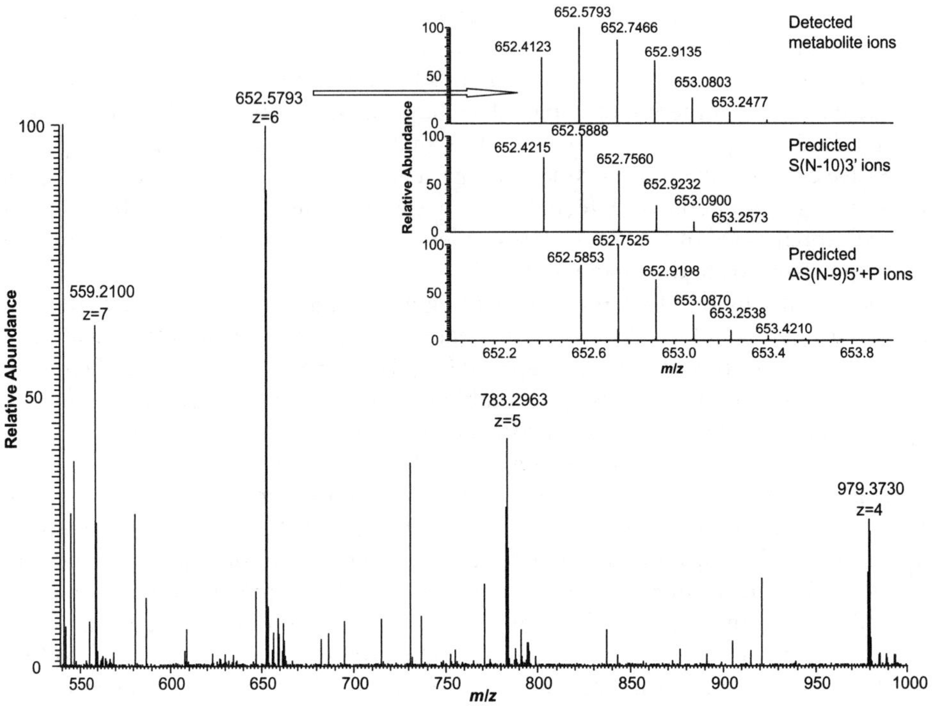

Figure 13.6 An example of siRNA metabolite identification using accurate mass measurements. The inset compares the experimentally detected isotope distribution of the –6 charge state of an unknown metabolite generated from rat liver microsomes with the predicted isotope distributions of two potential isobaric metabolites. The accurate mass spectra of the detected metabolite ions suggests that this metabolite is identified as S(N-10)3′, but not AS(N-9)5′+P. Reproduced from Zou et al. [36] with permission of John Wiley & Sons, Inc. (Copyright 2008).

metabolism are governed by a complex combination of their sequence potentially containing a nuclease substrate (e.g., UpA for an RNAse A-like endonuclease), their exposure to various biomatrices, and the placement of chemically modified nucleotides across their sequence.

13.10 DATA ANALYSIS

As oligonucleotide identification and sequencing by MS and MS/MS technologies continue to innovate and mature toward automated high-throughput screening (HTS) applications, the software tools necessary for the data analysis of such numerous and potentially complicated spectra must also advance. Without further development of automated or semi-automated software, data analysis becomes the bottleneck when identifying oligonucleotide therapeutics using mass spectrometry-based techniques. Currently, there is a dearth of oligonucleotide sequencing software from commercial vendors. Most programs and algorithms are developed and supplied through academic sources.

(a)

Retention time (min)	Peak code	Peak assignment	Sequence	Observed mass[a] (Da)	Calculated mass[a] (Da)
19.4/19.2/ 19.1	AS	AS (antisense strand)	5'-rArGrAmAmAmAfUfUmGmAmGmAmGmAmAmGfUfCfCmUmU-3'	6960.1	6960.5
18.3	A3	AS(N-3)3'	5'-rArGrAmAmAmAfUfUmGmAmGmAmGmAmAmGfUfC-3'	6013.1	6012.9
18.0	A4	AS(N-4)3'	5'-rArGrAmAmAmAfUfUmGmAmGmAmGmAmAmGfU-3'	5706.3	5705.7
18.0/17.9/ 17.8	S	S (sense strand)	5'-BGGAfCfUfUfCfUfCfUfCAAfUfUfUfUfCfUTTB-3'	6814.6	6814.0
16.6	S3	S(N-3)3'	5'-BGGAfCfUfUfCfUfCfUfCAAfUfUfUfUfCfU-3'	6026.1	6025.5
16.2	S4	S(N-4)3'	5'-BGGAfCfUfUfCfUfCfUfCAAfUfUfUfUfC-3'	5717.7	5717.3
15.9	S5	S(N-5)3'	5'-BGGAfCfUfUfCfUfCfUfCAAfUfUfUfU-3'	5410.3	5410.1
15.5	S6	S(N-6)3'	5'-BGGAfCfUfUfCfUfCfUfCAAfUfUfU-3'	5101.9	5102.0
15.1/15.0	S7	S(N-7)3'	5'-BGGAfCfUfUfCfUfCfUfCAAfUfU-3'	4793.7	4793.8
14.6	S8	S(N-8)3'	5'-BGGAfCfUfUfCfUfCfUfCAAfU-3'	4485.6	4485.7
14.2/14.3	S9	S(N-9)3'	5'-BGGAfCfUfUfCfUfCfUfCAA-3'	4177.4	4177.5
13.6	S10	S(N-10)3'	5'-BGGAfCfUfUfCfUfCfUfCA-3'	3864.7	3864.3
11.7	S11	S(N-11)3'	5'-BGGAfCfUfUfCfUfCfUfC-3'	3551.8	3551.1
9.3	S12	S(N-12)3'	5'-BGGAfCfUfUfCfUfCfU-3'	3243.8	3243.9

[a] The observed and calculated masses are average masses shown in ProMass processing reports.

(b)

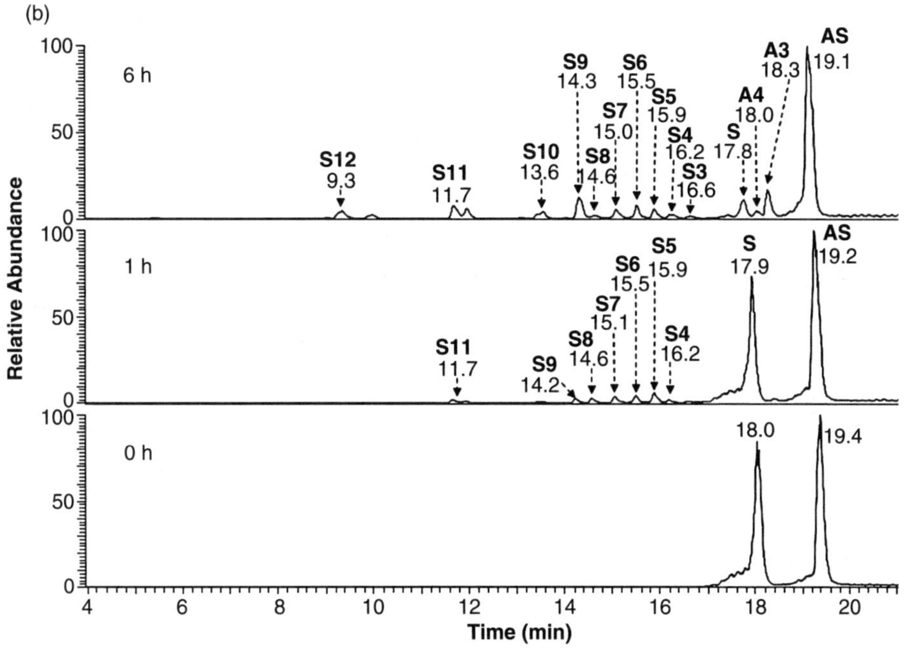

Figure 13.7 (a) Summary table of the identified metabolite sequences for the HBV263 siRNA duplex in rat liver microsomes incubations. (b) Total ion chromatograms of the HBV263 duplex incubated in rat liver micrsomes for 0, 1, and 6 h. The new metabolite peaks appear in the samples after 1 and 6 h incubations. See summary table for the LC peak assignments and sequences. Reproduced from Zou et al. [36] with permission of John Wiley & Sons, Inc. (Copyright 2008).

One of the few commercial programs available for oligonucleotide sequence characterization for LC-MS type data is ProMass from Novatia (Monmouth Junction, NJ), which is currently offered for Thermo (Waltham, MA) Xcalibur and Waters MassLynx platforms. When provided an oligonucleotide sequence, the software calculates the full length mass of the strand as well as the mass for every possible truncated sequence from the 5′ and 3′ ends and the combination of both. The truncated sequences represent the possible metabolite species originating from 5′ exo- and 3′ exonuclease degradation. Next, ProMass deconvolutes the masses from each defined experimental chromatography peak and matches the experimental masses with the theoretical full length and truncated sequences, and reports the identified sequences. Chemical modifications such as 2′-flu and 2′-O-Me nucleotides have been incorporated into the ProMass algorithm, facilitating the identification of chemically modified therapeutics, such as siRNAs.

As discussed previously, oligonucleotide sequencing by tandem mass spectrometry represents an important and crucial tool for elucidating siRNA strands. CID with modern instruments such as a QqTOF provides sufficient fragmentation coverage to sequence oligonucleotides up to the 20–25 mer level, ideally suited for siRNA identification. Unfortunately, the complexity of oligonucleotide tandem mass spectrometry spectra requires the use of software to examine the spectra for any time-constrained analysis. Rozenski and McCloskey developed user interactive software, Simple Oligonucleotide Sequencer (SOS), for rapid oligonucleotide sequencing from MS/MS type spectra [104]. The program rapidly inspects and assigns each step of the sequence mass ladder in both the 5′→3′ and 3′→5′ directions using multiple ion series with a graphical comparison of sequence homology and overlap. Chemical modifications of the base, sugar, or backbone are defined by the user and can be used in any combination, thus allowing SOS to be applicable to sequencing modified siRNA therapeutics.

Recently, Kretschmer et al. developed a data analysis program for sequencing chemically modified oligonucleotides from tandem mass spectrometry data [105]. The program deconvolutes the MS/MS spectra for fragment ions from up to three charge states and then evaluates these data sets against calculated theoretical fragment masses for all 5′ [a_n], [b_n], [c_n], [d_n], and [a_n-B] fragments and all 3′ [w_n], [x_n], [y_n], and [z_n] fragments. The algorithm successfully identified the sequence of two siRNA single strands chemically modified with 2′-O-Me and phosphorothioate linkages. The software is ideal for the sequence confirmation of chemically modified oligonucleotides for quality control-type purposes. The Kretschmer algorithm is not intended for *de novo* sequencing as the expected nucleotide sequence is needed for the analysis.

13.11 FUTURE TRENDS AND CONCLUSIONS

As the understanding of RNAi biology continues to flourish along with advances in safe and effective siRNA delivery modalities, the potential of siRNA to develop into the next biopharmaceutical is quickly becoming a reality. Since an increasing number of siRNAs are entering clinical studies, it is evident that dependable bioanalytical methods are needed for the comprehension of their metabolic profiles and PK parameters. Mass spectrometry offers a precise and robust analytical platform for all stages of the siRNA drug discovery process, such as the quality control analysis of synthetic drug products, quantitative bioavailability studies, and metabolic stability. Analysis of oligonucleotides by mass spectrometry is not trivial and is associated with many physical, chemical, and technological hurdles. As sample preparation, robust ionization, mass spectrometer sensitivity, novel tandem mass spectrometry techniques, and innovative software continue to develop, siRNA analysis will become more routine and approachable throughout the industry.

REFERENCES

[1] Napoli C, Lemieux C, Jorgensen R. Introduction of a chimeric chalcone synthase gene into petunia results in reversible co-suppression of homologous genes in trans. Plant Cell 1990;2:279–89.

[2] Romano N, Macino G. Quelling: transient inactivation of gene expression in Neurospora crassa by transformation with homologous sequences. Mol Microbiol 1992;6: 3343–53.

[3] Fire A, Xu S, Montgomery MK, Kostas SA, Driver SE, Mello CC. Potent and specific genetic interference by double-stranded RNA in *Caenorhabditis elegans*. Nature 1998; 391:806–11.

[4] Zamore PD, Tuschl T, Sharp PA, Bartel DP. RNAi: double-stranded RNA directs the ATP-dependent cleavage of mRNA at 21 to 23 nucleotide intervals. Cell 2000;101: 25–33.

[5] Elbashir SM, Harborth J, Lendeckel W, Yalcin A, Weber K, Tuschl T. Duplexes of 21-nucleotide RNAs mediate RNA interference in cultured mammalian cells. Nature 2001;411:494–8.

[6] Sepp-Lorenzino L, Ruddy M. Challenges and opportunities for local and systemic delivery of siRNA and antisense oligonucleotides. Clin Pharmacol Ther 2008;84: 628–32.

[7] Stanton MG, Colletti SL. Medicinal chemistry of siRNA delivery. J Med Chem 2010;53:7887–901.

[8] Li L, Shen Y. Overcoming obstacles to develop effective and safe siRNA therapeutics. Expert Opin Biol Ther 2009;9:609–19.

[9] Oh YK, Park TG. siRNA delivery systems for cancer treatment. Adv Drug Deliv Rev 2009;61:850–62.

[10] Kurreck J. RNA interference: from basic research to therapeutic applications. Angew Chem Int Ed Engl 2009;48:1378–98.

[11] Savas JN, Tanese N. A combined immunoprecipitation, mass spectrometric and nucleic acid sequencing approach to determine microRNA-mediated post-transcriptional gene regulatory networks. Brief Funct Genomics 2010;9:24–31.

[12] Peters L, Meister G. Argonaute proteins: mediators of RNA silencing. Mol Cell 2007;26:611–23.

[13] Liu J, Carmell MA, Rivas FV, Marsden CG, Thomson JM, Song JJ, et al. Argonaute2 is the catalytic engine of mammalian RNAi. Science 2004;305:1437–41.

[14] Ameres SL, Martinez J, Schroeder R. Molecular basis for target RNA recognition and cleavage by human RISC. Cell 2007;130:101–12.

[15] Filipowicz W. RNAi: the nuts and bolts of the RISC machine. Cell 2005;122:17–20.

[16] Gaynor JW, Campbell BJ, Cosstick R. RNA interference: a chemist's perspective. Chem Soc Rev 2010;39:4169–84.

[17] Watts JK, Deleavey GF, Damha MJ. Chemically modified siRNA: tools and applications. Drug Discov Today 2008;13:842–55.

[18] Deverre JR, Boutet V, Boquet D, Ezan E, Grassi J, Grognet JM. A competitive enzyme hybridization assay for plasma determination of phosphodiester and phosphorothioate antisense oligonucleotides. Nucleic Acids Res 1997;25:3584–9.

[19] Wei X, Dai G, Marcucci G, Liu Z, Hoyt D, Blum W, et al. A specific picomolar hybridization-based ELISA assay for the determination of phosphorothioate oligonucleotides in plasma and cellular matrices. Pharm Res 2006;23:1251–64.

[20] Raymond CK, Roberts BS, Garrett-Engele P, Lim LP, Johnson JM. Simple, quantitative primer-extension PCR assay for direct monitoring of microRNAs and short-interfering RNAs. RNA 2005;11:1737–44.

[21] Stratford S, Stec S, Jadhav V, Seitzer J, Abrams M, Beverly M. Examination of real-time polymerase chain reaction methods for the detection and quantification of modified siRNA. Anal Biochem 2008;379:96–104.

[22] Seitzer J, Zhang H, Koser M, Pei Y, Abrams M. Effect of biological matrix and sample preparation on qPCR quantitation of siRNA drugs in animal tissues. J Pharmacol Toxicol Methods 2011;63:168–73.

[23] Varkonyi-Gasic E, Wu R, Wood M, Walton EF, Hellens RP. Protocol: a highly sensitive RT-PCR method for detection and quantification of microRNAs. Plant Methods 2007;3:12.

[24] Duncan DD, Eshoo M, Esau C, Freier SM, Lollo BA. Absolute quantitation of microRNAs with a PCR-based assay. Anal Biochem 2006;359:268–70.

[25] Maroney PA, Chamnongpol S, Souret F, Nilsen TW. A rapid, quantitative assay for direct detection of microRNAs and other small RNAs using splinted ligation. RNA 2007;13:930–6.

[26] Liu WL, Stevenson M, Seymour LW, Fisher KD. Quantification of siRNA using competitive qPCR. Nucleic Acids Res 2009;37:e4.

[27] Overhoff M, Wunsche W, Sczakiel G. Quantitative detection of siRNA and single-stranded oligonucleotides: relationship between uptake and biological activity of siRNA. Nucleic Acids Res 2004;32:e170.

[28] Singh SK, Nielsen P, Koshkin AA, Wengel J. LNA (locked nucleic acids): synthesis and high-affinity nucleic acid recognition. Chem Commun 1998;4:455–6.

[29] Obika S, Nanbu D, Hari Y, Andoh J, Morio K, Doi T, et al. Stability and structural features of the duplexes containing nucleoside analogues with a fixed N-type conformation, 2′ -O,4′ -C-methyleneribonucleosides. Tetrahedron Lett 1998;39:5401–4.

[30] Jensen TB, Langkjaer N, Wengel J. Unlocked nucleic acid (UNA) and UNA derivatives: thermal denaturation studies. Nucleic Acids Symp Ser (Oxf) 2008;52:133–4.

[31] Whitehouse CM, Dreyer RN, Yamashita M, Fenn JB. Electrospray interface for liquid chromatographs and mass spectrometers. Anal Chem 1985;57:675–9.

[32] Karas M, Hillenkamp F. Laser desorption ionization of proteins with molecular masses exceeding 10,000 Daltons. Anal Chem 1988;60:2299–301.

[33] McCarthy SM, Gilar M, Gebler J. Reversed-phase ion-pair liquid chromatography analysis and purification of small interfering RNA. Anal Biochem 2009;390:181–8.

[34] Beverly M, Hartsough K, Machemer L, Pavco P, Lockridge J. Liquid chromatography electrospray ionization mass spectrometry analysis of the ocular metabolites from a short interfering RNA duplex. J Chromatogr B Analyt Technol Biomed Life Sci 2006;835:62–70.

[35] Bahr U, Aygun H, Karas M. Detection and relative quantification of siRNA double strands by MALDI mass spectrometry. Anal Chem 2008;80:6280–5.

[36] Zou Y, Tiller P, Chen IW, Beverly M, Hochman J. Metabolite identification of small interfering RNA duplex by high-resolution accurate mass spectrometry. Rapid Commun Mass Spectrom 2008;22:1871–81.

[37] Beverly MB. Applications of mass spectrometry to the study of siRNA. Mass Spectrom Rev 2011;30:979–98.

[38] Wheller R. The extraction and analysis of siRNA from rat plasma using mixed mode SPE and HPLC-MS/MS. Proceeding of the 58th ASMS Conference on Mass Spectrometry and Allied Topics. Salt Lake City, Utah, 2010.

[39] Ikuta S, Chattopadhyaya R, Dickerson RE. Reverse-phase polystyrene column for purification and analysis of DNA oligomers. Anal Chem 1984;56:2253–6.

[40] Huber CG, Oefner PJ, Bonn GK. High-performance liquid chromatographic separation of detritylated oligonucleotides on highly cross-linked poly-(styrene-divinylbenzene) particles. J Chromatogr 1992;599:113–8.

[41] Huber CG, Oefner PJ, Bonn GK. High-resolution liquid chromatography of oligonucleotides on nonporous alkylated styrene-divinylbenzene copolymers. Anal Biochem 1993;212:351–8.

[42] Huber CG, Oefner PJ, Preuss E, Bonn GK. High-resolution liquid chromatography of DNA fragments on non-porous poly(styrene-divinylbenzene) particles. Nucleic Acids Res 1993;21:1061–6.

[43] Gilar M, Fountain KJ, Budman Y, Neue UD, Yardley KR, Rainville PD, et al. Ion-pair reversed-phase high-performance liquid chromatography analysis of oligonucleotides: retention prediction. J Chromatogr A 2002;958:167–82.

[44] Null AP, Nepomuceno AI, Muddiman DC. Implications of hydrophobicity and free energy of solvation for characterization of nucleic acids by electrospray ionization mass spectrometry. Anal Chem 2003;75:1331–9.

[45] Beverly M, Hartsough K, Machemer L. Liquid chromatography/electrospray mass spectrometric analysis of metabolites from an inhibitory RNA duplex. Rapid Commun Mass Spectrom 2005;19:1675–82.

[46] Abrams MT, Koser ML, Seitzer J, Williams SC, DiPietro MA, Wang W, et al. Evaluation of efficacy, biodistribution, and inflammation for a potent siRNA nanoparticle: effect of dexamethasone co-treatment. Mol Ther 2010;18:171–80.

[47] Zhang G, Lin J, Srinivasan K, Kavetskaia O, Duncan JN. Strategies for bioanalysis of an oligonucleotide class macromolecule from rat plasma using liquid chromatography-tandem mass spectrometry. Anal Chem 2007;79:3416–24.

[48] Huang TY, Liu J, Liang X, Hodges BD, McLuckey SA. Collision-induced dissociation of intact duplex and single-stranded siRNA anions. Anal Chem 2008;80: 8501–8.

[49] Ivleva VB, Yu YQ, Gilar M. Ultra-performance liquid chromatography/tandem mass spectrometry (UPLC/MS/MS) and UPLC/MS(E) analysis of RNA oligonucleotides. Rapid Commun Mass Spectrom 2010;24:2631–40.

[50] Stults JT, Marsters JC. Improved electrospray ionization of synthetic oligodeoxynucleotides. Rapid Commun Mass Spectrom 1991;5:359–63.

[51] Potier N, Van Dorsselaer A, Cordier Y, Roch O, Bischoff R. Negative electrospray ionization mass spectrometry of synthetic and chemically modified oligonucleotides. Nucleic Acids Res 1994;22:3895–903.

[52] Shah S, Friedman SH. An ESI-MS method for characterization of native and modified oligonucleotides used for RNA interference and other biological applications. Nat Protoc 2008;3:351–6.

[53] Castleberry CM, Rodicio LP, Limbach PA. Electrospray ionization mass spectrometry of oligonucleotides. Curr Protoc Nucleic Acid Chem 2008;35:1–19.

[54] Limbach PA, Crain PF, Mccloskey JA. Molecular-mass measurement of intact ribonucleic-acids via electrospray-ionization quadrupole mass-spectrometry. J Am Soc Mass Spectrom 1995;6:27–39.

[55] Greig M, Griffey RH. Utility of organic bases for improved electrospray mass spectrometry of oligonucleotides. Rapid Commun Mass Spectrom 1995;9:97–102.

[56] Muddiman DC, Cheng XH, Udseth HR, Smith RD. Charge-state reduction with improved signal intensity of oligonucleotides in electrospray ionization mass spectrometry. J Am Soc Mass Spectrom 1996;7:697–706.

[57] Huber CG, Buchmeiser MR. On-line cation exchange for suppression of adduct formation in negative-ion electrospray mass spectrometry of nucleic acids. Anal Chem 1998;70:5288–95.

[58] Farand J, Beverly M. Sequence confirmation of modified oligonucleotides using chemical degradation, electrospray ionization, time-of-flight, and tandem mass spectrometry. Anal Chem 2008;80:7414–21.

[59] Jiang Y, Hofstadler SA. A highly efficient and automated method of purifying and desalting PCR products for analysis by electrospray ionization mass spectrometry. Anal Biochem 2003;316:50–7.

[60] Kharlamova A, Prentice BM, Huang TY, McLuckey SA. Electrospray droplet exposure to gaseous acids for reduction of metal counter-ions in nucleic acid ions. Int J Mass Spectrom 2011;300:158–66.

[61] McGinley M, Scott G, Hail M. Novel isolation and analysis method for oligonucleotide therapeutics and their metabolites form biological matrices by LC/MS. Proceedings of the 58th ASMS Conference on Mass Spectrometry and Allied Topics. Salt Lake City, Utah, 2010.

[62] Huber C, Oberacher H. Analysis of nucleic acids by on-line liquid chromatography-mass spectrometry. Mass Spectrom Rev 2001;20:310–43.

[63] Lin ZJ, Li W, Dai G. Application of LC-MS for quantitative analysis and metabolite identification of therapeutic oligonucleotides. J Pharm Biomed Anal 2007;44:330–41.

[64] Apffel A, Chakel JA, Fischer S, Lichtenwalter K, Hancock WS. Analysis of oligonucleotides by HPLC-electrospray ionization mass spectrometry. Anal Chem 1997;69: 1320–5.

[65] Huber CG, Krajete A. Analysis of nucleic acids by capillary ion-pair reversed-phase HPLC coupled to negative-ion electrospray ionization mass spectrometry. Anal Chem 1999;71:3730–9.

[66] Huber CG, Krajete A. Sheath liquid effects in capillary high-performance liquid chromatography-electrospray mass spectrometry of oligonucleotides. J Chromatogr A 2000;870:413–24.

[67] Premstaller A, Oberacher H, Huber CG. High-performance liquid chromatography-electrospray ionization mass spectrometry of single- and double-stranded nucleic acids using monolithic capillary columns. Anal Chem 2000;72:4386–93.

[68] Oberacher H, Parson W, Muhlmann R, Huber CG. Analysis of polymerase chain reaction products by on-line liquid chromatography-mass spectrometry for genotyping of polymorphic short tandem repeat loci. Anal Chem 2001;73:5109–15.

[69] Tremblay GA, Oldfield PR. Bioanalysis of siRNA and oligonucleotide therapeutics in biological fluids and tissues. Bioanalysis 2009;1:595–609.

[70] Landesman Y, Svrzikapa N, Cognetta A, 3rd, Zhang X, Bettencourt BR, Kuchimanchi S, et al. *In vivo* quantification of formulated and chemically modified small interfering RNA by heating-in-Triton quantitative reverse transcription polymerase chain reaction (HIT qRT-PCR). Silence 2010;1:16.

[71] Holzl G, Oberacher H, Pitsch S, Stutz A, Huber CG. Analysis of biological and synthetic ribonucleic acids by liquid chromatography-mass spectrometry using monolithic capillary columns. Anal Chem 2005;77:673–80.

[72] Taoka M, Yamauchi Y, Nobe Y, Masaki S, Nakayama H, Ishikawa H, et al. An analytical platform for mass spectrometry-based identification and chemical analysis of RNA in ribonucleoprotein complexes. Nucleic Acids Res 2009;37:e140.

[73] Willems AV, Deforce DL, Lambert WE, Van Peteghem CH, Van Bocxlaer JF. Rapid characterization of oligonucleotides by capillary liquid chromatography-nano electrospray quadrupole time-of-flight mass spectrometry. J Chromatogr A 2004;1052: 93–101.

[74] Thayer JR, Puri N, Burnett C, Hail M, Rao S. Identification of RNA linkage isomers by anion exchange purification with electrospray ionization mass spectrometry of automatically desalted phosphodiesterase-II digests. Anal Biochem 2010;399:110–7.

[75] Thomas B, Akoulitchev AV. Mass spectrometry of RNA. Trends Biochem Sci 2006;31:173–81.

[76] Sauer S. The essence of DNA sample preparation for MALDI mass spectrometry. J Biochem Biophys Methods 2007;70:311–8.

[77] Pieles U, Zurcher W, Schar M, Moser HE. Matrix-assisted laser desorption ionization time-of-flight mass spectrometry: a powerful tool for the mass and sequence analysis of natural and modified oligonucleotides. Nucleic Acids Res 1993;21:3191–6.

[78] Nordhoff E, Ingendoh A, Cramer R, Overberg A, Stahl B, Karas M, et al. Matrix-assisted laser desorption/ionization mass spectrometry of nucleic acids with wavelengths in the ultraviolet and infrared. Rapid Commun Mass Spectrom 1992;6: 771–6.

[79] Nordhoff E, Schurenberg M, Thiele G, Lubbert C, Kloeppel KD, Theiss D, et al. Sample preparation protocols for MALDI-MS of peptides and oligonucleotides using prestructured sample supports. Int J Mass Spectrom 2003;226:163–80.

[80] Smirnov IP, Roskey MT, Juhasz P, Takach EJ, Martin SA, Haff LA. Sequencing oligonucleotides by exonuclease digestion and delayed extraction matrix-assisted laser desorption ionization time-of-flight mass spectrometry. Anal Biochem 1996;238: 19–25.

[81] Gao H, Liu Y, Rumley M, Yuan H, Mao B. Sequence confirmation of chemically modified RNAs using exonuclease digestion and matrix-assisted laser desorption/ionization time-of-flight mass spectrometry. Rapid Commun Mass Spectrom 2009;23: 3423–30.

[82] Shaw JP, Kent K, Bird J, Fishback J, Froehler B. Modified deoxyoligonucleotides stable to exonuclease degradation in serum. Nucleic Acids Res 1991;19:747–50.

[83] Kawasaki AM, Casper MD, Freier SM, Lesnik EA, Zounes MC, Cummins LL, et al. Uniformly modified 2′-deoxy-2′-fluoro phosphorothioate oligonucleotides as nuclease-resistant antisense compounds with high-affinity and specificity for RNA targets. J Med Chem 1993;36:831–41.

[84] Egli M, Minasov G, Tereshko V, Pallan PS, Teplova M, Inamati GB, et al. Probing the influence of stereoelectronic effects on the biophysical properties of oligonucleotides: comprehensive analysis of the RNA affinity, nuclease resistance, and crystal structure of ten 2′-O-ribonucleic acid modifications. Biochemistry 2005;44:9045–57.

[85] Bahr U, Aygun H, Karas M. Sequencing of single and double stranded RNA oligonucleotides by acid hydrolysis and MALDI mass spectrometry. Anal Chem 2009;81: 3173–9.

[86] Farand J, Gosselin F. *De novo* sequence determination of modified oligonucleotides. Anal Chem 2009;81:3723–30.

[87] Wu J, McLuckey SA. Gas-phase fragmentation of oligonucleotide ions. International J Mass Spectrom 2004;237:197–241.

[88] Mcluckey SA, Vanberkel GJ, Glish GL. Tandem mass-spectrometry of small, multiply charged oligonucleotides. J Am Soc Mass Spectrom 1992;3:60–70.

[89] Taucher M, Rieder U, Breuker K. Minimizing base loss and internal fragmentation in collisionally activated dissociation of multiply deprotonated RNA. J Am Soc Mass Spectrom 2010;21:278–85.

[90] Gardner MW, Li N, Ellington AD, Brodbelt JS. Infrared multiphoton dissociation of small-interfering RNA anions and cations. J Am Soc Mass Spectrom 2010;21: 580–91.

[91] Tromp JM, Schurch S. Gas-phase dissociation of oligoribonucleotides and their analogs studied by electrospray ionization tandem mass spectrometry. J Am Soc Mass Spectrom 2005;16:1262–8.

[92] Smith SI, Brodbelt JS. Hybrid activation methods for elucidating nucleic acid modifications. Anal Chem 2011;83:303–10.

[93] Johnson JL, Guo W, Zang J, Khan S, Bardin S, Ahmad A, et al. Quantification of raf antisense oligonucleotide (rafAON) in biological matrices by LC-MS/MS to support pharmacokinetics of a liposome-entrapped rafAON formulation. Biomed Chromatogr 2005;19:272–8.

[94] Dai G, Wei X, Liu Z, Liu S, Marcucci G, Chan KK. Characterization and quantification of Bcl-2 antisense G3139 and metabolites in plasma and urine by ion-pair reversed phase HPLC coupled with electrospray ion-trap mass spectrometry. J Chromatogr B Analyt Technol Biomed Life Sci 2005;825:201–13.

[95] Murphy AT, Brown-Augsburger P, Yu RZ, Geary RS, Thibodeaux S, Ackermann BL. Development of an ion-pair reverse-phase liquid chromatographic/tandem mass spectrometry method for the determination of an 18-mer phosphorothioate oligonucleotide in mouse liver tissue. Eur J Mass Spectrom 2005;11:209–15.

[96] Kohler M, Thomas A, Walpurgis K, Schanzer W, Thevis M. Mass spectrometric detection of siRNA in plasma samples for doping control purposes. Anal Bioanal Chem 2010;398:1305–12.

[97] Geary RS. Antisense oligonucleotide pharmacokinetics and metabolism. Expert Opin Drug Metab Toxicol 2009;5:381–91.

[98] Griffey RH, Greig MJ, Gaus HJ, Liu K, Monteith D, Winniman M, et al. Characterization of oligonucleotide metabolism *in vivo* via liquid chromatography/electrospray tandem mass spectrometry with a quadrupole ion trap mass spectrometer. J Mass Spectrom 1997;32:305–13.

[99] Gaus HJ, Owens SR, Winniman M, Cooper S, Cummins LL. On-line HPLC electrospray mass spectrometry of phosphorothioate oligonucleotide metabolites. Anal Chem 1997;69:313–9.

[100] Wei X, Dai G, Liu Z, Cheng H, Xie Z, Marcucci G, et al. Metabolism of GTI-2040, a phosphorothioate oligonucleotide antisense, using ion-pair reversed phase high performance liquid chromatography (HPLC) coupled with electrospray ion-trap mass spectrometry. AAPS J 2006;8:E743–E755.

[101] Noll BO, McCluskie MJ, Sniatala T, Lohner A, Yuill S, Krieg AM, et al. Biodistribution and metabolism of immunostimulatory oligodeoxynucleotide CPG 7909 in mouse and rat tissues following subcutaneous administration. Biochem Pharmacol 2005;69: 981–91.

[102] Turner JJ, Jones SW, Moschos SA, Lindsay MA, Gait MJ. MALDI-TOF mass spectral analysis of siRNA degradation in serum confirms an RNAse A-like activity. Mol Biosyst 2007;3:43–50.

[103] Wilson C, Keefe AD. Building oligonucleotide therapeutics using non-natural chemistries. Curr Opin Chem Biol 2006;10:607–14.

[104] Rozenski J, McCloskey JA. SOS: a simple interactive program for *ab initio* oligonucleotide sequencing by mass spectrometry. J Am Soc Mass Spectrom 2002;13:200–3.

[105] Kretschmer M, Lavine G, McArdle J, Kuchimanchi S, Murugaiah V, Manoharan M. An automated algorithm for sequence confirmation of chemically modified oligonucleotides by tandem mass spectrometry. Anal Biochem 2010;405:213–23.

14

MASS SPECTROMETRY FOR METABOLOMICS

PETIA SHIPKOVA AND MICHAEL D. REILY

14.1 INTRODUCTION

Metabolomics, or often referred to as metabonomics, is the quantitative and qualitative analysis of endogenous small molecule metabolites. Metabolomics can be viewed as the small molecule complement to other "-omics" techniques such as genomics, transcriptomics, and proteomics, often collectively referred to as systems biology. Figure 14.1 shows the "-omics" cascade and reflects the complexity and connectivity of the biochemical continuum of a living organism. Metabolomics is the end step in the cascade that can be linked directly and quantitatively to a phenotype (i.e., wild type, knock-out) or to a treatment (i.e., dosing of a drug, fasting, change in diet). The metabolome was first defined as the "total complement of metabolites in a cell" [1] excluding xenobiotics and xenobiotic metabolites [2]. A more comprehensive, and perhaps relevant definition from a pharmaceutical perspective, expands the metabolome to include small molecules introduced and modified by diet, medication, environmental exposure, and coexisting organisms [3]. It is this broader definition of the metabolome and the measurement of how it changes that underlies the concept of metabonomics [4].

The metabolome reflects the overall biochemical state of an organism, and a comprehensive measurement of changes in the metabolome in response to external stressors such as drug treatment, genetic modification, and environmental change can lead to deducing the relationship between a perturbation and affected biochemical pathways, and, ultimately, to the discovery of biomarkers that report on a perturbation. The metabolome is generally considered to contain around 20,000 small molecule metabolites. Compared with the large number of individual RNAs

Mass Spectrometry for Drug Discovery and Drug Development, First Edition. Edited by Walter A. Korfmacher.

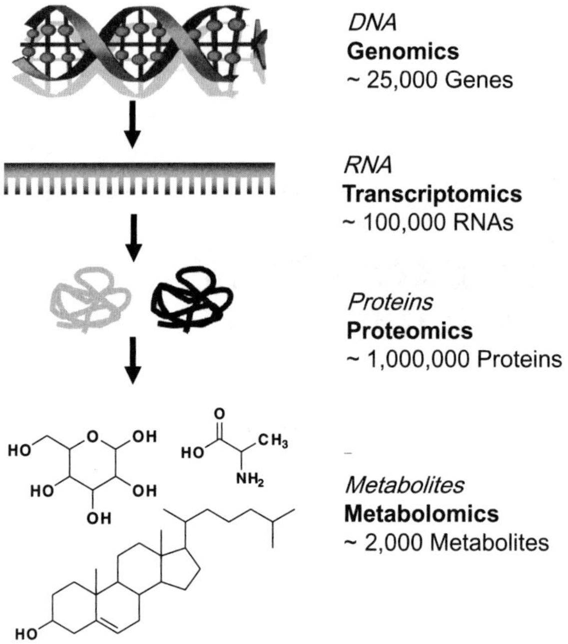

Figure 14.1 "-Omics" cascade: genomics, transcriptomics, proteomics, metabolomics.

(around 100,000) considered in transcriptomic profiling and what is thought to be around a million proteins in proteomics (Fig. 14.1), metabolomics includes a relatively small number of analytes, which makes it within reach for analytical characterization. However, even with only a few thousand metabolites, the accurate and comprehensive measurement of the metabolome is still an extremely complicated task. Some of the difficulties come from the great chemical diversity of the metabolome, from small polar hydrophilic molecules such as sugars and amino acids to various lipophilic components such as sterols, fatty acids, and phospholipids, all present at very different basal concentrations, from low picomolar to high millimolar levels. Conversely, within a class of biomolecules, there may be many chemically similar members and separation and characterization of these can be challenging. For example, there are hundreds of possible phosphatidyl cholines with combinations of different fatty acid chains, whose physical properties are similar enough to confound chromatographic separation. Metabolomic profiling varies from semi-quantitative (relative) to highly accurate concentration measurements (absolute) of a few to as many endogenous components as possible. Some of the analytical challenges as well as possible solutions for capturing endogenous components in a quantitative manner are discussed in this chapter.

14.2 METABOLOMICS IN DRUG DISCOVERY

Metabolomic applications are relatively new to drug discovery; however, even in the span of just a few years, they have received a lot of attention in the literature

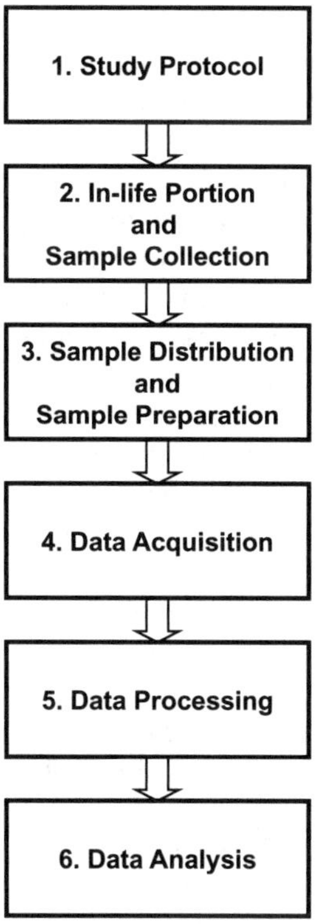

Figure 14.2 Typical 6-step procedure for metabolomic studies.

as outlined in a number of review articles ([5–8] and references therein). In drug discovery, metabolomics is usually employed in a comparative mode, where sample sets from two or more groups are directly compared. These groups typically include a normal state (i.e., healthy individuals or wild type) and a perturbed state (administration of a drug, change in lifestyle, knock-out, etc.), and they are analyzed for detecting qualitative or quantitative changes in the metabolome that can be related to the perturbation.

A typical metabolomic experiment consists of six distinct steps as highlighted in Figure 14.2. The process begins with careful consideration of the desired outcome of the investigation and the design of a study protocol (step 1) that would best enable answering the questions of interest. This protocol design typically includes choice of subjects (animal species, age, strain, gender, etc.), number of subjects per group (in typical rodent studies $n \geq 5$ is desirable), vehicle control, treatment regiment, time-points, tissues/samples of interest, and so on. Step 2 of the metabolomics study corresponds to the in-life portion of the experiment and sample collection. Specifics, such as choice of housing, diet, time, and procedures during sample collection, can play a major role in the outcome of the study [9]. Step 3 includes sample distribution, sample preparation, and storage. Typically, a number of different end points are included in a metabolomic study, such as standard clinical chemistry measurements, drug exposure, mass spectrometry (MS) and/or nuclear magnetic resonance (NMR) metabolomic measurements, and so on, and sample treatment at the collection point and subsequent distribution is extremely important to minimize potential degradation due to freeze–thaw cycles, transfer, and so on. Sample preparation is dependent on the nature of the desired analytical approach (liquid chromatography-MS [LC-MS], gas chromatography-MS [GC-MS], etc.), as well as the matrix of interest. There are numerous reports in the literature about possible approaches for sample preparation from different matrices, including plasma/serum, tissue homogenates, urine, and cerebrospinal fluid (CSF). [10–12].

Step 4 includes analytical data acquisition where emphasis is focused on elimination of signal variability due to analysis conditions. This is achieved by, where possible, using the same mobile phase and the same analytical column throughout a study, randomizing samples from different groups and including various internal standards, pooled-quality controls (QCs) [13], and so on. The success of a metabolomics study is highly dependent on ensuring that observed variability between samples in a group is due to biological differences and not analytical artifacts created during data collection. The fifth step includes data processing/chemometrics, which is typically done with commercially available or proprietary in-house built software packages. There are a number of choices of commercial products including Rosetta Biosoftware™ (supported by Ceiba Solutions, support@rosettabio.com), Genedata™ (GeneData AG, Basel, Switzerland) as well as MS vendor software by MarkerLynx™ (Waters, Milford, MA), Sieve™ (ThermoFisher Scientific, San Jose, CA), MarkerView™ (Sciex, Foster City, CA), Profiler™ (Phenomenome Discoveries, Saskatoon, Canada, and Shimadzu, Columbia, MD), and MassHunter™ (Agilent, Santa Clara, CA). The final and in some ways most complicated and time-consuming step includes data analysis and interpretation. This step is typically a joint effort across many functional groups including biology/toxicology, analytical chemistry, and chemometrics. Only when there is an integrated effort among these disciplines, can meaningful analytical findings fully lend themselves to the interpretation of biochemical mechanisms and the discovery of relevant biomarkers. Each one of these six steps is crucial for the acquisition of a robust, reproducible, and interpretable metabolomic data set.

14.3 ANALYTICAL APPROACHES FOR MEASURING THE METABOLOME

There are three main approaches in the qualitative and quantitative measurement of the metabolome: (1) fingerprinting, (2) nontargeted metabolomics, and (3) targeted metabolomics. Although these approaches differ technically and philosophically in the amount and type of information they capture, there is no absolute separation between the categories and, as a result, a certain level of overlap clearly exists. Each approach has its own unique characteristics and analytical workflows; however, in all three cases, a meaningful metabolomics outcome can only be achieved based on a carefully designed plan that includes the six steps described above. The ultimate goal is the generation of high-quality data that can be translated into a statistically significant biochemical snapshot of an organism's metabolome using sophisticated software tools.

The two most common MS techniques for measuring the metabolome include liquid chromatography-mass spectrometry (LC-MS) and gas chromatography-mass spectrometry (GC-MS), although there are numerous reports on other mass spectrometry-based approaches, including capillary electrophoresis (CE), matrix-assisted laser desorption/ionization (MALDI), and desorption electrospray ionization (DESI) coupled to MS as discussed later in this chapter. It should be mentioned that nuclear magnetic resonance (NMR) is still one of the most commonly used metabolomic techniques, and although discussion of NMR applications for metabolomics is beyond the scope of this chapter, NMR is complementary to LC-MS [14] and is often used in conjunction with MS approaches for a more complete characterization of the metabolome. All metabolomic approaches have their unique characteristics and particular advantages and disadvantages will be highlighted for fingerprinting, nontargeted and targeted metabolomics.

14.3.1 Fingerprinting

Fingerprinting profiling is analytically the simplest and most direct of the three metabolomic approaches. It typically involves generation of relatively simple analytical data sets from complex biological samples with minimally prepared peripheral biofluids or cell media, which allow direct comparison of samples based on spectroscopic, chromatographic, or spectrometric data. The generated data sets or "analytical fingerprints" reflect the composition of the samples under investigation, and are often treated as large multivariate data sets, with each analytical "feature" corresponding to some molecular component in the sample matrix. With the use of widely available statistical tools such as principal component analysis (PCA), these data sets can be used as multivariate input to directly compare and contrast samples within a study. When considering the entire sample set as a whole, this approach can distinguish different treatment groups, assess the degree of differentiation, and even suggest what molecular changes may be responsible for the separation. The main advantage of this approach is that annotation of the analytical data is not required, and therefore data analysis is quick and relatively simple. Figure 14.3 shows an example of a fingerprinting analysis of LC-MS data from male rat urine samples collected at 0 h, 4 h, 8 h, 24 h, and 48 h after dosing with 2-bromoethanamine, a known kidney toxicant. In this type of presentation, the combined LC-MS data from each sample is represented by a single dot and the distance between dots represents the multivariate difference between the samples. The samples from different time-points are color coded for recognition purposes (pretest samples are shown in yellow, 4 h in blue, 8 h in purple, 24 h in blue, and 48 h in green). The PCA plot clearly separates the different time-points and suggests a time-dependent biochemical trajectory with a maximal affect at 8 h and a subsequent return toward the predose (0 h) state by 48 h. The data analysis is carried out without annotation or knowledge of the identity of the individual components contributing to the PCA group separation.

There are numerous examples in the literature utilizing fingerprinting techniques using a variety of analytical approaches such as NMR spectroscopy [15–17], infrared (IR) and Raman spectroscopy [18–21], and CE-MS [22, 23], atmospheric pressure desorption-MS [24], direct injection [25–28], LC-MS [29–32], and GC-MS [33–35].

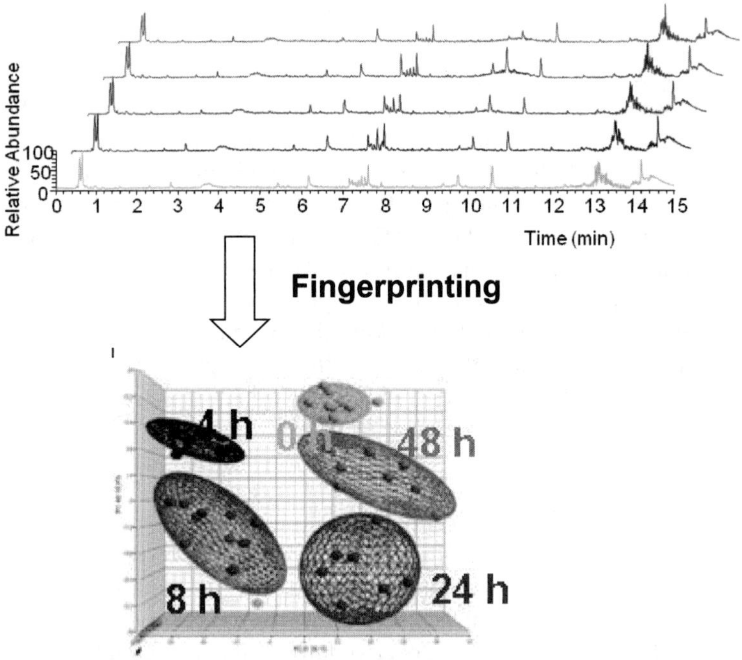

Figure 14.3 PCA plot of LC-MS data from male rat urine samples collected at 0 h, 4 h, 8 h, 24 h, and 48 h after dosing with 2-bromoethanamine, a known kidney toxicant. Pretest samples (0 h) are shown in yellow, 4 h in blue, 8 h in purple, 24 h in blue, and 48 h in green.

In general, chromatography-interfaced techniques such as LC-MS and GC-MS provide highly sensitive and selective analytical data, but are rarely used as fingerprinting techniques, since they introduce sources of variability at the chromatographic level (i.e., column deterioration, temperature, and matrix effects) and significantly increase the analysis time and data complexity. In contrast, fingerprinting approaches have been demonstrated to be remarkable reproducible for spectroscopic methods such as NMR [36] and IR [21, 37], which can quickly and reliably generate data on minimally manipulated samples, most notably direct analysis of serum and urine without the need of sample cleanup (i.e., protein precipitation, desalting, etc.) or derivatization.

14.3.2 Targeted Metabolomics

Targeted metabolomics is in some ways the exact opposite of the fingerprinting approach. Here, instead of searching for gross differences regardless of their identity, the analytes of interest are selected a priori, usually to address certain specific biological questions within a study. Targeted metabolomics is the most quantitative of the three approaches and, therefore, the generated data are the easiest to interpret. Since the analytes are known, it is possible and desirable to measure their absolute concentrations with appropriate use of stable-labeled standards and/or calibration

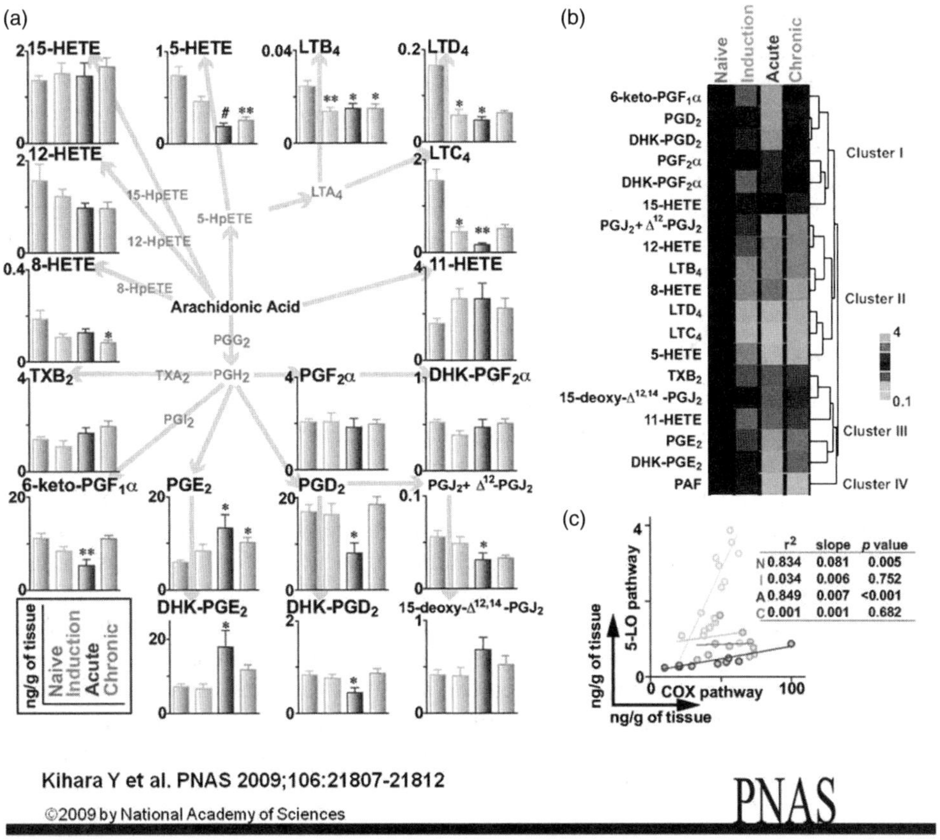

Figure 14.4 Example of targeted metabolomics: absolute quantitation of various eicosanoids and their role in mechanisms of inflammatory responses, pain, and fever. Reprinted from Kihara Y et al. PNAS 2009;106:21807–21812 with permission.

curves and optimized separation and detection methods. There are a number of publications in the literature illustrating excellent applications of targeted metabolomic analyses [38, 39]. Figure 14.4 shows one such approach for absolute quantitation of various eicosanoids and their role in mechanisms of inflammatory responses, pain, and fever [40]. The authors, Kihara et al., have successfully quantified various eicosanoid metabolites of arachidonic acid in cerebrospinal fluid (CSF) of naïve mice and experimental autoimmune encephalomyelitis (EAE) mice in the induction, acute, and chronic phase. The data clearly outline the effect of treatment on each group for each chosen specific arachidonic metabolites.

The targeted approach was recently enhanced further by the introduction of a commercially available kit allowing absolute quantitation of up to 186 metabolites from five metabolites classes (AbsoluteIDQ™p180, Biocrates Life Sciences AG, Innsbruck, Austria).

Sometimes targeted metabolomic analyses are erroneously equated to quantitative assays for measurement of endogenous biomarkers. The lines between the two analyses are indeed not well defined; however, in general, only multiplexed assays for tens to hundreds of different analytes are considered metabolomic assessments. The approach is not limited to molecules from a certain class of compounds with similar physicochemical properties; however, targeted metabolite assays often result from the desire to more accurately quantify specific sets of molecules, that is, phospholipids, bile acids, and so on.

The drawback of a targeted metabolomics approach is that unexpected observations are not monitored for and recorded and, therefore, discovery of novel biomarkers or mechanisms cannot be achieved.

14.3.3 Nontargeted Metabolomics

Nontargeted metabolomics is the most complex and the most challenging analysis of the three, and it is the richest in biochemical sample information. As its name suggests, there is no prior information of the compound classes of potential interest and, as a result, the analyses are focused on capturing quantitative information on as many different analytes as possible to maximize the chance of detecting statistically significant differences in altered biochemical pathways. Such analyses include dramatically different chemical classes of compounds from small polar hydrophilic components such as amino acids and sugars to larger lipophilic components such as sterols and fatty acids; therefore, optimization of elution and detection conditions for all analytes is incredibly complicated. These differences present numerous analytical challenges, including choice of reconstitution solvent, chromatography, column stationary phase, elution profile, polarity, and many other parameters. There are literature reports of nontargeted metabolomic approaches that utilize multiple chromatographic systems, such as hydrophilic interaction liquid chromatography (HILIC) and reversed-phase liquid chromatography (RPLC), to analyze samples and therefore capture different analytes, or alternatively, both GC-MS and LC-MS to provide complementary measurements.

In our lab we have chosen to combine the best of both worlds and capture most of the hydrophilic and lipophilic components, utilizing one 15-min long reversed-phase step gradient as shown in Figure 14.5. Positive and negative mode electrospray analyses are subject of separate injections/analyses. Although far from ideal, this is a good first-look at the biochemical makeup of different sample groups, and it can serve as an excellent guide to further focused studies. In fact, quite often, nontargeted metabolomics are used for biomarker discovery or in the search for altered biochemical pathways that are further refined and confirmed with targeted metabolomic or specific quantitative assays.

Nontargeted metabolomic analytical techniques typically include a combination of NMR spectroscopy and at least one MS-based approach. Of these, NMR is highly quantitative and very reproducible both longitudinally and across laboratories [36], although it suffers from poor sensitivity and thus has been largely limited to proton NMR [41]. The most commonly used hyphenated MS systems employ GC, LC, or CE as a front–end separation technique. General MS methodologies for metabolomics have been recently reviewed [3, 10, 42] as well as each individual approach: LC-MS [43, 44], GC-MS [34, 45, 46], and CE-MS [23, 47]. There have also been reports in the literature highlighting direct injection techniques such as DI [26, 27] and MALDI [48, 49]. While these approaches greatly simplify the data processing

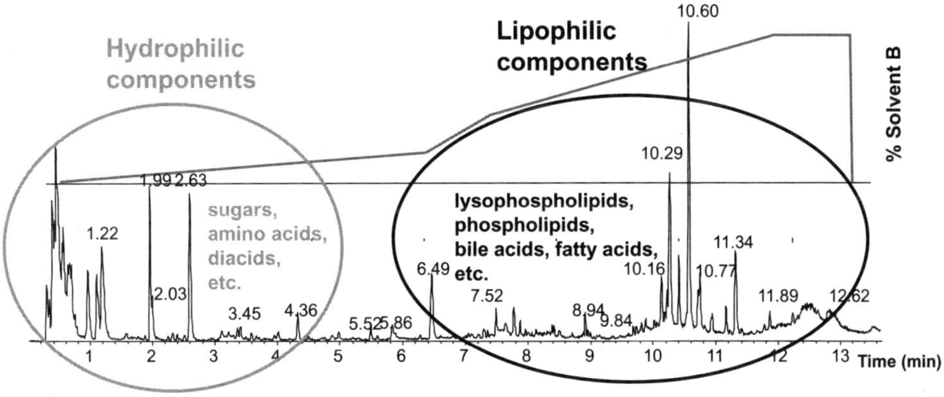

Figure 14.5 Typical LC-MS metabolomics profiling chromatogram capturing both hydrophilic and lipophilic components.

step, the major drawbacks include the potential for ion suppression or enhancement as well as the lack of separation of isobaric compounds.

Annotation is probably the most significant challenge in nontargeted metabolomics. The annotation step is typically done during data analysis using in-house built, free-access web-based libraries such as METLIN (http://metlin.scripps.edu), HMDB (http://www.hmdb.ca), and MMCD (http://mmcd.nmrfam.wisc.edu), or commercial libraries such as FIEHNLIB (http://fiehnlab.ucdavis.edu/Metabolite-Library-2007) and NIST (http://www.nist.gov). Often, the final structure confirmation is validated using commercially available, synthesized, or isolated standards. A common and widely recognized problem with metabolomic libraries, both free-access and commercial, is that they are not comprehensive and, as a result, most nontargeted analyses reveal many components that change in a statistically significant manner that frequently cannot be readily annotated. There are a number of consortium projects under way to extensively study and annotate the human metabolome, such as the Human Metabolome Project (results updated in HMDB) and the Human Plasma Metabolome (HuPMet). The individual steps in data annotation and analysis are illustrated in Figure 14.6.

A list of molecular components or metabolites of interest, showing change as a result of treatment, is generated as an end result of data processing. Clustering and trend analyses give further information in highlighting the components of interest. Annotation of these analytes is of utmost importance for the elucidation of novel biochemical pathways without which it would be virtually impossible to gather insights into the mechanism of action. The identification of unknowns revealed from a nontargeted metabolomics study is often the most significant bottleneck in the process since it is labor-intensive and often requires expertise from multiple analytical disciplines. However, the potential payoff for success is novel insight into biochemical mechanisms and biomarker discovery. The data processing and data analysis steps are major parts of any metabolomic analysis but are especially relevant in nontargeted metabolomics, and improving the accuracy and extent of data annotation will be crucial for more impactful metabolomics analyses.

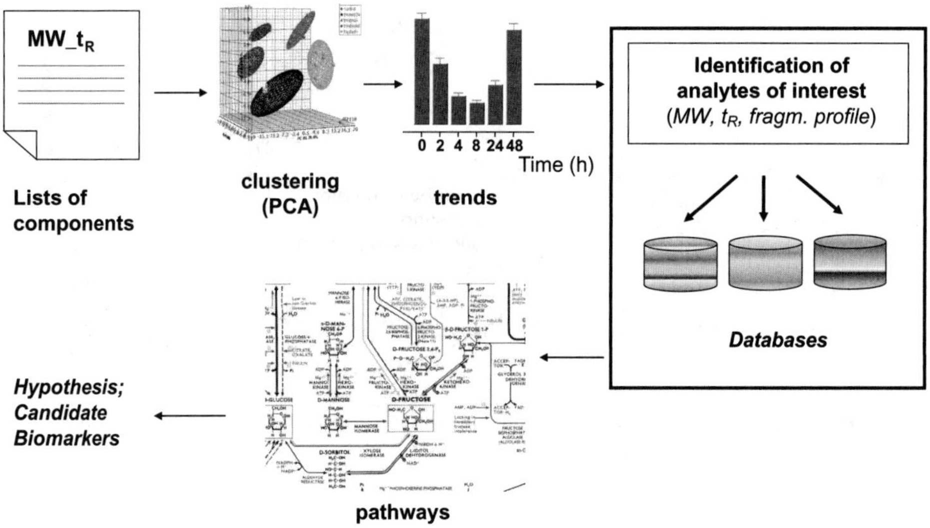

Figure 14.6 Data processing, data analysis, and elucidation of biochemical pathways for LC-MS-based metabolomics.

14.4 CHROMATOGRAPHIC AND NONCHROMATOGRAPHIC SOLUTIONS FOR MS METABOLOMICS

There are a number of significant advantages of using nonchromatographic approaches for MS metabolomics. Flow injection analysis (FIA) and direct injection/infusion (DI) are by far the simplest and fastest ways of sample introduction into a mass spectrometer [26, 27]. Additionally, data processing is dramatically simplified due to elimination of the need for chromatographic alignment and decrease in background provided by LC columns, solvents, and so on. This approach is most suitable for fingerprinting metabolomics approaches. Examples include direct infusion of yeast cell extracts [28], chip-based nanospray [50, 51], DESI [52], and many others. The main drawbacks of nonchromatographic techniques include poor detection of entire classes of compounds due to matrix incompatibility or ion suppression as well as lack of separation of isobaric compounds. In addition, both unique chemical entities and gas phase fragment ions are generated during sample ionization.

Chromatography adds an additional dimension in separation (in addition to *m/z*) and therefore greatly increases both the resolving power of an experiment and the complexity of the acquired data. Although the final detection is based on the *m/z* values, optimized chromatographic separation is crucial to obtaining high-quality data since it is directly related to peak definition and subsequent data processing. Robust and reproducible chromatographic separation minimizes matrix effects and decreases the potential of co-elution of isobaric components. However, the benefits come with the price of an increased data complexity and the need of sophisticated software data processing tools becomes even more pronounced.

Chromatographic separations are typically applicable for both targeted and nontargeted metabolomic analyses. Reversed-phase high-performance liquid chromatography chromatography (HPLC) is still the most commonly used approach, while ultra-high pressure liquid chromatography (UHPLC) is quickly becoming the chromatographic method of choice due to its superior resolution and speed of analysis compared with HPLC [31, 32, 53, 54]. Hydrophilic interaction liquid chromatography (HILIC) has also gained in popularity over the last few years [55, 56], and may provide an excellent orthogonal separation approach to traditional reversed-phase chromatography. Alternatives to LC-MS metabolomics include gas chromatography (GC) [33, 34, 45, 46], capillary electrophoresis (CE) [23, 47], or supercritical fluid chromatography (SFC) [57, 58]. GC-MS is the oldest and best established of all MS-based metabolomic approaches. It provides an excellent sensitivity and resolution and it has the advantage of using standardized electron impact (EI) based spectral libraries [59, 60] for analyte identification. The drawbacks of GC-MS include cumbersome sample preparation, long analysis times as well as complex data analysis.

14.5 MS INSTRUMENTATION FOR METABOLOMICS

The two main classes of mass spectrometers used for metabolomic experiments are triple quadrupoles and high-resolution accurate mass (HRAM) systems. Triple quadrupoles find their application as fast scanning and highly sensitive mass spectrometers mainly in targeted metabolomic analyses. There are reports in the literature utilizing tens of different MS/MS transitions, often including stable-labeled internal standards for quantitation. Examples include the quantification of 100 metabolites per sample in plant extracts [38] or in *Escherichia coli* [61]. Generally, the selected reaction monitoring (SRM) approach is preferred when a limited number of analytes (<100) are to be measured accurately [39, 43]. Commercial kits containing predetermined concentrations of stable-labeled internal standards (Sigma Aldrich, Biocrates, etc.) greatly enhance this approach.

HRAM mass spectrometers, including time-of-flight (TOF), Fourier transform-ion cyclotron resonance (FT-ICR) and Orbitrap systems, offer an excellent alternative to both targeted metabolomic experiments, where a large (>50 MS/MS transitions) set of analytes are monitored. However, the biggest advantage of HRAM instruments is in their ability to collect all ions within a predetermined mass range with excellent sensitivity and mass accuracy (typically <5 ppm) and sufficient dynamic range. This indiscriminative detection enables discovery of unknown or unexpected components changing with treatment or potential biomarkers.

14.6 SUMMARY AND FUTURE TRENDS

The value and place for metabolomics in drug discovery, from single-cell experiments to analysis of complex clinical studies, has been clearly established. The various analytical approaches highlighted in this chapter will all play a role in the quest for complete analysis of the metabolome, and metabolomics, combined with other "-omics" techniques, is and will continue to be an invaluable tool for defining the biochemical state of an organism and discovery of relevant biomarkers. Because MS is highly sensitive and delivers a fundamental molecular parameter (mass), it is in many ways the ultimate detection tool for expanding our reach into the metabolome. The current array of MS technologies discussed here will no doubt be greatly expanded in the near future to assist in this pursuit.

REFERENCES

[1] Tweeddale H, Notley-McRobb L, Ferenci T. Effect of slow growth on metabolism of *Escherichia coli*, as revealed by global metabolite pool ("metabolome") analysis. J Bacteriol 1998;180:5109–16.

[2] Beecher CWW. The human metabolome. In Harrigan GG, Goodacre R, editors. Metabolic Profiling: Its Role in Biomarker Discovery and Gene Function. Boston, MA: Kluwer Academic Publishers, 2003, pp. 311–19.

[3] Dunn WB. Current trends and future requirements for the mass spectrometric investigation of microbial, mammalian and plant metabolomes. Phys Biol 2008;5(1):011001.

[4] Nicholson JK, Lindon JC, Holmes E. Metabonomics: understanding the metabolic responses of living systems to pathophysiological stimuli via multivariate statistical analysis of biological NMR spectroscopic data. Xenobiotica 1999;29:1181–9.

[5] Wishart DS. Applications of metabolomics in drug discovery and development. Drugs R D 2008;9(5):307–22.

[6] Robertson DG, Watkins P, Reily MD. Metabolomics in toxicology: preclinical and clinical applications. Toxicol Sci 2010;120(S1):S146–70.

[7] Drexler DM, Reily MD, Shipkova PA. Advances in mass spectrometry applied to pharmaceutical metabolomics. Anal Bioanal Chem 2010;399:2645–53.

[8] Lu X, Zhao X, Bai C, Zhao C, Lu G, Xu G. LC-MS-based metabonomics analysis. J Chromatogr B 2008;866:64–76.

[9] Robertson DG, Reily MD, Lindon JC, Holmes E, Nicholson JK. Metabonomic technology as a tool for rapid throughput *in vivo* toxicity screening. In Vanden Heuvel JP, Perdew, GJ, Mattes WB, Greenlee WF, editors. Comprehensive Toxicology. Amsterdam: Elsevier Science, 2002, pp. 583–610.

[10] Villas-Bôas SG, Mas S, Åkesson M, Smedsgaard J, Nielsen J. Mass spectrometry in metabolome analysis. Mass Spectrom Rev 2005;24:613–46.

[11] Shipkova PA, Luk CE, Hnatyshyn S, Sanders M. Sample preparation approaches and data analysis for metabonomic profiling of plasma samples. Proceedings of the 55th ASMS; Indianapolis, IN, 2007.

[12] Bruce SJ, Jonsson P, Antti H, Cioarec O, Trygg J, Marklund SL, Moritz T. Evaluation of a protocol for metabolic profiling studies on human blood plasma by combined ultra-performance liquid chromatography/mass spectrometry: from extraction to data analysis. Anal Biochem 2008;372:237–49.

[13] Sangster T, Major H, Plumb R, Wilson AJ, Wilson ID. A pragmatic and readily implemented quality control strategy for HPLC-MS and GC-MS-based metabonomic analysis. Analyst 2006;131:1075–8.

[14] Robertson DG, Reily MD, Baker JD. Metabonomics in pharmaceutical discovery and development. J Proteome Res 2007;6:526–39.

[15] Robosky LC, Robertson DG, Baker JD, Rane S, Reily MD. *In vivo* toxicity screening programs using metabonomics. Comb Chem High Throughput Screen 2002;5:651–62.

[16] Shockor JP, Holmes E. Metabonomic applications in toxicity screening and disease diagnosis. Curr Top Med Chem 2002;2:35–51.

[17] Coen M, Holmes E, Lindon JC, Nicholson JK. NMR-based metabolic profiling and metabonomic approaches to problems in molecular toxicology. Chem Res Toxicol 2008;21:9–27.

[18] Ellis DI, Goodacre R. Metabolic fingerprinting in disease diagnosis: biomedical applications of infrared and Raman spectroscopy. Analyst 2006;131:875–85.

[19] Elliott GN, Worgan H, Broadhurst D, Draper J, Scullion J. Soil differentiation using fingerprint Fourier transform infrared spectroscopy, chemometrics and genetic algorithm-based feature selection. Soil Biol Biochem 2007;39:2888–96.

[20] Brown KL, Palyvoda OY, Thakur JS, Nehlsen-Cannarella SL, Fagoaga OR, Gruber SA, Auner GW. Raman spectroscopic differentiation of activated versus non-activated T lymphocytes: an in vitro study of an acute allograft rejection model. J Immunol Methods 2009;340:48–54.

[21] Pistorius AMA, Degrip WJ, Egorova-Zachernyuk TA. Monitoring of biomass composition from microbiological sources by means of FT-IR spectroscopy. Biotechnol Bioeng 2009;103:123–9.

[22] Allard E, Sjögren E, Lennernäs H, Sjöberg PJR, Bergquist J. Comparing capillary electrophoresis-mass spectrometry fingerprints of urine samples obtained after intake of coffee, tea, or water. Anal Chem 2008;80:8946–55.

[23] Garcia-Perez I, Earll ME, Angulo S, Barbas C, Legido-Quigley C. Metabolic fingerprinting with capillary electrophoresis. J Chromatogr A 2008;1204:130–9.

[24] Chen HW, Wortmann A, Zenobi RJ. Neutral desorption sampling coupled to extractive electrospray ionization mass spectrometry for rapid differentiation of biosamples by metabolomic fingerprinting. J Mass Spectrom 2007;42:1123–35.

[25] Brown SC, Kruppa G, Dasseux JL. Metabolomics applications of FT-ICR mass spectrometry. Mass Spectrom Rev 2005;24:223–31.

[26] Hansen MAE, Smedsgaard J. Automated work-flow for processing high-resolution direct infusion electrospray ionization mass spectral fingerprints. Metabolomics 2007;3:41–54.

[27] Beckmann M, Parker D, Enot D, Duval E, Draper J. High-throughput, nontargeted metabolite fingerprinting using nominal mass flow injection electrospray mass spectrometry. Nat Protoc 2008;3:486–504.

[28] Madalinski G, Godat E, Alves S, Lesage D, Genin E, Levi P, Labarre J, Tabet JC. Direct introduction of biological samples into a LTQ-Orbitrap hybrid mass spectrometer as a tool for fast metabolome analysis. Anal Chem 2008;80:3291–303.

[29] Wilson ID, Nicholson JK, Castro-Perez J, Granger JH, Johnson KA, Smith BW, Plumb RS. High resolution "ultra performance" liquid chromatography coupled to oa-TOF mass spectrometry as a tool for differential metabolic pathway profiling in functional genomic studies. J Proteome Res 2005;4:591–8.

[30] Gika HG, Theodoridis GA, Wingate JE, Wilson ID. Liquid chromatography and ultra-performance liquid chromatography-mass spectrometry fingerprinting of human urine: sample stability under different handling and storage conditions for metabonomics studies. J Chromatogr A 2008;1189:314–22.

[31] Plumb R, Jones M, Rainville P, Nicholson JA. Rapid simple approach to screening pharmaceutical products using ultra-performance LC coupled to time-of-flight mass spectrometry and pattern recognition. J Chromatogr Sci 2008;46:193–8.

[32] Michopoulos F, Lai L, Gika H, Theodoridis G, Wilson I. UPLC-MS-based analysis of human plasma for metabonomics using solvent precipitation or solid phase extraction. J Proteome Res 2009;8:2114–21.

[33] Jonsson P, Johansson A, Gullberg J, Trygg AJ, Grung B, Marklund S, Sjostrom M, Antti H, Moritz T. High-throughput data analysis for detecting and identifying differences between samples in GC/MS-based metabolomic analyses. Anal Chem 2005;77:5635–42.

[34] Fiehn O. Extending the breadth of metabolite profiling by gas chromatography coupled to mass spectrometry. Trends Anal Chem 2008;27:261–9.

[35] Vallejo M, García A, Tuñón J, García-Martínez D, Angulo S, Martín-Ventura JL. Plasma fingerprinting with GCMS in acute coronary syndrome. Anal Bioanal Chem 2009;394(6): 1517–24.

[36] Keun HC, Ebbels TM, Antti H, Bollard ME, Beckonert O, Schlotterbeck G. Analytical reproducibility in (1)H NMR-based metabonomic urinalysis. Chem Res Toxicol 2002;15:1380–6.

[37] Chen W, Yu F, Yuan Y. Application of FTIR and metabolomics analysis in high-throughput screening strains. Huagong Xuebao/Journal of Chemical Industry and Engineering (China) 2007;58:2336–40.

[38] Sawada Y, Akiyama K, Sakata A, Kuwahara A, Otsuki H, Sakurai T, Saito K, Hirai MY. Widely targeted metabolomics based on large-scale MS/MS data for elucidating metabolite accumulation patterns in plants. Plant Cell Physiol 2009;50:37–47.

[39] Lu W, Bennett BD, Rabinowitz JD. Analytical strategies for LCMS-based targeted metabolomics. J Chromatogr B 2008;871:236–42.

[40] Kihara Y, Matsushita T, Kita Y, Uematsu S, Akira S, Kira JI, Ishii S, Shimizu T. Targeted lipidomics reveals mPGES-1-PGE2 as a therapeutic target for multiple sclerosis. Proc Natl Acad Sci U S A 2009;106:21807–12.

[41] Reily MD, Lindon JC. NMR spectroscopy: principles and instrumentation. In Robertson DG, Holmes, E, Nicholson JK, editors. Metabonomics in Safety Assessment. New York: Taylor & Francis, 2005, pp. 75–104.

[42] Lokhov PG, Archakov AI. Mass spectrometry methods in metabolomics. Biochem (Moscow) Suppl Ser B Biomed Chem 2009;3:1–9.

[43] Chen C, Gonzalez FJ, Idle JR. LCMS-based metabolomics in drug metabolism. Drug Metab Rev 2007;39:581–97.

[44] Vogeser M, Seger C. A decade of HPLC-MS/MS in the routine clinical laboratory—goals for further developments. Clin Biochem 2008;41:649–62.

[45] Pasikanti KK, Ho PC, Chan EC. Development and validation of a gas chromatography/mass spectrometry metabonomic platform for the global profiling of urinary metabolites. Rapid Commun Mass Spectrom 2008;22:2984–92.

[46] Pasikanti KK, Ho PC, Chan EC. Gas chromatography/mass spectrometry in metabolic profiling of biological fluids. J Chromatogr B 2008;871:202–11.

[47] Monton MRN, Soga T. Metabolome analysis by capillary electrophoresis-mass spectrometry. J Chromatogr A 2007;1168:237–46.

[48] M'Koma AE, Blum DL, Norris JL, Koyama T, Billheimer D, Motley S, Ghiassi M, Ferdowsi N, Bhowmick I, Chang SS, Fowke JH, Caprioli RM, Bhowmick NA. Detection of pre-neoplastic and neoplastic prostate disease by MALDI profiling of urine. Biochem Biophys Res Commun 2007;353:829–34.

[49] Miura D, Fujimura Y, Tachibana H, Wariishi H. Highly sensitive matrix-assisted laser desorption ionization-mass spectrometry for high-throughput metabolic profiling. Anal Chem 2010;82:498–504.

[50] Juraschek R, Dulcks T, Karas M. Nanoelectrospray-more than just a minimized-flow electrospray ionization source. J Am Soc Mass Spectrom 1999;10:300–8.

[51] Schmidt A, Karas M, Dulcks R. Effect of different solution flow rates on analyte ion signals in nano-ESI MS, or when does ESI turn into nano-ESI? J Am Soc Mass Spectrom 2003;14:492–500.

[52] Takats Z, Wiseman JM, Gologan B, Cooks RG. Mass spectrometry sampling under ambient conditions with desorption electrospray ionization. Science 2004;306:471–3.

[53] Gika HG, Macpherson E, Theodoridis GA, Wilson ID. Evaluation of the repeatability of ultra-performance liquid chromatography-TOF-MS for global metabolic profiling of human urine samples. J Chromatogr B 2008;871:299–305.

[54] Zelena E, Dunn WB, Broadhurst D, Francis-McIntyre S, Carroll KM, Begley P, O'Hagan S, Knowles JD, Halsall A, Wilson ID, Kell DB. Development of a robust and repeatable UPLC-MS method for the long-term metabolomic study of human serum. Anal Chem 2009;81:1357–64.

[55] Gika HG, Theodoridis GA, Wilson ID. Hydrophilic interaction and reversed-phase ultra-performance liquid chromatography TOF-MS for metabonomic analysis of Zucker rat urine. J Sep Sci 2008;31:1598–608.

[56] Kamleh A, Barrett MP, Wildridge D, Burchmore RJS, Scheltema RA, Watson DG. Metabolomic profiling using Orbitrap Fourier transform mass spectrometry with hydrophilic interaction chromatography: a method with wide applicability to analysis of biomolecules. Rapid Commun Mass Spectrom 2008;22:1912–8.

[57] Shenar N, Cantel S, Martinez J, Enjalbal C. Comparison of inert supports in laser desorption/ionization mass spectrometry of peptides: pencil lead, porous silica gel, DIOS-chip and NALDI™ target. Rapid Commun Mass Spectrom 2009;23:2371–9.

[58] Bobeldijk I, Hekman M, De Vries-Van Der Weij J, Coulier L, Ramaker R, Kleemann R, Kooistra T, Rubingh C, Freidig A, Verheij E. Quantitative profiling of bile acids in biofluids and tissues based on accurate mass high resolution LC-FTMS: compound class targeting in a metabolomics workflow. J Chromatogr B 2008;871:306–13.

[59] Halket JM, Waterman D, Przyborowska AM, Patel RKP, Fraser PD, Bramley PM. Chemical derivatization and mass spectral libraries in metabolic profiling by GC/MS and LC/MS/MS. J Exp Bot 2005;56:219–43.

[60] Matsuda F, Shinbo Y, Oikawa A, Hirai MY, Fiehn O, Kanaya S, Saito K. Assessment of metabolome annotation quality: a method for evaluating the false discovery rate of elemental composition searches. PLoS ONE 2009;4(10):e7490.

[61] Bennett BD, Kimball EH, Gao M, Osterhout R, Van Dien SJ, Rabinowitz JD. Absolute metabolite concentrations and implied enzyme active site occupancy in *Escherichia coli*. Nat Chem Biol 2009;5:593–9.

15

QUANTITATIVE ANALYSIS OF PEPTIDES WITH MASS SPECTROMETRY: SELECTED REACTION MONITORING OR HIGH-RESOLUTION FULL SCAN?

LIEVE DILLEN AND FILIP CUYCKENS

15.1 INTRODUCTION

Pharmaceutical companies have predominantly focused on the development of new drug candidates on small organic molecules. However, in recent years, therapeutic proteins and peptides have become an increasingly large component of their drug portfolios [1, 2]. Moreover, besides the usefulness of peptides as therapeutic drugs, a variety of endogenous peptides are used as biomarkers for assessment of a disease state or a therapeutic intervention [3–5]. Therefore, quantitative analysis of peptides in biological matrices is crucial in support of pharmacokinetic and pharmacodynamic studies. In this chapter we define peptides as amino acid sequences with a molecular weight (MW) of less than 10,000 Da.

A large majority of preclinical and clinical drug development programs for peptides rely on immunological assays for quantitative analysis. These assays offer great sensitivity and throughput, but can have selectivity and specificity limitations and often show limited dynamic ranges for quantification [6]. Independent orthogonal techniques can be added to support the validity of the assay. Furthermore, the long development times required for immunoassays are not compatible with current flow charts, the large number of compounds, and the short cycle times required in

Mass Spectrometry for Drug Discovery and Drug Development, First Edition. Edited by Walter A. Korfmacher.

modern early drug discovery programs. New workflows and approaches are imperative.

High-performance liquid chromatography-mass spectrometry (HPLC-MS) as well as HPLC-tandem mass spectrometry (MS/MS) have the potential to offer an orthogonal technique to the immunoassay and can potentially address most of the shortcomings of immunoassays. Typically, for selective mass spectrometric methods, selected reaction monitoring (SRM) is preferred, where one or more peptide-specific fragmentation transitions are monitored. These approaches are familiar to the bioanalytical community as they are routinely applied for small molecules as well. Workflows optimizing the SRM transition to maximize sensitivity and selectivity can be largely automated for small molecules, and many vendors offer off-the-shelf software packages (DiscoveryQuant—ABSciex, Foster City, CA; QuanOptimize—Waters, Milford, MA; QuickQuan—ThermoFisher Scientific, Waltham, MA). However, for peptide SRM methods, these automated optimization workflows are less well developed. Under electrospray ionization (ESI), proteins and peptides are present as multiply charged ions. The distribution of the charge states is dependent on experimental conditions and on peptide concentration [7, 8]. In addition, the mechanisms of fragmentation of peptides are sensitive to small changes in collision energy (CE). Often, limited fragmentation is observed at low collision energies, while the optimal CE often results in many ladder sequence ions (*b* and *y* ions) [9], valuable for qualitative identification purposes, but which dilute the SRM signal in a quantification workflow. At higher collision energies, complete fragmentation can result in individual immonium ions that do not discriminate between peptides and therefore are not the preferred choice for development of an SRM method.

Many studies report the use of mass spectrometry-based approaches for quantification of proteins (MW > 10,000 Da) [10, 11]. Most of these quantitative workflows for larger proteins involve enzymatic digestion followed by liquid chromatography for separation of the resulting peptides. The quantification of a selected "signature peptide" as a surrogate for the intact protein is not without its shortcomings. Selectivity and specificity may be limited, as the tryptic peptide may not reveal all the different post-translation modifications of the intact protein precursor and the peptide may be a component of other proteins if not carefully selected. Ultimately, this approach for quantification of a protein through the quantification of a signature peptide also invokes an SRM-based approach as the standard method of choice for sensitive and selective analysis.

All the above-mentioned challenges in MS-based quantification of peptides have forced bioanalysts to expand their horizons and look for alternative strategies. Selected ion monitoring (SIM) omits the fragmentation step and can lead to higher absolute sensitivity, but insufficient selectivity is observed when analyzing in biological matrices. Pseudo-SRM (parent to parent transition) has not delivered the expected improvements. An interesting approach has been proposed by Klaassen et al. [12], who demonstrated that high field asymmetric waveform ion mobility mass spectrometry (FAIMS) is a selectivity-enhancing technique able to filter out matrix interferences. These authors succeeded in the validation of a problematic assay for a peptide with FAIMS-pseudo-SRM. New ion mobility mass spectrometry (IMS) developments have been introduced (e.g., differential ion mobility) and need further investigation to understand the consequences for peptide analysis.

With the focus on peptide quantification, two approaches have been proposed that take advantage of the multiple charge state of peptides. Targeted enhanced multiple charge scanning (tEMC) is carried out in a linear ion trap (LIT) and removes the singly charged background ions by optimizing the exit lens barrier [13]. Targeted ion parking (TIP) selectively traps several multiply charged peptide ions. In the LIT, reaction with negatively charged reagent ions reduces the charge state of the peptide (increasing the *m/z* value), and the higher *m/z* ion, which initially was not trapped, will now be monitored for quantification. This approach has shown potential for quantification of intact peptides/proteins [14].

Although ultra-performance liquid chromatography (UPLC) combined with MS/MS has now been recognized for years as the state-of-the-art bioanalytical tool for quantification, high-resolution mass spectrometry (HRMS) has steadily entered the bioanalytical territory, particularly in the drug discovery arena, where efficiency and fit-for-purpose quality are critical. The biotransformation community has for some time embraced HRMS technology as an indispensable tool for structure elucidation of unknown metabolites, and with the potential to combine quantitative and qualitative workflows. The bioanalytical community, on the other hand, has not generally been considering HRMS as a viable alternative to SRM quantification mainly due to lower sensitivity and limited dynamic ranges. However, improvements in the most common accessible high-resolution instruments (time-of-flight [TOF] and orbital trapping instruments) offer excellent mass resolution (>20,000) and mass accuracy (<5 ppm) and sufficient scan speed (>1 Hz) [15]. The new detectors have also addressed, to a large extent, the dynamic range and sensitivity gap. For small molecule quantification, full scan HRMS has been utilized for a broad variety of applications. In particular, when multiple analytes are to be quantified in a single injection, the SRM workflow has limitations and full scan HRMS has been proposed as a better alternative. Examples in food safety, forensic and environmental research, as well as toxicology and therapeutic drug monitoring have been described in the literature [16–20]. In HRMS experiments, selectivity is achieved through narrow width extracted ion chromatograms (EIC) achievable through a combination of high resolution and good mass accuracy. Full scan HRMS data contain data for all analytes across the selected mass range, each of which can, in principle, be extracted and quantitated. With respect to peptide quantification, examples of the use of full scan acquisition on HRMS instruments are scarce. We have recently published a comparative study on the quantification of six peptides on two MS platforms: on a triple quadrupole instrument with an SRM workflow and on a Q-TOF instrument using full scan HRMS [21].

After some general considerations on the mass spectrometric behavior of peptides, we will first discuss advantages and shortcomings of SRM-based approaches for peptide quantification. We will continue with considerations on full scan HRMS workflows for peptides including a discussion and explanation of the impact of resolution and mass accuracy in relation to the optimal extraction window to be chosen. Strategies related to combining multiple charge states and isotopic peaks as a function of the peptide mass will also be examined. Although we recognize the importance of sample preparation and chromatography for a good peptide quantification method, we consider these aspects out of scope for the current discussion, and instead refer to some excellent reviews [8, 22, 23].

15.2 IONIZATION AND FRAGMENTATION OF PEPTIDES

Peptides in a biological matrix need some separation from endogenous (background) and peptide-related (e.g., metabolites) molecules which is most often realized through HPLC. As such, the peptides are presented for MS analysis in the HPLC eluent and volatilization into a gaseous state is an essential step. The most popular atmospheric pressure ionization for peptides is ESI, in which charged droplets are formed when the sample solution is sprayed through a capillary held at a high voltage (2–5 kV). Since ESI is a very mild ionization technique polar, nonvolatile, high mass biomolecules are typically ionized as intact species. One particular characteristic of peptides is that they appear as multiply charged ions after ESI with mass-to-charge ratios (*m/z*) that allow detection of these molecules with both conventional quadrupole and ion trap analyzers. We focus in this chapter on Enfuvirtide (Fuzeon®, T20, obtained from Roche through a local pharmacy), a 36-amino acid peptide that functions as a human immunodeficiency virus (HIV) fusion inhibitor. Figure 15.1 shows the amino acid sequence and an ESI spectrum of enfuvirtide obtained in positive ion mode on a TOF instrument showing the generation of multiply charged $[M+4H]^{4+}$ and $[M+3H]^{3+}$ peaks. Deconvolution software allows the reconstitution of the singly charged spectrum and indicates very good mass accuracy (1.6 mDa or 0.35 ppm—theoretical monoisotopic mass $[M+H]^+$ 4490.1945, measured $[M+H]^+$ 4490.1919). The importance of good mass accuracy for quantitative analysis will be exemplified later on.

ESI can be performed both in positive and in negative ion modes. Since most peptides/proteins contain multiple basic amino acid residues, they are most often

Charge state	Calculated monoisotopic mass
1+	4490.1945
2+	2245.6011
3+	1497.4034
4+	1123.3045

Enfuvirtide: Ac-YTSLIHSLIEESQNQQEKNEQELLELDKWASLWNWF-NH_2

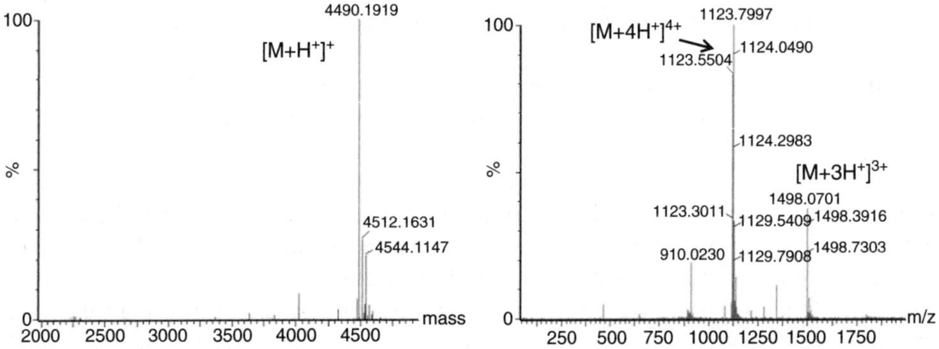

Figure 15.1 Electrospray spectrum for Enfuvirtide obtained in positive ion mode and its deconvoluted spectrum as a centroid mass spectrum.

analyzed in positive ion mode [24]. Nevertheless, examples of applications in negative ion mode have also been described [25] and will be addressed later in relation to SRM-driven quantification workflows. With respect to ionization, it is notable that the intensity of the most abundant ion in a mass spectrum is influenced by many different factors [7]. Minor changes in source parameters (orifice potential, capillary voltage) or changes in the composition of the mobile phase can have a significant impact on signal intensity. The ESI response can also be potentially enhanced by derivatization [26, 27]. However, in a bioanalytical workflow in early discovery, this approach is typically not practical because of the short cycle times and throughput required. At the point where a new drug candidate is selected for further development, a derivatization strategy can be valuable and considered if needed for enhancement of sensitivity. Enhancement of ionization efficiency at very low flow rates (nL/min to low μL/min) through nonpneumatically assisted electrospray conditions has been documented and rationalized [28]. Although commonly used and favored in proteomic investigations, the robustness of these micro- and nanoflow rate systems for analyzing high-throughput biological samples is of concern. Efforts are being directed toward chip-based "plug and play" formats (Agilent HPLC-chip, Santa Clara, CA, and Waters Trizaic format, Milford, MA). The next years will reveal whether these developments are sufficiently robust to be generally adopted in bioanalytical labs when there is a need for ultimate sensitivity or when only low volumes of samples are available.

Peptides typically fragment along their backbone at the peptide amide bond; the most commonly observed ions are the N-terminal *b*-ions as well as the C-terminal *y*-ions [9, 29]. Sequential series of these fragment ions, also called "ladder sequence ions," are often generated. These offer sequence-rich information that facilitates identification of unknown peptides but challenges the development of very sensitive assays for quantification since the product ion signal intensity is diluted over multiple peaks. The fragmentation of peptides is sensitive to small variations in the CE, often with very limited fragmentation at low CE. At higher CE complete fragmentation results in the appearance of the non-discriminatory immonium ions. In proteomic quantification workflows, algorithms have been proposed to predict the CE as a function of charge state for tryptic peptides [30]. However, the versatility of this tool for non-tryptic therapeutic and biomarker peptides has not yet been demonstrated. The introduction of novel peptide entities including unnatural amino acids and/or backbone modifications (cyclizations, end-blocking through acetylation) as well as conjugation (pegylation) further complicates the use of these prediction programs.

15.3 SRM FOR QUANTIFICATION OF PEPTIDES

15.3.1 Triple Quadrupole Instruments

Triple quadrupole mass spectrometers contain two quadrupole mass analyzers separated by a radiofrequency (RF)-only quadrupole collision cell. They are by far the most popular instruments for quantification by HPLC-MS/MS due to relative high duty cycle, which results in excellent sensitivity. After selection of the parent *m/z* in the first quadrupole, fragmentation in the collision cell, and monitoring of specific product ions in the third quadrupole, these instruments offer thus far unsurpassed selectivity and specificity. SRM allows ions of interest to pass through continuously, and theoretically 100% efficiency could be obtained if the parent ion is completely fragmented into one single product ion without any losses to other ions during collisional activation.

For peptides, however, the situation is less favorable. Depending on the number of amino acids, the parent ion signal is already spread over multiple charge states, resulting in lower ion intensities per peak (as shown in Fig. 15.1 for Enfuvirtide where the signal is spread over the 4^+ and the 3^+ ions). To mitigate this loss in sensitivity, signals from different charge states can be summed but at the expense of reducing the duty cycle per transition.

Quadrupole instruments typically operate at unit resolution (also defined as 0.7 Da full width at half maximum [FWHM]). Higher resolution is required when individual isotopic peaks of the multiply charged peaks need to be resolved. However, to maximize the signal intensity for multiply charged ions (3 charges or more), higher FWHM values have been applied for SRM selection [29, 30]. In addition to all these considerations, SRM-based method development for a peptide is guided by the following considerations and evaluations.

First, with multiply charged peaks present in the spectrum of peptides, one charge state can be the dominant species in the spectrum, and experimental conditions can sometimes be modified to enhance and/or favor the generation of one particular charge state. Even with one predominant charge state available, the evaluation should also include the fragmentation efficiency of the individual charge states. Although some papers mention that larger peptides require higher collision energies and thereby result in multiple fragments with lower sensitivity [31], the impact of higher charge states (e.g., 3^+ and higher charge states) on fragmentation efficiency has not been studied extensively, to the best of our knowledge. On the other hand, as mentioned earlier, prediction of fragmentation patterns and CE for tryptic peptides (most often 2^+ ions) has been documented [30]. Occasionally, we have observed that the most intense signal in the first quadrupole does not necessarily lead to the best yield of product ions (personal observation).

Second, the SRM approach able to deliver the best quantification strategy needs to be considered. Summation of multiple SRM transitions can increase the response, but this does not invariably lead to an improvement in signal-to-noise ratio (S/N). This approach can also reduce variability of the response since the relative charge state distributions of a peptide are a function of experimental conditions and the concentration of the peptide [8].Variability of the assay can potentially be reduced when SRM transitions from all the available charge states are used to build the SRM method.

Figure 15.2 shows an example for three β-amyloid peptides ($A\beta_{38}$, $A\beta_{40}$, $A\beta_{42}$). A method with a single SRM transition for the most abundant ion (5^- charged peak) has a lower limit of quantitation (LLOQ) of 0.5 ng/mL for each peptide. Including SRM transitions for the three observed charged states (adding 3^- and 4^-) lowers the LLOQ to 0.1 ng/mL. The SRM chromatograms for the single transition method for the three peptides show an S/N of 2–3 (Fig. 15.2A), while the chromatograms (Fig. 15.2B) that are the result of the summation of three transitions per peptide covering the three charge states show an S/N >5. However, with three transitions for three peptides, a total of nine transitions are monitored, and careful consideration and optimization of dwell times in relation to the chromatographic peak width is

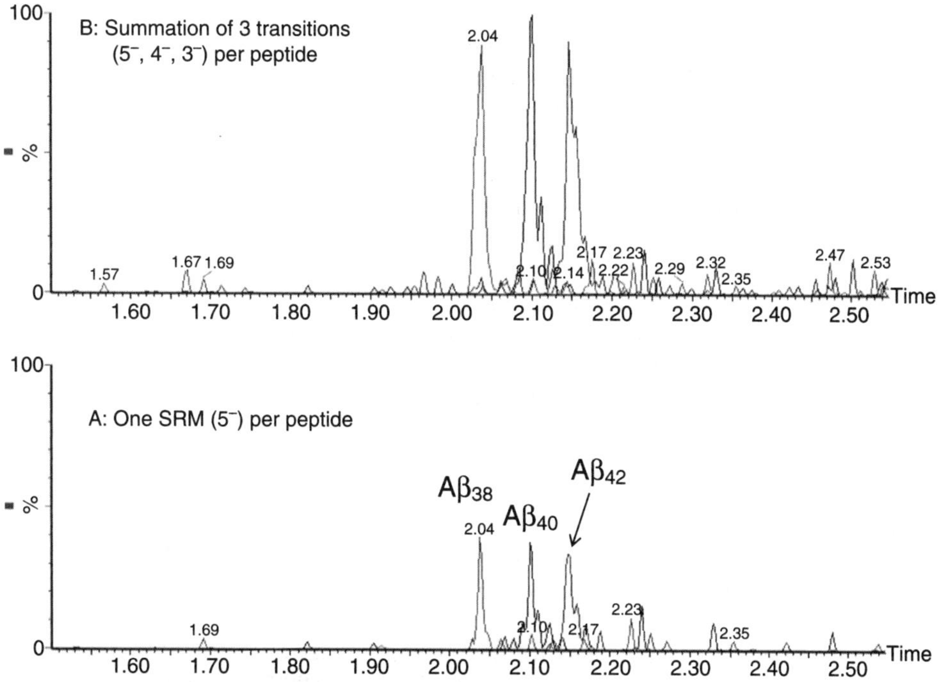

	UPLC-XevoTQS	
column	Acquity UPLC 300 A C_{18}, 1.7μm 2.1 x 50 mm PST	
flow rate	400 μl/min	
solvents	Solvent A: 0.2% NH_4OH - Solvent B: 0.2% NH_4OH in acetonitrile	
sample prep	4 volumes CSF + 1 volume 10% NH_4OH in acetonitrile	
injection volume	15 μL	
SRM	Ab_{38}	(5-) 825.3 → 821.6 (4-) 1031.8 → 1027.3 (3-) 1376.0 → 1370.1
	Ab_{40}	(5-) 864.9 → 861.3 (4-) 1081.4 → 1076.8 (3-) 1442.1 → 1436.1
	Ab_{42}	(5-) 901.8 → 898.1 (4-) 1127.5 → 1122.9 (3-) 1503.6 → 1497.6

Time (min)	%A	%B
0	90	10
0.5	90	10
2	40	60
2.5	5	95
2.8	90	10
3.5	90	10

Figure 15.2 Overlaid SRM LLOQ chromatograms (0.1 ng/mL cerebrospinal fluid) for three amyloid peptides ($A\beta_{38}$, $A\beta_{40}$, $A\beta_{42}$). (A) Result obtained with 1 SRM (5- peptide ion transition) per peptide; (B) transitions resulting from three different charge state peptide ions are summed. Experimental procedure included.

essential. For quantitative analysis, it is generally accepted that at least 15–20 points are needed for a good peak description allowing reliable quantification. With UPLC, peak widths are typically only a few seconds (2–3 s). In the example here, simultaneous analysis of three amyloid peptides with three transitions per peptides at a scan time of 10 ms results in a total scan time of 90 ms (including the interscan delay, this results in a total scan time of >100 ms). In the case of a chromatographic peak width of 2 s, a 100-ms scan time results in 20 points over the peak, which is close to the minimum number of data points for good practice quantification.

Adding more SRM transitions for simultaneous quantification of more peptides or for quantification with internal standards (IS) is then problematic. Split acquisition time periods per peptide (if chromatographic separation is achieved) or further reduction of dwell times are needed to mitigate the quantification challenge of sufficient data points for an adequate peak description. The most recent fast-scanning triple quadrupole instruments (e.g., API5500 [ABSciex], Vantage [ThermoFisher], and Xevo TQS [Waters)] claim dwell times down to 1 ms. Figure 15.3 shows the impact of dwell time on the reproducibility of peak areas for a small molecule assay. Dwell times were varied, but the total number of data points defining the peak was kept constant through introduction of dummy transitions. The SRM transition for the compound and its stable-isotope-labeled IS were both included. Six replicate injections at the LLOQ level for the compound (as defined in the validated method) and at a higher concentration level for the internal standard were analyzed. Dummy transitions with high mass differences differed at least 500 Da from the parent and daughter ion mass. On the older generation triple quadrupoles, dwell times of a few 100 ms up to 1 s were frequented, in line with peak widths obtained on regular HPLC systems. Figure 15.3A demonstrates, however, that dwell times below 50 ms clearly impact reproducibility of the obtained peak areas on the older triple quadrupole instruments. This effect is more pronounced for the lower peak areas at the LLOQ. Coefficients of variation (CVs) for the LLOQ peak area varied from 10% at 50 ms dwell time and raised to 77% with 5 ms dwell times. CVs for the peak areas of the IS were less affected due to the higher absolute peak areas obtained and were 3% at 50 ms up to 15% at 5 ms dwell time. Another interesting observation was the reduced peak area when dummy transitions were included that differed at least 500 mass units from the transition of the compound of interest. The improvements in the latest generation triple quadrupole experiments are huge when comparing Figure 15.3A and Figure 15.3B. CVs for the LLOQ peak areas varied from 1.4 % at 50 ms dwell time and increased to 8% with 5 ms dwell times. CVs for the peak areas of the IS varied from 0.8 % to 2.3% on this latest generation instrument. Although executed for small molecules, the conclusions of these experiments can be extended for peptide SRM methods. Low dwell times need careful consideration in order to obtain robust quantitative SRM assays.

Although the majority of published peptide/protein MS analyses have been performed in positive ion mode, occasional examples of analysis in negative ion mode have been reported [5, 32, 33]. Early reports described the use of trifluoroacetic acid (TFA) as an HPLC mobile phase modifier, and often the TFA adduct was observed. For cetrorelix, selected ion monitoring (SIM) of the TFA adduct in negative ion mode (at a high *m/z* value of 1543)—not including any MS/MS step—was used [25]. TFA has largely been replaced by MS-friendly alternatives (acetic acid or formic acid) to minimize the loss in sensitivity.

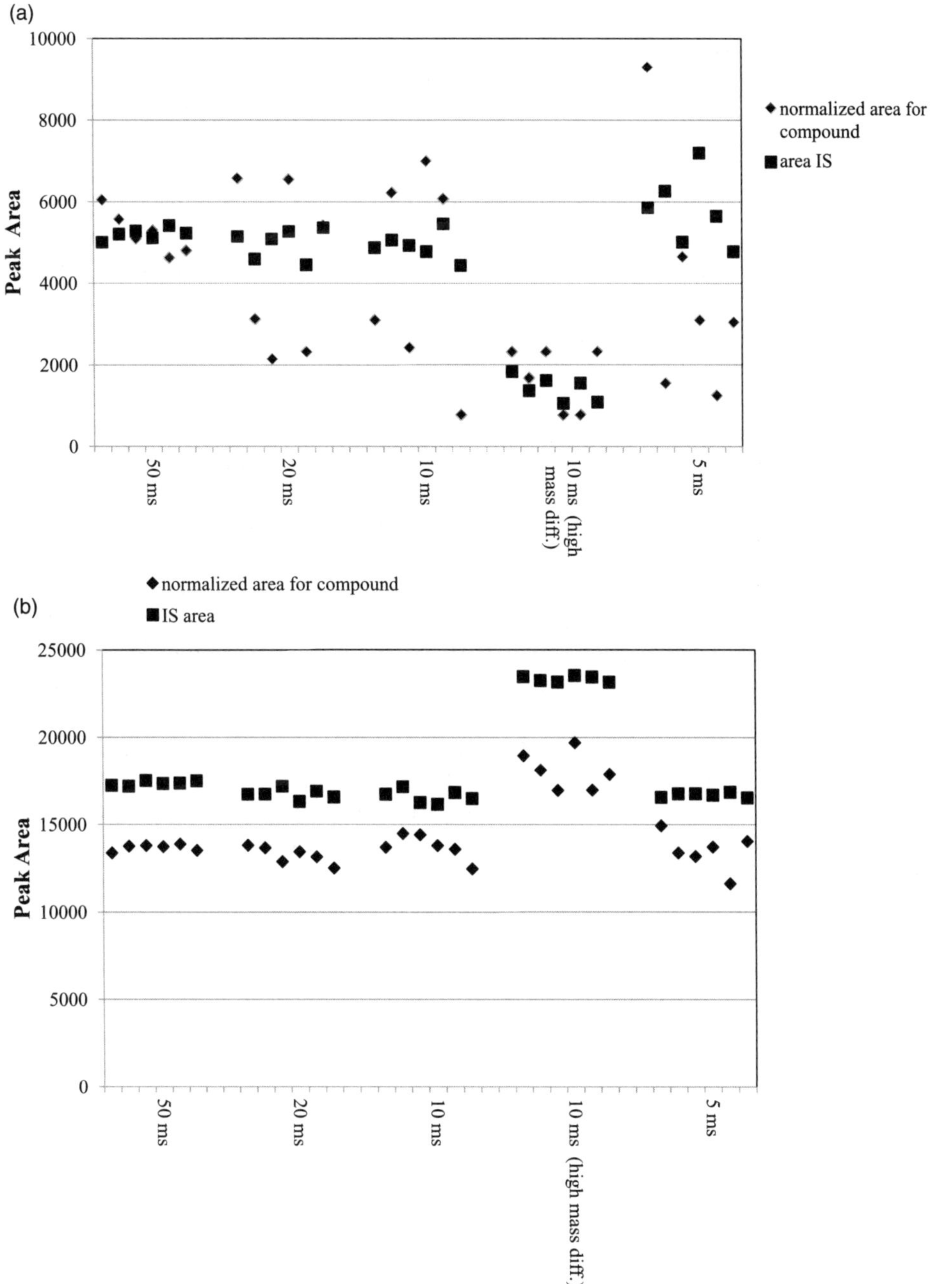

Figure 15.3 (A) Peak area in function of dwell times for six replicate injections at the LLOQ level of a small molecule drug and for the internal standard on a regular triple quadrupole instrument. (B) Peak area in function of dwell times for six replicate injections at the LLOQ level of a small molecule drug and for the internal standard on a latest generation triple quadrupole instrument.

Over the last two years, we have developed quantitative methods for over 30 peptide drug candidates. Many of the peptides had modifications such as N-terminal acetylation and/or C-terminal amidation. In addition, sequences containing unnatural amino acids (e.g., D-amino acids, aminobutyric acid) have been presented for analysis as well. Analysis in negative ion mode often gave the best bioanalytical performance, particularly (but not exclusively) when N-terminal acetylation and/or acidic amino acids (glutamic acid, aspartic acid) were included. Enfuvirtide (Fuzeon, T20), a 36-amino acid peptide with an MW of 4492 presents a good example (see Fig. 15.1 for structure). This peptidic HIV-fusion inhibitor is currently on the market for treatment of patients with HIV. HPLC-MS/MS methods for quantitative analysis of Enfuvirtide in plasma have been described by Chang et al. [34] and van den Broek et al. [30]. Both methods include the analysis in positive ion mode, the former working with an adapted protein precipitation method reaching LLOQs of 10 ng/mL and the latter using a solid-phase extraction (SPE) method with a reported LLOQ of 20 ng/mL (no concentration step was included since evaporation and reconstitution of the SPE eluate resulted in variable results). In the latter method, an 11-min run time is needed for chromatographic separation of T20 and M20. The M20 metabolite is formed by deamidation at the C-terminus, representing a mass increase of only 1 Da (NH_2 at the C-terminus is replaced by OH) and therefore chromatographic separation is desirable since the resolution available on triple quadrupole instruments is not able to distinguish between these entities. Attempts to use enzymatic digestion for Enfuvirtide as a tool to improve the sensitivity of the assay did not deliver the expected gain [35].

Formation of proteolytic fragments of a larger peptide through enzymatic digestion might result in better ionization and fragmentation since smaller peptides generally tend to have better MS properties. However, the enzymatic cleavage step introduces extra complexities and variabilities in the assay that need to be controlled. For this particular application, quantification of the five different chymotryptic fragments of Enfuvirtide did not result in a more sensitive assay compared with quantification of the intact peptide. Building a method for Enfuvirtide in negative ion mode proved to have some advantages: the formation of one major fragment ion (loss of acetyl) following collision-induced dissociation of the 4^- charged peak at low CE can offer a potentially more sensitive method. In positive ion mode, higher CEs were required to obtain fragmentation, and more fragment ions of equal intensity are observed. For negative ion MS, ammonium hydroxide (NH_4OH) was used as the mobile phase, resulting in a different chromatographic selectivity. The separation of the deamidated metabolite and the parent peptide C-terminal amide with NH_4OH-based mobile phase results in elution of the metabolite before the parent peptide, while in formic acid-based HPLC the elution order is reversed [36]. SRM chromatograms for Enfuvirtide and its metabolite in negative ion mode with a 5-min run time on an UHPLC system are presented in Figure 15.4. In protein-precipitated plasma (20 μL plasma + 50 μL H_2O + 200 μL acetonitrile) obtained from rats dosed intravenously with 2 mg/kg Enfuvirtide (four time-points shown), the rate of formation of the metabolite could be determined (Fig. 15.4).

Without doubt, positive ion mode is the preferred ionization mode for peptide analysis in most applications, but we are convinced that for selected applications, negative ion mode may present a better choice in terms of selectivity and sensitivity. For example, ammonium hydroxide has clearly proven to be the preferred HPLC

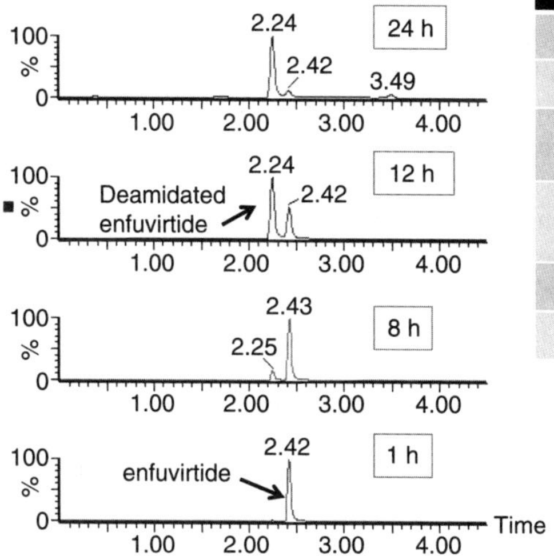

	UPLC-XevoTQS
Column	Acquity C_{18}, 2.1 x 50 mm,1.7 m
Flow rate	400 µl/min
solvents	Solvent A: 0.2% NH_4OH - Solvent B: 0.2% NH_4OH in acetonitrile
Sample prep	20 µL mouse plasma 50 µL water 200 µL CH_3CN
Injection volume	1 µL
SRM	897.3>882.5+897.3>888.5

Time (min)	%A	%B
0	99	1
0.5	99	1
3	60	40
3.5	5	95

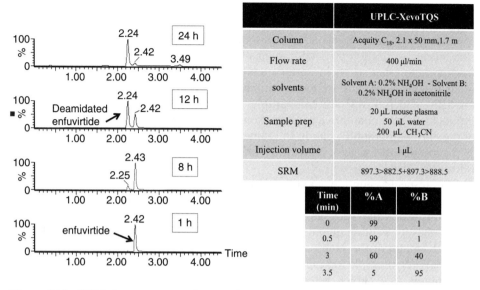

Figure 15.4 SRM chromatograms (in negative ion mode) of plasma obtained from mouse following oral administration of Enfuvirtide. Experimental procedure included.

solvent modifier in the application of HPLC-MS/MS quantification of amyloid peptides in cerebrospinal fluid [5, 33]. These peptides are extremely challenging, as the peptide containing 42 amino acids ($A\beta_{42}$) is very hydrophobic. Solubility and adsorption issues can be prevented by the use of ammonium hydroxide. In the latter two studies, SRM detection was performed in the negative ion mode. In these methods, the SRM transition measures the loss of between one and three water molecules. As such, the specificity of the assay is likely to be limited, particularly when applied to samples in complex biological matrices. In this case, however, the method yielded results that were fully validated and in line with the bioanalytical guidelines for method validation. Other investigators [37, 38] have kept the chromatography with NH_4OH but switched to positive ion mode for ESI-MS analysis. Similar sensitivity to the negative ion methods was realized through an optimized but extensive and laborious SPE sample preparation.

15.3.2 Ion Trap Instruments

Although triple quadrupoles are the preferred technology for quantitative analysis, ion trap MS can sometimes be a valuable alternative. Both three-dimensional and linear (two-dimensional) ion traps have claimed their territory in bioanalytical laboratories, predominantly for qualitative analyses. In ion traps, MS^n is performed “in time”; the trap is filled with all ions formed, precursor ions are selectively retained and subsequently excited, fragmented, and, finally, the resulting product ions are detected or further fragmented. Lower duty cycles are inherent to such a process, which, in combination with a smaller dynamic range, generally result in inferior performance compared with triple quadrupole MS. A more limited dynamic range of ion traps compared with triple quadrupole instruments is a consequence of the trapping process whereby only a limited number of ions can be trapped in the same space to avoid space charge effects. In some instances, however, the different detection mechanism of ion trap MS can also be beneficial. The different mechanism of fragmentation can result in better quantitative performance by the formation of other product ions that are more optimal, as described previously. The scan times for SRM and full MS^2 are comparable. This allows the summation of the major product ions or a qualitative confirmation of the identity of the compound or purity of the chromatographic peak in the same time frame. Lastly, the ability to perform MS^n experiments can have a favorable impact on S/N in cases where MS^2 selectivity is insufficient.

With respect to peptide quantification, several reports have been published which make use of ion trap technology [39–41], many including combinations of quadrupole and ion traps, such as the hybrid quadrupole LIT. The exploratory new approaches for peptide quantification are evolving and several proposed workflows take advantage of the many opportunities offered by the hybrid quadrupole LIT [13, 14]. When multiple MS platforms are available, the ion trap technology is generally not the first choice when selecting an instrument for quantitative analysis. However, for the quantification of peptide biomarkers in plasma, when the SRM method suffers from background interferences, the ion trap technology might offer solutions. For example, introducing MS^3 in a hybrid QTRAP instrument has resulted in enhanced selectivity and sensitivity [42].

15.4 HIGH RESOLUTION MS FOR PEPTIDE QUANTIFICATION

The two most popular HRMS instruments for quantitative bioanalysis today are the TOF and orbital trapping instruments. Therefore, all highlighted examples and discussion below will refer to these two platforms, and magnetic sector and Fourier transform-ion cyclotron resonance (FT-ICR) MS instruments will not be considered. TOF instruments show increasing resolution at higher *m/z* (on average, a resolution [R] of 20–40,000 and <5 ppm mass accuracy), while orbital trapping instruments offer superior mass accuracies and resolutions at the lower mass range (typically <2 ppm and R > 100,000 are possible), but resolution gradually drops with increasing *m/z*.

Qualitative analysis of drugs and their metabolites is increasingly being performed on high-resolution MS platforms. For small molecules, integrated workflows combining both quantification and identification in one analysis have been proposed [43–45]. It is expected that these combined approaches can add significant value to the experiments performed. Adoption of HRMS for quantification in the bioanalytical community requires no compromise on throughput, sensitivity, and dynamic range. LC separations should be developed with this purpose and should not be adapted for optimal separation of all potential metabolites. This statement is of course controversial, and the debate whether a compromise is acceptable if much greater value is added to the data is still ongoing.

The use of HRMS for the analysis of biological samples provides an alternative to SRM for differentiating the compound(s) of interest from background interferences. Selectivity and specificity are realized through narrow width *m/z* extraction.

Matrix-related isobaric background ions and analyte ions can be distinguished from drug-related components, depending largely on the mass resolution of the instrument used. In HPLC-MS/MS method development for a macrolide immunosuppressant, matrix effects prevented successful validation and introduction of HRMS/MS considerably reduced chemical noise [20]. A first example of a full validation of a bioanalytical method using HRMS was published by Fung et al. [46] for the quantification of prednisone and prednisolone in human plasma. Multiple studies have described quantification of small molecules using full scan HRMS. Full scan MS offers the advantage of analyzing all ions present over the selected mass range. Therefore, applications with simultaneous analysis of multiple analytes benefit to a large extent when analyzed with HRMS since the scanning capacity of the triple quadrupole instruments might not be sufficient (see previous discussion on SRM). The same advantage holds for peptides since multiple charge states and/or multiple fragment ions result in multiple SRM transitions for one peptide.

Peptide analysis using full scan MS allows detection of all available charge states in the spectrum and removes the fragmentation step from the workflow. Figure 15.5A shows the total ion current (TIC) chromatogram for a protein-precipitated plasma sample fortified with 2 μg/mL Enfuvirtide obtained on a Q-TOF MS system. In Figure 15.5B an EIC of the most abundant isotope of the 4^+ charged ion (with a

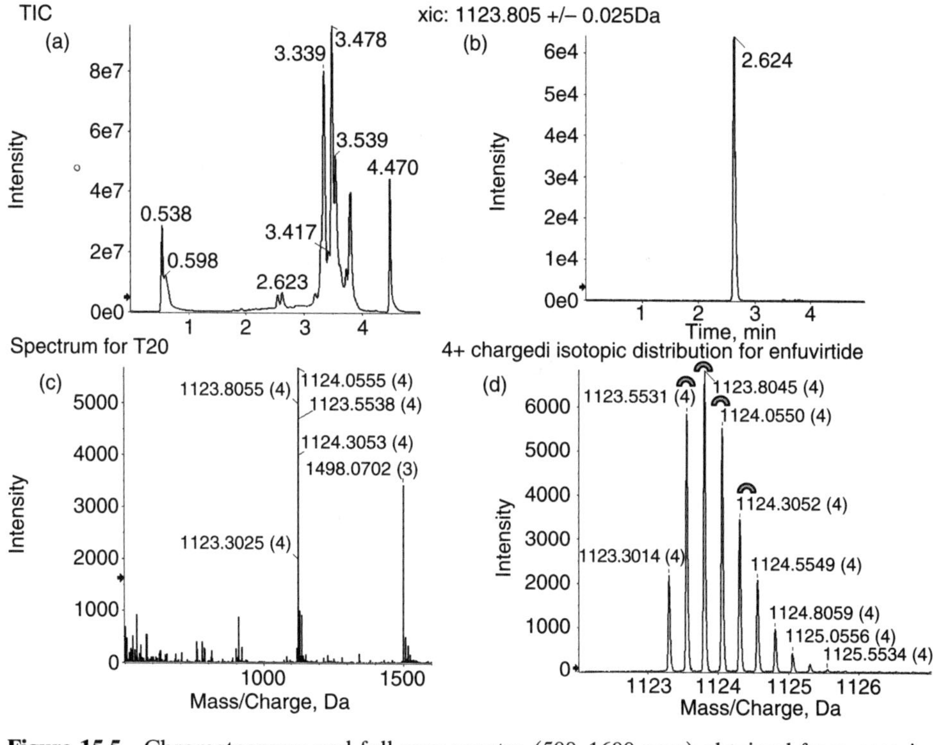

Figure 15.5 Chromatograms and full scan spectra (500–1600 amu) obtained from protein-precipitated plasma sample spiked with 2000 ng/mL Enfuvirtide. (A) Total ion chromatogram (TIC); (B) extracted ion chromatogram of the most abundant isotope of the 4+ charged ion (theoretical mass 1123.8054); (C) the full spectrum obtained at Rt 2.6 min with a zoomed view of the isotopic distribution of the 4+ charged peak. For method details, see Reference 21.

calculated *m/z* of 1123.8054) was generated with a 50-mDa mass extraction window (MEW). The mass spectrum with a zoomed view on the isotopic distribution of the 4^+ charged Enfuvirtide peak is visualized in Figure 15.5C,D. The arrows indicate the isotopes of this charge state that were included for quantification. This experimentally obtained distribution overlays perfectly with the theoretically calculated distribution (not shown). For Enfuvirtide, the monoisotopic peak only accounts for 7.4% of the total abundance of the 4^+ charged peak area. Therefore, combination of a number (or all) isotopic peaks can increase the sensitivity of the assay. Including all isotopic peaks for quantification is not always preferred, since at lower concentrations, lower abundance peaks may not be detected or fully discriminated, resulting in less reliable quantification. In a recent study [21], we evaluated the effect of summation of the four most abundant isotopic peaks per charge state for Enfuvirtide (indicated in Fig. 15.5D). The obtained LLOQs for six peptides on a TripleTof™ instrument (AB Sciex, Foster City, CA) ranged from 5 to 100 ng/mL in protein-precipitated plasma. Software-guided tools proposing the optimum choice of the charge states (taking into account the experimental conditions or based on an experimentally obtained mass spectrum) and the number of isotopes per charge state for optimal quantification would be helpful.

15.4.1 How Does Mass Resolving Power Influence Quantification?

To achieve the desired selectivity and sensitivity, the selection of the extraction window is very important. As discussed by Xia et al. [47], considerations on the choice of the MEW are related to the resolving power of the mass spectrometer and to the *m/z* of the analyte of interest. Resolving power is defined as R = M/FWHM (M = *m/z* value of the compound; FWHM = full width at half maximum). For HRMS data obtained in profile mode, covering the entire mass peak for quantification, the MEW should be at least two times the FWHM. Table 15.1 gives calculations for the maximum MEW for two resolving powers (30,000 and 100,000) for

TABLE 15.1 Maximum Theoretical Mass Extraction Window and Experimentally Obtained FWHM for Two Resolving Powers for the Most Abundant Isotopic Peak of Three Different Charge States for Enfuvirtide

Charge State	*m/z*	R (Instrument Settings)	Calculated FWHM mDa	Max MEW mDa	Measured Resolution on TripleTOF5600	Measured FWHM on TOF mDa
3+	1,498.0714	30,000	50	100	43,230	35
4+	1,123.8054	30,000	37	74	36,790	30
5+	899.2458	30,000	30	60	36,260	25
					Measured resolution on Orbitrap	Measured FWHM on Orbitrap
3+	1,498.0714	100,000	15	30	68,841	21
4+	1,123.8054	100,000	11	22	76,860	14
5+	899.2458	100,000	9	18	<LOD	<LOD

LOD, limit of detection.

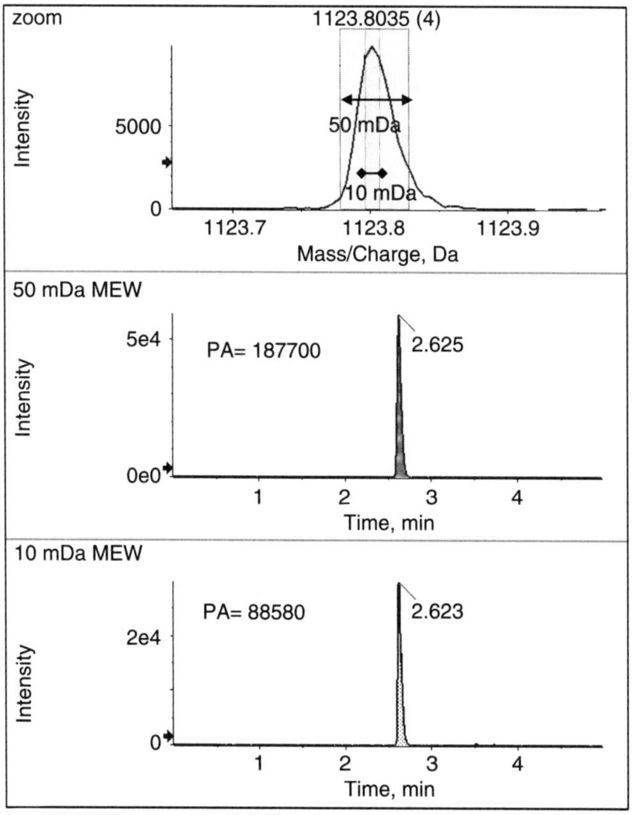

Figure 15.6 Extraction windows of 50 and 10 mDa applied on the most abundant isotope of the 4^+ peak of Enfuvirtide (theoretically *m/z* 1123.8054) PA = peak area. For method details, see Reference 21.

the most abundant isotope peak for three different charge states of Enfuvirtide. Experimentally obtained resolutions are also included. The MEWs need to be adapted as a function of the *m/z* value if the aim is to cover the same peak area for the different charge states. In practice, for quantitative analysis, a fraction of this maximum MEW will be applied as needed to exclude endogenous components. Figure 15.6 exemplifies the effect of the MEW on the peak intensity in an EIC. The maximum MEW (see Table 15.1) for the 4^+ charged isotopic peak (theoretically at *m/z* 1123.8054) is 74 mDa (±37 mDa). Both applied MEWs of 50 and 10 mDa in Figure 15.6 cover only a part of this maximum MEW. In Figure 15.7, two extracted LLOQ chromatograms (5 ng/mL) for Enfuvirtide with 50 and 10 mDa MEWs of the four most abundant isotopes of two charge states are given. The impact on both the signal intensity (peak height drops for Enfuvirtide from about 700 counts with 50 mDa MEW to 250 counts with 10 mDa MEW for the peak at 2.6 min) as well as on the matrix background can be noticed. In this example, matrix interference peaks relative to the drug peak do not improve substantially with a reduced MEW; only comparable reduction in signal intensity is observed.

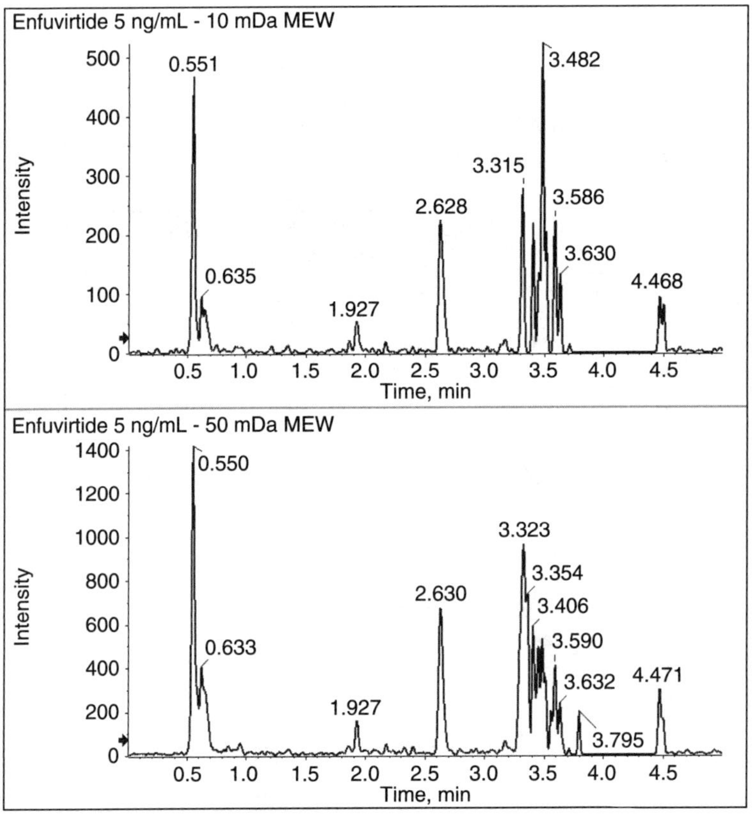

Figure 15.7 EIC obtained with two different MEWs applied on the four most abundant isotopes of two charge states (3^+ and 4^+) on a plasma sample spiked with 5 ng/mL Enfuvirtide. For method details see [21].

As argued by Xia et al. [47], there is no value in further narrowing the MEW if no S/N improvement can be obtained. Their investigation concluded that higher mass resolution in combination with narrow MEWs (10 ppm or lower) results in lower numbers of interfering peaks in the extracted chromatograms. With respect to peptide quantification, the chance of interfering background peaks appearing at identical mass increases as more isotopic peaks and charge states are included in the quantitative method but can be counteracted by very narrow MEWs at higher resolutions. This evaluation should be performed at the LLOQ.

An important difference between TOF and orbital trapping instruments relates to the impact of resolution on the sensitivity of the MS signal. TOF instruments sacrifice sensitivity when increased resolution is applied, while on orbital trapping instruments higher resolution is obtained through longer scanning time in the Orbitrap. In the latter case, no compromise on sensitivity is made. However, scan speed is reduced substantially; therefore, the chromatographic peak is defined by less data points.

A recent study by Ruan et al. [48] describes the bioanalysis of intact lysozyme (MW 15, 000) using HRMS on an Orbitrap instrument. The 10 most intense isotopic ions of the octuply protonated lysozyme were combined to obtain a linear dynamic range from 0.5 to 500 μg/mL. These investigators discussed the interrelation of three critical factors: mass resolving power, MW of the analyte, and MEW for quantification. They suggest that a minimal resolving power of at least four times the MW of the peptide/protein analyte is required for enhanced extraction with HRMS. If insufficient resolution is used, the isotopic signals are just baseline separated and the extracted chromatogram of the individual isotopic peaks obtained with a maximum MEW (2 × FWHM) will not show any advantage compared with one extracted chromatogram covering the whole cluster as would occur on low-resolution instruments.

In conclusion, HRMS quantification of peptides/proteins will depend on the resolving power of the mass spectrometer in relation to the MW of the analytes considered. Thus, peptides with a MW of 10,000 Da can be adequately quantified using an HRMS instrument that offers at least a resolution of 40,000 at the *m/z* of the multiply charged peptide peak. In a TOF instrument, this increased resolution results in reduced sensitivity, while on an Orbitrap instrument, the use of higher resolution results in increased scan times.

15.4.2 Transformation of Raw HRMS Spectral Data: Deconvoluted, Centroid, or Profile Data?

Multiply charged peaks can be deconvoluted to singly charged peaks. In most MS software packages, deconvolution software only works at the level of individual mass spectra. Biopharmalynx (Waters, Milford, MA) is a software package with built-in deconvolution tools (MaxEnt3) that was introduced for impurity profiling of large biomolecules. It allows control–analyte comparisons, returns deconvoluted chromatographic peaks, and identifies the peptide of interest and its degradation products. Such an approach could be adapted for state-of-the-art bioanalytical quantification of deconvoluted data but currently has not been implemented or explored. This workflow would greatly simplify the interpretation of the data.

Deconvolution software can only assess profile data, but in HRMS, data are often centroided, primarily to reduce the size of the data files generated. Both on-the-fly and post-acquisition processing are available and the mass spectral data are simplified to one line (or stick) per peak. The methodology used for centroiding is crucial and not always understood by the end user. In Masslynx (Waters), centroiding and mass measurement are executed based on the peak width at 80% of the height of the mass peak. The height of the produced stick corresponds to the peak area of the original profile data. Using centroid mode, Xia et al. [47] demonstrated that the apparent selectivity is increased as a consequence of this data processing.

In their global selectivity assessment, the total number of plasma endogenous component peaks was analyzed from acetonitrile-precipitated plasma in the EICs of 153 model compounds. At a resolving power of 20,000, a MEW of 50 ppm resulted in the detection of more than 400 endogenous background peaks (and 120 out of the 153 model compounds had at least 1 endogenous plasma component in the EIC) when profile data were used. Working with the same data set but with centroided data, the number of endogenous peaks was reduced to about 100 (and for 55 out of the 153 compound at least 1 endogenous plasma component was detected in the EIC). These authors propose smaller MEWs for profile data compared with centroid data, explaining that there is a reduced probability of missing data points when mass axis shifts in profile mode. However, with consistent mass accuracies, we would expect that narrower MEWs can be applied in centroid mode (and as such afford better selectivity without a compromise on sensitivity) compared with MEWs applied in profile mode where reduced MEWs have a negative impact on the extracted signal intensity (Fig. 15.6). No formal comparison of quantitative assay performance (sensitivity, accuracy, reproducibility) using centroid or profile data of the same sample set have been published (to our knowledge).

We have evaluated the quantitative performance for six peptides with full scan HRMS on a TripleTof 5600 instrument (ABSciex) [21]. On this instrument, acquisition in centroid mode is currently not offered and only post-acquisition centroiding is available, which is a cumbersome process not suited for batch analysis of many samples. A correct choice of the MEWs is essential for good sensitivity for data measured on peptides in profile mode. Larger MEWs will lead to higher intensity peaks and, thus, generally better sensitivity; however, as a consequence, the likelihood of co-extraction of endogenous interferences increases. With stable mass accuracy across a chromatographic peak (scan-to-scan reproducibility) as well as over the whole chromatographic run time, reduced MEWs applied on centroid data will result in increased S/N by enhancing specificity, without affecting peak intensity. Better understanding of the impact of the differences between profile and centroid data on typical bioanalytical parameters such as sensitivity, selectivity, specificity, and accuracy will result in preferred strategies for quantitative workflows.

15.4.3 Does HRMS SRM Offer a Valuable Strategy?

Many high-resolution configuration instruments allow HRMS SRM-based approaches (Q-TOF, LTQ-Orbitrap, Q-exactive). The precursor ion of interest is selected by the quadrupole or LTQ at unit resolution and fragmented in the collision cell similar to standard SRM. However, the product ion spectrum is scanned at high resolution instead of unit resolution. For small molecules, it has been reported that selection of product ions at high resolution can improve selectivity and specificity [20]. However, the challenges of the scan time paradigm discussed for SRM workflows are also relevant for HRMS SRM. Monitoring the full product ion spectrum and combining multiple product ions obtained in HRMS can potentially mitigate any loss in sensitivity. Selection of product ions can be done during data processing rather than before data acquisition, allowing the optimum choice of the product ion that provides the best sensitivity and lowest background interference for each analyte in the mixture. For peptide quantification, the SRM workflow will be faced with similar shortcomings to those observed in SRM workflows: the overall peptide signal is diluted over the different charge states and often unfavorable fragmentation is noticed. Therefore, HRMS SRM is unlikely to improve sensitivity for peptide quantification.

15.5 IMPACT OF THE DUTY CYCLES OF DIFFERENT INSTRUMENTATION

A fundamental difference between SRM- and TOF-based approaches relates to the intrinsic difference in duty cycles. The duty cycle is defined as the proportion of the time during which the system analyzes a specific *m/z*. SRM allows the ions of interest to pass through continuously. Theoretically, 100% detection efficiency is obtained if the parent ion is completely converted into one single product ion and the loss during collisional activation is considered negligible. The duty cycle is reduced in proportion to the number of SRM channels monitored in the experiment. With respect to peptide quantification on triple quadrupole instruments, duty cycles are somewhat compromised, because multiple SRM transitions need to be monitored to reduce the loss of ions spread over the multiple charge states and multiple fragments usually observed. Orthogonal acceleration TOF (oaTOF), on the other hand, pushes packets of ions into the flight tube. In the period where no ions are pulsed in the TOF, ions are wasted. Therefore, the duty cycle is reduced (to about 5–25% of maximal). Although reduced duty cycles are intrinsic to oaTOF, this is partially compensated by the simultaneous detection of all ions, which is particularly beneficial for peptides since all charge state ions can be summed. The application of an enhanced duty cycle through an optimized pusher frequency for a small *m/z* region and potential trapping of ions before TOF measurement can further improve the quantitative output of a TOF-based MS system. The latter is a common practice on Orbitrap instruments to improve the duty cycle. In parallel to the high-resolution scanning in the Orbitrap—the rate-limiting step—MS^n data can be generated in the LTQ ion trap (on LTQ-Orbitrap), or ions selected by a quadrupole can be collected and stored in the C-trap while waiting for the next Orbitrap scanning event (Q-Exactive). How all these considerations finally translate in a more sensitive quantification on one or the other platform is very much compound and matrix dependent.

15.6 CONCLUSION

HRMS is likely to play an increasingly important role in quantitative bioanalysis [49]. Only the time frame for broad adoption and the preferred approaches (full scan HRMS or HRMS-SRM) still need to be defined and will depend largely on the performance and cost of the instruments relative to triple quadrupole-based solutions. Newer HRMS instruments provide excellent resolution, mass accuracy, and dynamic range. Therefore, they become a valuable alternative to triple quadrupole instruments with the advantage of eliminating time-consuming tuning and optimization for optimal conditions for each SRM. This offers, particularly in early drug development, opportunities to increase efficiency. For peptide quantification, limited by multiple SRMs due to the multiple charge states in combination with narrow width chromatographic peaks and analytes with poor fragmentation efficiency, the advantages of HRMS quantification are evident. It is hoped that the sharing of experiences and results will convince the bioanalytical community to further explore these new approaches. New developments such as the combination of ion mobility separation with HRMS can further improve selectivity and offer valuable additions for the quantitative HRMS workflow.

ACKNOWLEDGMENTS

The authors are very grateful to Liesbeth Vereyken and Willy Cools for contributions to the experiments referred to in the chapter. We also acknowledge valuable discussions with Russell Mortishire–Smith, Ronald de Vries, and Hesham Ghobarah.

REFERENCES

[1] Vlieghe P, Lisowski V, Marinez J, Khrestchatisky M. Synthetic therapeutic peptides: science and market. Drug Discov Today 2010;15:40–56.

[2] Reichert J, Pechon P, Tartar A, Dunn M. Development trends for peptide therapeutics; a comprehensive quantitative analysis of peptide therapeutics in clinical development (report summary). Peptide Therapeutic Foundation: 2010.

[3] Van Den Broek I, Sparidans RW, Schellens JHM, Beijnens JH. Sensitive liquid chromatography-tandem mass spectrometry assay for absolute quantification of ITIH4-derived putative biomarker peptides in clinical serum samples. Rapid Commun Mass Spectrom 2010;24(13):1842–50.

[4] Van Den Broek I, Sparidans RW, Schellens JHM, Beijnens JH. Quantitative assay for six potential breast cancer biomarker peptides in human serum by liquid chromatography coupled to tandem mass spectrometry. J Chromatogr B 2010;878:590–602.

[5] Dillen L, Cools W, Vereyken L, Timmerman P. A screening UHPLC–MS/MS method for the analysis of amyloid peptides in cerebrospinal fluid of preclinical species. Bioanalysis 2011;3(1):45–55.

[6] Hoofnagle A, Wener M. The fundamental flaws of immunoassays and potential solutions using tandem mass spectrometry. J Immunol Methods 2009;347:3–11.

[7] Hewavitharana AK, Herath HMDR, Shaw PN, Cabot PJ, Kebarle P. Effect of solvent and electrospray mass spectrometer parameters on charge state distribution of peptides—a case study using liquid chromatography/mass spectrometry method development for beta-endorphin assay. Rapid Commun Mass Spectrom 2010;24:3510–4.

[8] Campbell JL, Le Blanc JCY. Peptide and protein drug analysis by MS: challenges and opportunities for the discovery environment. Bioanalysis 2011;3(6):645–57.

[9] MacLean B, Tomazela DM, Abbatiello SE, et al. Effect of collision energy optimization on the measurement of peptides by selected reaction monitoring (SRM) mass spectrometry. Anal Chem 2010;82:10116–24.

[10] Kuhn E, Wu J, Karl J, Liao H, Zolg W, Guild B. Quantification of C-reactive protein in the serum of patients with rheumatoid arthritis using multiple reaction monitoring mass spectrometry and 13C-labeled peptide standards. Proteomics 2004;4(4):1175–86.

[11] Hagman C, Ricke D, Ewert S, Bek S, Falchetto R, Bitsch F. Absolute quantification of monoclonal antibodies in biofluids by liquid chromatography-tandem mass spectrometry. Anal Chem 2008;80(4):491–8.

[12] Klaassen T, Swandt S, Kapron JT, Roemer A. Validated quantitation method for a peptide in rat serum using liquid chromatography/high field asymmetric waveform ion mobility spectrometry. Rapid Commun Mass Spectrom 2009;23:2301–6.

[13] Hao C, Campbell JL, Verkerk UH, Le Blanc JCY, Siu KWM. Performance and attributes of liquid chromatography—mass spectrometry with targeted charge separation in quantitative analysis of therapeutic peptides. J Am Soc Mass Spectrom 2011;22(1):67–74.

[14] Campbell JL, Le Blanc JCY. Targeted ion parking for the quantitation of biotherapeutic proteins: concepts and preliminary data. J Am Soc Mass Spectrom 2010;21(12):2011–22.

[15] Korfmacher W. High resolution mass spectrometry will dramatically change our drug-discovery bioanalysis procedures. Bioanalysis 2011;3(11):1169–71.

[16] Kaufmann A, Butcher P, Maden K, Walker S, Widmer M. Quantitative and confirmative performance of liquid chromatography coupled to high resolution mass spectrometry compared with tandem mass spectrometry. Rapid Commun Mass Spectrom 2011;25(7):979–92.

[17] Soler C, Hamilton B, Furey A, James KJ, Manes J, Pico Y. Comparison of four mass analyzers for determining carbosulfan and its metabolite in citrus by liquid chromatography/mass spectrometry. Rapid Commun Mass Spectrom 2006;20:2151–64.

[18] Kadar H, Veyrand B, Barbarossa A, et al. Development of an analytical strategy based on liquid chromatography-high resolution mass spectrometry for measuring perfluorinated compounds in human breast milk: application to the generation of preliminary data regarding perinatal exposure in France. Chemosphere 2011;85(3):473–80.

[19] Jiwan JLH, Wallemacq P, Herent MF. HPLC-high resolution mass spectrometry in clinical laboratory? Clin Biochem 2011;44:136–47.

[20] Taillon MP, Furtado M, Garofolo F. Challenges of developing a bioanalytical method for a macrolide immunosuppressant compound by LC-MS/MS. Bioanalysis 2011;3(1): 1201–15.

[21] Dillen L, Cools W, Vereyken L, Lorreyne W, Huybrechts T, de Vries R, Ghobarah H, Cuyckens F. Comparison of triple quadrupole and high resolution time of flight mass spectrometry for quantification of peptides. Bioanalysis 2012;4(3):565–79.

[22] Van Den Broek I, Sparidans R, Schellens J, Beijnen J. Quantitative bioanalysis of peptides by liquid chromatography coupled to tandem mass spectrometry. J Chromatogr B 2008;872:1–22.

[23] Ewles M, Goodwin L. Bioanalytical approaches to analyzing peptides and proteins by LC-MS/MS. Bioanalysis 2011;3(12):1379–97.

[24] Tamvakopoulos C. Mass spectrometry for the quantification of bioactive peptides in biological fluids. Mass Spectrom Rev 2007;26:389–402.

[25] Niwa M, Enomoto K, Yamashita K. Measurement of the novel decaeptide cetrorelix in human plasma and urine by liquid chromatography–electrospray ionization mass spectrometry. J Chromatogr B Biomed Sci Appl 1999;729(1–2):245–53.

[26] Mirzaei H, Regnier F. Enhancing electrospray ionization efficiency of peptides by derivatization. Anal Chem 2006;78(12):4175–83.

[27] Abian J, Oosterkamp AJ, Gelpí E. Comparison of conventional, narrow-bore and capillary liquid chromatography/mass spectrometry for electrospray ionization mass spectrometry: practical considerations. J Mass Spectrom 1999;34(4):244–54.

[28] Roepstorff P, Fohlman J. Proposal for a common nomenclature for sequence ions in mass spectra of peptides. Biomed Mass Spectrom 1984;11(11):601.

[29] Delinsky DC, Hill KT, White CS, Bartlett MG. Quantitation of the large polypeptide glucagon by protein precipitation and LC/MS. Biomed Chromatogr 2004;18(9):700–5.

[30] van den Broek I, Sparidans RW, Huitema AD, Schellens JH, Beijnen JH. Development and validation of a quantitative assay for the measurement of two HIV-fusion inhibitors, enfuvirtide and tifuvirtide, and one metabolite of enfuvirtide (M-20) in human plasma by liquid chromatography-tandem mass spectrometry. J Chromatogr B Analyt Technol Biomed Life Sci 2006;837(1–2):49–58.

[31] Shipkova P, Drexler DM, Langish R, Smalley J, Salyan ME, Sanders M. Application of ion trap technology to liquid chromatography/mass spectrometry quantitation of large peptides. Rapid Commun Mass Spectrom 2008;22(9):1359–66.

[32] Fierens C, Thienpont LM, Stöckl D, Willekens E, De Leenheer AP. Quantitative analysis of urinary C-peptide by liquid chromatography-tandem mass spectrometry with a stable isotopically labelled internal standard. J Chromatogr A 2000;896(1–2):275–8.

[33] Oe T, Ackermann BL, Inoue K. Garner CO, Gelfanova V, Dean RA, Siemers ER, Holtzman DM, Farlow MR, Blair IA. Quantitative analysis of amyloid β peptides in cerebrospinal fluid of Alzheimer's disease patients by immunoaffinity purification and stable isotope dilution liquid chromatography/negative electrospray ionization tandem mass spectrometry. Rapid Commun Mass Spectrom 2006;20:3723–35.

[34] Chang D, Kolis SJ, Linderholm KH, Julian TF, Nachi R, Dzerk AM, Lin PP, Lee JW, Bansal SK. Bioanalytical method development and validation for a large peptide HIV fusion inhibitor (Enfuvirtide, T-20) and its metabolite in human plasma using LC-MS/MS. J Pharm Biomed Anal 2005;38(3):487–96.

[35] van den Broek I, Sparidans RW, Schellens JH, Beijnen JH. Enzymatic digestion as a tool for the LC-MS/MS quantification of large peptides in biological matrices: measurement of chymotryptic fragments from the HIV-I fusion inhibitor enfuvirtide and its metabolite M-20 in human plasma. J Chromatogr B Analyt Technol Biomed Life Sci 2007; 854:245–59.

[36] Cuyckens F, Dillen L, Cools W, Bockx M, Vereyken L, de Vries R, Mortishire-Smith RJ. Extracting metabolite ions of peptide drugs from an *in vivo* matrix background. Bioanalysis 2012;4(3):595–604.

[37] Chambers E, Lame ME, Diehl DM, Zhang Y. Analytical challenges in the development of a quantitative SPE/LC/MS/MS assay for amyloid beta peptides in cerebrospinal fluid. Poster at the 58th ASMS Conference Salt Lake City, US 2010.

[38] Ismaiel OA, Yuan M, Jenkins RG, Karnes HG. Evaluation of 2-Dimensional Chromatographic Techniques for Eliminating Matrix Effects and Improving Bioanalysis of Peptides Using UPLC-MS/MS poster presentation at EBF symposium Barcelona 2010.

[39] Hatziieremia S, Kostomitsopoulos N, Balafas V, Tamvakopoulos C. A liquid chromatographic/tandem mass spectroscopic method for quantification of the cyclic peptide melanotan-II. Plasma and brain tissue concentrations following administration in mice. Rapid Commun Mass Spectrom 2007;21(15):2431–8.

[40] Bredehöft M, Schänzer W, Thevis M. Quantification of human insulin-like growth factor-1 and qualitative detection of its analogues in plasma using liquid chromatography/electrospray ionisation tandem mass spectrometry. Rapid Commun Mass Spectrom 2008;22(4):477–85.

[41] Vatansever B, Lahrichi SL, Thiocone A, Salluce N, Mathieu M, Grouzmann E, Rochat B. Comparison between a linear ion trap and a triple quadrupole MS in the sensitive detection of large peptides at femtomole amounts on column. J Sep Sci 2010;33: 2478–88.

[42] Fortin T, Salvador A, Charrier JP, Lenz C, Bettsworth F, Lacoux X, Choquet G, Kastylevsky F, Lemoine J. Multiple reaction monitoring cubed for protein quantification at the low nanogram/milliliter level in nondepleted human serum. Anal Chem 2009;81: 9343–52.

[43] O'Connor D, Mortishire-Smith R. High-throughput bioanalysis with simultaneous acquisition of metabolic route data using ultra-performance liquid chromatography coupled with time-of-flight mass spectrometry. Anal Bioanal Chem 2006;385:114–21.

[44] Zhang N, Yu S, Tiller P, Yeh S, Mahan E, Emary W. Quantitation of small molecules using high-resolution accurate mass spectrometers—a different approach for analysis of biological samples. Rapid Commun Mass Spectrom 2009;23(7):1085–94.

[45] Bateman K, Kellman M, Muenster H, Papp R, Taylor L. Quantitative-qualitative data acquisition using a benchtop Orbitrap mass spectrometer. J Am Soc Mass Spectrom 2009;20(8):1441–50.

[46] Fung E, Xia Y-Q, Aubry A-F, et al. Full scan high resolution accurate mass spectrometry (HRMS) in regulated bioanalysis: LC-HRMS for the quantitation of prednisone and prednisolone in human plasma. J Chromatogr B 2011;879:2919–27.

[47] Xia Y-Q, Lau J, Olah T, Jemal M. Targeted quantitative bioanalysis in plasma using liquid chromatography/high-resolution accurate mass spectrometry: an evaluation of global selectivity as a function of mass resolving power and extraction window, with comparison of centroid and profile modes. Rapid Commun Mass Spectrom 2011;25:2863–28.

[48] Ruan Q, Ji QC, Arnold ME, Humphreys WG, Zhu M. Strategy and its implications of protein bioanalysis utilizing high-resolution mass spectrometric detection of intact protein. Anal Chem 2011;83:8937–44.

[49] Ramanathan R, Jemal M, Ramagiri S, Xia YQ, Humprheys WG, Olah T, Korfmacher WA. It is time for a paradigm shift in drug discovery bioanalysis: from SRM to HRMS. J Mass Spectrom 2011;46:595–601.

INDEX

Mass Spectrometry for Drug Discovery and Drug Development, First Edition. Edited by Walter A. Korfmacher.

WILEY SERIES ON MASS SPECTROMETRY

Series Editors

Dominic M. Desiderio
Departments of Neurology and Biochemistry
University of Tennessee Health Science Center

Nico M. M. Nibbering
Vrije Universiteit Amsterdam, The Netherlands

John R. de Laeter • *Applications of Inorganic Mass Spectrometry*
Michael Kinter and Nicholas E. Sherman • *Protein Sequencing and Identification Using Tandem Mass Spectrometry*
Chhabil Dass • *Principles and Practice of Biological Mass Spectrometry*
Mike S. Lee • *LC/MS Applications in Drug Development*
Jerzy Silberring and Rolf Eckman • *Mass Spectrometry and Hyphenated Techniques in Neuropeptide Research*
J. Wayne Rabalais • *Principles and Applications of Ion Scattering Spectrometry: Surface Chemical and Structural Analysis*
Mahmoud Hamdan and Pier Giorgio Righetti • *Proteomics Today: Protein Assessment and Biomarkers Using Mass Spectrometry, 2D Electrophoresis, and Microarray Technology*
Igor A. Kaltashov and Stephen J. Eyles • *Mass Spectrometry in Structural Biology and Biophysics: Architecture, Dynamics, and Interaction of Biomolecules, Second Edition*
Isabella Dalle-Donne, Andrea Scaloni, and D. Allan Butterfield • *Redox Proteomics: From Protein Modifications to Cellular Dysfunction and Diseases*
Silas G. Villas-Boas, Ute Roessner, Michael A.E. Hansen, Jorn Smedsgaard, and Jens Nielsen • *Metabolome Analysis: An Introduction*
Mahmoud H. Hamdan • *Cancer Biomarkers: Analytical Techniques for Discovery*
Chabbil Dass • *Fundamentals of Contemporary Mass Spectrometry*
Kevin M. Downard (Editor) • *Mass Spectrometry of Protein Interactions*
Nobuhiro Takahashi and Toshiaki Isobe • *Proteomic Biology Using LC-MS: Large Scale Analysis of Cellular Dynamics and Function*
Agnieszka Kraj and Jerzy Silberring (Editors) • *Proteomics: Introduction to Methods and Applications*
Ganesh Kumar Agrawal and Randeep Rakwal (Editors) • *Plant Proteomics: Technologies, Strategies, and Applications*
Rolf Ekman, Jerzy Silberring, Ann M. Westman-Brinkmalm, and Agnieszka Kraj (Editors) • *Mass Spectrometry: Instrumentation, Interpretation, and Applications*
Christoph A. Schalley and Andreas Springer • *Mass Spectrometry and Gas-Phase Chemistry of Non-Covalent Complexes*
Riccardo Flamini and Pietro Traldi • *Mass Spectrometry in Grape and Wine Chemistry*
Mario Thevis • *Mass Spectrometry in Sports Drug Testing: Characterization of Prohibited Substances and Doping Control Analytical Assays*
Sara Castiglioni, Ettore Zuccato, and Roberto Fanelli • *Illicit Drugs in the Environment: Occurrence, Analysis, and Fate Using Mass Spectrometry*
Ángel Garciá and Yotis A. Senis (Editors) • *Platelet Proteomics: Principles, Analysis, and Applications*

Luigi Mondello • *Comprehensive Chromatography in Combination with Mass Spectrometry*
Jian Wang, James MacNeil, and Jack F. Kay • *Chemical Analysis of Antibiotic Residues in Food*
Walter A. Korfmacher (Editor) • *Mass Spectrometry for Drug Discovery and Drug Development*